AF781176

Molecular Biology Intelligence Unit

Immunobiology of Carbohydrates

Simon Y.C. Wong, Ph.D.
Carbohydrate Immunology Group
Edward Jenner Institute for Vaccine Research
Compton, Newbury, U.K.

Gemma Arsequell, Ph.D.
Unit of Glycoconjugate Chemistry
IIQAB-CSIC
Barcelona, Spain

Landes Bioscience / Eurekah.com
Georgetown, Texas
U.S.A.

Kluwer Academic / Plenum Publishers
New York, New York
U.S.A.

IMMUNOBIOLOGY OF CARBOHYDRATES

Molecular Biology Intelligence Unit

Eurekah.com / Landes Bioscience
Kluwer Academic / Plenum Publishers

Copyright ©2003 Eurekah.com and Kluwer Academic / Plenum Publishers

All rights reserved.
No part of this book may be reproduced or transmitted in any form or by any means, electronic or mechanical, including photocopy, recording, or any information storage and retrieval system, without permission in writing from the publisher, with the exception of any material supplied specifically for the purpose of being entered and executed on a computer system; for exclusive use by the Purchaser of the work.

Printed in the U.S.A.

Kluwer Academic / Plenum Publishers, 233 Spring Street, New York, New York, U.S.A. 10013
http://www.wkap.nl/

Please address all inquiries to the Publishers:
Eurekah.com / Landes Bioscience, 810 South Church Street
Georgetown, Texas, U.S.A. 78626
Phone: 512/ 863 7762; FAX: 512/ 863 0081
www.Eurekah.com
www.landesbioscience.com

Immunobiology of Carbohydrates, edited by Simon Y.C. Wong and Gemma Arsequell, Landes / Kluwer dual imprint / Landes series: Molecular Biology Intelligence Unit.

ISBN: 0-306-47844-7

While the authors, editors and publisher believe that drug selection and dosage and the specifications and usage of equipment and devices, as set forth in this book, are in accord with current recommendations and practice at the time of publication, they make no warranty, expressed or implied, with respect to material described in this book. In view of the ongoing research, equipment development, changes in governmental regulations and the rapid accumulation of information relating to the biomedical sciences, the reader is urged to carefully review and evaluate the information provided herein.

Library of Congress Cataloging-in-Publication Data

Immunobiology of carbohydrates / [edited by] Simon Y.C. Wong, Gemma Arsequell.
p. ; cm. -- (Molecular biology intelligence unit)
Includes bibliographical references and index.
ISBN 0-306-47844-7
1. Carbohydrates--Immunology. 2. Antigen-antibody reactions--Physiological effect.
[DNLM: 1. Carbohydrates--immunology. 2. Immunity, Cellular. 3. Receptors, Immunologic--physiology. QU 75 I33 2003] I. Wong, Simon Y. C. II. Arsequell, Gemma. III. Molecular biology intelligence unit (Unnumbered)
QR186.6.C37I47 2003
616.07'9--dc22 2003018772

Dedicated to
the memory of our special friend Dr. David Wing.

CONTENTS

EDITORS

Simon Y.C. Wong, Ph.D.
Carbohydrate Immunology Group
Edward Jenner Institute for Vaccine Research
Compton, Newbury, U.K.
simon.wong@jenner.ac.uk
Chapter 7

Gemma Arsequell, Ph.D.
Unit of Glycoconjugate Chemistry
IIQAB-CSIC
Barcelona, Spain
garqbp@iiqab.csic.es

CONTRIBUTORS

Shizuo Akira
Department of Host Defense
Research Institute for Microbial Diseases
Osaka University
Suita, Osaka, Japan
sakira@biken.osaka-u.ac.jp
Chapter 6

Vasso Apostolopoulos
The Austin Research Institute
Immunology and Vaccine Laboratory
Heidelberg, Australia
v.apostolopoulos@ari.unimelb.edu.au
Chapter 14

Gurdysal S. Besra
School of Biosciences
The University of Birmingham
Edgbaston
Birmingham, U.K.
gbesra@bham.ac.uk
Chapter 10

Gordon D. Brown
Sir William Dunn School of Pathology
Oxford University
Oxford, U.K.
Chapter 5

Paul R. Crocker
The Wellcome Trust Biocentre
at Dundee
University of Dundee
Dundee, U.K.
Chapter 8

Lynn G. Dover
School of Biosciences
The University of Birmingham
Edgbaston
Birmingham, U.K.
l.g.dover@bham.ac.uk
Chapter 10

Terry W. Du Clos
VA Medical Center
Research Service 151
Albuquerque, New Mexico, U.S.A.
tduclos@unm.edu
Chapter 3

Tim Elliott
Cancer Sciences Division
Southampton University
Southampton General Hospital
Southampton, U.K.
T.J.Elliott@soton.ac.uk
Chapter 11

Uri Galili
Departments of Cardiovascular Thoracic Surgery and Immunology
RUSH University
Chicago, Ilinois, U.S.A.
Uri-Galili@rush.edu
Chapter 15

Philippe Gasque
Brain Inflammation and Immunity Group and Complement Biology Group
Department of Medical Biochemistry and Immunology
University of Wales College of Medicine
Cardiff, U.K.
Gasque@cardiff.ac.uk
Chapter 4

Denise Golgher
Cancer Sciences Division
Southampton University
Southampton General Hospital
Southampton, U.K.
D.B.Golgher@soton.ac.uk
Chapter 11

Siamon Gordon
Sir William Dunn School of Pathology
Oxford University
Oxford, U.K.
Chapter 5

Mark Howarth
Cancer Sciences Division
Southampton University
Southampton General Hospital
Southampton, U.K.
Chapter 11

Tsuneyasu Kaisho
Department of Host Defense
Research Institute for Microbial Diseases
Osaka University
Suita, Osaka, Japan
Chapter 6

Reiji Kannagi
Research Institute Program of Molecular Pathology
Aichi Cancer Center
Nagoya, Japan
rkannagi@aichi-cc.jp
Chapter 1

A.A. Lindberg
Aventis (formerly Pasteur Merieux Connaught)
Marcy l'Etoile, France
Chapter 13

Ian C.M. MacLennan
MRC Centre for Immune Regulation
University of Birmingham Medical School
Birmingham, U.K.
i.c.m.maclennan@bham.ac.uk
Chapter 9

Luisa Martinez-Pomares
Sir William Dunn School of Pathology
Oxford University
Oxford, U.K.
Chapter 5

Ian McKenzie
The Austin Research Institute
Immunology and Vaccine Laboratory
Heidelberg, Victoria, Australia
Chapter 14

P. Moingeon
Aventis (formerly Pasteur Merieux Connaught)
Marcy l'Etoile, France
philippe.moingeon@aventis.com
Chapter 13

Carolyn Mold
University of New Mexico
School of Medicine
Department of Molecular Genetics and Microbiology
Albuquerque, New Mexico, U.S.A.
Chapter 3

M. Moreau
Aventis (formerly Pasteur Merieux Connaught)
Marcy l'Etoile, France
Chapter 13

B. Paul Morgan
Brain Inflammation and Immunity Group and Complement Biology Group
Department of Medical Biochemistry and Immunology
University of Wales College of Medicine
Cardiff, U.K.
Chapter 4

Lars Nitschke
Institute of Virology and Immunobiology
University of Wurzburg
Wurzburg, Germany
nitschke@vim.uni-wuerzburg.de
Chapter 8

Magdalena Plebanski
The Austin Research Institute
Immunology and Vaccine Laboratory
Heidelberg, Victoria, Australia
Chapter 14

Steven A. Porcelli
Department of Microbiology and Immunology
Albert Einstein College of Medicine
Bronx, New York, U.S.A.
porcelli@aecom.yu.edu
Chapter 10

Vince Pozsgay
National Institute of Child Health and Human Development
National Institutes of Health
Bethesda, Maryland, U.S.A.
vipo@helix.nih.gov
Chapter 12

Sally Rogers
Institute of Animal Health, Compton
Berkshire, U.K.
sarah.rogers@bbsrc.ac.uk
Chapter 7

Osamu Takeuchi
Department of Host Defense
Research Institute for Microbial Diseases
Osaka University
Suita, Osaka, Japan
Chapter 6

Philip R. Taylor
Sir William Dunn School of Pathology
Oxford University
Oxford, U.K.
ptaylor@molbiol.ox.ac.uk
Chapter 5

Carola G. Vinuesa
The Medical Genome Center
John Curtin School for Medical Research
Australia National University
Canberra, Australia
Chapter 9

Russell Wallis
Glycobiology Institute
Department of Biochemistry
University of Oxford
Oxford, U.K.
rwallis@glycob.ox.ac.uk
Chapter 2

PREFACE

Immune responses toward carbohydrate antigens could play critical roles in protection against microbial infections and tumour formation. Thus, carbohydrate antigens are candidates for vaccine development and immunotherapy. In addition, some microbial carbohydrates have stimulatory effects on the immune system, and as such could also be used as adjuvants to improve the immunogenicity of vaccines. In contrast anti-carbohydrate responses could also be undesirable as they prevent successful blood transfusion and organ transplantation and be contributing factors in autoimmune diseases. Modulation of these responses would therefore be of clinical benefit. Despite the clear importance of understanding the immunobiology of carbohydrates, this subject has been understudied and poorly appreciated by most immunologists for decades. There may be several reasons for this. One reason may be the lack of detailed information on cell surface carbohydrate structures and the regulatory mechanisms of glycosylation pathways that generate these structures in a time- and cell-dependent manner. Another reason may be the lack of biologically relevant reagents and assays to understand their interactions with carbohydrate binding proteins (or lectins) in the innate immune system. It is also possible that immune responses against carbohydrate antigens were generally thought to be less interesting and more difficult to modulate than responses toward protein antigens.

Recent advances in glycobiology and immunology are beginning to provide many more insights into the potential role of carbohydrate-protein interactions in the innate immune system and the immune responses against carbohydrate antigens. The purpose of this book is to promote the subject of carbohydrate immunology. It has three main themes: (1) Carbohydrate antigens/determinants and their recognition by proteins and cell surface receptors involved in the innate immune response or lymphocyte development and recruitment; (2) B and T cell responses toward carbohydrate antigens; and (3) generation and modulation of anti-carbohydrate responses in vaccine and immunotherapy development.

(1) Chapter 1 by R Kannagi introduces carbohydrate antigens as important determinants in immunological context. Chapters 2-4 focus on three different types of serum proteins that are involved in the innate recognition of microbial carbohydrates. The binding mechanism used by serum mannan binding protein (R Wallis), C-reactive protein (TW DuClos and C Mold) and complement proteins (P Gasque and BP Morgan) to differentiate self and non-self carbohydrate structures and their functions in innate and acquired immunity are reviewed. Chapters 5-6 describe pathogen recognition receptors on antigen-presenting cells. PR Taylor and colleagues focus on the biology of lectins on macrophages and dendritic cells while O Takeuchi and colleagues review toll-like receptors and their

ligand specificities and signalling mechanisms. To conclude the first section, S Rogers and SYC Wong summarise in Chapter 7 the structure and function of five groups of receptors belonging to the C-type lectin and lectin-like superfamily. In Chapter 8, L Nitschke and PR Crocker describe the Siglec family of receptors with specificity for sialic acid-containing carbohydrate structures.

(2) In Chapter 9, CG Vinuesa and ICM MacLennan cover B cell biology and special features of antibody responses to polysaccharides. In Chapter 10, SA Porcelli and colleagues describe glycoconjugate antigen presentation by the CD1 protein family and CD1-restricted T cell responses. To conclude the second section with Chapter 11, D Golgher and colleagues review studies on the MHC-restricted T cell recognition of glycopeptides.

(3) In Chapter 12, V Pozsgay provides a comprehensive review of bacterial carbohydrate synthesis. In Chapter 13, P Moingeon and colleagues illustrate the many challenges and opportunities in developing glycoconjugate vaccines against infectious diseases, and in Chapter 14 V Apostolopoulos review the potential of carbohydrate-based immunotherapeutics and vaccines for cancer and infectious diseases. Finally in Chapter 15, Galili presents the past and future approaches to modulate anti-gal antibody responses for the purpose of xenotransplantation.

There is no doubt that the number of topics in the immunobiology of carbohydrates will increase in the near future, but we believe that the current version will provide a valuable resource and a stimulus for many more studies in this relatively neglected area of immunological research.

Simon Y.C. Wong, Ph.D.
Gemma Arsequell, Ph.D.

ACKNOWLEDGEMENTS

We would like to thank all the authors for their enthusiasm for this project and their valuable contributions. We are indebted to Mr. Geoffrey Guile for his extreme care in proof reading all the chapters. We would like to acknowledge Professor Kurt Drickamer and many of our former colleagues at the Oxford Glycobiology Institute for educating us on the structure and function of carbohydrates, enzymes and lectins. We are grateful for the encouragement and support of Professor Raymond Dwek during our time in the Glycobiology Institute in Oxford. Finally Simon Wong would like to express his gratitude and appreciation of the support received from Professor Peter Beverley, the director of The Edward Jenner Institute for Vaccine Research.

Chapter 1

Carbohydrate Blood Group Antigens and Tumor Antigens

Reiji Kannagi

Abstract

Cell surface carbohydrates are recognized to be important antigens in protective immune responses against bacteria and viruses, as well as in adverse reactions in blood transfusion and organ transplantation. They also figure prominently in autoimmune diseases as autoantigens, and can contribute to diagnosis and possible immunotherapy for cancers. The monoclonal antibody approach has been responsible for the recent advances that led to the discovery of numerous hitherto-unknown carbohydrate determinants, some of which have been shown to perform crucial physiological functions. Carbohydrate determinants play the key roles in intercellular recognition and adhesion, and profoundly affect physiological behavior of cells. For instance, it was only recently found that they determine homing and distribution patterns of leukocyte subsets in the body, thus maintaining homeostasis of the immune system, and serving to regulate leukocyte recruitment in inflammatory responses. Significant progress has been achieved in the molecular biology of carbohydrate antigens: almost every glycosyltransferase controlling synthesis of major carbohydrate determinants has already been cloned and the regulatory mechanisms for the expression of individual carbohydrate antigens are currently under investigation.

ABO(H) Blood Group Antigens

ABO blood group antigens are carried by glycosphingolipids and glycoproteins, and widely distributed among various tissues. They are thus called histo-blood group antigens. The H antigen has the terminal structure Fucα1→2Galβ (Table 1). The A antigen has the terminal structure GalNAcα1→3(Fucα1→2)Galβ, and is synthesized by an α1→3 *N*-acetylgalactosaminyltransferase (A-enzyme) using UDP-GalNAc as a donor, and H antigen as an acceptor. The B antigen has the terminal structure Galα1→3(Fucα1→2)Galβ, and is synthesized by an α1→3 galactosyltransferase (B-enzyme) using UDP-Gal as a donor, and H antigen as an acceptor. ABO(H) antigens are carried by various carbohydrate core chains in glycolipids and glycoproteins. Those carried by glycolipids show considerable structural variation and are classified into four classes, the types 1–4 chains indicated in Table 2.

The apparent K_ms of A- and B-enzymes for UDP-GalNAc and UDP Gal are not very different from each other, and it has been proposed that the difference in k_{cat} explains the A- or B-specificity of the enzyme.[1–3] Cloning of the genes encoding these enzymes in 1990 revealed them to be highly homologous, and the enzymes encoded by the *A1* and *B* alleles differ in only four of their 354 amino acids, which are Arg176→Gly, Gly235→Ser, Leu266→Met and Gly268→Ala. The *O1* allele, the most frequent *O* allele, is characterized by deletion of nucleotide 261G. This results in a frame shift and the production of a truncated peptide with no enzymatic activity (Fig. 1). The molecular background of several variant ABO phenotypes has

Immunobiology of Carbohydrates, edited by Simon Y.C. Wong and Gemma Arsequell.
©2003 Eurekah.com and Kluwer Academic / Plenum Publishers.

Table 1. ABO(H) blood group determinants

Antigen	Structure
A antigen	GalNAcα1→3Galβ1→R 2 ↑ Fucα1
B antigen	Galα1→3Galβ1→R 2 ↑ Fucα1
H antigen	Galβ1→R 2 ↑ Fucα1

also been disclosed (for reviews see refs. 4,5). The *A2* allele has a deletion of the termination codon, which leads to the addition of 21 extra amino acids, and the *O2* allele has no deletion of 261G, but has a Gly268→Arg replacement, which is believed to abrogate enzymatic activity (Fig. 1). The type 3 and type 4 chains (Table 2) are poor substrates for the enzyme encoded by the *A2* allele, while that encoded by the *A1* allele produces a significant amount of type 3 and type 4 chain A antigens. Random combination of these common alleles (*A1*, *A2*, *B*, *O1* and *O2*) can result in 15 different genotype combinations, which can be still further increased by the addition of other less-frequent alleles.

The H determinant serves as the substrate for A- and B-transferases in individuals with blood types A, B and AB, while it is the end product of the synthetic pathway in blood type O individuals. The H determinants are synthesized by two α1→2 fucosyltransferases, and the genes for them were cloned as *FUT1* (*H*)[6,7] and *FUT2* (*Se*), respectively.[8,9] These enzymes are

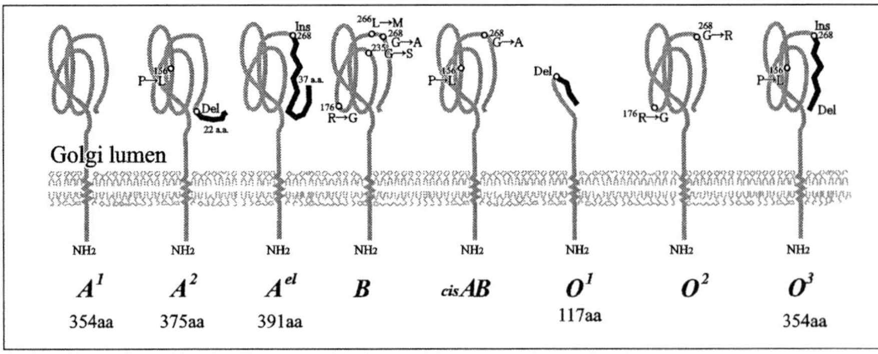

Figure 1. Schematic illustration of products of the major ABO alleles in the membrane of the Golgi apparatus. Golgi glycosyltransferases usually have a single transmembrane domain flanked by a short amino-terminal domain on the cytosolic side and a longer carboxy-terminal domain on the Golgi lumen side, thus forming the so-called type II transmembrane protein structure. Amino acid substitutions and nucleotide deletion (Del) or insertion (Ins) present in common ABO alleles are depicted. Adapted from references 167,168.

Table 2. Structural variation in blood group A-active determinants

Type 1 Chain

Type 1 Chain A

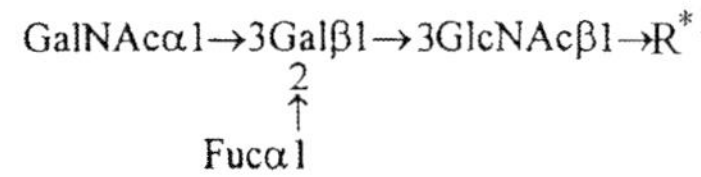

ALe^b^

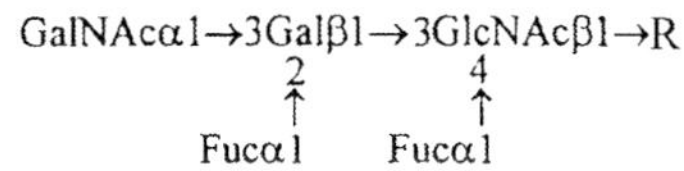

Type 2 Chain

A^a^

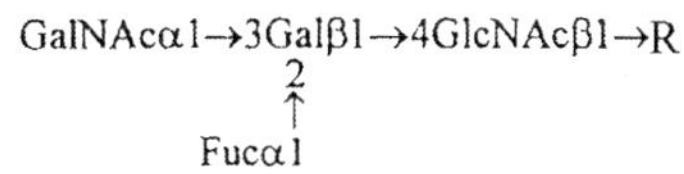

ALe^y^

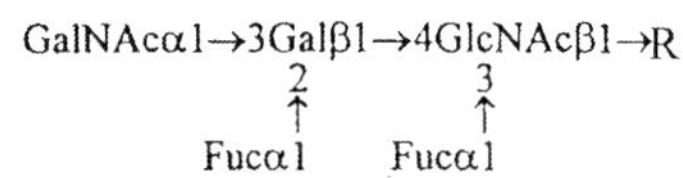

A^b^

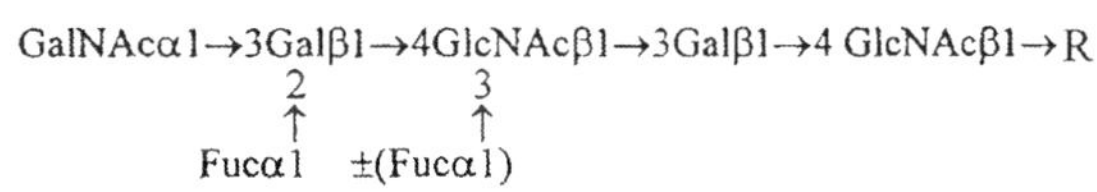

A^c^

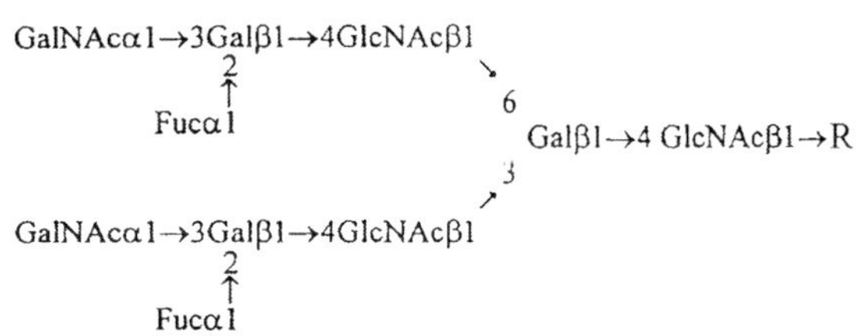

A^d^

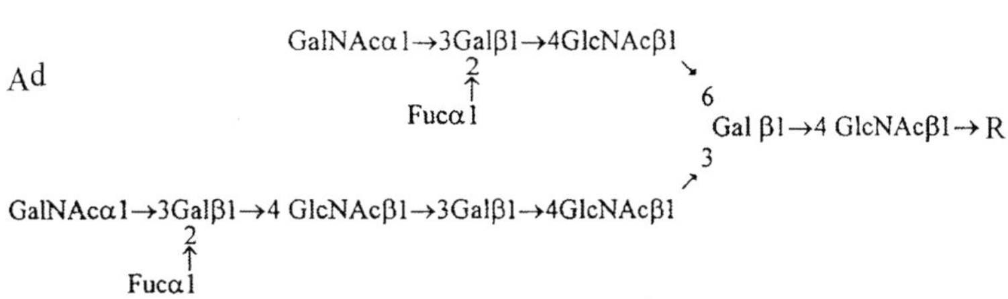

continued on next page

known to utilize both type 1 and type 2 chain substrates, and less efficiently the type 3 and type 4 chain substrates. Various mutation patterns have been reported in individuals deficient in these enzymes such as in the Bombay and Reunion island phenotypes.[10] *FUT1* is responsible for the synthesis of type 2 chain H determinants on erythrocytes, while *FUT2* is mainly involved in the synthesis of type 1 chain H and Le^b^ determinants on epithelial cells in the intestine. A striking difference between the α1→2 fucosyltransferases encoded by *FUT1* and *FUT2* is their apparent *K*m for GDP-fucose, which is 10–30 μM for *FUT1*, but as high as 0.1–0.2 mM for *FUT2*.[6,9,11]

Table 2. Structural variation in blood group A-active determinants (continued)

Type 3 Chain

A on A

GalNAcα1→3Galβ1→3GalNAcα1→3Galβ1→4 GlcNAcβ1→R
2 ↑ Fucα1 (on Galβ1→3GalNAc); 2 ↑ Fucα1 (on Galβ1→4 GlcNAc)

(H on A)

Galβ1→3GalNAcα1→3Galβ1→4 GlcNAcβ1→R
2 ↑ Fucα1 (on first Galβ1); 2 ↑ Fucα1 (on Galβ1→4 GlcNAc)

Type 4 Chain

Globo-A

GalNAcα1→3Galβ1→3GalNAcβ1→3Galα1→R
2 ↑ Fucα1 (on Galβ1)

(Globo-H)

Galβ1→3GalNAcβ1→3Galα1→R
2 ↑ Fucα1 (on Galβ1)

*1 R: Galb1Æ4Glcb1ÆCer in glycolipids, where these structures were elucidated. R could also be core structures composed of mannose residues and chitobiose in N-glycans, and core 1 ~ 4 in O-glycans for type 1 ~ 3 chain A determinants.

ABO(H) Determinants in Transfusion and Transplantation

The ABO system presents the most serious barrier to blood transfusion. In addition, it is involved in hemolytic disease of the newborn in ABO-incompatible pregnancy. As the incidence of Rh-incompatible hemolytic disease has decreased, ABO-incompatibility has become the major cause of hemolytic disease of the newborn. The ABO(H) determinants are also important transplantation antigens. ABO-incompatible solid organ transplantation is usually not performed except in cases of limited availability of cadaver donors, or for emergency transplantations. Although the removal of anti-A and -B natural antibodies by plasmapheresis and/or immunoadsorption before transplantation markedly reduces the frequency of hyperacute rejection crises, ABO-incompatibility is still the most significant risk factor affecting long-term allograft function, e.g. in ABO-incompatible living kidney transplantation.[12] ABO-incompatible allogeneic bone marrow transplantation is not uncommon, where anti-A or -B isoagglutinins sometimes cause pure red cell aplasia. Whether in ABO-incompatible organ transplantation, bone marrow transfusion or hemolytic disease of the newborn, IgG alloantibody is recognized to be more pathogenic than IgM antibody. Anti-A or -B alloantibody is usually IgM-dominant, but some patients have significant levels of IgG antibody. The mechanism for IgM/IgG switching in anti-A and -B alloantibody response remains to be elucidated. The distinction between *A1* and *A2* is implicated in incompatible organ transplantation.[13] *A1* determinant seems to be more immunogenic than *A2*, as kidney transplantation of *A2* kidneys is reported to cause an incompatibility reaction far less frequently than *A1* kidneys.[14]

Tumor-Associated Changes of ABO(H) Determinants

The expression of A and B determinants is considerably reduced in most cancer cells, thus paving the way to a relative increase of H antigen expression in cancer. This reduction is explained by the decrease in A- and/or B-enzyme in cancer, and this is attributed to transcriptional suppression either by some cancer-associated changes in some transcription factors, or epigenetic modification of the gene by DNA methylation.[4] An increase in α1→2 fucosyltransferase activity is known to occur in some cancers,[15] further accelerating the synthesis and expression of H-related carbohydrate determinants in cancer cells. Cancer-associated change in α1→2 fucosylation could be biphasic, as an increased expression of the Thomsen-Friedenreich (TF) antigen is noted in advanced cancers. This carries an exposed terminal galactose residue that remains unmodified by α1→2 fucosyltransferase.

Another interesting phenomenon in connection with the cancer-associated change of ABO(H) determinants is the appearance of illegitimate A or B determinants in cancers in patients with an irrelevant ABO blood group. Possible mistakes in blood subgroup typing or the cross-reactivity of diagnostic antibodies or lectins to A-like antigen such as Tn- or Forssman antigen may need to be taken into consideration in the interpretation of earlier results. However, the occurrence of illegitimate A determinants in cancer is repeatedly reported even in recent literature. It is mysterious that illegitimate A determinants occasionally appear in cancer tissues of type O individuals. The type *O2* individuals would have a greater chance of developing the illegitimate A or B antigen than *O1* individuals, since *O2* individuals have a full-length, although mutated, protein similar to the AB enzyme. Cancer tissues with illegitimate A determinant expression are reported to contain A-transferase-like protein as detected by an antibody specific for the A-enzyme, but most of the patients still have authentic *O1* genotype.[16]

The increase of H- and related determinants such as Le^y noted in some cancers is proposed to be associated with the remarkably increased transcription of *FUT1*.[15] On the other hand, some reports suggested that *FUT2* is mainly involved in Fucα1→2 modification in colon cancers.[17] In patients with blood type A, B or AB, another factor to be considered is the cancer-associated decrease of A- and B-antigen synthesis, which may result in a relative accumulation of H-antigen.

I and i Antigens

A low titer of antibodies that react with autologous red blood cells (RBCs) at cold temperature is present in the sera of essentially all adults. I antigen is one of the typical antigens recognized by such cold agglutinins (CA). This antigen is expressed on RBCs of essentially all adults except those with a rare inherited i-phenotype. CA having anti-i specificity react to RBCs of rare i-phenotype, and cord but not adult RBCs.

I and i antigens are both carbohydrate in nature, and are found in glycolipids and glycoproteins. I antigenicity is carried by the type 2 chain polylactosamines which contain GlcNAcβ1→6Galβ branching, while i antigen resides in the long straight type 2 chain polylactosamines without the branching structure (Table 3).[18–20]

The Ii antigens have physiological significance as differentiation antigens. Cord RBCs are mostly i antigen dominant, while adult RBCs are essentially all I antigen dominant. The iI-switching of RBCs is known to occur during the first year after birth. The beguiling hypothesis had been that the change in carbohydrate expression might somehow relate to the switching of fetal hemoglobin to adult hemoglobin. In the current knowledge, however, there is no direct relationship between i antigen expression and fetal hemoglobin production. It is known that the adult i-phenotype is sometimes associated with congenital cataract. This may be related to abnormal keratan sulfate synthesis in the lens of such individuals, since Ii antigens serve as core carbohydrate structures in keratan sulfate proteoglycans.

The genetic background of Ii antigen expression is complicated, because there is more than one enzyme that can synthesize the GlcNAcβ1→6Galβ branching of I antigen structures. At present two β1→6 *N*-acetylglucosaminyltransferases are known for I antigen synthesis,[21–23] and other isoenzymes yet to be identified may also be involved. It has been recently reported that the gene for one of the candidate enzymes is mutated in several individuals of rare i-phenotype.[24]

Table 3. Glycolipid antigens recognized by cold-reactive autoantibodies

Antigen recognized by cold agglutinin

I antigen

Galβ1→4GlcNAcβ1↘3
Galβ1→4GlcNAcβ1→3Galβ1→4Glcβ1→1Cer
Galβ1→4GlcNAcβ1↗6

i antigen

Galβ1→4GlcNAcβ1→3Galβ1→4GlcNAcβ1→3Galβ1→4Glcβ1→1Cer

Vo antigen (Sia-*l*1)

NeuAcα2→3Galβ1→4GlcNAcβ1→3Galβ1→4GlcNAcβ1→3Galβ1→4Glcβ1→1Cer

Gd antigen (Sia-*lb*1,2)

NeuAcα2→3Galβ1→4GlcNAcβ1→3Galβ1→4GlcNAcβ1→3Galβ1→4Glcβ1→1Cer

NeuAcα2→3Galβ1→4GlcNAcβ1↘3
Galβ1→4GlcNAcβ1→3Galβ1→4Glcβ1→1Cer
NeuAcα2→3Galβ1→4GlcNAcβ1↗6

Fl antigen (Sia-*b*1)

NeuAcα2→3Galβ1→4GlcNAcβ1↘3
Galβ1→4GlcNAcβ1→3Galβ1→4Glcβ1→1Cer
Fucα1→2Galβ1→4GlcNAcβ1↗6

Antigen recognized by Donath-Landsteiner antibody

P antigen

GalNAcβ1→3Galα1→4Galβ1→4Glcβ1→1Cer

Ii Antigens and Cold Agglutinin Disease

Ii antigens are known as autoantigens in CA disease. This is a hemolytic anemia sometimes accompanied by hemoglobinuria, which occurs in association with the presence of CA autoantibodies having a higher titer and greater thermal amplitude. CA disease develops as a complication in a variety of disorders, including mycoplasma pneumonia, infectious mononucleosis and malignant lymphoma, or for unknown reasons. Most of the autoantibodies have anti-I specificity. Autoantibodies with anti-i specificity are known to occur in CA disease secondary to infectious mononucleosis and malignant lymphoma.

Since Ii antigens serve as synthetic precursors for ABO(H) antigens as shown in Figure 2, the end products of the synthetic pathway contain both Ii-antigenic structures and ABO(H) antigens. It is interesting to note that autoantibodies are frequently directed to Ii-antigenic structures but not to the terminal ABO(H) antigenic structures. Antibodies of apparent anti-HI, anti-AI and anti-BI specificity reportedly occur, but only very rarely. Most autoantibodies are directed to Ii-antigenic structures or their sialylated derivatives (Table 3 and Fig. 2). An interesting example is the autoantibody having so called "Fl"-specificity, which is directed to the glycolipid molecule having sialic acid, I- and ABO(H) determinants, but it recognizes only the sialic acid, I- and αFuc portions common to these structures irrespective of the ABO blood group (Fig. 3). The A- or B-antigenic portion in the antigen molecules is not included in the Fl epitope.

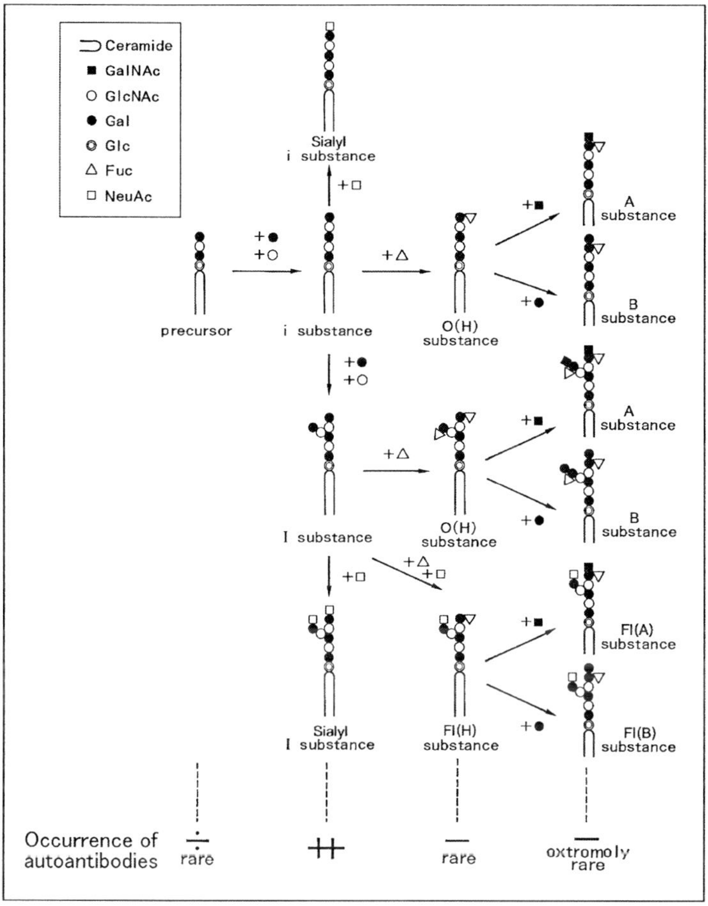

Figure 2. Ii antigens as synthetic precursors for ABO(H) determinants and auto-antigens for cold agglutinins. The autoantibodies are directed specifically to the precursor antigen epitopes, not to the strong alloantigenic epitopes at the terminus.

Characteristics of Anti-Ii Antibodies

High idiotypic connectivity is one of the characteristic features of cold-reactive anti-Ii antibodies. This was noticed as early as 1968 when polyclonal anti-idiotypic antibodies against anti-Ii antibodies were generated.[25] The molecular background of this idiotypic homogeneity is that these anti-Ii antibodies are encoded by the V_H4-34 heavy chain gene (the same gene is designated as V_H4.21 in earlier references). The V_H4-34 gene is found in virtually all cases of CA disease, whether in idiopathic CA diseases, lymphoproliferative disorders, or after infection with Epstein-Barr virus or *Mycoplasma pneumoniae.*[26–31] Examples of the first three cases of monoclonal CA are given in Table 4. Ii antigens were shown to bind to the framework region 1 (FR1) domain of immunoglobulin encoded by the V_H4-34 gene, and the antibodies encoded by V_H4-34 are proposed to be intrinsically autoreactive without requiring somatic mutation and independently of the associated light chains. The V_H4-34 gene is used in 5% of healthy adult B lymphocytes[32] and is frequently found in diffuse large-cell lymphoma, primary central nervous system lymphoma, B-chronic lymphocytic leukemia, and autoimmune disorders.[33] In addition to anti-Ii antibodies, the gene is frequently utilized in encoding anti-carbohydrate antibodies including those specific for the A-antigen[31] and some gangliosides.

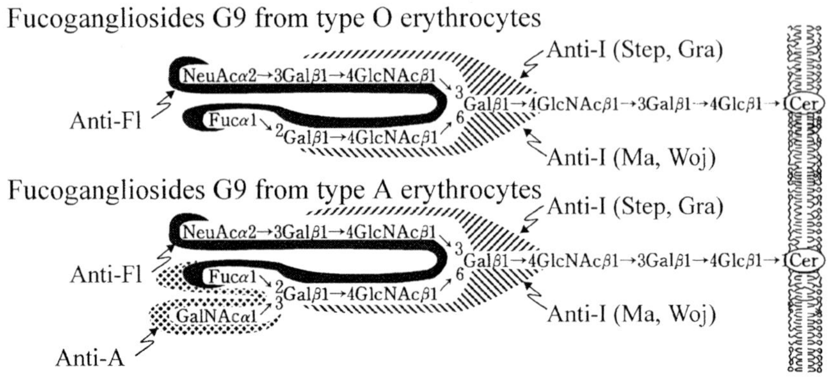

Figure 3. Antigenic epitope recognized by typical anti-I cold agglutinins and an "Fl"-type cold agglutinin. Typical antigenic epitope defined by anti-A alloantibody is shown for comparison. Step, Gra, Ma and Woj refer to well-characterized human monoclonal anti-I cold agglutinins. The epitope recognized by the "Fl"-type antibody is located nearest to the ABO(H) determinants among the epitopes defined by known autoantibodies. It recognizes the epitope composed of I antigen, sialic acid and fucose. Its epitope, however, does not include ABO(H) determinants on human erythrocyte fucoganglioside G9. The structures of fucogangliosides G9 prepared from type O (upper panel) and type A (lower panel) erythrocytes[169] are shown.

Mouse monoclonal antibodies directed to Ii and related antigens are also encoded by a very limited set of variable region genes. As shown in Table 4, V_H441 and V_HX24 (which is 98% homologous to V_H441) encoded all antibodies so far investigated. The V_H441 gene in mice is also known to encode other anti-carbohydrate antibodies, including essentially all antibodies directed to Lewis^x (Le^x) and galactosylgloboside (SSEA-3), both of which have β-galactose terminal as well as Ii antigens.[34]

Ii Antigen As Receptors for Pathogenic Microorganisms and Other Substances

B lymphocytes producing self-reactive antibodies are usually deleted by a stringent selection process. Most anti-Ii antibodies in the sera of healthy individuals do not react to their own RBCs at physiological body temperatures, but do so at cold temperatures. Thus they would not normally function as autoantibodies. This suggests that the immune selection mechanism for B lymphocytes with anti-Ii specificity would not be disturbed under normal physiological conditions.

It is interesting to note that I antigen serves as a receptor for some pathogenic microorganisms such as myxoviruses and mycoplasma. Binding of microorganisms to Ii antigens would be expected to limit the intramolecular motion of sugar residues in Ii determinants. This effect is similar to the application of cold temperatures, which enhances antibody binding to Ii. In this sense, anti-Ii antibodies may well be useful for self-defense. The physiological significance of antibodies encoded by V_H4–34 heavy chain gene is not entirely clear, but a role in surveying for any abnormality occurring on cell surface carbohydrates bearing the Ii- and related determinants is likely. The antibodies, however, become pathogenic when mutations introduced in their variable region confer reactivity against unaffected Ii antigen on intact RBCs at near body temperatures. Transient CA disease sometimes occurs secondary to mycoplasma infections.

Even some drugs and chemicals may eventually bind to Ii antigens. In some drug-induced immune hemolytic anemia, formation of a ternary complex of pathogenic antibody, causative drug and RBCs is known to be highly dependent on the presence of I antigen on the cells, as if the I antigen were a receptor for the causative drug or drug-antibody complex.[35–37]

Table 4. Usage of variable region genes encoding human and mouse anti-Ii and related antibodies

Antibody (Specificity)	Heavy Chain			Light Chain	
	VH	D	JH	Vκ	Jκ
Human					
KAU (anti-I)	VH4-34 (4.21)	*n.id.*[*1]	JH4	VκIIIb Hum/kv325/kIIIb	Jκ4
FS-1 (anti-I)	VH4-34 (4.21)	XP'1	JH3	VκII	Jκ2
FS-2 (anti-I)	VH4-34 (4.21)	Q52	JH4	VκIIIb HumIGKVQ	Jκ5
Mouse					
C-6 (anti-I)	VH441[*2]	DQ52	JH3	VκII (VκTEPC105)	Jκ4
F1α87 (anti-I)	VHX24[*2]	DSP2.2	JH3	*n.id.*	*n.id.*
Tan13 (anti-i)	VHX24	DSP2.8	JH2	VκVI (VκX24)	Jκ5
1B2 (anti-PG, i)[*3]	VH441	DSP2.8	JH1	*n.id.*	*n.id.*
F1H11 (anti-PG, i)[*3]	VH441	*n.id.*	JH2	VκVI (VκOX-1)	Jκ2

[*1] Encoding germline gene was not identified.
[*2] VH441 and VHX24 both belong to VH4 family and are 98% homologous to each other.
[*3] Reactive to both paragloboside (neolactotetraosylceramide) and i-antigen.

P Blood Group Antigens

The P blood group system consists of three major antigens, P, P^k and P^1, which are known to be on glycolipids. P (globoside, Gb_4[*1]) and P^k (ceramide trihexoside, CTH, or Gb_3) antigens are expressed on RBCs of essentially all individuals, and deficient only on RBCs of rare individuals of P^k phenotype who lack globoside, or those of *p* phenotype who lack both globoside and CTH on their RBCs.

[*1] Abbreviations according to the IUPAC-IUB Joint Commission on Biochemical Nomenclature (JCBN) Nomenclature of glycolipids, recommendations 1997,[166] where Gb refers to globo-series, Lc to lacto-series, nLc to neolacto-series, and Gg to ganglio-series glycolipids.

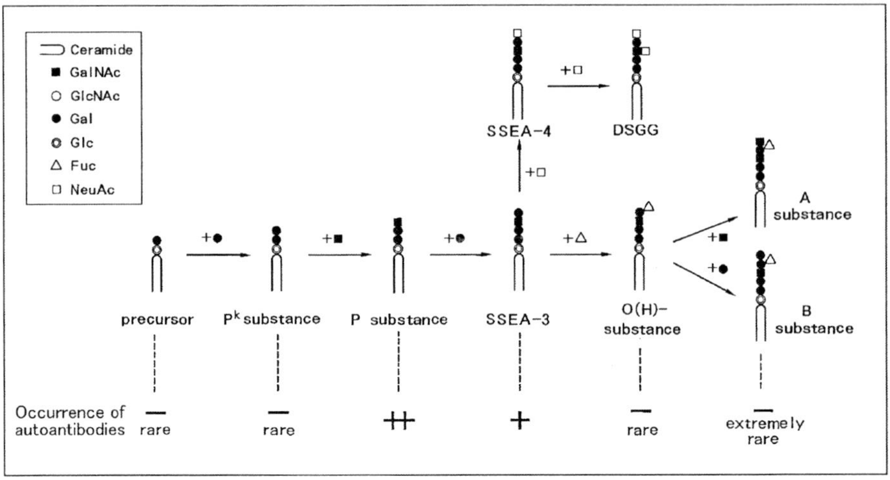

Figure 4. P antigen as synthetic precursor for ABO(H) determinants and autoantigens for Donath-Landsteiner antibodies. The autoantibodies are directed specifically to the parecursor antigen epitopes, and not to the strong alloantigenic epitopes at the terminus.

Globoside and CTH belong to the globo-series glycolipids (Table 5). As globoside is the major neutral glycolipid in human erythrocytes, it was long believed to be the final product in globo-series glycolipid synthesis. In some cells and tissues, however, globoside turned out to be a precursor of more complex glycolipids such as galactosylgloboside (SSEA-3), sialosyl galactosylgloboside (SSEA-4), disialosyl galactosylgloboside (DSGG) and type 3 chain H and A determinants (Table 5).[38–43] These substances are present on RBCs as well; DSGG is one of the major gangliosides of human RBCs, and type 3 chain H and A antigens are significant components of H or A antigens carried by RBCs. Genes for an $\alpha 1 \rightarrow 4$ galactosyltransferase which synthesizes CTH,[44,45] and $\beta 1 \rightarrow 3$ *N*-acetylgalactosaminyltransferase which synthesizes globoside[46] have been cloned recently. A gene for $\beta 1 \rightarrow 4$ galactosyltransferase which synthesizes galactosylgloboside was also cloned,[47] and its relation to the P blood group system is the subject of ongoing studies.[48–50]

These findings indicate that the P antigen, globoside Gb_4, is an intermediate metabolite in the synthetic pathway of globo-series glycolipids, and that the final products of the pathway are ABO(H) determinants. The entire synthetic pathway is shown in Figure 4. It is very similar but not identical to the synthetic pathway of neolacto-series ABO(H) antigen, where Ii antigens serve as synthetic precursors for AB(H) antigen synthesis (Fig. 2). Since these globo-series ABO(H) antigens are synthesized through a pathway different from that for the synthesis of neolacto-series ABO(H) antigens, the internal core sugar sequence is entirely different from that of neolacto-series ABO(H) antigens.

Physiological Significance of Antigens in P Blood Group System

The incidence of individuals lacking P antigen, such as type *p* or P^k individuals, is very low at 5.8 or less per 1,000,000. Since the P antigen has been shown to express in embryos in a stage-specific manner, the antigen could be regarded as a differentiation or developmental antigen in most of the population. In fact, early-stage embryos are very rich in globo-series glycolipids, and nearly all of them including P^k and P antigens appear and disappear sequentially during the development of early-stage embryos. The extended globo-series glycolipids such as SSEA-3 and -4 are good markers for human embryonal teratocarcinoma cells and pre-implantation embryos. While expression of SSEA-1 is rather restricted to murine embryos, SSEA-3 and -4 are expressed in humans, and both of these antigens have been utilized to

Table 5. Extended globo-series glycolipid antigens related to P blood group system

Antigen	Structure	Synonyms
CTH	Galα1→4Galβ1→4Glcβ1→1Cer	P^k-antigen, Gb_3, CD77
Globoside	GalNAcβ1→3Galα1→4Galβ1→4Glcβ1→1Cer	P-antigen, Gb_4
Galactosylgloboside	Galβ1→3GalNAcβ1→3Galα1→4Galβ1→4Glcβ1→1Cer	SSEA-3, (Gb_5)
MSGG	NeuAcα2→3Galβ1→3GalNAcβ1→3Galα1→4Galβ1→4Glcβ1→1Cer	SSEA-4
DSGG	NeuAcα2↘6 NeuAcα2→3Galβ1→3GalNAcβ1→3Galα1→4Galβ1→4Glcβ1→1Cer	
Sulfated Gb_5	SO_3^-–3Galβ1→3GalNAcβ1→3Galα1→4Galβ1→4Glcβ1→1Cer	
Globo-H	Galβ1→3GalNAcβ1→3Galα1→4Galβ1→4Glcβ1→1Cer Fucα1↗2	Type 4 chain H
Globo-A	GalNAcα1→3Galβ1→3GalNAcβ1→3Galα1→4Galβ1→4Glcβ1→1Cer Fucα1↗2	Type 4 chain A
Forssman	GalNAcα1→3GalNAcβ1→3Galα1→4Galβ1→4Glcβ1→1Cer	(Gb_5)

characterize human pluripotent or embryonic stem cells or related cell lines.[51-53] Globo-series glycolipids are differentiation antigens also in blood cells.[54-56] They also appear on some cancer cells such as kidney cancers, where DSGG is supposed to play a significant role in the adhesion of cancer cells to endothelial cells during the course of hematogenous metastasis.[57-60] The counter receptor on endothelial cells for DSGG has not been identified, but recently Siglec7, a member of the family of sialic acid binding proteins, was shown to bind to DSGG.[61]

Globo-series glycolipids are known to participate in carbohydrate-carbohydrate interaction. Globoside or galactosylgloboside interacts with paragloboside (lactoneotetraosylceramide, nLc4), Gg_3,[62] or G_{M3}.[63] CTH is also known as CD77, the antigen specifically appearing on cells from Epstein Barr virus-induced B cell leukemia.[54,55] An impressive historical example of cancer association of P-blood group substance is the P-like antigen in a tumor of a patient with blood type *p*.[64-66] The patient developed high titers of the antibody directed to the P-blood group antigen in the serum and survived without any evidence of recurrence long after surgical treatment. This case represents the first reported example showing the possibility of cancer immunotherapy targeted to the carbohydrate determinants at the surface of the cancer cells.

Paroxysmal Cold Hemoglobinuria and P Antigen

The P antigen is also known as the auto-antigen recognized by the so-called Donath-Landsteiner antibody in patients with paroxysmal cold hemoglobinuria (PCH). This autoantibody is a biphasic antibody; it binds to human RBCs at cold temperatures, and activates complement to cause hemolysis at or near body temperature. Besides patients with idiopathic PCH, there are those who develop it secondary to syphilis. Infantile transient PCH sometimes occurs secondary to common viral infections.

Among the globo-series glycolipids of human erythrocytes, the Donath-Landsteiner autoantibodies in PCH are specifically directed to the P antigen, the precursor molecule, and the autoantibodies directed to the globo-series ABO(H) antigens, a final product, are very rare (Fig. 4). Some Donath-Landsteiner antibodies recognize the intermediate structures such as SSEA-3 or -4 antigens, but such antibodies are less frequent than those directed to the P antigen. A very similar situation is seen in CA disease. Although the differentiation antigens, such as Ii and P antigens, and strong alloantigenic epitopes such as ABO(H) determinants, are simultaneously present on the RBCs, the autoimmune reaction is directed specifically to the differentiation antigen epitopes only. Similar to Ii antigens, P antigen is also known to serve as a receptor for some bacteria, which infect the intestine and urinary bladder such as *E coli*, and some bacterial toxins.[67] This implies that the immune system is more prone to autoreactivity towards self differentiation antigens rather than strong alloantigens, and that the dysfunction of autoreactive B cell clones could be triggered by infection by some viruses and bacteria.

One of the striking differences between the autoantibodies in PCH and those in CA disease is that the former are mostly IgG, while the latter are almost entirely IgM. The globo-series glycolipids seem to elicit a significant IgG response in humans; even alloantibodies appearing in the sera of rare *p* and P^k phenotypes contain a significant amount of IgG. The precise reason for this peculiar antigen-dependent IgM/IgG switching in antibody production is not known at present, but internal α-galactose residue in the structure may well be involved. It is tempting to speculate that the production of IgG antibodies directed to A- or B- antigens in patients with ABO-incompatible pregnancy may be related to the response to type 3 chain and/or type 4 chain A- or B- antigen. Some T-lymphocytes are known to recognize α-galactosylceramide,[68] but it is not known if they have helper activity or not.

Lewis Blood Group Antigens

The Lewis blood group system is composed of two major carbohydrate antigens, Le^a and Le^b. Two minor antigens, Le^c and Le^d, were found from detailed serological studies, and several sialylated or sulfated forms of antigens were identified mainly by monoclonal antibodies. All determinants are based on the type 1 chain lactosamine structure as shown in Table 6.

Table 6. Structures of type 1 chain antigens related to Lewis blood group

Antigenicity	Structure
Le^a	Galβ1→3GlcNAcβ1→R Fucα1↗4
Le^b	Galβ1→3GlcNAcβ1→R Fucα1↗2 Fucα1↗4
Le^c	Galβ1→3GlcNAcβ1→R
Le^d (type 1 chain H)	Galβ1→3GlcNAcβ1→R Fucα1↗2
Elongated Le^a	Galβ1→3GlcNAcβ1→3Galβ1→4GlcNAcβ1→R Fucα1↗4
Le^a on Le^x	Galβ1→3GlcNAcβ1→3Galβ1→4GlcNAcβ1→R Fucα1↗4 Fucα1↗3
Repetitive Le^a	Galβ1→3GlcNAcβ1→3Galβ1→3GlcNAcβ1→R Fucα1↗4 Fucα1↗4
Sialyl Le^a	NeuAcα2→3Galβ1→3GlcNAcβ1→R Fucα1↗4
Sialyl Le^c	NeuAcα2→3Galβ1→3GlcNAcβ1→R
Disialyl Le^c	NeuAcα2↘6 NeuAcα2→3Galβ1→3GlcNAcβ1→R
Disialyl Le^a	NeuAcα2↘6 NeuAcα2→3Galβ1→3GlcNAcβ1→R Fucα1↗4
3'-Sulfo Le^a	SO_3^-– 3Galβ1→3GlcNAcβ1→R Fucα1↗4

The α-fucose residue linked to penultimate βGlcNAc through 1→4 linkage in Table 6 is synthesized by an α1→4 fucosyltransferase encoded by *Le* or *Fuc-T III* (*FUT3*) gene. This enzyme has a capability to utilize both type 1 chain and type 2 chain substrates and is termed α1→4/3 fucosyltransferase, although it has a strong preference for type 1 chain substrates. The α-fucose residue linked to terminal β-galactose through 1→2 linkage in Table 6 is synthesized by the α1→2 fucosyltransferase encoded by *FUT2* (*Se*), which was described in the section on ABO(H) antigens. Individuals with both *Le* and *Se* genes have Le^a and Le^b antigens in their saliva and Le^{a-b+} RBCs, while those without *Se* gene have only Le^a antigen in their saliva and Le^{a+b-} RBCs, and those without *Le* gene have Le^{a-b-} RBCs. In Europeans and white Americans, Le^{a-b+} comprises 72% of the population, followed by Le^{a+b-} (22%), and Le^{a-b-} (6%), while the percentage of Le^{a-b-} is as high as 22% in Afro-Americans.[69] The Le^{a+b+} phenotype is frequently encountered in people in East Asia and the Pacific rim region, due to the presence of *Se* genes encoding less efficient α1→2 fucosyltransferase.

Le^a and Le^b antigens do not express well in cord blood cells. Le^a antigen starts to appear soon after birth, but Le^b develops much later. Expression of Le^b is known to reach the adult level at 6 years of age, explaining why the RBCs of most children show Le^{a+b+} phenotype. It has long been known that Le^a and Le^b glycolipid antigens are mainly synthesized by intestinal epithelial cells, secreted into the blood stream, and adsorbed at the surface of RBCs.[70] This process is sometimes affected by abnormalities in serum lipoprotein composition in pregnancy or malignant disorders, and this eventually results in a prominent decrease of Le^a and Le^b antigen expression on RBCs.

Lewis Blood Group Antigens in Malignant and Non-Malignant Disorders

Monoclonal antibodies raised against cancer-associated antigens by immunizing human cancer cells are quite often found to be specific for carbohydrate determinants. N19-9, one of such antibodies obtained by immunization of human colon cancer cells, recognizes carbohydrate antigen preferentially expressed on cancers of the digestive organs including pancreas, biliary tract and colon. The antigen thus termed carbohydrate antigen 19-9 or CA19-9. The biochemical entity of the antigenic epitope defined by this antibody turned out to be a sialylated form of Le^a antigen, where sialic acid is attached to terminal β-galactose through a 2→3 linkage (Table 6). Many other monoclonal antibodies raised against human cancer cells, including SPan1, C-50 and KMO-1, were later shown to have similar specificity to that of CA19-9 in that they recognize the sialyl Le^a antigen. All antibodies are useful in detecting the sialyl Le^a antigen in cancer tissues in immunohistochemical examination, and also in detecting soluble forms of the antigen in the sera of patients. The positive occurrence of the antigen in the sera of patients with pancreatic cancers is reported to be as high as 80–90%.

An antibody raised against human pancreatic cancer cells, DU-PAN-2, was shown to recognize sialyl Le^c antigen (Table 6). Expression of the sialyl Le^a antigen is dependent on the Lewis blood group, and the antigen does not appear on tumors in patients with the Le^{a-b-} phenotype. Sialyl Le^c frequently appears on such tumors, but less frequently than sialyl Le^a in tumors from patients with the Le^{a-b+} or Le^{a+b-} phenotypes.

Possible Mechanisms for the Enhanced Expression of Sialyl Le^a in Cancers

The exact reason why expression of sialyl Le^a antigen is enhanced in cancers remains unclear to date. Immunohistochemical examination of cancer tissue specimens of the digestive organs indicates an increased expression of sialyl Le^a antigen in cancer cells compared to corresponding non-malignant epithelial cells, while Le^a antigen is equally expressed on both types of cells. This finding suggests that a change in sialic acid modification of Le^a is responsible for the increased expression of sialyl Le^a in cancer cells. Sialyl Le^a antigen is synthesized sequentially by, first, the formation of the type 1 chain precursor through the action of a β1→3 galactosyltransferase, followed by its modification with α2→3 sialyltransferase and α1→4 fucosyltransferase (Fuc-T III). The enzymatic activity of the β1→3 galactosyltransferase for synthesis of type 1 chain precursors is known to be decreased,[71] and the α1→4 fucosyltransferase

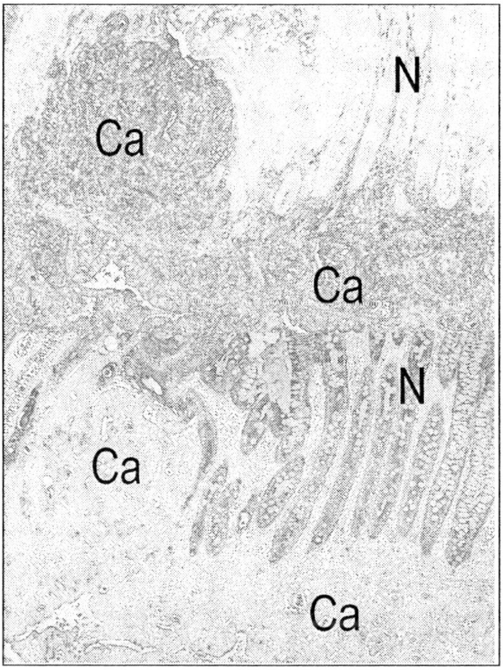

Figure 5. Typical distribution pattern of sialyl Le^a (upper panel) and disialyl Le^a (lower panel) antigens in colon cancer tissues. Ca, cancer cells; N, non-malignant epithelial cells. Note that α2→3 monosialyl Le^a antigen is preferentialaly expressed on cancer cells, while α2→3, 2→6 disialyl Le^a antigen is specifically localized in non-malignant epithelial cells. Adapted from reference 77.

activity shows no significant change between cancer cells and non-malignant epithelial cells in the digestive organ.[72] The only enzyme among them that shows a significant increase in cancer cells is a α2→3 sialyltransferase activity towards type 1 chain substrates.[73] These findings were confirmed by the recent comparison of transferase mRNA content between cancer and non-malignant epithelial cells.[74,75] The preferential expression of sialyl Le^a antigen in these types of cancers may, at least partly, attributable to an increased expression of α2→3 sialyltransferase for type 1 chain substrates.

The expression of the disialyl Le^a antigen, which has an additional sialic acid residue attached to the C6 position of penultimate *N*-acetylglucosamine (Table 6), is strongly expressed on non-malignant epithelial cells in the digestive organs, whereas its expression is significantly decreased in cancer cells (Fig. 5).[76,77] These findings suggest that a cancer-associated decrease of α2→6 sialylation would result in an increase in sialyl Le^a in cancers. Here a change in sialic acid modification is again involved in the increased expression of sialyl Le^a in cancer cells. The expression of 3'-sulfo Le^a antigen (Table 6) is also known to decrease upon malignant transformation of the same cell types. This would result in the accumulation of non-sialylated Le^a antigen, and eventually in sialyl Le^a antigen due to the action of α2→3 sialyltransferase.

Sialyl Le^a and Cancer Metastasis

Sialyl Le^a antigen on cancer cells serves as a ligand for E-selectin, a cell adhesion molecule in the selectin family that is expressed on vascular endothelial cells, and which mediates adhesion of cancer cells to vascular beds (Fig. 6).[78] This adhesion is proposed to be involved in the hematogenous metastasis of cancer cells.[79] Therefore, sialyl Le^a antigen is not merely a marker for cancers, but a functional molecule involved in the malignant cell progression. The patients whose cancer cells strongly express sialyl Le^a antigen have an increased risk of developing hematogenous metastasis, and tend to have a poorer prognosis than patients having cancer cells which do not or only weakly express this antigen. This is most noticeable in the statistics of patients with colon cancers (Fig. 7),[80–82] where hematogenous metastasis to the liver is the main factor determining the prognosis.

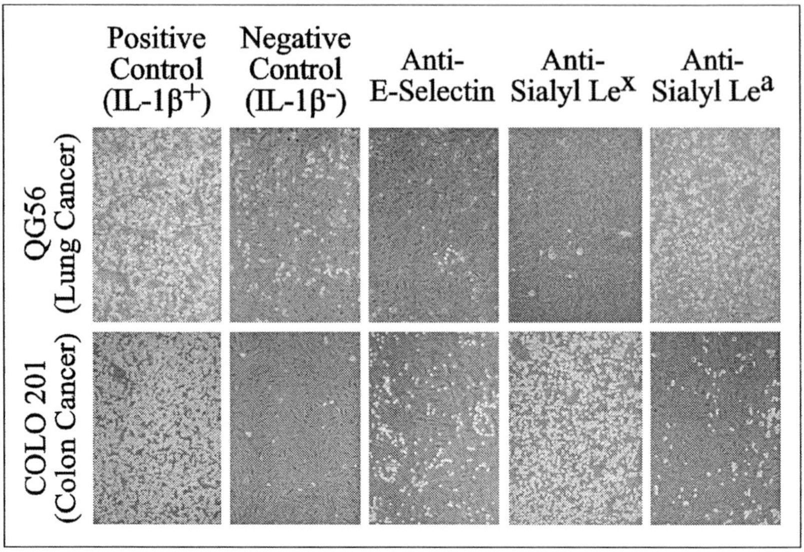

Figure 6. E-selectin-mediated adhesion of human cancer cells to endothelial cells. Typical examples of non-static monolayer cell adhesion experiments indicating adhesion of cultured human cancer cells to IL-1β-stimulated human umbilical vein endothelial cells. Note that adhesion of a cultured human lung cancer cell line QG56 is inhibited by anti-sialyl Lewis^x antibody, while that of the cultured human colon cancer cell line COLO201 is inhibited by anti-sialyl Lewis^a antibody. Adhesion of both lines is completely inhibited by anti-E-selectin antibody. Adapted from reference 78.

Differentiation and Developmental Carbohydrate Determinants, Lewis^x, Lewis^y, Sialyl Lewis^x and Related Antigens

Le^x (CD15), Le^y and sialyl Le^x (CD15s) antigens are isomers of Le^a, Le^b and sialyl Le^a antigens, respectively (Table 7). They have type 2 chain lactosamine as their backbone structure, and are synthesized by glycosyltransferases partly overlapped, but mostly different from, those for Le^a, Le^b and sialyl Le^a antigens.

The α-fucose residue linked to penultimate βGlcNAc through 1→3 linkage is synthesized by various α1→3 fucosyltransferases each of which has a distinct substrate specificity. Enzymes encoded by *Fuc-T IV* (*FUT4*), *Fuc-T VI* (*FUT6*), *Fuc-T III* (*Se*, *FUT3*) and *FUT9* are involved in the synthesis of Fucα1→3 linkage in Le^x and Le^y, while *FUT1* and *FUT2* are involved in the synthesis of Fucα1→2 linkage at the terminal galactose moiety. Fucosyltransferases involved in the synthesis of Fucα1→3 linkage in sialyl Le^x antigen include those encoded by *Fuc-T VII* (*FUT7*), *Fuc-T VI*, *Fuc-T III* and to a lesser extent *Fuc-T IV*. ST3Gal4 is proposed as the main sialyltransferase that provides sialylated substrates for these fucosyltransferases in its synthesis. Thus, expression of these antigens is, although not completely unrelated to, largely independent of the Lewis blood phenotype of the individuals. These antigens do not appear on RBCs and are not regarded as blood group antigens.

A developmental antigen specifically expressed in the murine pre-implantation embryos called SSEA-1 is the Le^x antigen that is carried by a long straight polylactosamine having the i antigenic structure (Table 7).[83] SSEA-1 is proposed to be involved in cell adhesion and cell compaction in pre-implantation embryos, but to date its counter-receptor proteins have not been identified. The cell adhesive activity of Le^x antigen is proposed to depend on carbohydrate-carbohydrate interaction among Le^x antigens.[84]

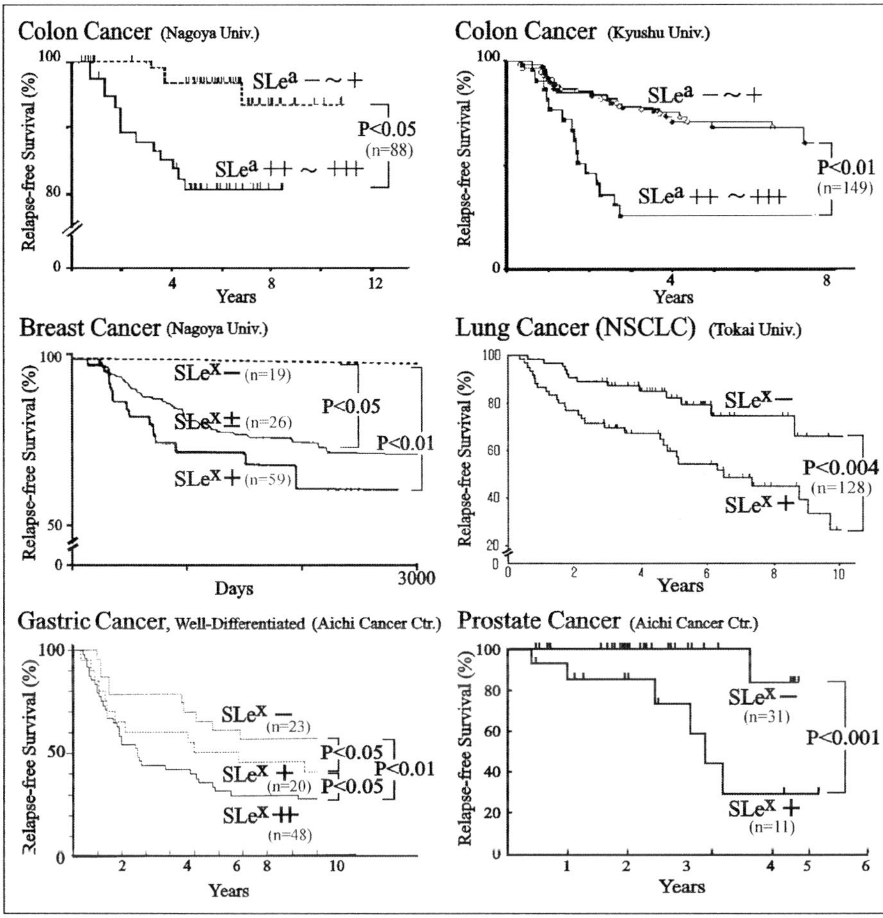

Figure 7. Correlation between the prognosis of patients with cancers and the degree of expression of sialyl Le^a or sialyl Le^x determinants in cancer tissues. Adapted from references 80–82, 170–172.

Sialyl Le^x and Sialyl 6-sulfo Le^x As Ligands for Cell Adhesion Molecules of the Selectin Family

Sialyl Le^x antigen is expressed on human leukocytes, and provides the specific ligands for the cell adhesion molecules of the selectin family that are expressed on vascular endothelial cells. The interaction of leukocytes and endothelial cells mediated by this cell adhesion system is the first step in the extravasation of leukocytes into the inflammatory lesion. Sialyl Le^x antigen is constitutively expressed on neutrophils and monocytes, but is not expressed on most resting lymphocytes in the peripheral blood of healthy individuals.[85,86] Its expression is strongly induced on lymphocytes upon activation, and this is accompanied by the transcriptional induction of the *Fuc-T VII* gene (Fig. 8).[87,88] Small subsets of lymphocytes such as natural killer and helper memory T lymphocytes are known to constitutively express sialyl Le^x-like antigens. The antigen called CLA (cutaneous lymphocyte antigen), which is defined by a monoclonal antibody HECA-452, and proposed to be specific to a skin-homing subset of helper memory T cells, is also a sialyl Le^x-like antigen.[89,90] Activated T helper 1 lymphocytes are known to express this antigen more strongly than T helper 2 lymphocytes.[91]

Table 7. Type 2 chain antigens

Antigen	Structure
Le^x	Galβ1→4GlcNAcβ1→R Fucα1↗3
Le^y	Galβ1→4GlcNAcβ1→R Fucα1↗2 Fucα1↗3
Type 2 chain H	Galβ1→4GlcNAcβ1→R Fucα1↗2
Sialyl Le^x	NeuAcα2→3Galβ1→4GlcNAcβ1→R Fucα1↗3
SSEA-1 (Le^x-i)[*1]	Galβ1→4GlcNAcβ1→3Galβ1→4GlcNAcβ1→R Fucα1↗3
Sialyl Le^x-i	NeuAcα2→3Galβ1→4GlcNAcβ1→3Galβ1→4GlcNAcβ1→R Fucα1↗3
Sialyl difucosyl antigen	NeuAcα2→3Galβ1→4GlcNAcβ1→3Galβ1→4GlcNAcβ1→R Fucα1↗3 Fucα1↗3
Sialyl 6-sulfo Le^x	SO_3^-↘6 NeuAcα2→3Galβ1→4GlcNAcβ1→R Fucα1↗3
6-Sulfo Le^x	SO_3^-↘6 Galβ1→4GlcNAcβ1→R Fucα1↗3
3'-Sulfo Le^x	SO_3^-– 3Galβ1→4GlcNAcβ1→R Fucα1↗3

[*1] Minimal structure required by anti-SSEA-1 antibody.

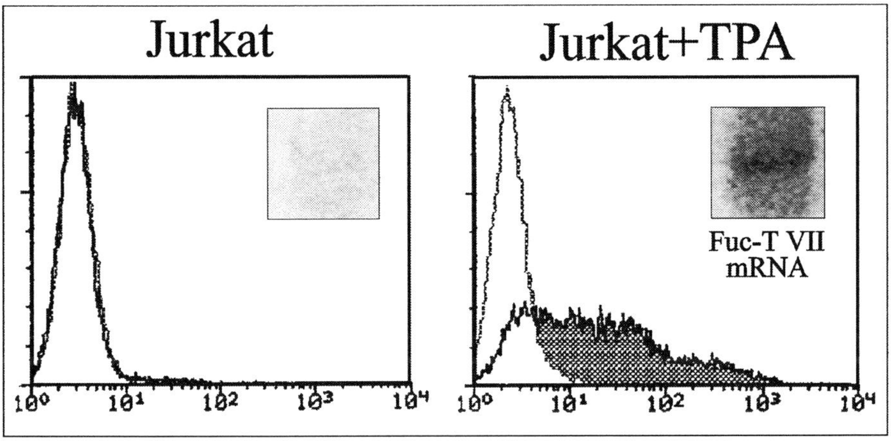

Figure 8. Induction of sialyl Le^x determinant on cultured human T lymphoid cells Jurkat, stimulated with TPA. Left and right insets, Northern blot for *Fuc-T VII*, indicating a marked induction of transcription associated with the surface appearance of sialyl Le^x.

There is good evidence showing that some sialylated and fucosylated determinants other than sialyl Le^x also serve as ligands for selectins.[92–95] The penultimate *N*-acetylglucosamine is fucosylated in sialyl Le^x, while in such determinants, fucosylation is noted at more internal *N*-acetylglucosamine residues. These determinants found in cells of myeloid lineage are termed "myeloglycan",[95] and are most probably synthesized by Fuc-T IV.

The presence of sialyl Le^x or these sialylated and fucosylated carbohydrate determinants on ligand molecules is sufficient for their interaction with E-selectin, while P-selectin requires these carbohydrate determinants to be expressed on a specific core protein, PSGL-1 (P-selectin glycoprotein ligand-1). PSGL-1 is expressed on virtually all human leukocytes, and the sulfation of tyrosine residues near its amino terminal region is a prerequisite for P-selectin to interact properly with the ligands.

L-selectin is unique in that it mediates the homing of lymphocytes, even in the absence of inflammation, as a routine homeostatic process for maintaining the proper distribution of lymphocytes. High endothelial venules (HEVs) of peripheral lymph nodes are known to serve as an entrance for homing lymphocytes, and the carbohydrate ligands expressed on human HEV are identified as sialyl 6-sulfo Le^x,[96–98] while some 6'-sulfated or 6,6'-disulfated antigens have been noted in the HEVs of rodents.[99–101] Sialyl 6-sulfo Le^x is also expressed on a subset of leukocytes, where it serves as a ligand for E- and P-selectin.[102] The sialyl 6-sulfo Le^x antigen (also termed as 6-sulfo sialyl Le^x in earlier reports[*2]) has an additional sulfate residue at the C6 position of the penultimate *N*-acetylglucosamine (Table 7). The 6-sulfation is mediated by the action of GlcNAcβ 6-sulfotransferases, and several genes for the candidate enzymes involved in sialyl 6-sulfo Le^x synthesis have been cloned.[103–105] Selectin ligand activity of sialyl 6-sulfo Le^x is regulated by a unique post-translational modification of terminal sialic acid, which is *N*-deacetylated and cyclized by a sialic acid cyclase.[106, 107] The cyclic sialyl 6-sulfo Le^x antigen thus formed has no selectin binding activity, and is found on various subsets of human leukocytes.

[*2] Sialyl 6-sulfo Le^x was previously termed 6-sulfo sialyl Le^x in our previous reports as well as those from other laboratories. Later we noticed the term is confusing, since it appears as if the sulfated residue is carried somewhere in the sialic acid moiety. The situation is even more complicated when the sialic acids carry further modification,[106] such as de-*N*-acetyl sialyl 6-sulfo Le^x. Confusion would thus result if it were called 6-sulfo de-*N*-acetyl sialyl Le^x. For these reasons, we have adopted the notation "sialyl 6-sulfo Le^x" for this compound since 1999.

Circulating human leukocytes strongly express sialyl Le^x antigen, but it is hardly detectable on leukocytes of other animal species such as mouse and rat, when the same specific monoclonal antibodies are used for detection of the antigen.[108] The nature of selectin ligands on animal leukocytes is still enigmatic. Since fucosyltransferase Fuc-T VII-KO or Fuc-T VII/IV-KO mice have significantly reduced selectin ligands on their leukocytes,[109,110] it is evident that the ligands in mice are also the products of the same set of glycosyltransferases as in humans. Nevertheless, the most common anti-sialyl Le^x antibodies fail to react with mouse leukocytes.

Expression of Le^x, Le^y and Sialyl Le^x Antigens in Cancers

These type 2 chain antigens are frequently expressed in various human cancer tissues. Sialyl Le^x also appears in the sera of patients, and is a useful marker for the serum diagnosis of cancers.[111,112] Frequent occurrence of the antigen is noted in cancers originating in various organs including the lung, breast and ovary. The antigen is also expressed in cancers of the digestive organs, but in these organs the positive frequency of sialyl Le^x is lower than that of sialyl Le^a. This is most probably due to a difference in the distribution of synthetic precursors for these antigens: the type 2 chain substrates are ubiquitously present, while the type 1 chain substrates are abundant and predominant in the digestive organs.

Expression of sialyl Le^x is significantly increased in cancer cells compared to non-malignant epithelial cells. Increased expression of Le^x and Le^y antigens is also noted in cancers. These carbohydrate antigens are carried by glycolipids, *N*- and *O*-linked carbohydrate side chains of glycoproteins. Most anti-sialyl Le^x antibodies detect sialyl Le^x carried by glycoproteins as well as by glycolipids. A unique monoclonal antibody called NCC-ST-439 directed to cancer-associated antigens recognizes the sialyl Le^x antigen carried by *O*-linked carbohydrate side chains of glycoproteins, but is not reactive to conventional sialyl Le^x glycolipids.[113]

Mechanism for Increased Expression of Sialyl Le^x Antigen in Cancers

The exact mechanism that leads to the increased expression of sialyl Le^x in cancers is not clear at this moment. Assays for fucosyltransferase activity in cancers have frequently revealed no significant difference in the overall enzymatic activity between cancer and non-malignant epithelial cells. Among the fucosyltransferase candidates involved, Fuc-T VII and IV are known to be dynamically regulated at the transcriptional level, and Fuc-T VII is clearly responsible for the increased expression of sialyl Le^x in malignant cells in some cases, especially in certain leukemias.[114,115] Transcription of other fucosyltransferases usually remains unchanged between these types of cells. Likewise, the sialyltransferase activity for type 2 chain substrates and mRNA for ST3Gal4, the gene encoding for the enzyme, are not always increased in cancers.

The sialyl 6-sulfo Le^x antigen, recently identified as a major ligand for L-selectin in human peripheral lymph nodes, is also expressed on some epithelial cells. In contrast to the conventional sialyl Le^x, which is preferentially expressed in cancer tissues rather than non-malignant epithelia, the sialyl 6-sulfo Le^x antigen was found to be expressed preferentially in the nonmalignant colonic epithelia rather than cancer cells in human colorectal cancer tissues (Fig. 9).[116] A non-sialylated determinant, 6-sulfo Le^x (Table 7), was also preferentially localized in the non-malignant epithelia.[116] These findings suggest that a significant decrease in 6-sulfation occurs upon malignant transformation of colonic epithelial cells, and this would be at least partly responsible for the accumulation in cancer cells of conventional non-sulfated sialyl Le^x, the cancer-associated antigen. Another sulfated form, 3'-sulfo Le^x, is also expressed on some human cancer tissues and cultured cancer cell lines,[116,117] but it remains to be determined whether its expression shows a significant change between cancerous and non-malignant epithelial cells.

The synthesis of complex carbohydrate determinants on normal epithelial cells tends to be impaired upon malignant transformation. This phenomenon is termed "incomplete synthesis" of carbohydrate determinants in cancer, and comprises one of the major principal mechanisms for tumor-associated alteration of carbohydrate determinants. The increased expression of sialyl

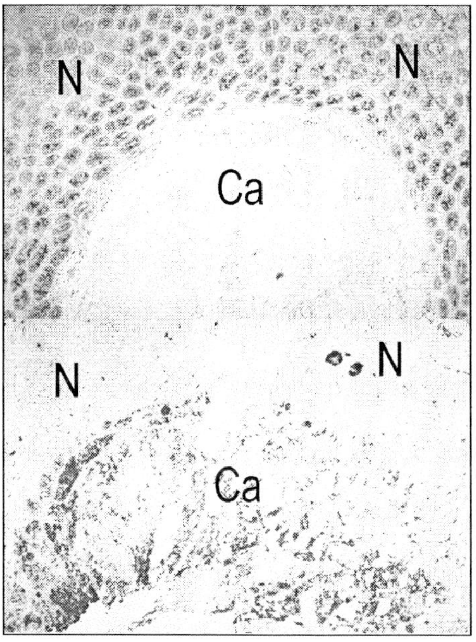

Figure 9. Typical distribution pattern of sialyl Le^x (upper panel) and sialyl 6-sulfo Le^x (lower panel) antigens in colon cancer tissues. Note that sialyl Le^x antigen is preferentially expressed on cancer cells, while sialyl 6-sulfo Le^x antigen is specifically localized in non-malignant epithelial cells. Ca, cancer cells; N, non-malignant epithelial cells. Adapted from reference 116.

Le^x by the suppression of 6-sulfation, and the enhanced expression of sialyl Le^a by the suppression of 6-sialylation as proposed above, could be regarded as typical examples of this mechanism as schematically shown in Figure 10.

Sialyl Le^x Antigen and Cancer Metastasis

Sialyl Le^x antigen expressed on cancer cells mediates their adhesion to vascular endothelial cells and promotes hematogenous metastasis of cancers as well as sialyl Le^a antigen (Fig. 6). Patients with cancer cells strongly expressing sialyl Le^x tend to have a poor prognosis, as observed in cases of cancers of lung, breast, ovary, prostate and stomach (Fig. 7). Generally sialyl Le^a plays a more predominant role than sialyl Le^x in cancers of the digestive organs, but in gastric cancers the contribution of sialyl Le^x is comparable to that of sialyl Le^a. Cancers of the liver and kidney also express sialyl Le^x, but in these cancers no correlation was observed between expression of the antigen and prognosis. It has been proposed that other carbohydrate determinants play significant roles in the adhesion of kidney cancer cells to vascular endothelial cells.[57,60] The sialylated and internally fucosylated antigens are also expressed in cancer cells, and serve as ligands for selectins. E-selectin is known to be critically involved in the adhesion of cancer cells to endothelial cells. P-selectin is less frequently involved, since cancer cells usually lack the expression of PSGL-1,[118] the core protein indispensable for P-selectin binding. When P-selectin is involved, CD24 or some other proteins may substitute for PSGL-1 in the adhesion process.[119,120] Malignant cells that express L-selectin may infiltrate or metastasize into lymph nodes via L-selectin dependent pathway.[121] In fact, cells expressing L-selectin are frequently found in malignant lymphoma and other hematological malignancies.

Other Carbohydrate Antigens Associated with Cancers

Ganglio-series glycolipids also appear in a differentiation- and development-dependent manner. Some determinants among them are regarded as cancer-associated, including G_{D3}, G_{D2}, G_{M2} in melanoma and neuroblastoma, and fucosyl G_{M1} in small cell carcinoma of the lung. They appear mostly in the malignant cells of neuroendocrine origin, while G_{M2} and extended G_{M2} (Cad blood group) are also found in cancers of the digestive organs or in non-small cell carcinoma of the lung. Extended G_{M2} determinants are on ganglio-series glycolipids such

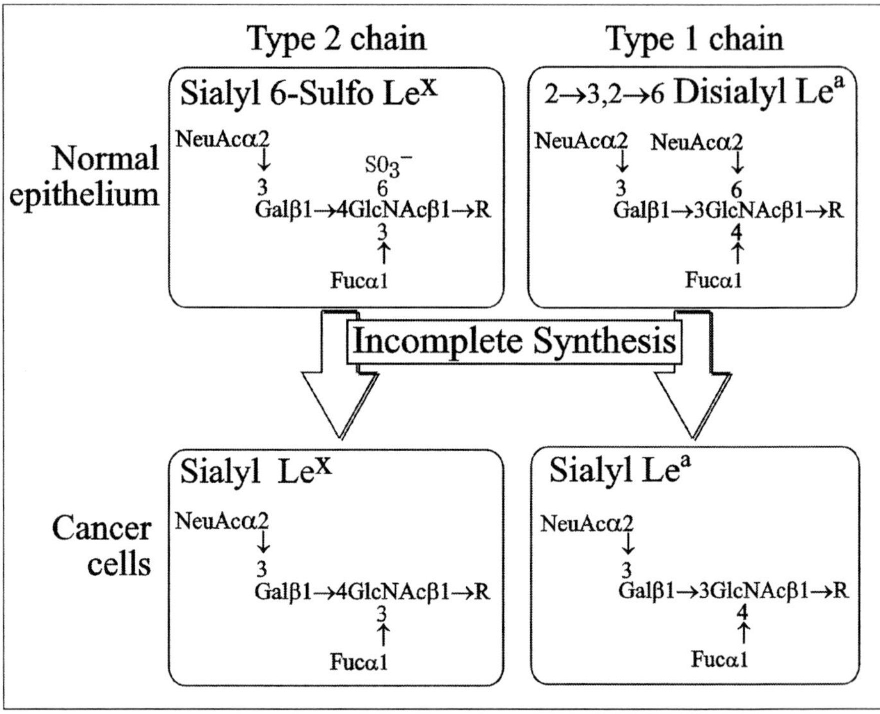

Figure 10. Schema illustrating the induction of sialyl Le^a or sialyl Le^x expression in cancers as a result of "incomplete synthesis" of more complex carbohydrate determinants, disialyl Le^a or sialyl 6-sulfo Le^x.

as GalNAc-G_{D1a}, as well as neolacto- and lacto-series[60] glycolipids (Table 8). The same terminal structure carried by *O*-glycan was recently shown to be a major carbohydrate side chain in non-malignant colonic mucin.[122]

Thomsen-Friedenreich (TF or T), Tn and sialyl Tn antigens are relatively short-chain carbohydrate antigens on *O*-glycans, and are important cancer-associated antigens (Table 9). These antigens are expressed in cancer cells originating in various organs such as the breast, ovary, colon and urinary bladder, and are proposed to be good targets for immunotherapy. Sialyl Tn antigen is the target molecule recognized by several monoclonal antibodies generated against cancer-associated antigens. These include the B72.3 and CC49 antibodies. The sialyl Tn antigen has also been found in the sera of patients with these types of cancers, especially ovarian cancers. This antigen is used as a serum tumor marker.

Some candidate genes for the glycosyltransferases responsible for the synthesis of these determinants have been cloned, including a β1→3 galactosyltransferase for TF antigen,[123] α1→3 *N*-acetylgalactosaminyltransferases for Tn antigen,[124–128] and α2→3 sialyltransferases for the synthesis of sialyl Tn antigen.[129,130] It is not clear at present whether or not alteration in the expression of these genes in cancerous cells is responsible for the appearance of the TF, Tn or sialyl Tn antigen. Since *O*-glycans on normal epithelial mucins have more complex carbohydrate structures, the appearance of these simple antigens in cancers has been ascribed to the "incomplete synthesis" of carbohydrate antigens rather than an increase in the synthesis of shorter chains. TF antigen is one of the major carriers and synthetic precursors for ABO(H) determinants, and its appearance in cancers must be partly related to the cancer-associated reduction of ABO(H) determinants.

Table 8. Cancer associated ganglio-series glycolipids and related antigens

Antigen	Structure
GD3	Galβ1→4Glc1→Cer NeuAcα2→8NeuAcα2↗3
GD2	GalNAcβ1→4Galβ1→4Glc1→Cer NeuAcα2→8NeuAcα2↗3
GM2	GalNAcβ1→4Galβ1→4Glc1→Cer NeuAcα2↗3
GalNAc-GD1a	GalNAcβ1→4Galβ1→3GalNAcβ1→4Galβ1→4Glc1→Cer NeuAcα2↗3 NeuAcα2↗3
Sd^a (CAD) glycolipid (neolacto-series)	GalNAcβ1→4Galβ1→4GlcNAcβ1→3Galβ1→4Glc1→Cer NeuAcα2↗3
Sd^a glycoproteins	
Glycophorin A (*O*-glycan)	GalNAcβ1→4Galβ1→3GalNAcα1→Ser/Thr NeuAcα2↗3
Tamm-Horsfall protein (*N*-glycan)	GalNAcβ1→4Galβ1→3GlcNAcβ1→3Galβ1→R NeuAcα2↗3
GalNAcDSLc$_4$ (lacto-series)	GalNAcβ1→4Galβ1→3GlcNAcβ1→3Galβ1→4Glc1→Cer NeuAcα2↗3 NeuAcα2↗6

Another factor to be taken into consideration is the biochemical properties of the enzymes involved in the synthesis of these antigens. For instance, the β1→3 galactosyltransferase responsible for TF antigen synthesis is known to have a very high *K*m value (i.e., 630 μM) for UDP-galactose.[131] A similarly high *K*m value for the same substrate is also known for a β1→3 galactosyltransferase responsible for the production of type 1 chain precursor, the substrate for the synthesis of all type 1 chain based Lewis blood group substances including the important cancer-associated antigen, sialyl Le^a. It has also been suggested that the rate-limiting step is not the activity of the transferase, but that of the UDP-galactose transporter. This transporter transfers UDP-galactose from the cytosol to the Golgi apparatus. Indeed, introduction of a gene for UDP-galactose transporter is known to confer TF antigen expression and increase the expression of sialyl Le^a on transfected cells.[132]

A unique GlcNAcα1→4Galβ1→R determinant is known to be on *O*-glycans found in the stomach, pancreas and biliary tract.[133–135] It has been suggested that this determinant is a receptor for, and even to have some bactericidal activity against *Helicobacter pylori*, an important human pathogen which causes both gastric and duodenal ulcers and is associated with gastric cancer and lymphoma. The gene for the enzyme catalyzing the synthesis of this structure, α1→4-*N*-acetylglucosaminyltransferase, has been cloned,[136] and shown to belong to the same family as the gene for α1,4-galactosyltransferase involved in the synthesis of CTH (the P^k antigen).

Table 9. Thomsen-Friedenreich (TF), Tn and sialyl Tn antigens

Antigen	Structure
TF antigen	Galβ1→3GalNAcα1→Ser/Thr
Tn antigen	GalNAcα1→Ser/Thr
Sialyl Tn antigen	NeuAcα2→6GalNAcα1→Ser/Thr

Heterophile Antigens

Forssman Antigen, a Typical Heterophile Antigen

The term heterophile antigen refers to an antigen commonly encountered among various animal species. Its distribution is almost random, and not necessarily related to the evolutionary relationship of a given species. For instance, Forssman antigen, the most typical heterophile antigen, is expressed in erythrocytes and other tissues of "Forssman-positive" animals, such as guinea-pig, horse, cat, dog, sheep, goat, mouse, pigeon, chicken and turtle, while not expressed in "Forssman-negative" animals, such as rabbit, rat, cow, pig, human, monkey and frog. The antigen belongs to the globo-series glycolipids, and has the terminal structure GalNAcα1→3GalNAcβ1→Galα (Table 10). Variants carrying the same terminal structure based on gala-series or isoglobo-series glycolipids are also recorded.[137,138] Strict regulation of its expression is evident during differentiation or development in Forssman-positive animals. For instance, murine teratocarcinoma cells and pre-implantation embryos frequently express the antigen until upon cell differentiation or embryo development.[139–141] In contrast, expression of another heterophile antigen, α1→3Gal, is known to appear upon differentiation of murine teratocarcinoma cells.[142]

A gene for the canine α1→3 *N*-acetylgalactosaminyltransferase responsible for the synthesis of Forssman antigen has been cloned.[143] Interestingly, this enzyme has 42% amino acid identity to the A- and B-enzymes of the ABO(H) blood group system, and is regarded as having the same evolutionary origin as these enzymes. The recombinant canine enzyme has an ability to synthesize Forssman antigen, but no ability to utilize H antigen as substrate. In humans a full-length protein having 83% homology to that of the canine enzyme is produced in various tissues and organs, but it has no α1→3 *N*-acetylgalactosaminyltransferase activity because of the introduction of a few amino acid substitutions in the putative catalytic domain. This is in contrast to another heterophile antigen, αGal antigen, where the gene encoding the enzyme contains multiple frame shift mutations and premature termination in human cells, and is not effectively transcribed.[144,145] Although normal human cells and tissues do not contain Forssman antigen, its occurrence is frequently reported in human cancer tissues. It is not clear whether this involves the activation of the dormant enzyme, or the induction of some other enzymes having a similar enzymatic activity. Interestingly, but probably unrelated to cancer, the Forssman antigen is found to have a remarkable anti-depressive activity in a mouse model of depression.[146,147]

Table 10. Heterophile antigens

Antigen	Structure
Forssman antigen	GalNAcα1→3GalNAcβ1→3Galα1→4Galβ1→3Glcβ1→1Cer
a gala-series variant	GalNAcα1→3GalNAcβ1→3Galα1→4Galβ1→1Cer
an isoglobo-series variant	NeuAcα2↘3 GalNAcβ1→4Galβ1→3GlcNAcβ1↘6 GalNAcα1→3GalNAcβ1→3Galα1→3Galβ1→4Glcβ1→1Cer
αGal antigen	Galα1→3Galβ1→(3GlcNAcβ1→)R
H-D antigen	NeuGcα2→R

αGal Antigen and Xenotransplantation

αGal antigen, Galα1→3Galβ1→(3GlcNAcβ1→)R, is widely expressed in a variety of mammalian species, with the exception of man, apes, and Old World monkeys. This antigen was initially described as a pentasaccharide glycolipid antigen on rabbit RBCs having a weak blood group B-like activity. Forssman antigen is in contrast known as a typical antigen expressed on sheep RBCs. The gene for the α1→3 galactosyltransferase responsible has a high homology to A- and B-transferases in ABO(H) blood group system, and also to the α1→3 *N*-acetylgalactosaminyltransferase for Forssman antigen synthesis. The corresponding orthologue in human genome contains multiple mutations and is not effectively transcribed.[144,145] In animals having αGal antigen, its expression is differentiation-dependent; expression is induced during the differentiation of macrophages or teratocarcinoma cells. A species with no αGal antigen frequently develops natural anti-αGal antibodies in sera, which protect the host from pathogenic microorganisms and viruses.

Virus particles produced in αGal-positive animals express αGal epitopes on their envelope glycoproteins, and are inactivated by such natural anti-αGal antibodies. The αGal antigen is known to elicit at least partly a T lymphocyte-dependent immune response.[148] Anti-αGal antibodies are composed of both IgG and IgM antibodies,[149] and are encoded by a very limited set of V_H genes, which is either IGHV3-11 or 3-74 in humans.[150,151] The anti-αGal antibodies produced in α1→3 galactosyltransferase-KO mice are also encoded by a limited set of heavy chain genes, namely V_H441,[152] the same gene encoding murine anti-Ii antibodies according to some reports, or V_H22.1 according to others.[153] Since humans develop a strong immune response to the αGal epitope, its application for cancer therapy is being devised from autologous tumor cells artificially derivatized to express the αGal epitope.[154–156] Since αGal antigen is present on porcine tissues, it is the major target molecule for the hyperacute rejection of porcine organs xenotransplanted to primate,[157] and various strategies have been proposed to avoid rejection (for reviews see refs. 158,159).

Hanganutziu-Deicher (H-D) Antigen

A heterophile antigen that causes serum sickness in humans receiving a therapeutic injection of animal antitoxin serum such as anti-diphtheria toxin and anti-tetanus toxin was described independently by Hanganutziu and Deicher. The antigen is detectable in most animals including sheep, cow, horse, pig, rabbit, and guinea pig, but not in humans and chicken.

The antigen was identified as glycoconjugates having a terminal *N*-glycolyl sialic acid (NeuGc). NeuGc is formed by hydroxylation of *N*-acetyl sialic acid (NeuAc). There are three possible pathways for the conversion of NeuAc to NeuGc; from free NeuAc to NeuGc, from CMP-NeuAc to CMP-NeuGc, and lastly the direct modification of glycoconjugate-bound NeuAc residues to form NeuGc residues. The major pathway is thought to be the conversion of CMP-NeuAc to CMP-NeuGc by hydroxylation. The murine gene for the CMP-NeuAc hydroxylase that is responsible has been cloned,[160] and the hydroxylation reaction was shown to be mediated by an enzyme complex composed of CMP-NeuAc hydroxylase, cytochrome b5, and NADH cytochrome b5 reductase in the presence of NADH. Analysis of the human orthologue revealed a 92 base pair deletion in the CMP-NeuAc hydroxylase gene, which leads to the production of a short polypeptide and completely abrogates the enzymatic activity.[161-163] In H-D positive species, the antigen is distributed in various tissues and organs including blood cells and vascular endothelial cells, while the brain expresses predominantly NeuAc and a far lower amount of NeuGc.

Similarly to αGal antigen, H-D antigen is expressed in pig tissues, and is presumed to exert some adverse effect in xenotransplantaion. As baboons and monkeys also express H-D antigens, pig-to-non-human primate experimental models are not useful to address the question of whether H-D antigens can elicit a serious adverse reaction in xenotransplantation of pig organs to humans.

Some human malignant cells such as melanoma and colon cancers are reported to express H-D antigen, and thus the antigen has been proposed to be one of the target molecules for immunotherapy of human cancers.[164] Even some non-malignant human tissues are known to contain a small amount of NeuGc. It is not clear why cancers in humans having only the enzymologically inert polypeptide can express H-D antigen on their surface. One possibility is the presence of another conversion pathway in cancers, such as conversion at the level of free sialic acid, or direct modification of glycoconjugate-bound sialic acid residues (for a review see ref. 165). Glycolyl-CoA is proposed to be a possible donor of the glycolyl residue. Another possibility would be an abnormality in sialic acid transport. It is well known that cells take up free sialic acid from extracellular fluid and utilize it for the synthesis of sialoglycoconjugates. When the extracellular fluid contains much NeuGc derived from some dietary source or other milieu, cultured human cells are known to acquire it and express a significant amount of sialoglycoconjugates containing NeuGc. The mechanism for cellular uptake of sialic acid is not well characterized, but it is conceivable that human cancer cells, or even certain non-malignant tissues, may have a faster uptake rate of extracellular sialic acid, which may be related to the expression of the "illegitimate" *N*-glycolyl determinants in such tissues.

References

1. Seto NO, Palcic MM, Compston CA et al. Sequential interchange of four amino acids from blood group B to blood group A glycosyltransferase boosts catalytic activity and progressively modifies substrate recognition in human recombinant enzymes. J Biol Chem 1997; 272:14133-14138.
2. Seto NO, Compston CA, Evans SV et al. Donor substrate specificity of recombinant human blood group A, B and hybrid A/B glycosyltransferases expressed in Escherichia coli. Eur J Biochem 1999; 259:770-775.
3. Kamath VP, Seto NO, Compston CA et al. Synthesis of the acceptor analog αFuc(1→2)αGal-O$(CH_2)_7CH_3$: a probe for the kinetic mechanism of recombinant human blood human blood group B glycosyltransferase. Glycoconjugate J 1999; 16:599-606.
4. Hakomori S. Antigen structure and genetic basis of histo-blood groups A, B and O: their changes associated with human cancer. Biochim Biophys Acta 1999; 1473:247-266.
5. Yamamoto F. Cloning and regulation of the ABO genes. Transfus Med 2001; 11:281-294.

6. Rajan VP, Larsen RD, Ajmera S et al. A cloned human DNA restriction fragment determines expression of a GDP-L-fucose: β-D-galactoside 2-α-L-fucosyltransferase in transfected cells. Evidence for isolation and transfer of the human H blood group locus. J Biol Chem 1989; 264:11158-11167.
7. Larsen RD, Rajan VP, Ruff MM et al. Isolation of a cDNA encoding a murine UDP galactose:β-D-galactosyl-1,4-N-acetyl-D-glucosaminide α-1,3-galactosyltransferase: expression cloning by gene transfer. Proc Natl Acad Sci USA 1989; 86:8227-8231.
8. Rouquier S, Lowe JB, Kelly RJ et al. Molecular cloning of a human genomic region containing the H blood group α(1,2)fucosyltransferase gene and two H locus-related DNA restriction fragments. Isolation of a candidate for the human Secretor blood group locus. J Biol Chem 1995; 270:4632-4639.
9. Kelly RJ, Rouquier S, Giorgi D et al. Sequence and expression of a candidate for the human Secretor blood group α(1,2)fucosyltransferase gene (FUT2). Homozygosity for an enzyme-inactivating nonsense mutation commonly correlates with the non-secretor phenotype. J Biol Chem 1995; 270:4640-4649.
10. Fernandez-Mateos P, Cailleau A, Henry S et al. Point mutations and deletion responsible for the Bombay H null and the Reunion H weak blood groups. Vox Sang 1998; 75:37-46.
11. Sarnesto A, Kohlin T, Hindsgaul O et al. Purification of the secretor-type β-galactoside α1→2-fucosyltransferase from human serum. J Biol Chem 1992; 267:2737-2744.
12. Toma H, Tanabe K, Tokumoto T. Long-term outcome of ABO-incompatible renal transplantation. Urol Clin North Am 2001; 28:769- 13.
13. Rydberg L, Breimer ME, Brynger H et al. ABO-incompatible kidney transplantation (A2 to O). Qualitative and semiquantitative studies of the humoral immune response against different blood group A antigens. Transplantation 1990; 49:954-960.
14. Rydberg L. ABO-incompatibility in solid organ transplantation. Transfus Med 2001;11:325-342.
15. Sun J, Thurin J, Cooper HS et al. Elevated expression of H type GDP-L-fucose:β-D-galactoside α-2-L-fucosyltransferase is associated with human colon adenocarcinoma progression. Proc Natl Acad Sci USA 1995; 92:5724-5728.
16. David L, Leitao D, Sobrinho-Simoes M et al. Biosynthetic basis of incompatible histo-blood group A antigen expression: anti-A transferase antibodies reactive with gastric cancer tissue of type O individuals. Cancer Res 1993; 53:5494-5500.
17. Fujitani N, Liu Y, Toda S et al. Expression of H type 1 antigen of ABO histo-blood group in normal colon and aberrant expressions of H type 2 and H type 3/4 antigens in colon cancer. Glycoconjugate J 2000; 17:331-338.
18. Okada Y, Kannagi R, Levery SB et al. Glycolipid antigens with blood group I and i specificities from human adult and cord erythrocytes. J Immunol 1984; 133:835-842.
19. Niemann H, Watanabe K, Hakomori S. Blood group i and I activities of "lacto-N-norhexaosylceramide" and its analogues: the structural requirements for i-specificities. Biochem Biophys Res Commun 1978; 81:1286-1293.
20. Watanabe K, Hakomori SI, Childs RA et al. Characterization of a blood group I-active ganglioside. Structural requirements for I and i specificities. J Biol Chem 1979; 254:3221-3228.
21. Bierhuizen MFA, Mattei M-G, Fukuda M. Expression of the developmental I antigen by a cloned human cDNA encoding a member of a β-1,6-N-acetylglucosaminyltransferase gene family. Genes Dev 1993; 7:468-478.
22. Schwientek T, Nomoto M, Levery SB et al. Control of O-glycan branch formation. Molecular cloning of human cDNA encoding a novel β1,6-N-acetyl glucosaminyl transferase forming core 2 and core 4. J Biol Chem 1999; 274:4504-4512.
23. Yeh JC, Ong E, Fukuda M. Molecular cloning and expression of a novel β-1,6-N-acetylglucosaminyltransferase that forms core 2, core 4, and I branches. J Biol Chem 1999; 274.3215-3221.
24. Yu LC, Twu YC, Chang CY et al. Molecular basis of the adult i phenotype and the gene responsible for the expression of the human blood group I antigen. Blood 2001; 98:3840-3845.
25. Willims RC,Jr., Kunkel HG, Capra JD. Antigenic specificities related to the cold agglutinin activity of gamma M globulins. Science 1968; 161:379-381.
26. Li YC, Spellerberg MB, Stevenson FK et al. The I binding specificity of human V_H4-34 (V_H4-21) encoded antibodies is determined by both V_H framework region 1 and complementarity determining region 3. J Mol Biol 1996; 256:577-589.
27. Smith G, Spellerberg M, Boulton F et al. The immunoglobulin VH gene, VH4-21, specifically encodes autoanti-red cell antibodies against the I or i antigens. Vox Sang 1995; 68:231-235.
28. Riboldi P, Gaidano G, Schettino EW et al. Two acquired immunodeficiency syndrome-associated Burkitt's lymphomas produce specific anti-i IgM cold agglutinins using somatically mutated V_H4-21 segments. Blood 1994; 83:2952-2961.

29. Pascual V, Victor K, Lelsz D et al. Nucleotide sequence analysis of the V regions of two IgM cold agglutinins. Evidence that the V_H4-21 gene segment is responsible for the major cross-reactive idiotype. J Immunol 1991; 146:4385-4391.
30. Silberstein LE, Jefferies LC, Goldman J et al. Variable region gene analysis of pathologic human autoantibodies to the related i and I red blood cell antigens. Blood 1991; 78:2372-2386.
31. Thompson KM, Sutherland J, Barden G et al. Human monoclonal antibodies against blood group antigens preferentially express a V_H4-21 variable region gene-associated epitope. Scand J Immunol 1991; 34:509- 32.
32. Brezinschek HP, Brezinschek RI, Lipsky PE. Analysis of the heavy chain repertoire of human peripheral B cells using single-cell polymerase chain reaction. J Immunol 1995; 155:190-202.
33. Rettig MB, Vescio RA, Cao J et al. VH gene usage is multiple myeloma: complete absence of the VH4.21 (VH4-34) gene. Blood 1996; 87:2846-2852.
34. Snyder JG, Dinh Q, Morrison SL et al. Structure-function studies of anti-3-fucosyllactosamine (Le^x) and galactosylgloboside antibodies. J Immunol 1994; 153:1161-1170.
35. Duran-Suarez JR, Martin-Vega C, Argelagues E et al. The I antigen as an immune complex receptor in a case of haemolytic anaemia induced by an antihistaminic agent. Br J Haematol 1981; 49:153-154.
36. Duran-Suarez JR, Martin-Vega C, Argelagues E et al. Red cell I antigen as immune complex receptor in drug-induced hemolytic anemias. Vox Sang 1981; 41:313-315.
37. Habibi B, Basty R, Chodez S et al. Thiopental-related immune hemolytic anemia and renal failure. Specific involvement of red-cell antigen I. N Engl J Med 1985; 312:353-355.
38. Kannagi R, Levery SB, Ishigami S et al. New globoseries glycosphingolipids in human teratocarcinoma reactive with the monoclonal antibody directed to a developmentally-regulated antigen, SSEA-3. J Biol Chem 1983; 258:8934-8942.
39. Kannagi R, Cochran NA, Ishigami F et al. Stage-specific embryonic antigens (SSEA-3 and -4) are epitopes of a unique globo-series ganglioside isolated from human teratocarcinoma cells. Eur Mol Biol Orgn J 1983; 2:2355-2361.
40. Levery SB, Salyan MEK, Steele SJ et al. A revised structure for the disialosyl globo-series gangliosides of human erythrocytes and chicken skeletal muscle. Arch Biochem Biophys 1994; 312:125-134.
41. Kannagi R, Levery SB, Hakomori S. Blood group H antigen with globo-series structure: Isolation and characterization from human blood group O erythrocytes. FEBS Lett 1984; 175:397-401.
42. Bremer EG, Levery SB, Sonnio S et al. Characterization of a glycosphingolipid antigen defined by the monoclonal antibody MBr1 expressed in normal and neo-plastic epithelial cells of human mammary gland. J Biol Chem 1984; 259:14773-14777.
43. Clausen H, Watanabe K, Kannagi R et al. Blood Group A glycolipid (Ax) with globo-series structure which is specific for blood group A_1 erythrocytes: One of the chemical bases for A_1 and A_2 distinction. Biochem Biophys Res Commun 1984; 124:523-529.
44. Okajima T, Nakamura Y, Uchikawa M et al. Expression cloning of human globoside synthase cDNAs. Identification of β3Gal-T3 as βUDP-N-acetyl galactosamine :globotriaosylceramide β1,3-N-acetylgalactosaminyltransferase. J Biol Chem 2000; 275:40498-40503.
45. Keusch JJ, Manzella SM, Nyame KA et al. Cloning of Gb_3 synthase, the key enzyme in globo-series glycosphingolipid synthesis, predicts a family of α1,4-glycosyltransferases conserved in plants, insects, and mammals. J Biol Chem 2000; 275:25315-25321.
46. Kojima Y, Fukumoto S, Furukawa K et al. Molecular cloning of globotriaosyl ceramide /CD77 synthase, a glycosyltransferase that initiates the synthesis of globo-series glycosphingolipids. J Biol Chem 2000; 275:15152-15156.
47. Zhou D, Henion TR, Jungalwala FB et al. The β1,3-galactosyltransferase β3GalT-V is a stage-specific embryonic antigen-3 (SSEA-3) synthase. J Biol Chem 2000; 275:22631-22634.
48. Koda Y, Soejima M, Sato H et al. Three-base deletion and one-base insertion of the α(1, 4)galactosyltransferase gene responsible for the P phenotype. Transfusion 2002; 42:48-51.
49. Furukawa K, Iwamura K, Uchikawa M et al. Molecular basis for the p phenotype. Identification of distinct and multiple mutations in the α1,4-galactosyltransferase gene in Swedish and Japanese individuals. J Biol Chem 2000; 275:37752-37756.
50. Hellberg A, Poole J, Olsson ML. Molecular basis of the globoside-deficient P^k blood group Phenotype.Identification of four inactivating mutations in the UDP-N-acetylgalactosamine:globotriaosylceramide 3-β-N-acetylgalactosaminyltransferase gene. J Biol Chem 2002;
51. Thomson JA, Odorico JS. Human embryonic stem cell and embryonic germ cell lines. Trends Biotechnol 2000; 18:53-57.
52. Thomson JA, Itskovitz-Eldor J, Shapiro SS et al. Embryonic stem cell lines derived from human blastocysts. Science 1998; 282:1145-1147.

53. Shamblott MJ, Axelman J, Wang S et al. Derivation of pluripotent stem cells from cultured human primordial germ cells. Proc Natl Acad Sci USA 1998; 95:13726-13731.
54. Miyamoto D, Ueno T, Takashima S et al. Establishment of a monoclonal antibody directed against Gb_3Cer/CD77: a useful immunochemical reagent for a differentiation marker in Burkitt's lymphoma and germinal centre B cells. Glycoconjugate J 1997; 14:379-388.
55. Nudelman E, Kannagi R, Hakomori S et al. A glycolipid antigen associated with Burkitt lymphoma defined by a monoclonal antibody. Science 1983; 220:509- 511.
56. Kannagi R, Levery SB, Hakomori S. Sequential change of carbohydrate antigens associated with the differentiation of murine Leukemia cells, M1: iI-antigenic conversion and shifting of glycolipid synthesis. Proc Natl Acad Sci USA 1983; 80:2833-2848.
57. Satoh M, Handa K, Saito S et al. Disialosyl galactosylgloboside as an adhesion molecule expressed on renal cell carcinoma and its relationship to metastatic potential. Cancer Res 1996; 56:1932-1938.
58. Satoh M, Nejad FM, Ohtani H et al. Association of renal cell carcinoma antigen, disialylgalactosylgloboside, with c-Src and Rho A in clustered domains at the surface membrane. Int J Oncol 2000; 16:529-536.
59. Satoh M, Nejad FM, Nakano O et al. Four new human renal cell carcinoma cell lines expressing globo-series gangliosides. Tohoku J Exp Med 1999; 189:95-105.
60. Ito A, Levery SB, Saito S et al. A novel ganglioside isolated from renal cell carcinoma. J Biol Chem 2001; 276:16695-16703.
61. Ito A, Handa K, Withers DA et al. Binding specificity of siglec7 to disialogangliosides of renal cell carcinoma: possible role of disialogangliosides in tumor progression. FEBS Lett 2001; 504:82-86.
62. Song Y, Withers DA, Hakomori S. Globoside-dependent adhesion of human embryonal carcinoma cells, based on carbohydrate-carbohydrate interaction, initiates signal transduction and induces enhanced activity of transcription factors AP1 and CREB. J Biol Chem 1998; 273:2517-2525.
63. Kojima N, Hakomori S. Cell adhesion, spreading, and motility of G_{M3}-expressing cells based on glycolipid-glycolipid interaction. J Biol Chem 1991; 266:17552-17558.
64. Levine P, Bobbit OB, Waller RK. Isoimmunization by a new blood group factor in tumor cells. Proc Soc Exp Biol Med 1951; 77:403-405.
65. Kannagi R, Levine P, Watanabe K et al. Recent studies of glycolipid and glycoprotein profiles and characterization of the major glycolipid antigen in gastric cancer of a patient of blood group genotype pp(Tja^-) first studied in 1951. Cancer Res 1982; 42:5249-5254.
66. Hattori H, Uemura K, Ogata H et al. Characterization of glycolipids from the gastric cancer of a patient of p, O, Le^{a-b+} blood type: presence of incompatible blood group antigens in tumor tissues. Cancer Res 1987; 47:1968-1972.
67. Arab S, Lingwood CA. Intracellular targeting of the endoplasmic reticulum nuclear envelope by retrograde transport may determine cell hypersensitivity to verotoxin via globotriaosyl ceramide fatty acid isoform traffic. J Cell Physiol 1998; 177:646-660.
68. Kawano T, Cui J, Koezuka Y et al. Natural killer-like nonspecific tumor cell lysis mediated by specific ligand-activated Vα14 NKT cells. Proc Natl Acad Sci USA 1998; 95:5690-5693.
69. Daniels G. Human blood groups. Oxford: Blackwell Science Ltd., 1995.
70. Marcus DM, Cass LE. Glycosphingolipids with Lewis blood group activity: uptake by human erythrocytes. Science 1969; 164:553-555.
71. Seko A, Ohkura T, Kitamura H et al. Quantitative differences in GlcNAc:β1→3 and GlcNAc:β1→4 galactosyltransferase activities between human colonic adenocarcinomas and normal colonic mucosa. Cancer Res 1996; 56:3468-3473.
72. Dohi T, Hashiguchi M, Yamamoto S et al. Fucosyltransferase-producing sialyl Le^a and sialyl Le^x carbohydrate antigen in benign and malignant gastrointestinal mucosa. Cancer 1994; 73:1552-1561.
73. Akamatsu S, Yazawa S, Tachikawa T et al. α2→3Sialyltransferase associated with the synthesis of CA 19-9 in colorectal tumors. Cancer 1996; 77 Suppl:1694-1700.
74. Ito H, Hiraiwa N, Sawada-Kasugai M et al. Altered mRNA expression of specific molecular species of fucosyl- and sialyltransferases in human colorectal cancer tissues. Int J Cancer 1997; 71:556-564.
75. Petretti T, Kemmner W, Schulze B et al. Altered mRNA expression of glycosyl transferases in human colorectal carcinomas and liver metastases. Gut 2000; 46:359-366.
76. Itai S, Arii S, Tobe R et al. Significance of 2-3 and 2-6 sialylation of Lewis A antigen in pancreas cancer. Cancer 1988; 61:775- 77.
77. Itai S, Nishikata J, Yoneda T et al. Tissue distribution of sialyl 2-3 and 2-6 Lewis a antigens and the significance of serum 2-3/2-6 sialyl Lewis a antigen ratio for the differential diagnosis of malignant and benign disorders of the digestive tract. Cancer 1991; 67:1576-1587.
78. Takada A, Ohmori K, Yoneda T et al. Contribution of carbohydrate antigens sialyl Lewis A and sialyl Lewis X to adhesion of human cancer cells to vascular endothelium. Cancer Res 1993; 53:354-361.

79. Kannagi R. Carbohydrate-mediated cell adhesion involved in hematogenous metastasis of cancer. Glycoconjugate J 1997; 14:577-584.
80. Nakayama T, Watanabe M, Katsumata T et al. Expression of sialyl Lewis^a as a new prognostic factor for patients with advanced colorectal carcinoma. Cancer 1995; 75:2051-2056.
81. Shimono R, Mori M, Akazawa K et al. Immunohistochemical expression of carbohydrate antigen 19-9 in colorectal carcinoma. Am J Gastroenterol 1994; 89:101-105.
82. Kiriyama K, Watanabe T, Sakamoto J et al. [Expression and clinical significance of type-1 blood group antigens (Le^a, Le^b, CA19-9) in colorectal cancer—comparison with CEA]. Nippon Geka Gakkai Zasshi 1991; 92:320-330.
83. Kannagi R, Nudelman E, Levery SB et al. A series of human erythrocyte glyco sphingolipids reacting to the monoclonal antibody directed to developmentally-regulated antigen, SSEA-1. J Biol Chem 1982; 257:14865-14874.
84. Eggens I, Fenderson B, Toyokuni T et al. Specific interaction between Le^x and Le^x determinants. A possible basis for cell recognition in preimplantation embryos and in embryonal carcinoma cells. J Biol Chem 1989; 264:9476-9484.
85. Ohmori K, Yoneda T, Shigeta K et al. Sialyl SSEA-1 antigen as a carbohydrate marker of human natural killer cells and immature lymphoid cells. Blood 1989; 74:255-261.
86. Ohmori K, Takada A, Ohwaki I et al. A distinct type of sialyl Lewis X antigen defined by a novel monoclonal antibody is selectively expressed on helper memory T cells. Blood 1993; 82:2797-2805.
87. Knibbs RN, Craig RA, Natsuka S et al. The fucosyltransferase FucT-VII regulates E-selectin ligand synthesis in human T cells. J Cell Biol 1996; 133:911-920.
88. Wagers AJ, Stoolman LM, Kannagi R et al. Expression of leukocyte fucosyltransferases regulates binding to E-selectin—relationship to previously implicated carbohydrate epitopes. J Immunol 1997; 159:1917-1929.
89. Berg EL, Yoshino T, Rott LS et al. The cutaneous lymphocyte antigen is a skin lymphocyte homing receptor for the vascular lectin endothelial cell-leukocyte adhesion molecule 1. J Exp Med 1991; 174:1461-1466.
90. Berg EL, Robinson MK, Mansson O et al. A carbohydrate domain common to both sialyl Le^a and sialyl Le^x is recognized by the endothelial cell leukocyte adhesion molecule ELAM-1. J Biol Chem 1991; 266:14869-14872.
91. Austrup F, Vestweber D, Borges E et al. P- and E-selectin mediate recruitment of T-helper-1 but not T-helper-2 cells into inflamed tissues. Nature 1997; 385:81-83.
92. Handa K, Stroud MR, Hakomori S. Sialosyl-fucosyl poly-LacNAc without the sialosyl-Le^x epitope as the physiological myeloid cell ligand in E-selectin-dependent adhesion: Studies under static and dynamic flow conditions. Biochemistry 1997; 36:12412-12420.
93. Stroud MR, Handa K, Salyan ME et al. Monosialogangliosides of human myelogenous leukemia HL60 cells and normal human leukocytes. 1. Separation of E-selectin binding from nonbinding gangliosides, and absence of sialosyl-Le^x having tetraosyl to octaosyl core. Biochemistry 1996; 35:758-769.
94. Stroud MR, Handa K, Salyan ME et al. Monosialogangliosides of human myelogenous leukemia HL60 cells and normal human leukocytes. 2. Characterization of E-selectin binding fractions, and structural requirements for physiological binding to E-selectin. Biochemistry 1996; 35:770-778.
95. Stroud MR, Handa K, Ito K et al. Myeloglycan, a series of E-selectin-binding polylactosaminolipids found in normal human leukocytes and myelocytic leukemia HL60 cells. Biochem Biophys Res Commun 1995; 209:777-787.
96. Mitsuoka C, Kawakami-Kimura N, Kasugai-Sawada M et al. Sulfated sialyl Lewis X, the putative L-selectin ligand, detected on endothelial cells of high endothelial venules by a distinct set of anti-sialyl Lewis X antibodies. Biochem Biophys Res Commun 1997; 230:546-551.
97. Mitsuoka C, Sawada-Kasugai M, Ando-Furui K et al. Identification of a major carbohydrate capping group of the L-selectin ligand on high endothelial venules in human lymph nodes as 6-sulfo sialyl Lewis x. J Biol Chem 1998; 273:11225-11233.
98. Kimura N, Mitsuoka C, Kanamori A et al. Reconstitution of functional L-selectin ligands on a cultured human endothelial cell line by co-transfection of $\alpha 1 \rightarrow 3$ fucosyltransferase VII and newly cloned GlcNAcβ: 6-sulfotransferase cDNA. Proc Natl Acad Sci USA 1999; 96:4530-4535.
99. Hemmerich S, Rosen SD. 6'-sulfated sialyl Lewis x is a major capping group of GlyCAM-1. Biochemistry 1994; 33:4830-4835.
100. Hemmerich S, Leffler H, Rosen SD. Structure of the O-glycans in GlyCAM-1, an endothelial-derived ligand for L-selectin. J Biol Chem 1995; 270:12035-12047.
101. Tangemann K, Bistrup A, Hemmerich S et al. Sulfation of a high endothelial venule-expressed ligand for L-selectin: effects on tethering and rolling of lymphocytes. J Exp Med 1999; 190:935-941.

102. Kannagi R, Mitsuoka C, Kanamori A et al. Expression and selectin-binding activity of sialyl 6-sulfo Lewis X determinant on human helper memory T lymphocytes. In: Mason DY, ed. Leukocyte Typing VII. Oxford: Oxford University Press, 2002:39-41.
103. Uchimura K, Muramatsu H, Kadomatsu K et al. Molecular cloning and characterization of an N-acetylglucosamine-6-O-sulfotransferase. J Biol Chem 1998; 273:22577-22583.
104. Bistrup A, Bhakta S, Lee JK et al. Sulfotransferases of two specificities function in the reconstitution of high-endothelial-cell ligands for L-selectin. J Cell Biol 1999; 145:899-910.
105. Hemmerich S, Bistrup A, Singer MS et al. Sulfation of L-selectin ligands by an HEV-restricted sulfotransferase regulates lymphocyte homing to lymph nodes. Immunity 2001; 15:237-247.
106. Mitsuoka C, Ohmori K, Kimura N et al. Regulation of selectin binding activity by cyclization of sialic acid moiety of carbohydrate ligands on human leukocytes. Proc Natl Acad Sci USA 1999; 96:1597-1602.
107. Kannagi R. Regulatory roles of carbohydrate ligands for selectins in homing of lymphocytes. Curr Opin Struct Biol 2002; 12:in press.
108. Ito K, Handa K, Hakomori S-I. Species-specific expression of sialosyl-Le^x on polymorphonuclear leukocytes (PMN), in relation to selectin-dependent PMN responses. Glycoconjugate J 1994; 11:232-237.
109. Homeister JW, Thall AD, Petryniak B et al. The α(1,3)fucosyltransferases FucT-IV and FucT-VII exert collaborative control over selectin-dependent leukocyte recruitment and lymphocyte homing. Immunity 2001; 15:115-126.
110. Smithson G, Rogers CE, Smith PL et al. Fuc-TVII is required for T helper 1 and T cytotoxic 1 lymphocyte selectin ligand expression and recruitment in inflammation, and together with Fuc-TIV regulates naive T cell trafficking to lymph nodes. J Exp Med 2001; 194:601-614.
111. Kannagi R, Fukushi Y, Tachikawa T et al. Quantitative and qualitative characterization of cancer-associated serum glycoprotein antigens expressing fucosyl or sialosyl-fucosyl type 2 chain polylactosamine. Cancer Res 1986; 46:2619-2626.
112. Chia D, Terasaki PI, Suyama N et al. Use of monoclonal antibodies to sialylated Lewis^x and sialylated Lewis^a for serological test of cancer. Cancer Res 1985; 45:435-437.
113. Kumamoto K, Mitsuoka C, Izawa M et al. Specific detection of sialyl Lewis x determinant carried on the mucin GlcNAcβ1→6GalNAcα core structure as tumor-associated antigen. Biochem Biophys Res Commun 1998; 247:514-517.
114. Kannagi R. Transcriptional regulation of expression of carbohydrate ligands for cell adhesion molecules in the selectin family. Adv Exp Med Biol 2001; 491:267-278.
115. Hiraiwa N, Hiraiwa M, Kannagi R. Human T-cell leukemia virus-1 encoded Tax protein transactivates α1→3 fucosyltransferase Fuc-T VII, which synthesizes sialyl Lewis X, a selectin ligand expressed on adult T-cell leukemia cells. Biochem Biophys Res Commun 1997; 231:183-116.
116. Izawa M, Kumamoto K, Mitsuoka C et al. Expression of sialyl 6-sulfo Lewis x is inversely correlated with conventional sialyl Lewis x expression in human colorectal cancer. Cancer Res 2000; 60:1410-1416.
117. Capon C, Wieruszeski JM, Lemoine J et al. Sulfated Lewis X determinants as a major structural motif in glycans from LS174T-HM7 human colon carcinoma mucin. J Biol Chem 1997; 272:31957-31968.
118. Handa K, White T, Ito K et al. P-selectin-dependent adhesion of human cancer cells: Requirement for co-expression of a 'PSGL-1-like' core protein and the glycosylation process for sialosyl-Le^x or sialylsyl-Le^a. Int J Oncol 1995; 6:773-781.
119. Friederichs J, Zeller Y, Hafezi-Moghadam A et al. The CD24/P-selectin binding pathway initiates lung arrest of human A125 adenocarcinoma cells. Cancer Res 2000; 60:6714-6722.
120. Aigner S, Sthoeger ZM, Fogel M et al. CD24, a mucin-type glycoprotein, is a ligand for P-selectin on human tumor cells. Blood 1997; 89:3385-3395.
121. Qian F, Hanahan D, Weissman IL. L-selectin can facilitate metastasis to lymph nodes in a transgenic mouse model of carcinogenesis. Proc Natl Acad Sci USA 2001; 98:3976-3981.
122. Capon C, Maes E, Michalski JC et al. Sd^a-antigen-like structures carried on core 3 are prominent features of glycans from the mucin of normal human descending colon. Biochem J 2001; 358:657-123.
123. Ju T, Brewer K, D'Souza A et al. Cloning and expression of human core 1 β1,3-galactosyltransferase. J Biol Chem 2002; 277:178-186.
124. Hassan H, Reis CA, Bennett EP et al. The lectin domain of UDP-N-acetyl-D-galactosamine: Polypeptide N-acetylgalactosaminyltransferase-T4 directs its glycopeptide specificities. J Biol Chem 2000; 275:38197-38205.

125. Bennett EP, Hassan H, Mandel U et al. Cloning of a human UDP-N-acetyl-α-D-galactosamine: polypeptide N-acetylgalactosaminyltransferase that complements other GalNAc-transferases in complete O-glycosylation of the MUC1 tandem repeat. J Biol Chem 1998; 273:30472-30481.
126. Sutherlin ME, Nishimori I, Caffrey T et al. Expression of three UDP-N-acetyl-α-D-galactosamine: polypeptide GalNAc N-acetylgalactosaminyltransferases in adeno carcinoma cell lines. Cancer Res 1997; 57:4744-4748.
127. Wandall HH, Hassan H, Mirgorodskaya E et al. Substrate specificities of three members of the human UDP-N-acetyl-α-D-galactosamine: Polypeptide N-acetyl galactosaminyl transferase family, GalNAc-T1, -T2, and -T3. J Biol Chem 1997; 272:23503-23514.
128. White T, Bennett EP, Takio K et al. Purification and cDNA cloning of a human UDP-N-acetyl-α-D-galactosamine:polypeptide N-acetylgalactosaminyltransferase. J Biol Chem 1995; 270:24156-24165.
129. Kurosawa N, Kojima N, Inoue M et al. Cloning and expression of Galβ1,3GalNAc-specific GalNAc α2,6-sialyltransferase. J Biol Chem 1994; 269:19048-19053.
130. Kono M, Tsuda T, Ogata S et al. Redefined substrate specificity of ST6GalNAc II: a second candidate sialyl-Tn synthase. Biochem Biophys Res Commun 2000; 272:94-97.
131. Ju T, Cummings RD, Canfield WM. Purification, characterization, and subunit structure of rat core 1 β1,3-galactosyltransferase. J Biol Chem 2002; 277:169-177.
132. Kumamoto K, Goto Y, Sekikawa K et al. Increased expression of UDP-galactose transporter mRNA in human colon cancer tissues and its implication in synthesis of Thomsen-Friedenreich antigen and sialyl Lewis A/X determinants. Cancer Res 2001; 61:4620-4627.
133. Hotta K, Goso K, Kato Y. Human gastric glycoproteins corresponding to paradoxical concanavalin A staining. Histochemistry 1982; 76:107-134.
134. Ishihara K, Kurihara M, Goso Y et al. Peripheral α-linked N-acetylglucosamine on the carbohydrate moiety of mucin derived from mammalian gastric gland mucous cells: Epitope recognized by a newly characterized monoclonal antibody. Biochem J 1996; 318:409-416.
135. Nakamura N, Ota H, Katsuyama T et al. Histochemical reactivity of normal, metaplastic, and neoplastic tissues to α-linked N-acetylglucosamine residue-specific monoclonal antibody HIK1083. J Histochem Cytochem 1998; 46:793-136.
136. Nakayama J, Yeh JC, Misra AK et al. Expression cloning of a human α1→4-N-acetylglucosaminyltransferase that forms GlcNAcα1→4Galβ1→R, a glycan specifically expressed in the gastric gland mucous cell-type mucin. Proc Natl Acad Sci USA 1999; 96:8991-8996.
137. Bouhours D, Liaigre J, Richard C et al. Forssman penta- and tetraglycosylceramide are xenoantigens of ostrich kidney and liver. Glycobiology 1999; 9:875-886.
138. Yamamoto H, Iida-Tanaka N, Kasama T et al. Isolation and characterization of a novel Forssman active acidic glycosphingolipid with branched isoglobo-, ganglio-, and neolacto-series hybrid sugar chains. J Biochem (Tokyo) 1999; 125:923-930.
139. Clark GF, Gorbea CM, Cummings RD et al. Decreased biosynthesis of Forssman glycolipid after retinoic acid-induced differentiation of mouse F9 teratocarcinoma cells. Lectin-affinity chromatography of the glycolipid-derived oligosaccharide. Carbohydr Res 1991; 213:155-168.
140. Krupnick JG, Damjanov I, Damjanov A et al. Globo-series carbohydrate antigens are expressed in different forms on human and murine teratocarcinoma-derived cells. Int J Cancer 1994; 59:692-698.
141. Willison KR, Karol RA, Suzuki A et al. Neutral glycolipid antigens as developmental markers of mouse teratocarcinoma and early embryos: an immunologic and chemical analysis. J Immunol 1982; 129:603-609.
142. Cho SK, Yeh J, Cho M et al. Transcriptional regulation of α1,3-galactosyltransferase in embryonal carcinoma cells by retinoic acid. Masking of Lewis X antigens by α-galactosylation. J Biol Chem 1996; 271:3238-3246.
143. Haslam DB, Baenziger JU. Expression cloning of Forssman glycolipid synthetase: A novel member of the histo-blood group ABO gene family. Proc Natl Acad Sci USA 1996; 93:10697-10702.
144. Joziasse DH, Shaper JH, Jabs EW et al. Characterization of an α-1→3-galactosyltransferase homologue on human chromosome 12 that is organized as a processed pseudogene. J Biol Chem 1991; 266:6991-6998.
145. Koike C, Luppi P, Sharma SB et al. Molecular basis of evolutionary loss of α1,3-galactosyltransferase gene in higher primates. J Biol Chem 2002; 277:10114-10120.
146. Masuda Y, Sugiyama T. The effect of globopentaosylceramide on a depression model, mouse forced swimming. Tohoku J Exp Med 2000; 191:47-54.
147. Masuda Y, Ohnuma S, Sugiyama T. α2-adrenoceptor activity induces the antidepressant-like glycolipid in mouse forced swimming. Methods Find Exp Clin Pharmacol 2001; 23:19-21.
148. Chiang TR, Fanget L, Gregory R et al. Anti-Gal antibodies in humans and 1, 3α-galactosyltransferase knock-out mice. Transplantation 2000; 69:2593-2600.

149. Baquerizo A, Mhoyan A, Kearns-Jonker M et al. Characterization of human xeno reactive antibodies in liver failure patients exposed to pig hepatocytes after bioartificial liver treatment: an ex vivo model of pig to human xenotransplantation. Transplantation 1999; 67:5-18.
150. Kearns-Jonker M, Swensson J, Ghiuzeli C et al. The human antibody response to porcine xenoantigens is encoded by IGHV3-11 and IGHV3-74 IgVH germline progenitors. J Immunol 1999; 163:4399-4412.
151. Wang L, Radic MZ, Galili U. Human anti-Gal heavy chain genes. Preferential use of VH3 and the presence of somatic mutations. J Immunol 1995; 155:1276-1285.
152. Nozawa S, Xing PX, Wu GD et al. Characteristics of immunoglobulin gene usage of the xenoantibody binding to Gal-α(1,3)Gal target antigens in the Gal knockout mouse. Transplantation 2001; 72:147-155.
153. Chen ZC, Radic MZ, Galili U. Genes coding evolutionary novel anti-carbohydrate antibodies: studies on anti-Gal production in α1,3galactosyltransferase knock out mice. Mol Immunol 2000; 37:455-154.
154. LaTemple DC, Henion TR, Anaraki F et al. Synthesis of α-galactosyl epitopes by recombinant α1,3galactosyl transferase for opsonization of human tumor cell vaccines by anti-galactose. Cancer Res 1996; 56:3069-3074.
155. Chen ZC, Tanemura M, Galili U. Synthesis of α-gal epitopes (Galα1-3Galβ1-4GlcNAc-R) on human tumor cells by recombinant α1,3galactosyltransferase produced in Pichia pastoris. Glycobiology 2001; 11:577-586.
156. Galili U, Chen ZC, Manches O et al. Preparation of autologous leukemia and lymphoma vaccines expressing α-gal epitopes. J Hematother Stem Cell Res 2001; 10:501-511.
157. Cooper DK, Good AH, Koren E et al. Identification of α-galactosyl and other carbohydrate epitopes that are bound by human anti-pig antibodies: relevance to discordant xenografting in man. Transpl Immunol 1993; 1:198-205.
158. Joziasse DH, Oriol R. Xenotransplantation: the importance of the Galα1,3Gal epitope in hyperacute vascular rejection. Biochim Biophys Acta 1999; 1455:403-418.
159. Galili U. The α-gal epitope (Galα1-3Galβ1-4GlcNAc-R) in xenotransplantation. Biochimie 2001; 83:557-563.
160. Kawano T, Koyama S, Takematsu H et al. Molecular cloning of cytidine monophospho-N-acetylneuraminic acid hydroxylase. Regulation of species- and tissue-specific expression of N-glycolylneuraminic acid. J Biol Chem 1995; 270:16458-16463.
161. Irie A, Koyama S, Kozutsumi Y et al. The molecular basis for the absence of N-glycolylneuraminic acid in humans. J Biol Chem 1998; 273:15866-15871.
162. Chou HH, Takematsu H, Diaz S et al. A mutation in human CMP-sialic acid hydroxylase occurred after the Homo-Pan divergence. Proc Natl Acad Sci USA 1998; 95:11751-11756.
163. Varki A. N-glycolylneuraminic acid deficiency in humans. Biochimie 2001; 83:615-622.
164. Nakarai H, Chandler PJ, Kano K et al. Hanganutziu-Deicher antigen as a possible target for immunotherapy of melanoma. Int Arch Allergy Immunol 1990; 91:323-328.
165. Malykh YN, Schauer R, Shaw L. N-Glycolylneuraminic acid in human tumours. Biochimie 2001; 83:623-634.
166. Chester MA. IUPAC-IUB Joint Commission on Biochemical Nomenclature (JCBN). Nomenclature of glycolipids—recommendations 1997. Eur J Biochem 1998;257:293-298.
167. Clausen H, Bennett EP, Grunnet N. Molecular genetics of ABO histo-blood groups. Transfus Clin Biol 1994; 1:79-89.
168. Grunnet N, Steffensen R, Bennett EP et al. Evaluation of histo-blood group ABO genotyping in a Danish population: frequency of a novel O allele defined as O2. Vox Sang 1994; 67:210-215.
169. Kannagi R, Roelcke D, Peterson KA et al. Characterization of an epitope determinant structure in a developmentally-regulated glycolipid antigen defined by a cold agglutinin "FI"; recognition of α-sialosyl and α-fucosyl groups in a branched structure. Carbohydr Res 1983; 120:143-157.
170. Narita T, Funahashi H, Satoh Y et al. Association of expression of blood group-related carbohydrate antigens with prognosis in breast cancer. Cancer 1993; 71:3044-3053.
171. Ogawa J, Tsurumi T, Yamada S et al. Blood vessel invasion and expression of sialyl Lewis^x and proliferating cell nuclear antigen in stage I non-small cell lung cancer: Relation to postoperative recurrence. Cancer 1994; 73:1177-1183.
172. Futamura N, Nakamura S, Tatematsu M et al. Clinicopathologic significance of sialyl Le^x expression in advanced gastric carcinoma. Br J Cancer 2000; 83:1681-1687.

CHAPTER 2

Structural Basis for Mannose-Binding Protein Function in Innate Immunity

Russell Wallis

Summary

Mannose-binding protein (MBP) is the first component of the lectin pathway of the complement cascade. It targets pathogens by binding to arrays of sugars on cell surfaces and neutralizes them in an antibody-independent manner. Characterization of the interactions of MBP with carbohydrate ligands has provided detailed understanding of how exogenous sugars are selectively recognized. Recent studies of the interactions between MBP and its downstream targets have provided insight into the events that lead to pathogen neutralization. Separate point mutations in the gene encoding human MBP lead to a common immunodeficiency. Increased susceptibility to infections in individuals with mutant alleles is caused by a lack of functional MBP in serum, arising through structural changes in the encoded proteins.

Introduction

The complement system plays a central role in host defense against invasion by pathogens.[1] It comprises a series of serum proteins and enzymes that recognize microorganisms and target them for destruction (Fig. 1). Foreign cells are cleared following activation of a lytic pathway, comprising proteins that interact to form pores in cell walls. Cells are also removed by host leukocytes that are stimulated to phagocytose particles tagged by proteins or protein fragments derived from complement components.

Three different pathways initiate the effector functions of complement (Fig. 1). In the classical pathway, component C1q binds to immune complexes and triggers complement fixation through the step-wise activation of associated serine proteases C1r and C1s. This pathway is one of the major mechanisms for neutralization of pathogens following engagement by the adaptive immune system. The alternative pathway and the lectin pathway function in an antibody-independent manner and thus provide an initial line of immune defense. In the alternative pathway, spontaneous low-level hydrolysis of component C3 leads to the deposition of protein fragments onto the surface of cells. Foreign cells are destroyed following amplification of the complement cascade through a positive-feedback loop, whilst regulatory proteins avert activation on host cells, thus protecting against self-damage. In the lectin pathway, mannose-binding protein (MBP or mannose-binding lectin) binds to sugars present on pathogenic cells and triggers complement fixation by activating a protease called MBP-associated serine protease-2 (MASP-2), a homologue of C1r and C1s of the classical pathway.[2,3]

The complement system serves to direct and stimulate an effective adaptive immune response as well as neutralizing microorganisms directly.[4,5] Specific receptors present on B lymphocytes bind to fragments of complement proteins on the surface of antigens and reduce the threshold for B cell activation by antigen. Similar receptors are present on follicular dendritic

Immunobiology of Carbohydrates, edited by Simon Y.C. Wong and Gemma Arsequell.
©2003 Eurekah.com and Kluwer Academic / Plenum Publishers.

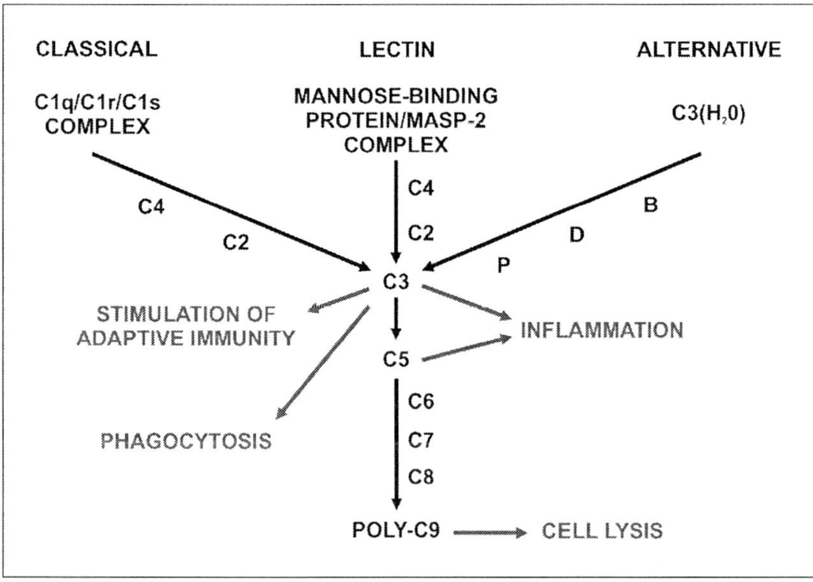

Figure 1. Complement cascade. Major components of the complement cascade are shown. Functions of the components are indicated in grey.

cells in germinal centers where they are believed to play an important role in maintaining immunological memory.

The complement system is a potent stimulator of inflammation. Peptides generated by cleavage of components C3, C4 and C5 are anaphylotoxins that bind to receptors on a variety of cells including monocytes, macrophages, granulocytes, endothelial cells and platelets. These peptides promote inflammation by inducing changes in vascular permeability, smooth muscle contraction, histamine release, chemotaxis and superoxide production.

Increasing evidence emphasizes the important role of the lectin pathway in immune function. This chapter focuses on the structural features that enable MBP to target and neutralize foreign cells and the molecular basis of a common immunodeficiency caused by a lack of functional MBP in serum.

Structural Organization of MBP

MBP is a member of the collectin family of animal lectins.[6] Polypeptides fold into four distinct domains: a short N-terminal cysteine-rich domain followed by a collagenous domain, an α-helical portion called the neck and a C-terminal carbohydrate-recognition domain (CRD) (Fig. 2). Three identical polypeptides associate to form subunits. Interactions between CRDs and the α-helices that form the neck initiate formation of the collagen triple helix, which assembles in a C- to N-terminal direction.[7] Disulfide bonds formed between cysteine residues in the N-terminal domains link polypeptides together. The pattern of disulfide bonds is heterogeneous and asymmetrical.

MBP subunits oligomerize to form larger structures through interactions between cysteine-rich domains and the first part of the collagenous domains (Fig. 2).[8] Disulfide bonds formed between cysteine residues near the N-terminus of each polypeptide link separate subunits together to stabilize these oligomers. Prior to secretion, attachment of small sulfhydryl-containing groups to free cysteine residues blocks further self-association of subunits, so that once secreted, oligomers no longer interact with each other. MBPs consist of a mixture of different oligomers. Human MBP comprises dimers to octamers of subunits of

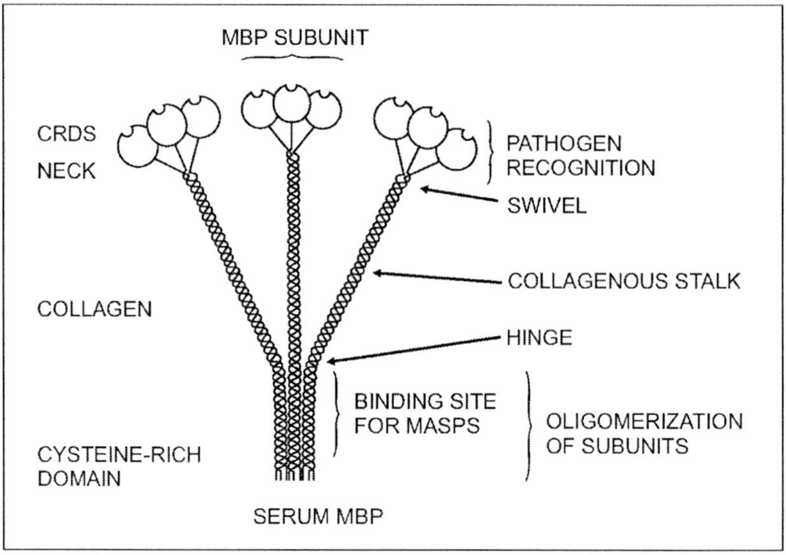

Figure 2. Domain organisation of serum MBP. An MBP trimer is shown. Dimers and tetramers of subunits have similar structural organisations. Specific functions of portions of MBP are indicated.

which dimers, trimers and tetramers are the predominant forms,[9,10] while rat MBP consists of monomers to tetramers of subunits.[11] Dimers, trimers and tetramers of rat MBP all activate complement but trimers and tetramers have higher activities than dimers.[8] Single MBP subunits have very low activity compared to the larger oligomers.

The collagenous domains of MBP resemble vertebrate collagens in terms of their structures and post-translational modifications.[12] Hydrodynamic analysis has shown that oligomers of rat MBP are asymmetrical, reflecting the fact that the collagenous regions form highly extended structures[8]. 4-Hydroxyproline and glucosylgalactosyl-5-hydroxylysine residues have been detected within the consensus sequences Hyp-Gly-Xaa and Hyl-Gly-Xaa in rat and human MBPs. Hydroxyproline residues are thought to stabilize the collagen triple helix, while the modified hydroxylysine residues may modulate subunit formation and oligomerization.[13]

MBP oligomers are bouquet-like structures in images obtained using rotary shadowing electron microscopy.[9] Measurements based on these images reveal that the point at which individual collagenous stems diverge from each other corresponds to an interruption in the Gly-Xaa-Yaa repeat of the collagen-like domain. This region, known as the hinge, is likely to be flexible (Fig. 2). A second potentially flexible region called the swivel is located at the junction between the collagen-like domain and the neck.[7] As MBP circulates in serum, flexibility of the hinge and swivel regions probably allows movement of the collagenous stalks relative to each other, enabling MBP to scan its local environment for target ligands. The swivel probably orientates the CRDs when MBP binds to a target cell.

Gene Organization of MBP

Two MBP genes have been identified in mammals.[14] In rats, mice and most other mammals, both genes are transcribed and translated and encode proteins called MBP-A and MBP-C. MBP-A is a serum protein, while MBP-C is found mainly in liver. In humans and chimpanzees, only MBP-C is produced.[15,16] It is found in serum and has structural and functional properties similar to rat MBP-A. The human orthologue of MBP-A is transcribed into mRNA, but encodes a truncated product that is not functional.

The human MBP gene consists of four exons, separated by three introns.[17] The first exon encodes the cysteine-rich domain and the first part of the collagenous domain, while the second, third and fourth exons encode the remainder of the collagenous domain, the neck region and the CRD, respectively. Three allelic variants have been identified that are associated with immunodeficiency[18-20] (see below). Each allele contains a separate point mutation in exon 1 of the gene (within codons 52, 54 and 57) that gives rise to a single amino acid substitution in the first part of the collagenous domain of the encoded protein (Arg^{32}→Cys, Gly^{34}→Asp and Gly^{37}→Glu). The frequencies of the alleles differ markedly in different populations.[2] For example, the codon 54 mutation is the most common in British Caucasians, occurring in approximately 25% of the population, while the codon 57 mutation is found in up to 50% of many sub-Sahara African populations. Based on the allelic frequencies in different populations, together with the patterns of human colonization, estimates have been made of when the mutations arose. For example, the codon 54 mutation is believed to have occurred 20,000-50,000 years ago, after colonization of North America but before the first human settlement in Australia.[21]

The promoter region of human MBP contains several responsive elements that modulate transcription levels.[17,22] Included in these elements are those commonly associated with acute-phase proteins. Indeed, serum levels of MBP are raised after malarial infection and after trauma, although the increases are relatively modest (< 3-fold) compared to those measured in other acute-phase proteins.[23] Several MBP promoter polymorphisms have been identified and designated as H/L, X/Y and P/Q.[24,25] Four different promoter haplotypes are commonly found and these are associated with different serum MBP levels. Haplotypes HYP and LYQ give rise to higher MBP levels while LYP and LXP are associated with lower levels. The three mutant MBP alleles appear to be linked with haplotypes that yield high or intermediate levels of gene expression, implying that low serum MBP levels in immunodeficient individuals are due to the mutations in the MBP gene rather than in the promoter.

Sugar Recognition by MBP

MBPs selectively bind to foreign cells through multiple CRD-sugar interactions.[3,26] Each CRD interacts with low affinity to a single carbohydrate residue and provides part of the binding energy that is necessary to form a stable complex capable of triggering complement fixation.[27] Crystal structures of CRDs of rat MBP in complex with mannose-containing sugars reveal that a ternary complex is formed between the CRD, a Ca^{2+} and the terminal mannose residue of the sugar.[28-30] Amino acid residues within the binding site form hydrogen bonds to the 3- and 4-OH groups of the sugar and also make co-ordination bonds to the Ca^{2+}. Binding specificity for monosaccharides is determined largely by the configuration of the hydroxyl groups on the carbohydrate moiety, so that residues that have hydroxyl groups equivalent to the equatorial 3- and 4-OH groups of mannose, such as N-acetylglucosamine and fucose, are also ligands.

Crystal structures of trimeric fragments of rat and human MBP reveal that the CRDs are configured in an ideal arrangement to bind to ligands projecting from a flat surface such as a bacterial or fungal cell.[31,32] The CRDs are maintained in a fixed geometry through hydrophobic packing between the CRD of one polypeptide and part of the α-helix that forms the neck in an adjacent polypeptide. Binding sites are 54 Å apart in rat MBP, which is too widely separated for multiple CRD-ligand interactions between a MBP subunit and a single high-mannose oligosaccharide. Because carbohydrates terminating in mannose are relatively rare on mammalian cells, MBPs will not bind to host cells but will instead target microorganisms that are covered in high-density arrays of mannose-like sugars.

MBP interacts with an extremely broad range of bacteria, fungi and parasitic cells, many of which are human pathogens.[33] For example, MBP has been shown to bind to and activate complement on clinical strains of *Burkholderia cepacia*, an important pathogen in patients with cystic fibrosis.[34] MBP was also found to interact and promote C4 deposition on numerous pathogens isolated from immunocompromised children.[35] In this study, MBP bound to species of

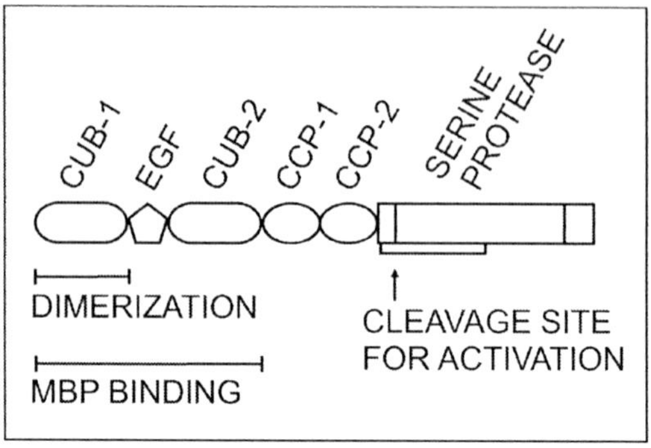

Figure 3. Domain organisation of MASPs. Regions involved in MASP dimerisation and in binding to MBP are shown.

Candida, Aspergillus fumigatus, Staphylococcus aureus, Escherichia coli, Klebsiella and strains of *Haemophilus influenzae* type b. Interestingly, in another study, MBP bound relatively poorly to species of anaerobic bacteria most commonly identified as causing disease, implying that there may be an inverse relationship between pathogenicity and the level of MBP binding.[36] This suggestion is consistent with the finding that patients with low levels of functional MBP appear to be more susceptible to infections caused by organisms that only rarely cause severe disease in immunocompetent individuals. For example, individuals with mutant alleles are more vulnerable to bacterial meningitis caused by Neisseria meningitidis.[37]

MBP/MASP Complexes

MBP circulates in complexes with three different MASPs (MASP-1, -2 and -3) and a protein comprising the two N-terminal domains of MASP-2, called MBP-associated protein 19 (MAP19 or sMAP).[38-42] MAP19 is an alternatively spliced product from the MASP-2 gene, whereas MASP-1 and MASP-3 are produced following alternative splicing of mRNAs encoded by the MASP-1/-3 gene. All three MASPs have the same domain organization: two N-terminal CUB domains (domains found in complement components C1r/C1s, Uegf and bone morphogenic protein) separated by an epidermal growth factor (EGF)-like domain, followed by two complement control protein (CCP) modules and a serine protease domain. The EGF-like module contains the features characteristic of a Ca^{2+}-binding EGF-like domain.

MBP circulates as complexes with zymogen forms of all three MASPs. However, MASP-2 alone is believed to trigger complement activation.[43] When MBP binds to a target cell, auto-activation of MASP-2 occurs through hydrolysis of the polypeptide chain at a specific cleavage site near the N-terminal end of the serine protease domain (Fig. 3). The activated serine protease remains attached to the rest of the molecule through a single disulfide bond. MASP-2 then cleaves component C4, exposing a reactive thioester group on the larger C4b fragment, which in turn reacts with an exposed hydroxyl or amino acid side chain on the target cell. Component C2 binds to the immobilized C4b fragment and is also cleaved by activated MASP-2. The resulting C4b2a complex subsequently cleaves and activates component C3, the next element in the pathway (Fig. 1).

MASPs are Ca^{2+}-independent homodimers stabilized through interactions between the N-terminal CUB modules.[44,45] The first three modules mediate binding to MBP in a Ca^{2+}-dependent manner, in which both CUB domains probably contact MBP. MASPs bind to the first portion of the collagen-like domain of MBP. Each protomer of the MASP dimer probably binds to a separate MBP subunit, so that MBP dimers bind to single MASP dimers, while MBP trimers and tetramers bind up to two MASP dimers. A complex between a MBP

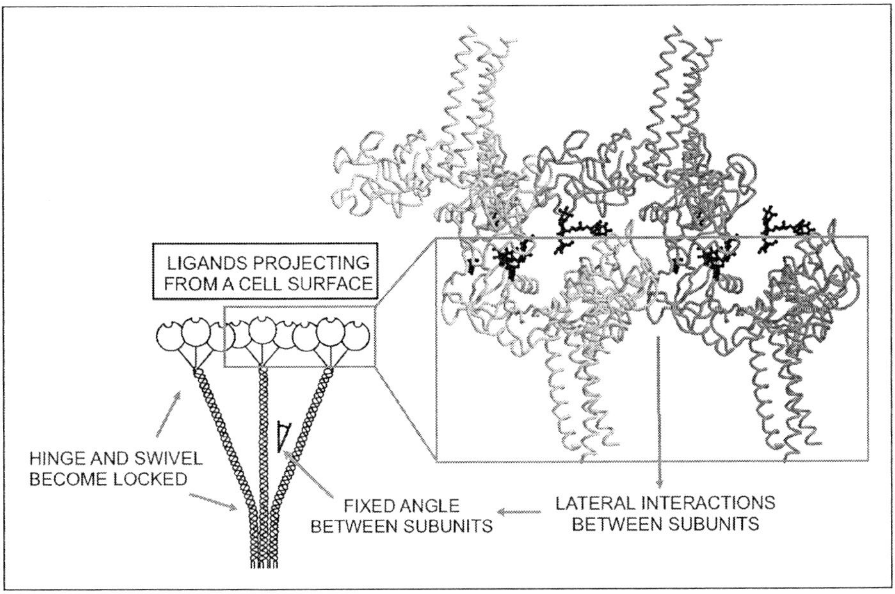

Figure 4. Structure of fragments of rat MBP in complex with high-mannose oligosaccharides reveals lateral interactions between CRDs of separate subunits. Fragments of MBP are arranged in layers cross-linked by sugar ligands in the crystal lattice. Lateral interactions occur between CRDs of adjacent subunits.[30] A similar arrangement of subunits would impart a fixed angle between the collagenous stalks when MBP binds to a target cell and might represent the activated conformation that triggers MASP auto-activation.

dimer and a single MASP dimer is sufficient to form a functional unit that is capable of becoming activated when MBP binds to a microorganism.

The roles of MASP-1, MASP-3 and MAP19 are not known. MASP-1 can cleave complement component C3, in vitro, but this activity is very low and is unlikely to be biologically important.[43] MASPs compete for binding sites on MBP.[45] Because the larger MBP oligomers can bind two MASPs, complexes containing more than one type of MASP probably circulate in serum. MASP-2 might activate MASP-1 or MASP-3. However, the zymogen forms of MASP-1 and MASP-2 do not interact with each other, so either MASP-1 or MASP-3 probably auto-activates. MAP19 binds to MBP, although with lower affinity than MASP-2.[46] Because MAP19 lacks the serine protease domain, it cannot activate complement directly. It may form part of some of the larger MBP/MASP-2 complexes that activate complement or it might compete for MASP-2 binding sites, thereby modulating complement activation.

Mechanism of Complement Activation by MBP

The mechanism of complement activation by the lectin pathway is not known. However, the crystal structure of a trimeric fragment of rat MBP in complex with an oligosaccharide ligand provides an indication of how activation might be initiated.[30] In this structure, MBP fragments are arranged in sheets cross-linked by oligosaccharides in the crystal lattice (Fig. 4). Lateral interactions occur between the CRDs of adjacent subunits and probably stabilize this arrangement. The CRDs in one layer present sugar ligands to the CRDs of the opposite sheet in the form of a surface of high-density carbohydrate structures, as might be displayed from a cell surface. Such an arrangement of subunits would lock the configuration of the hinge and swivel in a defined geometry, in which there is a relatively acute angle between the collagenous stalks of adjacent subunits. Because MASPs bind near the hinge, they could detect the conformation of the collagenous domains, leading to activation of the protease domain.

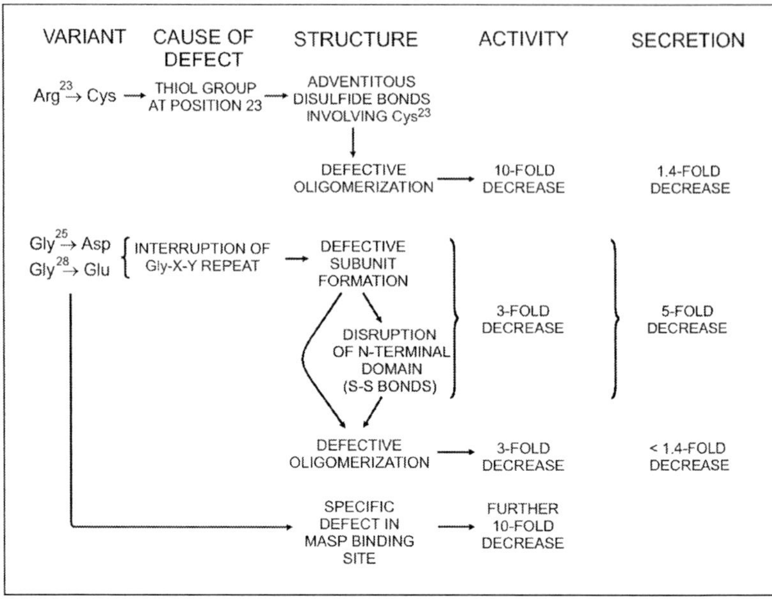

Figure 5. Molecular defects that give rise to reduced complement fixation and reduced secretion rates in mutant MBPs. Changes in complement-fixing activities and secretion rates relative to wild-type MBP were determined using recombinant rat protein containing mutations corresponding to the known human variants.[13, 51]

MBP-Associated Immunodeficiency

Immunodeficiency caused by three separate mutations in the MBP gene is associated with increased susceptibility to infections caused by a wide range of microorganisms.[2,18,35,36,47] Reduced levels of MBP in the serum of affected individuals compound low complement-fixing activity due to structural defects in the N-terminal regions of the protein. Individuals are most susceptible to infections in early childhood before the adaptive immune system is fully developed. Vulnerability is also increased when the adaptive immune system is compromised for example during HIV infection or following chemotherapy.[48,49] Immunodeficiency has also been described in adults with no other obvious predisposing disorders[50], implying that MBP plays an important immunological protective role throughout life.

The molecular basis of immunodeficiency has been examined using a model system in which the human mutations have been recreated in rat serum MBP.[51] An Arg^{23}→Cys mutation (corresponding to the Arg^{32}→Cys mutation in human MBP) leads to a lower proportion of trimers and tetramers of subunits and a corresponding increase in monomers of subunits. Because the smaller oligomers have lower complement-fixing activities than trimers and tetramers, the overall activity is reduced by approximately 10-fold.

Defective oligomerization of subunits in the Arg^{23}→Cys mutation is caused by adventitious disulfide bond formation during biosynthesis (Fig. 5).[51] Interchain disulfide bonds formed between Cys^{23} residues within separate subunits probably disrupt the local structure of the collagenous domain. Other Cys^{23} residues form disulfide bonds with other material, such as free cysteine or glutathione. Because the first part of the collagenous domain is involved in oligomerization of subunits, the structural changes in this region probably block self-association during biosynthesis, leading to secretion of smaller oligomers.

The molecular defect in rat MBP with the Gly^{25}→Asp and Gly^{28}→Glu mutations (corresponding to the Gly^{34}→Asp and Gly^{37}→Glu mutations in human MBP) is also due in part to a decrease in the amounts of the larger MBP oligomers (Fig. 5).[51] However, defective interactions with MASP-2 caused by disruption to the collagen-like domain contribute to the reduced activities of the mutant proteins. The Gly^{28}→Glu mutation leads to a 10-fold lower activity than the Gly^{25}→Asp mutation, implying that Gly^{28} is nearer to the MASP binding site.

The molecular defect in the glycine mutants is due to the disruption of the Gly-Xaa-Yaa repeat in the collagenous domain.[51] Replacement of glycine residues with bulky acidic groups probably destabilizes the collagen helix as has been found when these residues are introduced into synthetic collagen peptides. Because MBP polypeptides assemble in a C- to N-terminal direction, the mutations disrupt the N-terminal part of the collagenous domain and the cysteine-rich domain. Reduced affinities for the MASPs and defective oligomerization of subunits are both caused by the disruption to these regions of MBP.

The protein deficit in immunodeficient individuals probably arises through two mechanisms: decreased secretion and increased turnover. Rat MBPs containing either the Gly^{25}→Asp or the Gly^{28}→Glu mutations are secreted at rates five-fold lower than the rate of wild-type protein, implying that defective assembly during biosynthesis leads to reduced secretion levels.[13] The mechanism by which MBP is removed from serum is not known. However, the mutant proteins are structurally perturbed and might be cleared more rapidly either through the natural mechanism or by some other process such as degradation by endogenous proteases.

MBP-associated immunodeficiency is a dominant disorder.[2] It has been hypothesized that the defect in heterozygous individuals arises because the presence of even one mutant polypeptide in a MBP subunit would be disruptive. This suggestion is supported by recent studies using rat MBP.[52] MBPs, generated by co-expression of mutant and wild-type polypeptides, contained both mutant and wild-type chains and had the same structural and functional defects as the fully mutant proteins.

Complement Activation and Disease

The complement system is tightly regulated so that untimely activation is usually stopped before damage is done to host tissues. However, under certain circumstances these regulatory processes break down or are overwhelmed. Such occasions can be very damaging to the host. In some infectious disorders, such as meningococcal disease, the most severe disease appears to be correlated with high levels of inflammatory activity,[33] implying that excessive stimulation of the inflammatory response might contribute to the disease process. A link between MBP-associated immunodeficiency and increased susceptibility to meningococcal disease has been demonstrated.[37] In this study, although individuals with wild-type alleles were less vulnerable to infection, there was some evidence that the disease was more severe once such subjects were affected. A similar relationship between disease susceptibility and severity has been observed in other inherited complement defects. For example, deficiencies in the terminal components of complement are associated with mild recurrent attacks of meningococcal disease.[53]

A number of studies have examined possible links between MBP deficiency and inflammatory disorders such as rheumatoid arthritis[54-56] and systemic lupus erythmatosus.[57-60] Associations between increased frequencies of the mutant MBP alleles or protein deficiency and systemic lupus erythmatosus have been described in a number of different populations, although the role of MBP in the disease pathogenesis is not known. Well-defined relationships between this disorder and other genetic defects of the complement system have also been demonstrated, implying that defective complement function plays a key role in the disease. A link between mutant MBP alleles and rheumatoid arthritis is more controversial. Data from different groups appear to be contradictory and a clear association has not been established.

Some microorganisms exploit the complement system as a means of infecting a host. For example, *Mycobacteria* and other intracellular parasites utilize the complement-opsonization pathway for entry into host cells. Interestingly, the presence of mutant MBP alleles appears to be associated with increased resistance to such infections. In a recent study of infections caused by *Mycobacterium tuberculosis*, the frequency of the codon 54 mutation ($Gly^{34}\rightarrow Asp$) in an African population was found to be significantly higher amongst tuberculosis negative-controls compared to patients with pulmonary tuberculosis or tuberculosis meningitis.[61] Therefore, it seems likely that a dampened down lectin pathway may be beneficial in protecting against infection by these microorganisms. Such a selective advantage would explain, at least in part, why mutant MBP alleles are so prevalent in the population.

Another situation where complement activation may be injurious is when tissues starved of oxygen and nutrients are reperfused, for example following ischemia or infarction. Under these conditions acute stimulation of complement occurs, resulting in the influx of large numbers of neutrophils and extensive deposition of lytic components. Specific inhibitors of complement have been found to reduce infarct size in animal models of reperfusion injury, suggesting that complement contributes to tissue damage. The lectin pathway might be involved in this process because antibodies specific for MBP reduce significantly the degree of damage in a rat model system of myocardial reperfusion injury.[62,63]

MBP and Opsonization

Activation of any one of the three pathways of complement stimulates phagocytosis by host leukocytes through the deposition of complement-derived fragments onto the surfaces of microorganisms. Receptors on phagocytic cells bind to these fragments and activate the phagocytic pathway. MBP is also thought to function directly as an opsonin.[64] Several proteins and receptors have been proposed as MBP receptors, although none has been confirmed in this role as yet. It could be envisaged that binding of immobilized MBP to a receptor on phagocytic cells stimulates phagocytosis directly. Alternatively, binding might enhance phagocytosis mediated through an established mechanism such as complement- or immunoglobulin-mediated phagocytosis.[33]

MBP As a Therapeutic Agent

MBP does not bind to most mammalian cells because the sugars that are expressed on cell surfaces usually terminate in sialic acid, which is not a ligand. However, the glycosylation patterns on tumor cells are often different from those of normal cells and some tumor cell lines are recognized by MBP. Tumors derived from one such cell line were found to stop growing and decrease in size upon injection of vaccinia virus carrying the wild-type human MBP gene.[65] The mechanism of cytotoxicity in these experiments has not been established. However, the process is probably complement independent because similar effects were observed when the $Gly^{34}\rightarrow Asp$ mutant was used in place of wild-type MBP. These studies suggest that MBP could be used as an anti-cancer agent against certain types of tumors. Furthermore, MBP could potentially be used to target a wider range of tumor types by changing the binding specificity of the CRDs for sugar ligands expressed on such cells.

Replacement therapy for MBP-deficient individuals is another possible clinical use for MBP. Treatment with MBP might preclude many infections at an early stage by neutralizing the microorganisms before they become established within the host. MBP-replacement therapy would also have the advantage that it stimulates the host's natural immune response and it targets a wide range of pathogens, including bacteria, fungi and parasites. Furthermore, it is unlikely that MBP would be immunogenic because low levels of protein are produced in individuals with mutant alleles. Because MBP would almost certainly have to be administered intravenously and repeatedly, therapy might be particularly useful in circumstances where individuals are vulnerable for a short period. One such example is treatment following chemotherapy. Infection is a major cause of mortality in children following chemotherapy and recent studies suggest that individuals with low serum MBP are particularly vulnerable.[49] Thus, MBP infusions might represent a useful new therapeutic approach.

Acknowledgements

I thank Kurt Drickamer for critical reading of the manuscript and for many helpful discussions. I also thank Hadar Feinberg for providing the structure of MBP shown in Fig. 3. Funding is provided by grant 041845 from the Wellcome Trust.

References

1. Porter RR, Reid KBM. The biochemistry of complement. Nature 1978; 275:699-704.
2. Turner MW. Mannose-binding lectin: the pluripotent molecule of the innate immune system. Immunol Today 1996; 17:532-540.
3. Weis WI, Taylor ME, Drickamer K. The C-type lectin superfamily in the immune system. Immunol Rev 1998; 163:19-34.
4. Fearon DT, Carter RH. The CD19/CR2/TAPA-1 complex of B-lymphocytes: linking natural to acquired immunity. Annu Rev Immunol 1995; 13:127-149.
5. Fearon DT, Locksley RM. Elements of immunity—The instructive role of innate immunity in the acquired immune response. Science 1996; 272:50-54.
6. Drickamer K, Taylor ME. Biology of animal lectins. Annu Rev Cell Biol 1993; 9:237-264.
7. Wallis R, Drickamer K. Asymmetry adjacent to the collagen-like domain in rat liver mannose-binding protein. Biochem J 1997; 325:391-400.
8. Wallis R, Drickamer K. Molecular determinants of oligomer formation and complement fixation in mannose-binding proteins. J Biol Chem 1999; 274:3580-3589.
9. Lu J, Thiel S, Wiedemann H et al. Binding of the pentamer/hexamer forms of a mannan-binding protein to zymosan activates the proenzyme $C1r_2C1s_2$ complex of the classical pathway of complement, without involvement of C1q. J Immunol 1990; 144:2287-2294.
10. Lipscombe RJ, Sumiya M, Summerfield JA et al. Distinct physiochemical characteristics of human mannose-binding protein expressed by individuals of differing genotype. Immunol 1995; 85:660-667.
11. Yokota Y, Arai T, Kawasaki T. Oligomeric structures required for complement activation of serum mannan-binding proteins. J Biochem (Tokyo) 1995; 117:414-419.
12. Kadler K. Extracellular matrix 1: fibril-forming collagens. Protein Profile 1995; 2:491-619.
13. Heise CT, Nicholls JR, Leamy CE et al. Impaired secretion of rat mannose-binding protein resulting from mutations in the collagen-like domain. J Immunol 2000; 165:1403-1409.
14. Drickamer K, Dordal MS, Reynolds L. Mannose-binding proteins isolated from rat liver contain carbohydrate-recognition domains linked to collagenous tails. J Biol Chem 1986; 261:1034-1046.
15. Mogues T, Ota T, Tauber AI et al. Characterization of two mannose-binding protein cDNAs from rhesus monkey (Macaca mulatta): structure and evolutionary implications. Glycobiology 1996; 6:543-550.
16. Guo N, Mogues T, Weremowicz S et al. The human ortholog of rhesus mannose-binding protein-A gene is an expressed pseudogene that localizes to Chromosome 10. Mamm. Genome 1998; 9:246-249.
17. Taylor ME, Brickell PM, Craig RK et al. Structure and evolutionary origin of the gene encoding a human serum mannose-binding protein. Biochem J 1989; 262:763-771.
18. Sumiya M, Super M, Tabona P et al. Molecular basis of opsonic defect in immunodeficient children. Lancet 1991; 337:1569-1570.
19. Lipscombe RJ, Sumiya M, Hill AVS et al. High frequencies in African and non-African populations of independent mutations in the mannose binding protein gene. Hum Mol Genet 1992; 1:709-715.
20. Madsen HO, Garred P, Kurtzhals JA et al. A new frequent allele is the missing link in the structural polymorphism of the human mannan-binding protein. Immunogenetics 1994; 40:37-44.
21. Turner MW, Dinan L, Heatley S et al. Restricted polymorphism of the mannose-binding lectin gene of indigenous Australians. Hum Mol Genet 2000; 9:1481-1486.
22. Naito H, Ikeda A, Hasegawa K et al. Characterization of the human serum mannan-binding protein promoter. J Biochem 1999; 126:1004-1012.
23. Thiel S, Holmskov U, Hviid L et al. The concentration of the C-type lectin, mannan-binding protein, in human plasma increases during an acute phase response. Clin Exp Immunol 1992; 90:31-35.
24. Madsen HO, Satz ML, Hogh B et al. Different molecular events result in low protein levels of mannan-binding lectin in populations from Southeast Africa and South America. J Immunol 1998; 161:3169-3175.
25. Madsen HO, Garred P, Thiel S et al. Interplay between promoter and structural gene variants control basal serum level of mannan-binding protein. J Immunol 1995; 155:3013-3020.

26. Weis WI, Drickamer K. Structural basis of lectin-carbohydrate interaction. Annu Rev Biochem 1996; 65:441-473.
27. Iobst ST, Wormald MR, Weis WI et al. Binding of sugar ligands to Ca^{2+}-dependent animal lectins: I. Analysis of mannose binding by site-directed mutagenesis and NMR. J Biol Chem 1994; 269:15505-15511.
28. Weis WI, Drickamer K, Hendrickson WA. Structure of a C-type mannose-binding protein complexed with an oligosaccharide. Nature 1992; 360:127-134.
29. Ng KK-S, Drickamer K, Weis WI. Structural analysis of monosaccharide recognition by rat liver mannose-binding protein. J Biol Chem 1996; 271:663-674.
30. Ng K-S, Kolatar AR, Park-Snyder S et al. Orientation of bound ligands in mannose-binding proteins: implications for multivalent ligand recognition. J Biol Chem 2002; 277:16088-16095.
31. Sheriff S, Chang CYY, Ezekowitz RAB. Human mannose-binding protein carbohydrate-recognition domain trimerizes through a triple α-helical coiled coil. Nature Struct Biol 1994; 1:789-794.
32. Weis WI, Drickamer K. Trimeric structure of a C-type mannose-binding protein. Structure 1994; 2:1227-1240.
33. Jack DL, Klein NJ, Turner MW. Mannose-binding lectin: targetting the microbial world for complement attack and opsonophagocytosis. Immunol Rev 2001; 180:86-99.
34. Davies J, Neth O, Alton E et al. Differential binding of mannose-binding lectin to respiratory pathogens in cystic fibrosis. Lancet 2000; 355:1885-1886.
35. Neth O, Jack DL, Dodds AW et al. Mannose-binding lectin binds to a range of clinically relevant microorganisms and promotes complement deposition. Infect Immun 2000; 68:688-693.
36. Townsend R, Read RC, Turner MW et al. Differential recognition of obligate anaerobic bacteria by human mannose-binding lectin. Clin. Exp. Immunol. 2001; 124:223-228.
37. Hibberd ML, Sumiya M, Summerfield JA et al. Association of variants of the gene for mannose-binding lectin with susceptibility to meningococcal disease. Lancet 1999; 353:1049-1053.
38. Dahl MD, Thiel S, Matsushita M et al. MASP-3 and its association with distinct complexes of the mannan-binding lectin complement activation pathway. Immunity 2001; 15:127-135.
39. Takayama Y, Takada F, Takahashi A et al. A 100-kDa protein in the C4-activating component of Ra-reactive factor is a new serine protease having module organization similar to C1r and C1s. J Immunol 1994; 152:2308-2316.
40. Thiel S, Vorup-Jensen T, Stover CM et al. A second serine protease associated with mannan-binding lectin that activates complement. Nature 1997; 386:506-510.
41. Stover CM, Thiel S, Lynch NJ et al. Two constituents of the initiation complex of the mannan-binding lectin activation pathway of complement are encoded by a single structural gene. J Immunol 1999; 162:3481-3490.
42. Takahashi M, Endo Y, Fujita T et al. A truncated form of mannose-binding lectin-associated serine protease (MASP)-2 expressed by alternative polyadenylation is a component of the lectin complement pathway. Int Immunol 1999; 11:859-863.
43. Rossi V, Cseh S, Bally I et al. Substrate specificities of recombinant mannan-binding lectin-associated serine proteases-1 and -2. J Biol Chem 2001; 276:40880-40887.
44. Wallis R, Dodd RB. Interaction of mannose-binding protein with associated serine proteases: effects of naturally occurring mutations. J Biol Chem 2000; 275:30962-30969.
45. Chen C-B, Wallis R. Stoichiometry of complexes between mannose-binding protein and its associated serine proteases. Defining functional units for complement activation. J Biol Chem 2001; 276:25894-25902.
46. Thielens NE, Cseh S, Thiel S et al. Interaction properties of human mannan-binding lectin (MBL)-associated serine proteases-1 and -2, MBL-associated protein 19, and MBL. J Immunol 2001; 166:5068-5077.
47. Super M, Thiel S, Lu J et al. Association of low levels of mannan-binding protein with a common defect of opsonization. Lancet 1989; 2:1236-1239.
48. Garred P, Madsen HO, Balslev U et al. Susceptibility to HIV infection and progression of AIDS in relation to variant alleles of mannose-binding lectin. Lancet 1997; 349:236-240.
49. Neth O, Hann I, Turner MW et al. Deficiency of mannose-binding lectin and burden of infection in children with malignancy: a prospective study. Lancet 2001; 358:614-618.
50. Summerfield JA, Ryder S, Sumiya M et al. Mannose-binding protein gene mutations associated with unusual and severe infections in adults. Lancet 1995; 349:886-889.
51. Wallis R, Cheng JYT. Molecular defects in variant forms of mannose-binding protein associated with immunodeficiency. J Immunol 1999;163:4953-4959.
52. Wallis R. Dominant effects of mutations in the collagenous domain in mannose-binding protein. J Immunol 2002; 168:4553-4558.

53. Ross SC, Rosenthal PJ, Berberich HM, Densen P. Killing of Neisseria meningitidis by human neutrophils: implications for normal and complement-deficient individuals. J Infect Dis 1987; 155:1266-1275.
54. Graudal NA, Madsen HO, Tarp U et al. The association of variant mannose-binding lectin genotypes with radiographic outcome in rheumatoid arthritis. Arthritis Rheum 2000; 43:515-521.
55. Garred P, Madsen HO, Marquart H et al. Two edged role of mannose binding lectin in rheumatoid arthritis: a cross sectional study. J Rheumatol 2000; 27:26-34.
56. Ip WK, Lau YL, Chan SY et al. Mannose-binding lectin and rheumatoid arthritis in southern Chinese. Arthritis Rheum 2000; 43:1679-1687.
57. Davies EJ, Snowden N, Hillarby MC et al. Mannose-binding protein gene polymorphism in systemic lupus erythematosus. Arthritis Rheum 1995; 38:110-114.
58. Davies EJ, Teh LS, Ordi-Ros J et al. A dysfunctional allele of the mannose binding protein gene associates with systemic lupus erythematosus in a Spanish population. J Rheumatol 1997; 24:485-488.
59. Lau YL, Lau CS, Chan SY et al. Mannose-binding protein in Chinese patients with systemic lupus erythematosus. Arthritis Rheum 1996; 39:706-708.
60. Sullivan KE, Wooten C, Goldman D, Petri M. Mannose-binding protein genetic polymorphisms in black patients with systemic lupus erythematosus. Arthritis Rheum 1996; 39:2046-2051
61. Hoal-Van Helden EG, Epstein J, Victor TC et al. Mannose-binding protein B allele confers protection against tuberculous meningitis. Pediatr Res 1999; 45:459-464.
62. Jordan JE, Montalto MC, Stahl GL. Inhibition of mannose-binding lectin reduces postischemic myocardial reperfusion injury. Circulation 2001; 104:1413-1418.
63. Collard CD, Vakeva A, Morrissey MA et al. Complement activation after oxidative stress: Role of the lectin complement pathway. Am J Pathol 2000; 156:1549-1556.
64. Kuhlman M, Joiner K, Ezekowitz RAB. The human mannose-binding protein functions as an opsonin. J Exp Med 1989; 169:1733-1745.
65. Ma Y, Uemura K, Oka S. Antitumor activity of mannan-binding protein in vivo as revealed by a virus expression system: Mannan-binding protein dependent cell-mediated cytotoxicity. Proc Natl Acad Sci 1999; 96:371-375.

CHAPTER 3

C-Reactive Protein:
Structure, Synthesis and Function

Terry W. Du Clos and Carolyn Mold

Abstract

C-reactive protein (CRP) is an acute phase serum protein in man and a member of the pentraxin family of proteins. The pentraxins are conserved on an evolutionary basis having shared structural features across invertebrate and vertebrate species. The pentraxins share a novel cyclic pentameric structure that is resistant to heat and protease attack. Each of the five identical subunits has a shallow binding pocket on the binding face that binds phosphocholine and other ligands in a calcium-dependent manner. The other face of the pentraxin is responsible for receptor and C1q binding. CRP shares many properties with the adaptive immune system mediator, immunoglobulin. CRP binds to bacterial polysaccharides and glycolipids, damaged membranes and exposed nuclear antigens. This leads to binding of C1q and activation of the classical complement cascade resulting in the fixation of prophagocytic complement split products. CRP also has been shown to bind to the Fc receptors and to enhance phagocytosis of particulate antigens and microorganisms. The interaction of CRP with receptor-bearing cells is a stimulus for cytokine production. CRP is therefore one of the prototypic members of the innate immune system providing a recognition molecule for altered self material and foreign invaders. Through fixation of complement and interaction with Fc receptors clearance of effete cells or foreign invaders is enhanced in a noninflammatory manner. Thus CRP is a multifunctional protein whose roles in inflammation, host defense and autoimmunity are being intensively explored.

Introduction

C-reactive protein is a member of the pentraxin family of proteins. These proteins evolved very early on presumably as defense molecules. The oldest known members of this family are found in *Limulus polyphemus*, the horseshoe crab, although there may be homologues in even more primitive species. The pentraxin family is characterized by a cyclic planar structure consisting of five or, in the case of Limulus CRP, six subunits. Another characteristic of the pentraxin family is calcium-dependent binding to ligands including microbial polysaccharides. Two major groups of pentraxins are found in mammals. These are C-reactive protein (CRP) and serum amyloid P component (SAP). This review will be focused on CRP and mostly on human CRP although mention of important features of SAP will be made.

History

CRP was first discovered by physicians who were studying patients with severe acute infections with *Streptococcus pneumoniae*.[1] These studies were done prior to the availability of antibiotics and the patients were essentially untreated. Many of the patients died of their infection. On examining the blood of these patients over the course of their illness a substance was detected that was capable of agglutinating the organisms. This substance appeared at very high

Immunobiology of Carbohydrates, edited by Simon Y.C. Wong and Gemma Arsequell.
©2003 Eurekah.com and Kluwer Academic / Plenum Publishers.

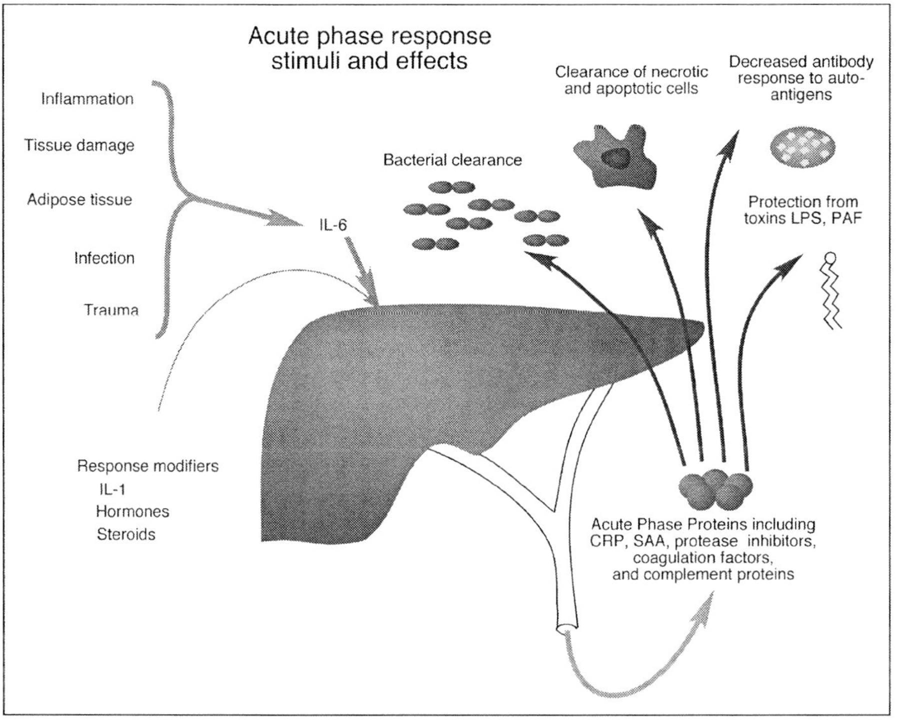

Figure 1. CRP synthesis. CRP is synthesized primarily by hepatocytes in response to the proinflammatory cytokine IL-6 but is influenced by IL-1 and other factors. CRP is released rapidly into the blood and travels to sites of tissue damage and/or inflammation.

levels very early in the course of the infection. In the patients who survived the infection the substance disappeared within a few days after clinical improvement. These changes all occurred before the development of specific IgG antibodies. The substance was named C-reactive protein because it reacted with the cell wall C polysaccharide (PnC) of the pneumococcus. The reaction with PnC and the bacteria was calcium-dependent and it was subsequently determined that the primary moiety recognized on the polysaccharide was phosphocholine (PC).[2]

Synthesis

CRP is synthesized primarily in the liver by hepatocytes[3] in response to a variety of inflammatory conditions including trauma, inflammation and infection (Fig. 1). The synthesis of these proteins is a part of the acute phase response and CRP is an acute phase reactant (proteins that are synthesized in increased amounts during the acute phase response). CRP is considered to be the classical acute phase protein as its blood concentration increases from less than about 1 μg/ml to as high as 600-1000 μg/ml during the height of an acute phase response. Interleukin-6 is the primary stimulus for CRP synthesis but levels of mRNA may also be influenced by interleukin-1, corticosteroids and other hormones.[4-6] As CRP is synthesized as a pentamer, assembly occurs in the endoplasmic reticulum. A unique feature of its synthesis is the regulation of secretion. CRP in the resting state is retained in the endoplasmic reticulum by binding to carboxylesterases.[7] Upon stimulated synthesis the binding of CRP to these sites is decreased and the transit time from the endoplasmic reticulum to secretion is markedly decreased.[8] The result of this enhanced secretion and high level of synthesis is that high levels of CRP can be seen in the blood very shortly after an acute phase stimulus.

Structure

As noted above, human CRP is a cyclic pentamer composed of five identical subunits. This finding was originally made by electron microscopy[9] but has been confirmed more recently using X-ray crystallography.[10,11] These subunits are noncovalently associated but the association is not reversible under physiological conditions. Each of the subunits (or protomers) contains two molecules of calcium that are bound to one face of the pentamer (Fig. 2A). These calcium ions are bound with moderate affinity and promote the stability of the pentamer. The opposite face of the pentamer has binding sites for C1q and Fc receptors (Fig. 2B).

The pentamer is remarkably resistant to thermal and enzymatic degradation.[12] The molecule undergoes no known posttranslational modifications unlike its nearest pentraxin family member SAP, which is glycosylated.

Ligand Binding

As noted above, the ligand first described for CRP was the cell wall polysaccharide of the pneumococcus or *S. pneumoniae*, PnC. The binding of CRP to PnC was shown to be calcium-dependent. Many of the ligands, but not all, share this reliance on calcium as calcium forms part of the binding pocket on CRP. Another characteristic of CRP binding to various ligands is inhibition by PC. This moiety is present at high density in the pneumococcal polysaccharide PnC.

There are a variety of ligands to which CRP binds in a noncalcium dependent manner. These are the polycationic, highly charged polymers such as poly-L-lysine, poly-L-arginine and a variety of related molecules.[13] The actual site of binding on CRP for polycations is probably close to the PC binding site. It has been suggested recently that the polycation-binding site and the PC binding site are nonoverlapping.[14] The PC-binding site on CRP is now well defined and is based on the solution of the CRP structure as a complex with PC.[11] The positively charged choline moiety reacts with a negatively charged glutamic acid on CRP. The negatively charged phosphate moiety reacts with two Ca^{2+} ions.

CRP also binds to a group of proteins found in the nuclei of mammalian cells. These proteins are bound to nucleic acids and are often the target of autoantibodies in various autoimmune diseases but especially systemic lupus erythematosus (SLE). These proteins are typically positively charged or contain short regions of positively charged residues. Robey et al[15] were the first to describe the interaction of CRP with nuclear antigens. These investigators found that CRP would react with chromatin and nucleosome core particles. They did not detect binding to histones for technical reasons and concluded that CRP bound to a neoantigen created by the interaction of histones and DNA. It was later determined that CRP reacted with histones in chromatin especially H1 and to a lesser extent H2A.[16,17] However, in intact cells CRP primarily interacts with small nuclear ribonucleoproteins (snRNPs).[18] Robey also showed staining of nuclei of cells in a "spotty pattern" and suggested that this was due to interaction with chromatin. However, this staining pattern appeared to be typical of the interaction of antibodies from patients with SLE with the snRNPs. The snRNPs are responsible for the splicing of high molecular weight RNA into mature mRNA. It was subsequently determined that CRP actually reacted with two of the major proteins in the snRNPs recognized by autoantibodies: the SmD protein and the 70kD protein.[18,19] Further examination of the determinants recognized on snRNP proteins and histones revealed a common motif of alternating glycine (G)/alanine-lysine/arginine (R). The GR repeat has been described in snRNP proteins D1 and D2 as well as B/B' (Fig. 3). The binding of CRP to these sites was calcium-dependent and inhibited by PC.[17]

A variety of microorganisms express PC on their cell wall teichoic acids and glycolipids including *S. pneumoniae*, *Salmonella typhimurium*, *Haemophilus influenzae*, *Pseudomonas aeruginosa*, *Neisseria*, *Clostridia*, *Lactococcus* and *Bacillus* species. Of these PC-expressing organisms *S. pneumoniae* and *H. influenzae* have been demonstrated to bind CRP. CRP binding to phosphorylated carbohydrates has also been well documented.[20] Phosphorylated carbohydrates are found on the surface of a variety of microorganisms including Leishmania.

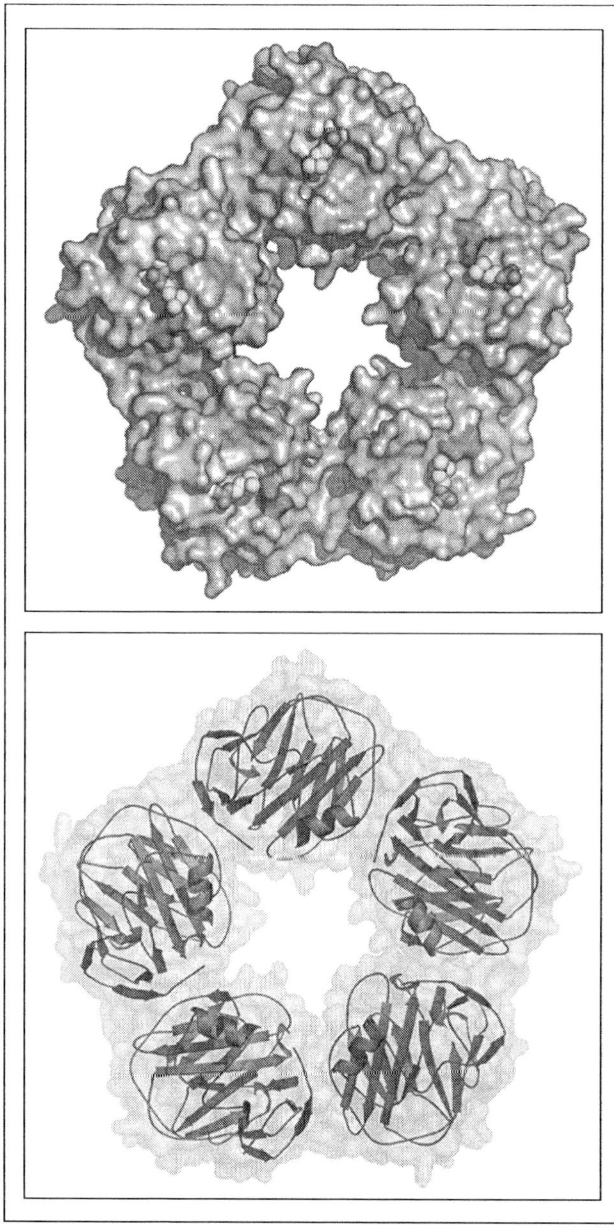

Figure 2. Structure of CRP. (A). The binding face of CRP is shown with the prototypic ligand PC. Calcium is shown as two spheres, which interact with the negatively charged oxygens of the phosphate moiety of PC. The choline interacts with the negatively charged glutamic acid residue 61 and perhaps with other residues as well. The figure is based on the crystal structure published by Thompson et al[11] (B). The receptor/C1q binding face is shown as a ribbon diagram.

HISTONE H2A 1-15	SGRGKQGGKARAKAK
HISTONE H1 177-191	AKKAKSPKKAAKAGRPKKAAKSPAKAKAVKPKAAK
D ANTIGEN M3 87-110	KSKKREAVAGRGRGRGRGRGRGRG

Figure 3. A determinant motif that is recognized by CRP in nuclear antigens. Three defined determinants on three nuclear antigens to which CRP bound were determined. Alignment of these sequences revealed a binding pattern characterized by repeating G/A,R/K amino acids. This highly charged, basic amino acid repeat serves as a marker for CRP recognition of exposed nuclear antigens, histone/chromatin and snRNPs.

Although membrane phospholipids sphingomyelin and phosphatidylcholine contain PC as their polar head groups CRP does not bind to intact membranes. The first indication that CRP may recognize damaged membranes was the finding that CRP interacted with liposomes that had incorporated lysophosphatidylcholine.[21] The authors concluded that CRP may be capable of recognizing damaged membranes. Support for this conclusion included the findings that CRP bound to cells treated with phospholipase[22] and to cells damaged by complement.[23] These results suggest that CRP may be an important recognition molecule for damaged self, especially during the acute phase response, which is invariably associated with tissue damage. Figure 4 shows the various types of ligands to which CRP binds.

The related molecule, SAP, has similar binding specificities that complement those of CRP. For example, whereas CRP binds to snRNPs in cell nuclei, SAP binds predominantly to chromatin[24] through interactions with histones[25] and DNA.[26] On cell membranes, CRP reacts with exposed phosphatidylcholine, and SAP binds to phosphatidylethanolamine, an inner leaflet phospholipid that is exposed when cells are damaged.[27] SAP also binds to a variety of sulfated and phosphorylated polysaccharides found on microbial polysaccharides and to matrix components including heparin, fibronectin and laminin.[28-31] SAP binding to ligands is also calcium-dependent.

Receptor Interactions

CRP is capable of interacting with the immune system through binding to receptors on phagocytic cells. It has been known for many years that CRP can act as an opsonin for bacteria.[32,33] These early studies showed that CRP could induce the uptake of various bacteria to which it bound. It was subsequently demonstrated that complement activation was responsible for at least part of the opsonic effect.[34] However, a direct role for CRP was also established, suggesting that a receptor for CRP must exist on phagocytic cells. Efforts to identify and characterize this receptor were fraught with several problems. However, several investigators suggested that the CRP-receptor was in some way associated with the receptors for immunoglobulin.[35,37] At that time it was unknown that three different classes of receptors for Ig were actually present on leukocytes, further confusing the picture. It was concluded however, that the CRP receptor and the Fc receptors were distinct, as efforts to cross inhibit were only partially successful and inhibitory monoclonal antibodies that blocked IgG binding were largely unable to block CRP binding to FcγR.[38,39,36]

Our laboratory was the first to demonstrate that CRP could interact directly with the high affinity receptor for IgG, FcγRI.[40,41] These finding have recently been confirmed by Bodman-Smith et al who established the binding affinity by surface plasmon resonance.[41a] Although CRP bound to FcγRI with moderate affinity it did not account for all the binding to phagocytic cells, since resting PMN and cell lines that lacked FcγRI, such as K562, also bound CRP. As CRP binding to a variety of cell lines appeared to correlate with the presence of FcγRII it was decided to directly test the binding of CRP to cells transfected with FcγRII.[42] In fact CRP

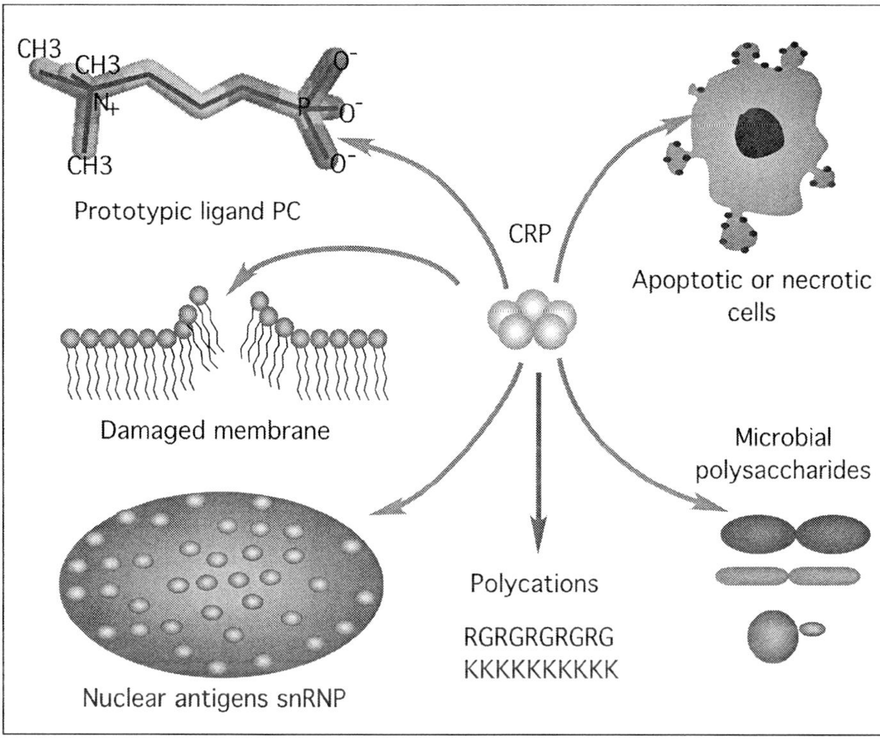

Figure 4. CRP ligands. The prototypic ligand is phosphocholine but CRP binds to a variety of self and foreign ligands. Interaction of CRP with damaged membranes occurs through the exposed PC group on phospholipids. Interaction of CRP with microbial polysaccharides may occur through PC groups, phosphorylated carbohydrates or other ligands. CRP interacts with proteins, including nuclear antigens through highly positively charged amino acids. CRP also interacts strongly certain polycations like poly-L-lysine or poly-L-arginine.

bound directly to these cells and could enhance the binding of opsonized zymosan to them. It was further shown that CRP binding to human FcγRIIa was preferential for the R over the H variant of an allelic polymorphism at amino acid 131.[43] This polymorphism has previously been shown to influence the binding of IgG2 to FcγRIIa.[44]

The generation of mice deficient in various FcγR allowed for the direct testing of the interaction of CRP with FcγR. Mice deficient in the γ-chain of FcεRI are unable to express either FcγRI or FcγRIII. These mice have decreased levels of CRP binding to macrophages. Mice deficient in FcγRII also showed decreased CRP binding and mice deficient in all three FcγRs showed no detectable binding of CRP. Based on binding studies using different cell types from normal and deficient mice, it was determined that CRP binds to FcγRI and FcγRII in the mouse.[45] These results have been confirmed using phagocytosis of a model particle, zymosan, by mouse macrophages.[46]

Cell signaling by CRP has been recently examined in a model of PMN signaling. Chi et al[47] used HL-60 cells differentiated to granulocytic morphology with DMSO to study the interaction of CRP with FcγRIIa. They reported that CRP induced phosphorylation events in FcγRIIa, phospholipase Cγ2 and phosphatidylinositol-3-kinase consistent with signaling through FcγRIIa. Experiments using inhibitors to block uptake of CRP-opsonized zymosan by human PMN are also consistent with signaling through FcγRIIa.[48]

As was the case for ligand binding, SAP is similar but not identical to CRP with respect to its receptors on phagocytic cells. Binding and phagocytosis experiments using human SAP have determined that SAP binds to human FcγRI, FcγRIIa and FcγRIII.[48] The predominant receptor used by PMN for phagocytosis of SAP-coated zymosan is FcγRIII.[48] In the mouse SAP opsonization may be mediated by FcγRI or FcγRIII.[46]

CRP and the Complement System

One of the primary effector mechanisms of CRP is its ability to activate the classical pathway of the complement system.[49] Early studies revealed that CRP bound C1q and that like IgG, close spacing of CRP on a surface or aggregation was required to activate C1.[50] Further studies showed that activation of complement by CRP occurs exclusively through the classical pathway. CRP activation of complement is efficient for the early part of the classical pathway with minimal activation of the terminal components.[51] Consequently, activation of complement by CRP generates cell bound fragments of C3 that are recognized by receptors for complement on phagocytic cells. The result is that CRP enhances clearance without generating proinflammatory complement split products. It appears that the lack of full activation of the classical pathway is due to the capacity of CRP to bind the complement regulatory protein factor H.[52] In fact CRP binding of factor H can inhibit complement activation by the alternative and lectin pathways.[53,54] Figure 5 illustrates the activation of complement by CRP binding to a surface.

The interaction of CRP with both C1q and factor H has been studied at the molecular level. Site-directed mutagenesis has been used to determine residues in CRP that are important for C1q binding.[55] On the basis of these results the shallow end of the cleft near the central pore on the effector face of CRP was proposed as the C1q binding site. These authors further suggest that CRP binding occurs through the globular head groups of C1q as is seen for IgG and IgM. However, binding and inhibition studies localized the CRP binding site to a peptide region of the collagen-like tails of C1q.[56] Two separate CRP binding sites have been identified on factor H. One is located in the short consensus/complement repeats (SCRs) 11-13 of factor H and the other in SCR 7.[57] The binding site in SCR 7 has been shown to overlap with the heparin and M protein binding sites in that region of factor H. Giannakis et al[57a] used alanine replacement mutagenesis to identify a cluster of positively charged residues in SCR 7 that are required for binding to heparin, M protein and CRP. The corresponding sites on CRP that interact with factor H have not been defined.

The role of complement in several in vivo activities of CRP has been tested using mice depleted of complement by cobra venom factor treatment. Both CRP and complement were required for clearance of CRP-coated erythrocytes,[59] for CRP-mediated opsonophagocytosis of *S. pneumoniae*,[60] and for CRP-mediated bactericidal activity against *H. influenzae*.[61] However, complement was not required for CRP and SAP effects on chromatin clearance.[62] In mice infected with *S. pneumoniae*, both CRP and complement contribute to protection.[63,64,64a]

CRP and Infection

The best-defined in vivo activity of CRP is it ability to protect animals from infection with *S. pneumoniae*. Very early studies suggested that CRP interacted with the bacteria directly and could enhance phagocytosis of bacteria in vitro.[32,33] However, the ability of CRP to actually protect animals from infection was not determined until much later.[65,66] These findings have been verified in transgenic mice expressing human CRP.[67] The protection of mice from infection is dependent on activation of the complement system by CRP.[63,64] The interaction of CRP with PC on the cell wall of these bacteria is thought to be important for recognition and protection. It has been determined that other pathogenic bacteria including *H. influenzae* express PC and may be recognized by CRP. Thus CRP was shown to induce killing of *H. influenzae* in vitro.[61] Protection of mice from infection with *Salmonella* species has been demonstrated recently.[68] It was also shown that CRP transgenic mice had increased antibody responses after

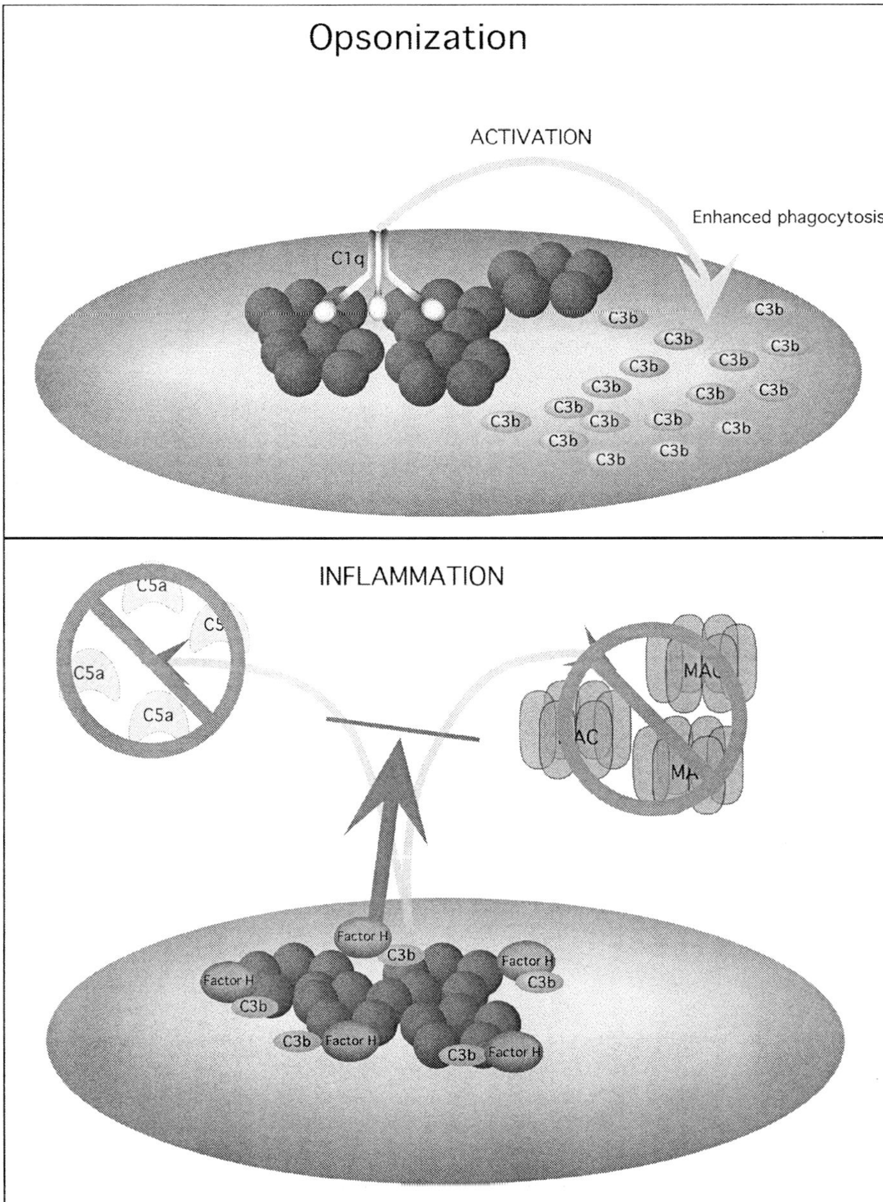

Figure 5. Complement activation by CRP. (A). The initial phase of complement activation is similar to IgG. CRP binds to a surface and the deposition of two CRP molecules closely spaced leads to C1q binding. The binding of C1q leads to activation of the early components of complement and deposition of C3b. (B). The next phase of complement activation by CRP is unlike IgG. CRP deposition on a surface leads to binding of the regulatory component factor H. Factor H inhibits the generation of the membrane attack complex and the generation of the complement split product C5a. Thus CRP-mediated complement activation is a less inflammatory process than is IgG-mediated activation.

immunization with an avirulent strain of *S. typhimurium*. The role of CRP in host defense against infection was recently reviewed.[69]

CRP and Autoimmunity

The first indication that CRP might interact with nuclear antigens may have been made in studies of rheumatoid synovium. It was shown that scattered cells in the rheumatoid synovium stained strongly for CRP and this staining appeared in the nuclei.[70] Other studies showed staining for CRP at sites of acute inflammation.[71,72] In these studies, CRP appeared to localize to sites containing infiltrating PMN. The first direct indication that CRP could interact with nuclear antigens was made by Robey et al.[15] These investigators showed that CRP reacted in a spotty pattern with nuclei. They also reported a direct interaction with chromatin and suggested that the reaction with chromatin produced the spotty staining of nuclei. It was subsequently determined that CRP reacted strongly with the splicing particles composed of small ribonucleoprotein complexes or snRNPs and was probably responsible for the staining of nuclei.[18] Figure 6 demonstrates the binding of CRP to fixed Hep-2 cells by immunofluorescence analysis. Analysis of the binding of CRP to snRNPs showed that CRP reacts with a lysine- or arginine-rich motif found in snRNPs and in histones.[17] The significance of this finding is still not completely explored but CRP had earlier been demonstrated to specifically inhibit antibody responses to antigens that it binds (see below). One study has demonstrated that CRP can prevent the accelerated autoimmunity induced in autoimmune mice by injections of chromatin.[73] In that study CRP produced a transient decrease in antibody responses and delayed mortality. More recently Szalai et al reported that CRP expressed as a transgene in NZB X NZW F1 mice delayed proteinuria and enhanced survival of these mice.[73a]

More recently it was demonstrated that CRP may interact with apoptotic cells to enhance their clearance without generating an immune response.[74] In these studies the authors demonstrated binding of CRP to apoptotic Jurkat T cells and augmentation of complement activation that was limited to the early complement components. These findings were consistent with the previously described interaction of CRP with damaged membranes and the known interaction of CRP with factor H. However, the authors demonstrated that CRP mediated opsonization of apoptotic cells by macrophages was associated with the expression of the anti-inflammatory cytokine, transforming growth factor β. Thus CRP, in the presence of complement can lead to clearance of apoptotic cells and perhaps further decrease the inflammation at the site of tissue damage.

Another possible mechanism for inhibition of immune responses to self-antigens is the ability of CRP to inhibit the immune response to epitopes to which it binds. It was shown that administration of CRP to mice prior to immunization with type 3 *S. pneumoniae* blocked the response to the major determinant, PC, by 90%.[75] This inhibition was shown to be specific for the PC determinant and no blocking of the response to the capsular polysaccharide was seen. Similarly immunization with PC-conjugated horse red blood cells (HRBC) in the presence of CRP led to decreased antibody responses to PC and not HRBC. Thus, CRP may inhibit antibody responses to antigens it binds by epitope blocking. As CRP binds to damaged membranes, nuclear antigens and other components of damaged cells it may serve to prevent autoantibody responses.[75]

Intriguing results have been reported by two groups suggesting a role for pentraxins in preventing autoimmunity.[58,76] In both reports mice made deficient in SAP by targeted gene disruption were found to have increased levels of anti-nuclear antibodies. In one strain of mice anti-chromatin antibodies were predominant and female SAP-deficient mice developed severe glomerulonephritis.[76] SAP is the primary pentraxin in mice where it is an acute phase reactant. Mouse CRP is found at very low levels.[77] One mechanism by which the pentraxins may prevent immunization by nuclear antigens is to alter the clearance of chromatin. SAP and to a lesser degree CRP are capable of slowing chromatin clearance and directing it toward the liver over the spleen and kidneys.[62]

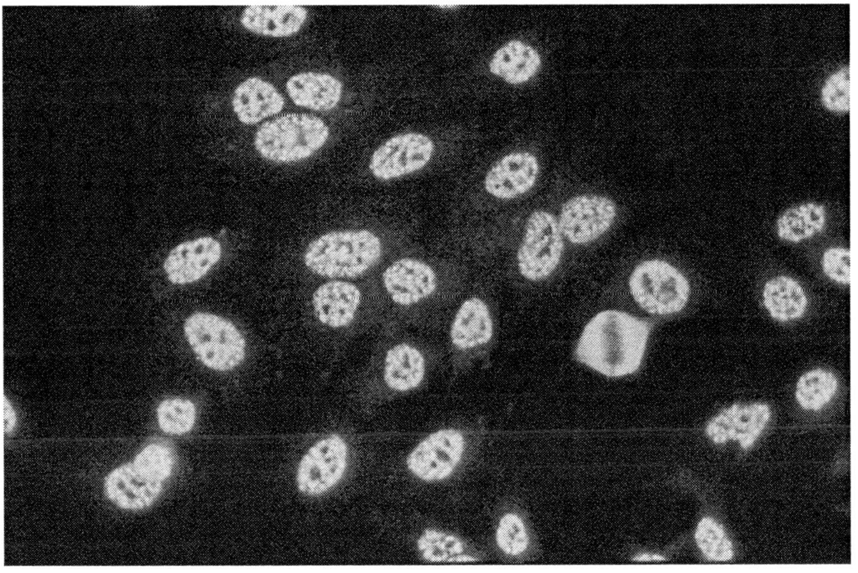

Figure 6. Immunofluorescent staining of nuclear antigens on fixed human cell line Hep-2. CRP shows string binding to the nuclear speckles, which were determined to be the small nuclear ribonucleoproteins (snRNPs), which are the target of autoantibodies in patients with systemic lupus erythematosus.

CRP and Cardiovascular Disease

Perhaps the first indication that CRP was involved in myocardial injury was the finding that it became localized in the necrotic myocardium of animals following ligation of the coronary artery.[78] The presence of CRP in the intima of atherosclerotic vessels was also noted very early on.[79] CRP has also been demonstrated, along with activated complement, in the damaged myocardium 24 hours after myocardial infarction.[80] These findings did not receive the attention of clinicians for many years, as the relationship of these findings to clinical disease in humans was unknown. Recently there has been a great interest in the role of CRP in the diagnosis and pathogenesis of atherosclerotic vascular disease. These studies have shown a clear association between levels of CRP and the risk of certain cardiovascular events.[81-83] Whether the relationship between CRP and these events is simply due to the known relationship between CRP and inflammation is unknown. Recently, CRP levels have been found to correlate with abdominal girth and body mass index suggesting that CRP may be simply a surrogate for another risk factor (reviewed in ref. 84). In addition CRP levels are higher in smokers than nonsmokers.

One theory of how CRP may be involved in atherosclerotic disease is that it is involved in the processing of lipids in the blood. It has been demonstrated that CRP binds to low-density lipoprotein (LDL) particles.[85,86] It has also been suggested that CRP may actually promote atherogenesis by the enhancement of uptake of LDL by macrophages. CRP, through an interaction with FcγRIIa, was shown to enhance the uptake of LDL by macrophages in vitro.[87]

There has been one study done in rats which suggests that CRP, through activation of the complement cascade, may directly enhance tissue destruction in experimentally induced myocardial infarction.[88] The effect of CRP was shown to be dependent on activation of the classical complement cascade. It is likely that the high CRP levels seen in acute myocardial events associated with complement activation act very differently from the low persistent CRP elevations seen in individuals with subacute inflammatory disease.

In conclusion, although CRP may not be a direct risk factor for the development of cardiovascular events it may nevertheless provide a useful, easily measured indicator of this risk. As far as the effect of CRP on the progression of an acute myocardial event is concerned, the data thus far suggest that CRP may promote inflammatory damage through activation of the complement cascade.

CRP and Inflammation

Despite many years of investigation in many laboratories the role of CRP in inflammation is somewhat controversial. Attempts to classify CRP as a pro- or anti-inflammatory molecule miss the point. The ability of CRP to react with different receptors on different cells to activate complement and to bind to various ligands would suggest that the effect of CRP on different inflammatory responses would vary. However, one direct way that CRP may act as an anti-inflammatory mediator is through its ability to downregulate the response to certain biological mediators. CRP was shown to block the response of platelets to platelet activating factor (PAF).[89,90] This inhibition is thought to be mediated by its binding to the PC moiety of PAF as CRP can inhibit the binding of PAF to platelets.[91]

In transgenic mice, CRP can protect the host from systemic and local inflammatory injury. The ability of CRP to protect against endotoxin shock was first described by Xia and Samols.[92] Mice expressing CRP were also protected from several mediators of septic shock including PAF and a combination of tumor necrosis factor (TNF-α) and interleukin-1 (IL-1β). These findings were confirmed by Chae et al[93] who demonstrated that CRP would completely protect mice from the hepatic dysfunction and mortality associated with *Vibrio vulnificus* lipopolysaccharide. Protection occurred even though the induced cytokine response was not affected. These studies suggest that CRP does not directly inactivate endotoxin, but rather decreases damage from the inflammatory response. The mechanism for this protective effect of CRP is unknown. However, the protection mediated by CRP is dependent on the presence of Fc receptors in these mice.[93a]

It has also been reported that high concentrations of CRP inhibit neutrophil responses to chemotactic factors[94,95] and neutrophil influx in chemoattractant-induced alveolitis.[96] Inhibition of migration and activation of neutrophils can decrease tissue injury in systemic inflammation. It is not yet clear how CRP signaling through FcγRII activation could interfere with chemotactic responses through G-protein coupled receptors. However, Zhong et al[94] found that CRP enhanced phosphatidylinositol-3 (PI-3) kinase and mitogen-activated protein (MAP) kinase activity consistent with FcγRII activation. Since the chemotactic response also requires PI-3 kinase activation, it is possible that the two pathways compete for signaling intermediates.

The interaction of CRP with platelets may be important in the inflammatory process as well. As noted above one effect of CRP is to block the interaction of proinflammatory molecules like PAF with platelets. However, there are several reports that CRP in an aggregated form or when complexed with certain ligands may enhance platelet aggregation. For example heat-aggregated CRP or CRP-ligand complexes induced platelet activation in a similar manner to heat aggregated IgG.[97] It was also demonstrated that the treatment of platelets with chymotrypsin increased the response to aggregated IgG and aggregated CRP but not collagen.[98] The activation of FcγRIIa by enzymatic treatment was later demonstrated by others.[99] These results suggest that CRP complexes like IgG complexes may activate platelets through FcγRII.

Despite earlier reports suggesting that CRP did not deposit in human tissues,[100] others have shown CRP localization in a variety of inflammatory sites.[70,72,78,101,102] These sites are often areas of tissue necrosis and may be related to PMN infiltration. Deposition of CRP in tissues generally leads to complement activation and in the case of infiltrating phagocytes to activation of Fcγ receptors. In these instances CRP may actually enhance the inflammatory process.

CRP and Cytokine Production

There are numerous reports on the ability of CRP to induce cytokine production primarily by monocytes and macrophages. Some of the results are in apparent conflict but these differences may be related to the cell types examined and the conditions employed. Using normal human monocytes stimulated with human CRP, Ballou and Lozanski reported that CRP was capable of inducing the production of the proinflammatory cytokines IL-1β, IL-6 and TNF-α.[103] The response was dose-dependent with maximal responses seen at about 50 μg/ml, a level that is seen in the blood of patients with a moderate inflammatory response e.g., active rheumatoid arthritis. Similarly, CRP induced IL-1β and TNF-α in human alveolar macrophages.[104] These reports suggested that in man CRP induces proinflammatory cytokines and could enhance the inflammatory response at tissue sites of inflammation. However, Tilg et al[105] reported that human peripheral blood mononuclear cells produced an excess of IL-1RA over IL-1 in response to CRP suggesting that CRP should actually downregulate the inflammatory response through stimulation of the anti-inflammatory cytokine IL-1RA. As these investigators did not examine purified cell types it is not easy to compare these results with the previous studies on mononuclear cells. Subsequent studies by others examined the effect of CRP on human monocytes and alveolar macrophages. Consistent with previous studies CRP induced IL-1β and IL-1RA by human monocytes but in alveolar macrophages CRP actually inhibited synthesis of these cytokines. These investigators also demonstrated that CRP could enhance the response of monocytes to LPS. A report by Cermak et al[106] showed that CRP could induce the production of tissue factor (TF) by monocytes. TF is a membrane bound glycoprotein that initiates the extrinsic pathway of coagulation. Elevation of TF is expected to enhance coagulation and thrombosis. These investigators found that, unlike LPS, CRP did not induce TF production by endothelial cells. This is consistent with the presence of FcγR on monocytes, but not endothelial cells.

The effects of CRP on cytokine production by mononuclear cells and macrophages are similar to those produced by aggregated IgG. However, CRP does not appear to bind to FcγRIII, which is important for some of the proinflammatory cytokine production by IgG.[107] Thus CRP is expected to produce a more modest inflammatory response than IgG and perhaps, as suggested by Tilg et al, the net effect of CRP is anti-inflammatory due to the stimulation of IL-1RA.

Summary

CRP is a molecule secreted by the liver in response to a perceived threat. The stimuli include damaged tissue, inflammation and infectious agents. CRP recognizes a variety of structures or determinants that are either foreign (bacterial polysaccharides) or normally not exposed (nuclear antigens, apoptotic or necrotic cells or damaged membranes). The interaction of CRP with these molecules tags them for interaction with other members of the innate immune system, macrophages, complement and PMNs. The result is clearance back to the liver for disposal in a noninflammatory manner. CRP may also deposit in the tissue where it may actually enhance the inflammatory response locally through stimulation of cytokines and deposition of complement. When the threat is over and the stimuli are no longer present CRP is cleared to hepatocytes. Thus there is a kind of CRP cycle in which CRP is released from the liver and later cleared by it.

References

1. Tillett WS, Francis Jr T. Serological reactions in pneumonia with a nonprotein fraction of pneumococcus. J Exp Med 1930; 52:561-571.
2. Volanakis JE, Kaplan MH. Specificity of C-reactive protein for choline phosphate residues of pneumococcal C-polysaccharide. Proc Soc Exp Biol Med 1971; 136:612-614.
3. Hurlimann J, Thorbecke GJ, Hochwald GM. The liver as the site of C-reactive protein formation. J Exp Med 1966; 123:365-378.

4. Li S-P, Liu T-Y, Goldman ND. cis-Acting elements responsible for interleukin-6 inducible C-reactive protein expression. J Biol Chem 1990; 265:4136-4142.
5. .Weinhold B, Rüther U. Interleukin-6-dependent and -independent regulation of the human C-reactive protein gene. Biochem J 1997; 327:425-429.
6. Mackiewicz A, Speroff T, Ganapathi MK et al. Effects of cytokine combinations on acute phase protein production in two human hepatoma cell lines. J Immunol 1991; 146:3032-3037.
7. Macintyre S, Samols D, Dailey P. Two carboxylesterases bind C-reactive protein within the endoplasmic reticulum and regulate its secretion during the acute phase response. J Biol Chem 1994; 269:24496-24503.
8. Macintyre SS, Kushner I, Samols D. Secretion of C-reactive protein becomes more efficient during the course of the acute phase response. J Biol Chem 1985; 260:4169-4173.
9. Osmand AP, Friedenson B, Gewurz H et al. Characterization of C-reactive protein and the complement subcomponent C1t as homologous proteins displaying cyclic pentameric symmetry (pentraxins). Proc Natl Acad Sci USA 1977; 74:739-743.
10. Emsley J, White HE, O'Hara BP et al. Structure of pentameric human serum amyloid P component. Nature 1994; 367:338-345.
11. Thompson D, Pepys MB, Wood SP. The physiological structure of human C-reactive protein and its complex with phosphocholine. Structure 1999; 7:169-177.
12. Kinoshita CM, Gewurz AT, Siegel JN et al. A protease-sensitive site in the proposed Ca^{2+}-binding region of human serum amyloid P and other pentraxins. Protein Sci 1992; 1:700-709.
13. DiCamelli R, Potempa LA, Siegel J et al. Binding reactivity of C-reactive protein for polycations. J Immunol 1980; 125:1933-1938.
14. Lee RT, Takagahara I, Lee YC. Mapping the binding areas of human C-reactive protein for phosphorylcholine and polycationic compounds. Relationship between the two types of binding sites. J Biol Chem 2002; 277:225-232.
15. Robey FA, Jones KD, Tanaka T et al. Binding of C-reactive protein to chromatin and nucleosome core particles. A possible physiological role of C-reactive protein. J Biol Chem 1984; 259:7311-7316.
16. Du Clos TW, Zlock LT, Rubin RL. Analysis of the binding of C-reactive protein to histones and chromatin. J Immunol 1988; 141:4266-4270.
17. Du Clos TW, Zlock L, Marnell LL. Definition of a C-reactive protein binding determinant on histones. J Biol Chem 1991; 266:2167-2171.
18. Du Clos TW. C-reactive protein reacts with the U1 small nuclear ribonucleoprotein. J Immunol 1989; 143:2553-2559.
19. Jewell WS, Marnell LL, Rokeach LA et al. C-reactive protein binding to the Sm-D protein of snRNPs. Identification of a short polypeptide binding region. Mol Immunol 1993; 30:701-708.
20. Culley FJ, Bodman-Smith KB, Ferguson MA et al. C-reactive protein binds to phosphorylated carbohydrates. Glycobiology 2000; 10:59-65.
21. Volanakis JE, Wirtz KWA. Interaction of C-reactive protein with artificial phosphatidylcholine bilayers. Nature 1979; 281:155-157.
22. Narkates AJ, Volanakis JE. C-reactive protein binding specificities: Artificial and natural phospholipid bilayers. Ann NY Acad Sci 1982; 389:172-181.
23. Li YP, Mold C, Du Clos TW. Sublytic complement attack exposes C-reactive protein binding sites on cell membranes. J Immunol 1994; 152:2995-3005.
24. Pepys MB, Booth SE, Tennent GA et al. Binding of pentraxins to different nuclear structures: C-reactive protein binds to small nuclear ribonucleoprotein particles, serum amyloid P component binds to chromatin and nucleoli. Clin Exp Immunol 1994; 97:152-157.
25. Hicks PS, Saunero-Nava L, Du Clos TW et al. Serum amyloid P component binds to histones and activates the classical complement pathway. J Immunol 1992; 149:3689-3694.
26. Pepys MB, Butler PJG. Serum amyloid P component is the major calcium-dependent specific DNA binding protein of the serum. Biochem Biophys Res Commun 1987; 148:308-313.
27. Schwalbe RA, Dahlbäck B, Coe JE et al. Pentraxin family of protein interacts specifically with phosphorylcholine and/or phosphorylethanolamine. Biochemistry (Mosc) 1992; 31:4907-4915.
28. Hind CRK, Collins PM, Renn D et al. Binding specificity of serum amyloid P component for the pyruvate acetal of galactose. J Exp Med 1984; 159:1058-1069.
29. Zahedi K. Characterization of the binding of serum amyloid P to laminin. J Biol Chem 1997; 272:2143-2148.
30. Kubak BM, Potempa LA, Anderson B et al. Evidence that serum amyloid P component binds to mannose-terminated sequences of polysaccharides and glycoproteins. Mol Immunol 1988; 25:851-858.

31. Loveless RW, Floyd-O'Sullivan G, Raynes JG et al. Human serum amyloid P component is a multispecific adhesive protein whose ligands include 6-phosphorylated mannose and the 3-sulphated saccharides galactose, N-acetylglucosamine and glucuronic acid. EMBO J 1992; 11:813-819.
32. Kindmark C-O. Stimulating effect of C-reactive protein on phagocytosis of various species of pathogenic bacteria. Clin Exp Immunol 1971; 8:941-948.
33. Hokama Y, Coleman MK, Riley RF. In vitro effects of C-reactive protein on phagocytosis. J Bact 1962; 83:1017-1024.
34. Mortensen RF, Osmand AP, Lint TF et al. Interaction of C-reactive protein with lymphocytes and monocytes: Complement-dependent adherence and phagocytosis. J Immunol 1976; 117:774-781.
35. Mortensen RF, Duszkiewicz JA. Mediation of CRP-dependent phagocytosis through mouse macrophage Fc-receptors. J Immunol 1977; 119:1611-1616.
36. Zeller JM, Kubak BM, Gewurz H. Binding sites for C-reactive protein on human monocytes are distinct from IgG Fc receptors. Immunol 1989; 67:51-55.
37. Müller H, Fehr J. Binding of C-reactive protein to human polymorphonuclear leukocytes: Evidence for association of binding sites with Fc receptors. J Immunol 1986; 136:2202-2207.
38. Zahedi K, Tebo JM, Siripont J et al. Binding of human C-reactive protein to mouse macrophages is mediated by distinct receptors. J Immunol 1989; 142:2384-2392.
39. Tebo JM, Mortensen RF. Characterization and isolation of a C-reactive protein receptor from the human monocytic cell line U-937. J Immunol 1990; 144:231-238.
40. Crowell RE, Du Clos TW, Montoya G et al. C-reactive protein receptors on the human monocytic cell line U-937. Evidence for additional binding to FcγRI. J Immunol 1991; 147:3445-3451.
41. Marnell LL, Mold C, Volzer MA et al. C-reactive protein binds to FcgRI in transfected COS cells. J Immunol 1995; 155:2185-2193.
41a. Bodman-Smith KB, Melendez AJ, Cambell I et al. C-reactive protein-mediated phagocytosis and phospholipase D signalling through the high affinity receptor of immunoglobulin G (FcγRI). *Immunol* 2002; 107:252-260.
42. Bharadwaj D, Stein MP, Volzer M et al. The major receptor for C-reactive protein on leukocytes is FcγRII. J Exp Med 1999; 190:585-590.
43. Stein MP, Edberg JC, Kimberly RP et al. C-reactive protein binding to FcγRIIa on human monocytes and neutrophils is allele specific. J Clin Invest 2000; 105:369-376.
44. Van der Pol WL, van de Winkel JGJ. IgG receptor polymorphisms: Risk factors for disease. Immunogenetics 1998; 48:222-232.
45. Stein MP, Mold C, Du Clos TW. C-reactive protein binding to murine leukocytes requires Fcγ receptors. J Immunol 2000; 164:1514-1520.
46. Mold C, Gresham HD, Du Clos TW. Serum amyloid P component (SAP) and C-reactive protein (CRP) mediate phagocytosis through murine Fcγ receptors. J Immunol 2001; 166:1200-1205.
47. Chi M, Tridandapani S, Zhong W et al. C-Reactive protein induces signaling through FcγRIIa on HL-60 granulocytes. J Immunol 2002; 168:1413-1418.
48. Bharadwaj D, Mold C, Markham E et al. Serum amyloid P component binds to Fc gamma receptors and opsonizes particles for phagocytosis. J Immunol 2001; 166:6735-6741.
49. Kaplan MH, Volanakis JE. Interactions of C-reactive protein with the complement system. I. Consumption of human complement associated with the reaction of C-reactive protein with pneumococcal polysaccharide and with the choline phosphatides, lecithin and sphingomyelin. J Immunol 1974; 112:2135-2147.
50. Claus DR, Siegel J, Petras K et al. Interactions of C-reactive protein with the first component of human complement. J Immunol 1977; 119:187-192.
51. Berman S, Gewurz H, Mold C. Binding of C-reactive protein to nucleated cells leads to complement activation without cytolysis. J Immunol 1986; 136:1354-1359.
52. Mold C, Gewurz H, Du Clos TW. Regulation of complement by C-reactive protein. Immunopharmacology 1999; 42:23-30.
53. Suankratay C, Mold C, Zhang Y et al. Complement regulation in innate immunity and the acute-phase response: Inhibition of mannan-binding lectin-initiated complement cytolysis by C-reactive protein (CRP). Clin Exp Immunol 1998; 113:353-359.
54. Mold C, Gewurz H. Inhibitory effect of C-reactive protein on alternative C pathway activation by liposomes and Streptococcus pneumoniae. J Immunol 1981; 127:2089-2092.
55. Agrawal A, Shrive AK, Greenhough TJ et al. Topology and structure of the C1q-binding site on C-reactive protein. J Immunol 2001; 166:3998-4004.
56. Jiang H, Robey FA, Gewurz H. Localization of sites through which C-reactive protein binds and activates complement to residues 14-26 and 76-92 of the human C1q A chain. J Exp Med 1992; 175:1373-1379.

57. Jarva H, Jokiranta TS, Hellwage J et al. Regulation of complement activation by C-reactive protein: Targeting the complement inhibitory activity of factor H by an interaction with short consensus repeat domains 7 and 8-11. J Immunol 1999; 163:3957-3962.
57a. Giannakis E, Jokiranta TS, Male DA et al. A common site within factor H SCR 7 responsible for binding heparin, C-reactive protein and streptococcal M protein. *Eur J Immunol* 2003; 33:962-969.
58. Soma M, Tamaoki T, Kawano H et al. Mice lacking serum amyloid P component do not necessarily develop severe autoimmune disease. Biochem Biophys Res Comm 2001; 286:200-205.
59. Nakayama S, Mold C, Gewurz H et al. Opsonic properties of C-reactive protein in vivo. J Immunol 1982; 128:2435-2438.
60. Edwards KM, Gewurz H, Lint TF et al. A role for C-reactive protein in the complement-mediated stimulation of human neutrophils by type 27 Streptococcus pneumoniae. J Immunol 1982; 128:2493-2496.
61. Weiser JN, Pan N, McGowan KL et al. Phosphorylcholine on the lipopolysaccharide of *Haemophilus influenzae* contributes to persistence in the respiratory tract and sensitivity to serum killing mediated by C-reactive protein. J Exp Med 1998; 187:631-640.
62. Burlingame RW, Volzer MA, Harris J et al. The effect of acute phase proteins on clearance of chromatin from the circulation of normal mice. J Immunol 1996; 156:4783-4788.
63. Nakayama S, Gewurz H, Holzer TJ et al. The role of the spleen in the protective effect of C-reactive protein in *Streptococcus pneumoniae* infection. Clin Exp Immunol 1983; 54:319-326.
64. Szalai AJ, Briles DE, Volanakis JE. Role of complement in C-reactive protein-mediated protection of mice from *Streptococcus pneumoniae*. Infect Immun 1996; 64:4850-4853.
64a. Mold C, Rodic-Polic B, Du Clos TW. Protection from *Streptococcus pneumoniae* infection by C-reactive protein and natural antibody requires complement but not Fc gamma receptors. *J Immunol* 2002; 168:6375-6381.
65. Mold C, Nakayama S, Holzer TJ et al. C-reactive protein is protective against *Streptococcus pneumoniae* infection in mice. J Exp Med 1981; 154:1703-1708.
66. Yother J, Volanakis JE, Briles DE. Human C-reactive protein is protective against fatal *Streptococcus pneumoniae* infection in mice. J Immunol 1982; 128:2374-2376.
67. Szalai AJ, Briles DE, Volanakis JE. Human C-reactive protein is protective against fatal *Streptococcus pneumoniae* infection in transgenic mice. J Immunol 1995; 155:2557-2563.
68. Szalai AJ, VanCott JL, McGhee JR et al. Human C-reactive protein is protective against fatal Salmonella enterica Serovar Typhimurium infection in transgenic mice. Infect Immun 2000; 68:5652-5656.
69. Du Clos TW, Mold C. The role of C-reactive protein in resolution of bacterial infection. Curr Opin Infect Dis 2001; 14:289-293.
70. Gitlin JD, Gitlin JI, Gitlin D. Localization of C-reactive protein in synovium of patients with rheumatoid arthritis. Arthritis Rheum 1977; 20:1491-1499.
71. Parish WE. Studies on vasculitis. I. Immunoglobulins, IC, C-reactive protein, and bacterial antigens in cutaneous vasculitis lesions. Clin Allerg 1971; 1:97-109.
72. Du Clos TW, Mold C, Paterson PY et al. Localization of C-reactive protein in inflammatory lesions of experimental allergic encephalomyelitis. Clin Exp Immunol 1981; 43:565-573.
73. Du Clos TW, Zlock L, Hicks PS et al. Decreased autoantibody levels and enhanced survival of (NZB X NZW) F_1 mice treated with C-reactive protein. Clin Immunol Immunopath 1994; 70:22-27.
73a. Szalai AJ, Weaver CT, McCrory MA et al. Delayed lupus onset in (NZB X NZW)F1 mice expressing a human C-reactive protein transgene. *Arth Rheum* 2003; 48:1602-1611.
74. Gershow D, Kim S, Brot N et al. C-reactive protein binds to apoptotic cells, protects the cells from assembly of the terminal complement components, and sustains an antiinflammatory innate immune response: Implications for systemic autoimmunity. J Exp Med 2000; 192:1353-1363.
75. Nakayama S, Du Clos TW, Gewurz H et al. Inhibition of antibody responses to phosphocholine by C-reactive protein. J Immunol 1984; 132:1336-1340.
76. Bickerstaff MCM, Botto M, Hutchinson WL et al. Serum amyloid P component controls chromatin degradation and prevents antinuclear autoimmunity. Nat Med 1999; 5:694-696.
77. Pepys MB, Baltz ML, Gomer K et al. Serum amyloid P component is an acute phase reactant in mouse. Nature 1979; 278:259-261.
78. Kushner I, Rakita L, Kaplan MH. Studies of acute-phase protein. II Localization of Cx-reactive protein in heart in induced myocardial infarction in rabbits. J Clin Invest 1963; 42:286-292.
79. Intorp HW, Milgrom F, Witebsky E. Antigens of normal and atherosclerotic intima. J Immunol 1969; 102:1404-1410.
80. Lagrand WK, Niessen HWM, Wolbink GJ et al. C-reactive protein colocalizes with complement in human hearts during acute myocardial infarction. Circulation 1997; 95:97-103.

81. Liuzzo G, Biasucci LM, Gallimore JR et al. The prognostic value of C-reactive protein and serum amyloid A protein in severe unstable angina. N Engl J Med 1994; 331:417-424.
82. Ridker PM, Cushman M, Stampfer MJ et al. Inflammation, aspirin and the risk of cardiovascular disease in apparently healthy men. N Engl J Med 1997; 336:973-979.
83. Ridker P, Buring JSJ, Matiaws M et al. Prospective study of C-reactive protein and the risk of future cardiovascular events among apparently healthy women. Circulation 1998; 98:731-733.
84. Lemieux I, Pascot A, Prud'homme D et al. Elevated C-reactive protein another component of the atherothrombotic profile of abdominal obesity. Arterioscler Thromb Vasc Biol 2001; 21:961-967.
85. de Beer F, Soutar A, Baltz M et al. Low density lipoprotein and very low density lipoprotein are selectively bound by aggregated C-reactive protein. J Exp Med 1982; 156:230-242.
86. Cabana VG, Siegel JN, Sabesin SM. Effects of the acute phase response on the concentration and density distribution of plasma lipids and apolipoproteins. J Lipid Res 1989; 30:39-49.
87. Zwaka TP, Hombach V, Torzewski J. C-reactive protein-mediated low density lipoprotein uptake by macrophages: Implications for atherosclerosis. Circulation 2001; 103:1194-1197.
88. Griselli M, Herbert J, Hutchinson WL et al. C-reactive protein and complement are important mediators of tissue damage in acute myocardial infarction. J Exp Med 1999; 190:1733-1739.
89. Vigo C. Effect of C-reactive protein on platelet-activating factor-induced platelet aggregation and membrane stabilization. J Biol Chem 1985; 260:3418-3422.
90. Kilpatrick JM, Virella G. Inhibition of platelet activating factor by rabbit C-reactive protein. Clin Immunol Immunopathol 1985; 37:276-281.
91. Filep JG, Herman F, Kelemen E et al. C-reactive protein inhibits binding of platelet-activating factor to human platelets. Thromb Res Suppl 1991; 61:411-421.
92. Xia D, Samols D. Transgenic mice expressing rabbit C-reactive protein are resistant to endotoxemia. Proc Natl Acad Sci USA 1997; 94:2575-2580.
93. Chae MR, Park BH, Kim JS et al. Protective effect of C-reactive protein against the lethality induced by *Vibrio vulnificus* lipopolysaccharide. Microbiol Immunol 2000; 44:335-340.
93a. Mold C, Rodriguez W, Rodic-Polic B et al. C-reactive protein mediates protection from lipopolysaccharide through interactions with Fc gamma R. *J Immunol* 2002; 169:7019-7025.
94. Zhong S, Zen Q, Tebo J et al. Effect of human C-reactive protein on chemokine and chemotactic factor-induced neutrophil chemotaxis and signaling. J Immunol 1998; 161:2533-2540.
95. Heuertz RM, Tricomi SM, Ezekieli UR et al. C-reactive protein inhibits chemotactic peptide-induced p38 mitogen-activated protein kinase activity and human neutrophil movement. J Biol Chem 1999; 275:17968-17974.
96. Ahmed N, Thorley R, Xia DY et al. Transgenic mice expressing rabbit C-reactive protein exhibit diminished chemotactic factor-induced alveolitis. Am J Respir Crit Care Med 1996; 153:1141-1147.
97. Fiedel BA, Simpson RM, Gewurz H. Activation of platelets by modified C-reactive protein. Immunol 1982; 45:439-447.
98. Fiedel BA, Siegel JN, Gewurz H et al. Comparison of the enzymatic sensitivities of the platelet receptor for human C-reactive protein and its functional relationship to the platelet IgG Fc receptor. Clin Exp Immunol 1982; 50:215-222.
99. van de Winkel JGJ, van Ommen R, Huizinga TWJ et al. Proteolysis induces increased binding affinity of the monocyte type II FcR for human IgG. J Immunol 1989; 143:571-578.
100. Vigushin DM, Pepys MB, Hawkins PN. Metabolic and scintigraphic studies of radioiodinated human C-reactive protein in health and disease. J Clin Invest 1993; 91:1351-1357.
101. Kushner I, Kaplan MH. Studies of acute phase protein. I An immunohistochemical method for the localization of Cx-reactive protein in rabbits: Association with necrosis in inflammatory lesions. J Exp Med 1961; 114:961-973.
102. Parish WE. Studies in vasculitis. VII. C-reactive protein as a substance perpetuating chronic vasculitis. Occurrence in lesions and concentrations in sera. Clin Allerg 1976; 6:543-550.
103. Ballou SP, Lozanski G. Induction of inflammatory cytokine release from human monocytes by C-reactive protein. Cytokine 1992; 4:361-368.
104. Galve-de Rochemonteix B, Wiktorowicz K, Kushner I et al. C-reactive protein increases production of IL-1 alpha, IL-1 beta, and TNF-alpha, and expression of mRNA by human alveolar macrophages. J Leuk Biol 1993; 53:439-445.
105. Tilg H, Vannier E, Vachino G et al. Antiinflammatory properties of hepatic acute phase proteins: Preferential induction of interleukin-1 (IL-1) receptor antagonist over IL-1β synthesis by human peripheral blood mononuclear cells. J Exp Med 1993; 178:1629-1636.
106. Cermak J, Key NS, Bach RR et al. C-reactive protein induces human peripheral blood monocytes to synthesize tissue factor. Blood 1993; 82:513-520.
107. Marsh CB, Wewers MD, Tan LC et al. Fc γ receptor cross-linking induces peripheral blood mononuclear cell monocyte chemoattractant protein-1 expression. J Immunol 1997; 158:1078-1084.

Chapter 4

Complement: A Major Humoral Effector System in Innate and Acquired Immunity

Philippe Gasque and B. Paul Morgan

Abstract

The complement cascade is an essential component of the phylogenetically ancient innate immune response and is crucial to our natural ability to prevent and combat infection. For instance, complement is involved in host defense by triggering the generation of a membranolytic complex (C5b9, membrane attack complex) at the surface of the pathogen and complement fragments (referred to as opsonins, eg. C3b, iC3b) interact with cell surface receptors (CR1, CR3 and CR4) to promote phagocytosis and a local proinflammatory response that ultimately contributes to the protection and healing of the host. Knowledge of the unique molecular and cellular innate immunological interactions that occur in the development and resolution of pathology should facilitate the design of effective therapeutic strategies to fight against pathogens.

Molecular Elements of the Innate Immune Response Involved in the Recognition of Pathogens and Toxic Cell Debris: A Common Ancestral Scavenging System

Innate immune systems use proteins encoded in the germ-line to identify potentially noxious substances. These proteins, whether they are cell surface receptors or soluble, function most commonly by recognizing carbohydrate structures on targets. From the concept originally presented by Medzhitov and Janeway,[1] it is now well established that soluble and membrane defense molecules of the innate immune system recognize pathogen-associated molecular patterns (PAMPs). PAMPs are expressed by microbes at sites of infection and inflammation, and shared by large groups of pathogens as conserved products not subject to antigenic variability. PAMPs are microbial structures that, upon interaction with elements of the host innate immune system, trigger the initiation of host protective responses with the clearance of the pathogen by phagocytic cells. In this process, macrophages are essential in the first line of defense against infections.[2]

More recently, attention has focused on the roles of phagocytosis and innate immunity in health and development. New and exciting findings highlight the role of phagocytosis as an important process in tissue homeostasis. It is well established that programmed cell death, or apoptosis, is an integral part of the development of all metazoans and that dying cells are removed very early on.[3,4] Remarkably, the cellular receptors and products (proteins, lipids, carbohydrates) that are involved in the recognition and subsequent phagocytosis of apoptotic cells are very similar to those utilized to recognize pathogens. The emerging paradigm is that innate immunity plays a role in the clearance of potentially dangerous and toxic self-derived entities such as apoptotic cells. In analogy to PAMPs, we and others have proposed that innate

Immunobiology of Carbohydrates, edited by Simon Y.C. Wong and Gemma Arsequell.
©2003 Eurekah.com and Kluwer Academic / Plenum Publishers.

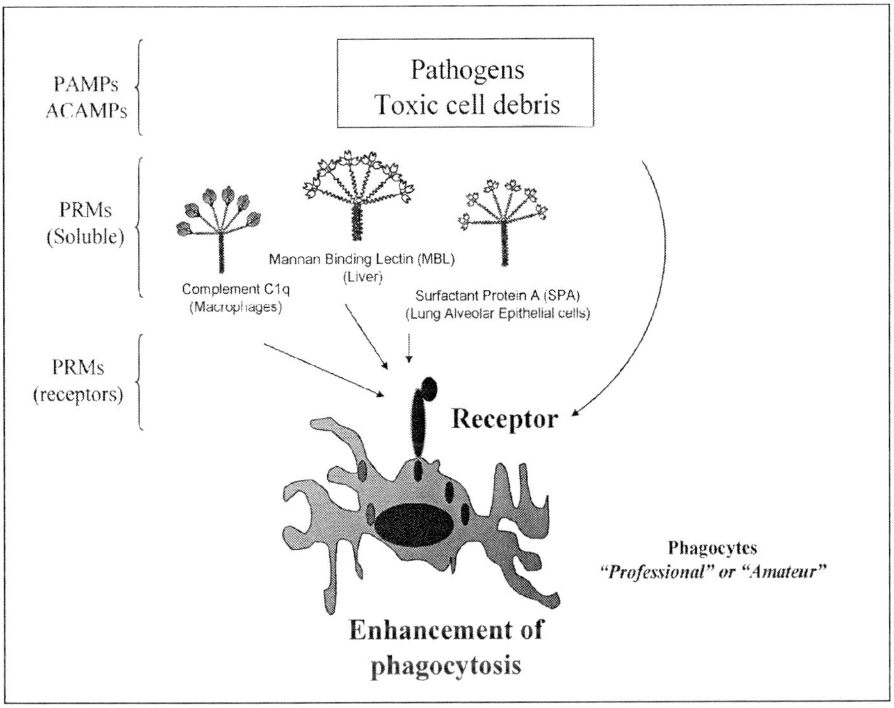

Figure 1. Defense collagens recognize PAMPs and ACAMPs.

immune molecules recognize apoptotic cell-associated molecular patterns (ACAMPs) expressed de novo by cells undergoing programmed cell death.[5,6] Ultimately, "professional" and "amateur" phagocytes recognize PAMPs and ACAMPs through specific soluble and membrane-bound pattern-recognition molecules (PRMs; e.g., collectins, phagocytic receptors), which lead to the clearance of various target cells. Thus, just as the adaptive immune response is known to be highly selective, innate immunity too is extremely effective at distinguishing between innocuous self cells and potentially noxious substances ranging from pathogens to toxic cell debris, a division based essentially upon recognition of specific "non self" signatures.

One group of molecules that have been shown to recognize PAMPs and ACAMPs has been collectively named "defense collagens" (Fig. 1). In the majority of these, globular carboxy-terminal domains recognize relatively broad categories of molecules, and collagen-like amino-terminal domains link the invading organism to powerful effectors of the immune system (PRMs of the humoral and cellular effector systems in blood and tissues). Soluble members of the defense collagens include C1q (the recognition component of the classical pathway of complement) and the collectins such as mannose binding lectin (MBL) and pulmonary surfactant protein A, (SPA).

The Complement System and Complement-Associated Proteins: Routes of Activation against Pathogens

The complement (C) system consists of some 30 fluid-phase and cell-membrane proteins and plays an important role in innate immunity by recognizing and targeting for killing pathogens such as bacteria, virus infected cells and parasites while preserving normal 'self' cells (for a review see ref. 7). Recent studies have demonstrated a marked conservation of the C system between invertebrates and mammals which points to a common ancestry of this system in host

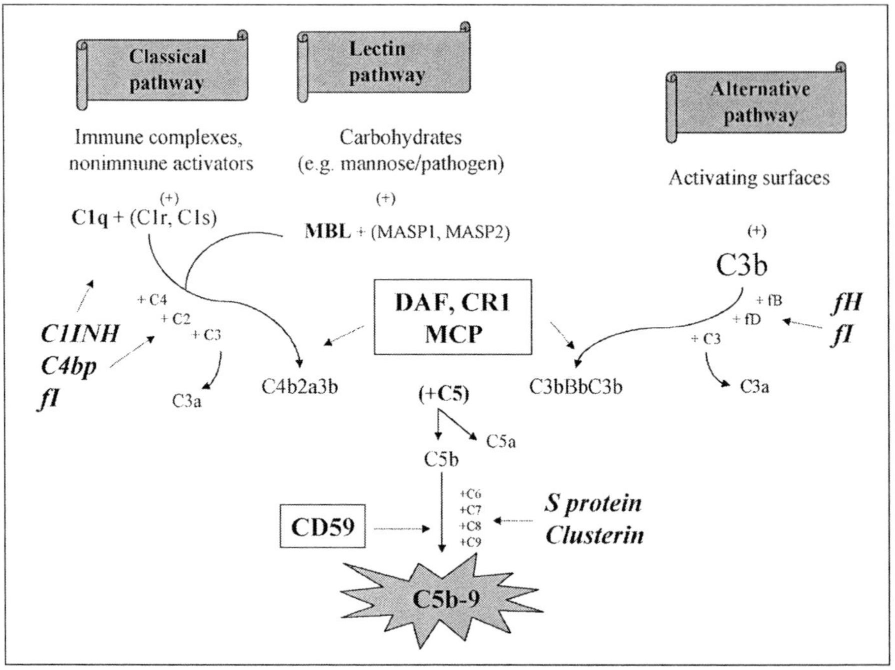

Figure 2. The C system is tightly regulated by a number of soluble (in bold italics) and membrane associated proteins (in bold, boxed).

defense and raises the paradigm of a critical role of C in tissue homeostasis.[8–10] In invertebrates (e.g., insects), the C system is very simple and is composed of two or three components. Surprisingly, these are not produced in the fat body, the functional equivalent of the liver where the majority of mammalian C components are produced, but are expressed by phagocytes.[11,12] In mammals, the hepatocyte is the major source of most C proteins, with the exception of C1q, factor D and C7. Interestingly, C1q mRNA expression is predominant in spleen, thymus and heart, although expression in tissue macrophages and some epithelial cells is seen in other organs.[13] Although the hepatocyte is the source of most of the serum C proteins, many other cell types, including monocytes, fibroblasts, epithelial and endothelial cells, can also synthesize most of the C components.[14] The relevance of this extrahepatic synthesis of C is only now being recognized.

C can be activated essentially by three distinct routes: the classical, alternative and lectin pathways (CP, AP, and LP respectively) (see Fig. 2). The AP is the most ancient of these pathways. The initiation of the AP (involving C3, factor B (fB), factor D (fD) and properdin (P)) does not depend upon the presence of immune complexes but is initiated following interactions with carbohydrate-rich particles lacking sialic acid (for a review see ref. 15). The AP is, for instance, activated by a diverse set of "natural" substances, including yeast cell walls, bacterial cell walls and damaged self cells. Activation leads to the deposition of C3 fragments on the target cells. The CP, involving C1q, C1r, C1s, C4, C2 and C3 components, is activated primarily by the interaction of C1q with antibody-antigen immune complexes. However, activation can also be achieved after interaction of C1q with nonimmune molecules such as polyanions (e.g., bacterial lipopolysaccharides, DNA and RNA), certain small polysaccharides, viral membranes, C reactive protein (CRP), serum amyloid P component (SAP), and more importantly, some poorly characterized bacterial, fungal and virus membrane components (for a review see ref. 16). CRP in mammals is an acute phase protein that can interact directly with microorganisms in a calcium-dependent manner to bring about C activation through the CP. It has been suggested that CRP-medicated C activation does not initiate an efficient terminal pathway with the formation of C5a and the membranolytic complex C5b9.[17] CRP and SAP are members of the pentraxin family of proteins which are unrelated to other known proteins, but are stably conserved in vertebrate and invertebrates (e.g., *Limulus polyphemus*).[18–20] In vertebrates CRP binds to a plethora of microbial polysaccharides while SAP interacts with carbohydrate moieties. Pentraxins are also know to bind to cell components (e.g., fibronectin, chromatin) in damaged tissues to aid in their removal through interactions with the opsonic C system as well as directly stimulating macrophages through binding to type I and type II Fc receptors (for reviews see refs. 17, 21).

MBL is a member of the family of calcium-dependent lectins, the collectins (collagenous lectins) and is homologous in structure to C1q. MBL is probably the most remarkable PRM of the innate immune system, owing to its selective binding to arrays of terminal mannose groups on a variety of bacteria.[22,23] MBL activates C by interacting with two serine proteases called MASP1 and MASP2 (MASP = MBL-associated serine protease). MASP2 cleaves and activates C4 and C2, and MASP1 may cleave C3 directly. These proteins of the C system comprise the lectin pathway (Fig. 2).[24–26]

The ultimate goal for the activation of the C system is the formation, through the C terminal pathway (TP, involving C5, C6, C7, C8 and C9 components), of a membrane attack complex (MAC, also called the C5b-9 complex) which disrupts and forms a pore in the phospholipid bilayer to lyse the target cell.[27] However, the C system, when activated at an inappropriate site and/or to an inappropriate extent, is remarkably effective at damaging host tissues and causing pathology as seen in degenerative disorders of the CNS.[28] To avoid this self-destructive tendency, host cells are protected by a battery of regulatory molecules (C inhibitors) which inhibit either the assembly of the C3-cleaving enzymes or the formation of the MAC. C1 inhibitor (C1-INH), C4b binding protein (C4bp), factor H (fH), factor I (fI), S protein (Sp) and clusterin are all soluble C inhibitors secreted by cells and are present in the plasma. The other C inhibitors are expressed on the cell membrane and include membrane cofactor protein (MCP, CD46), decay accelerating factor (DAF, CD55), and CD59 (Fig. 2).[29]

A target cell coated with C opsonins (C1q, C3b and iC3b) will be specifically recognized and phagocytozed by macrophages bearing C receptors (CR1, CR3, CR4, see Table 1). The recognition steps result both in the movement of phagocyte membrane (guided by the actin cytoskeleton) around the target and the formation of the phagosome which then fuses to lysosomes to initiate digestion. Of note, many organisms take advantage of opsonization by the C system to enhance their virulence. Opsonized pathogens bind to C receptors to gain entry into host cells so that they are sheltered from the innate and adaptive immune responses. Activation of host C deposits C3b on the surface, enabling the pathogen to subvert host cell C3 fragment receptors (CR1, CR2) and to enter the cell. Organisms as diverse as the human immunodeficiency virus (HIV) and pathogenic mycobacteria exploit this mechanism very efficiently.[30]

The C System and C-Associated Proteins: Clearance of Apoptotic Cells

Several recent lines of evidence suggest that C1q plays an important role in the clearance of apoptotic cells. Three independent studies have shown that C1q can bind directly and specifically to surface blebs of UV-induced apoptotic cells (keratinocytes and T cells) leading to the activation of the CP of C.[31–34] Moreover, it has been reported that C1q knockout mice showed a profound impairment in the clearance of apoptotic cells which then accumulated in the kidney leading to glomerulonephritis with immune deposits.[35] Mice lacking C2 and fB (i.e., with no functional CP or AP) did not develop glomerulonephritis nor autoantibodies, whereas C1q-deficient mice that also lacked C2 and fB developed glomerulonephritis without glomerular C3 deposition.[36] These observations support the hypothesis that C1q serves as a vital opsonin in the efficient recognition and physiological clearance of apoptotic cells. Collectins

Table 1. Soluble and membrane-bound pattern recognition molecules (PRMs) of the complement innate immune system and involved in host defence and scavenging of toxic cell debris (e.g., apoptotic cells)

Molecule	Structure	Location	Ligands	Function
Soluble				
MBL	Collectin, contains globular domains and collagen tail	Primary synthesis in liver	Microbial cell wall saccharides (mannose) Saccharides expressed de novo by apoptotic cells	Activates complement (opsonic and lytic activities) Promotes phagocytosis (through CR1)
C1q	Globular domains and collagen tail	Monocyte/macrophages and some epithelial cells	Immune complexes Toxic debris (DNA, βA4 amyloid) Epitopes on apoptotic cells	Activates complement Stimulates myeloid cells (through CR1) Promotes phagocytosis
C3	Disulfide-linked heterodimer, thiol ester containing protein	liver, myeloid and non-myeloid cells	Forms ester linkage to OH-groups on carbohydrates and proteins	Attachment of ligand for receptors to promote phagocytosis (CR3) or cell stimulation (CR2, CR1)
Receptors				
CR1 (CD35)	30 SCRs (Short Consensus Repeats)	Myeloid cells	C3b, C4b, C1q, MBL	Promotes phagocytosis Controls complement activation
CR2 (CD21)	15 SCRs	B cells and follicular dendritic cells	C3d	Augments B cell activation by antigen
CR3 (CD11b/CD18)	β2 integrin	Monocytes	iC3b, β-glucan, LPS, Fibrinogen, ICAM-1	Promotes phagocytosis, LPS clearance Non-phlogystic cell stimulation Promotes adhesion
CR4. p150,95 (CD11c/CD18)	β2 integrin	Macrophages	iC3b, LPS Fibrinogen, ICAM-1, CD23	As for CR3

Table adapted from Fearon and Locksley, Science, 1996, Ref 71.

have also been implicated in the binding and clearance of apoptotic cells. MBL was found to bind preferentially to apoptotic Jurkat T cells (UV irradiated) and aged neutrophils.[34] This binding seemed to involve the carbohydrate recognition domain of MBL interacting with yet uncharacterized cell surface carbohydrates. The binding of MBL to apoptotic cells was reduced following mannosidase treatment or by competition with high concentrations of mannose.

The Proposed C1q Receptor Involved in Phagocytosis and/or Signaling Events

One cell-surface molecule has emerged as a candidate defense collagen receptor for C1q, MBL and SPA. The first indication that such a receptor existed emerged from studies of monocyte phagocytic activity. Monocytes that had adhered to surfaces coated with C1q (or MBL or SPA) displayed a 4 to 10 fold enhancement of ingestion of targets opsonized with IgG or C.[37] Monoclonal antibodies, selected for their ability to inhibit the C1q-mediated enhancement of phagocytosis, were used to identify a cDNA clone encoding the cell-surface transmembrane glycoprotein designated as "C1q receptor that enhances phagocytosis" (C1qRp).[38] C1qRp was then found to be the analog of the rodent fetal stem cell marker AA4 which is involved in cell-cell interactions during hematopoietic and vascular development.[39–42] AA4 was abundantly expressed on endothelial cells and microglia and recent studies in human replicate this expression pattern, suggesting that C1qRp/AA4 is involved in cell signaling to promote phagocytosis and adhesion.[43] Recent studies have not supported the suggestion that C1qRp is a receptor for C1q, and the observed interactions are likely to be nonspecific, due to the heavily charged nature of C1q.[44] Moreover, C1qRp has been shown to be the antigen recognized by a pro-adhesive monoclonal antibody called mNI-11 and several antibodies against CD93, leading to a change in terminology for this molecule.[44–47] Our current understanding of the cellular and molecular properties of this receptor is incomplete and certainly warrants further investigation.

Other C1q and Defense Collagen Receptors (CR1, Cc1qr, CD91, Gc1qr)

CR1 (CD35) is found on circulating monocytes and neutrophils but the major site of expression is B lymphocytes. CR1 is a multifunctional receptor both in its ligand specificity and in the C regulatory activities.[48] As a receptor, CR1 binds to C opsonins (C4b, C3b, iC3b, C1q) and MBL, and as such has been implicated in phagocytic activities.[49] Several studies support a role for a cell surface receptor for the collagenous stalk of C1q (cC1qR) that binds the collagenous tails of C1q and MBL attached to apoptotic cells or pathogens to mediate phagocytosis.[50] The cC1qR is identical with calreticulin (CRT), a cytoplasmic protein that is a member of the family of heat-shock proteins, the most abundant and ubiquitous soluble intracellular proteins. Although CRT does not have a transmembrane domain, it appears to be expressed at the membrane through association with CD91 (also known as the $\alpha 2$ macroglobulin ($\alpha 2$m) receptor of LDL receptor related protein, LRP1) on the macrophage cell surface.[34] Yet another receptor for C1q is a 33-kDa protein that interacts with the globular heads of C1q and, logically, has been termed gC1qR. This protein is broadly expressed in tissues but, surprisingly, is located in mitochondria, suggesting that gC1qR is not a true cell-surface receptor. However, several lines of evidence show that this protein can be associated with the extracellular surface of the membrane, possibly through interactions with integrins or other transmembrane proteins, yet to be characterized.[51]

C Receptors Involved in Phagocytosis and Cell Recruitment (Chemotaxis)

C3, when activated on a cell surface becomes covalently bound (opsonized) as C3b which is subsequently cleaved to yield a very stable fragment iC3b. There is well documented evidence that CR3 (CD11b/CD18) and CR4 (CD11c/CD18 also known as the p150,95 antigen) are involved in the phagocytosis of targets opsonized with C3b and iC3b fragments (for reviews see refs. 52, 53). The capacity of CR3 to recognize natural microbial surface components such as β-glucan, lipopolysaccharide, lipophosphoglycan and other yet undefined structures is, perhaps, the critical event to stimulate macrophages leading to the elimination of pathogens, toxic debris and apoptotic cells.[52] Although CR3 is abundantly expressed by circulating monocytes, neutrophils and NK cells, CR4 is the most abundant C3 receptor on tissue macrophages.[54] Interestingly, in mouse, CR4 is abundantly expressed by dendritic cells but its role remains poorly defined. The diverse functions and activities ascribed to CR3 would appear to confer on this receptor the status of a master regulator of leukocytes, particularly with regard to innate immune responses. Indeed, CR3 is now considered as a key and ancestral receptor in innate immunity. As further evidence of its ancient roles, the α chain of a CR3-like molecule has recently been described in invertebrates.[55] CR3 thus legitimately joins the ranks of host PRMs alongside CD14 and the macrophage mannose receptor.

Both CR3 and CR4 belong to the β2 subgroup of the integrin superfamily and binding of CR3 to other ligands on the endothelium (i.e., ICAM-1) may be a necessary step in the migration of leukocytes into extravascular tissues.[56] Most, and perhaps all, of these protein ligands bind to a specialized inserted (I) domain in the α subunit of CR3. The β2 integrins were brought to prominence when it was realized that a clinical syndrome characterized by recurrent life-threatening infections, referred to as leukocyte adhesion deficiency, resulted from reduced or absent surface expression of β2 integrins or from expression of a nonfunctional form of β2 integrin. The syndrome results from mutations in the common CD18 subunit, and hence affects all of the β2 integrins.[57] The pathology was similarly reproduced in CD11b knockout mice in which neutrophils demonstrated impaired adhesion, phagocytosis, and oxygen radical generation in response to an immune challenge. Surprisingly, it would appear that CR3 in isolation is not the most potent activating receptor on phagocytic cells despite its role as an important component of innate immune responses.[58] Phagocytosis mediated by CR3 alone does not generate an oxidative burst, and ligation of CR3 suppresses the secretion of IL-12 and other proinflammatory signals.

It has been proposed that the binding of microbial or apoptotic motifs to CR3 or other PRM (e.g., CD14, mannose receptor, and scavenger receptor) is involved as early warning signals that trigger rapid recognition of and response to noxious substances. The presence of low levels of C3-opsonized particles or of LPS does not necessarily constitute a serious threat to the individual, hence it is essential that the response is measured and in proportion to the magnitude of the threat. Sustained macrophage activation therefore would require additional signals, which may include one or more of the following: (i) receptor clustering; (ii) costimulation by chemokines and chemoattractants (C3a, C5a) to allow switches between low- and high-affinity states, or (iii) cooperation with another type of PRM, which indicates the presence of more than one foreign ligand and hence raises the level of innate defense activity.

One aspect of the C system that has received consistent attention is the function and mechanism of action of the biologically active small fragments derived from the homologous C proteins C3 and C5. These C anaphylatoxins (C3a and C5a) released in the fluid phase after enzymatic cleavage of C3 and C5 respectively, are important proinflammatory molecules involved in the stimulation and chemotaxis of myeloid cells with specific anaphylatoxin receptors (C3aR and C5aR). These receptors belong to the rhodopsin family of seven transmembrane-spanning G-protein coupled receptors.[59] C5a is regarded as the most potent chemoattractant for neutrophils and monocytes, although C3a has also been shown to regulate inflammatory functions.[60,61] Early evidence indicated that C3aR was present only on myeloid cells such as macrophages, eosinophils and mast cells.[62-64] However, the demonstration that C3aR mRNA was expressed throughout the body and particularly in the adrenal gland, pituitary, and in the central nervous system, is consistent with C3aR having a much greater role in the pathogenesis of inflammatory and autoimmune diseases than previously suspected.[65]

Several investigators have found a close relationship between elevated plasma levels of C3a and its "inactive" metabolite C3adesArg in patients with septic shock and at risk of developing either adult respiratory distress syndrome or multi-organ failure.[66,67] C3a mediates numerous proinflammatory activities, including release of lysosomal enzymes from leukocytes, secretion of histamine from mast cells, smooth muscle contraction and chemoattraction of eosinophils and mast cells. These activities are in accordance with the generally accepted view that C3a participates positively in inflammatory reactions.[59] However, recent reports have strongly suggested that C3a can exhibit anti-inflammatory properties by suppressing LPS-induced TNF-α, IL1-β and IL-6 secretion from isolated peripheral blood mononuclear cells (PBMCs) and attenuating TNF-α and IL-6 secretion from lymphocytes.[68,69] The most recent addition to our understanding of the many potential activities of C3a has come from the characterization of genetic deletion of the C3a receptor in mice. This elegant and pioneering study demonstrated an important protective role for the C3aR in endotoxin shock, notably by attenuating LPS-induced proinflammatory cytokine production.[70]

CR2(CD21): The Link Between Innate and Acquired Immune Responses

In the past innate immunity has been considered as a stop-gap to provide rapid but inefficient antimicrobial host defense until the slower, but specific and efficient, acquired immune response develops.[71] This concept is no longer tenable as it is now established that C-derived fragments play an important role in shaping the antibody responses of acquired immunity. The role of C was shown first almost 30 years ago by Pepys, who demonstrated that the formation of antibodies against T-cell dependent antigens was reduced in animals in which C3 had been depleted.[72] The mechanism for this phenomenon has been elucidated more recently. On the plasma membrane of B cells the B cell receptor (BCR; membrane immunoglobulin) is associated with a complex of two proteins, CD19, a component of the acquired immune response system and CD21 (CR2), the receptor for C3d. It has been proposed that C activation on a newly encountered antigen is triggered via binding to "natural" IgM, CRP, collectins or alternative pathway components, leading to deposition of C3 fragments, including C3d, at the surface of the pathogen. Binding of the antigen-C3d complex to the BCR and CR2, respectively, and signaling through CD19 in the complex, induces a sustained activation of B cells. Indeed, when an antigen was coupled to C3d molecules, much less antigen was required to evoke a given level of antibody than was the case for the native antigen.[73] C1q can also act to bridge the innate and adaptive immune systems following binding to natural IgM complexed to antigens, including microbial antigens and certain autoantigens.[74]

The Opsonic Ancestral Element

From the foregoing sections, it is evident that C3 is central both to the activation of C and to its functions. Therefore, one must conclude that this component was the nucleus around which the C system has evolved. It has been proposed that the ancestral C system would have resembled a simplified AP comprising a C3-like and a fB-like component together with a receptor on a phagocytic cell (for a review see ref. 8). The major function of such a system would therefore have been opsonic (marking for phagocytosis) rather than lytic. However, the recent finding of a MBL-like molecule in invertebrates has raised the possibility that the lectin pathway might have predated the development of the AP, and hence be more ancient than has hitherto been assumed.[75]

C and Other Innate Immune Signaling Pathways: An Ancestral Innate Immune Signaling Pathway

The application of *Drosophila* genetics to decipher the mechanisms involved in host defense and development has generated insights into innate immunity and uncovered similarities with mammalian immune responses.[76] For instance, analysis of immunocompromised flies has demonstrated that the Toll signaling pathway, previously characterized as a regulator of dorsal-ventral polarity in developing embryos, also mediates defense against pathogens. In mammals, Toll-like receptors (TLRs) regulate mammalian innate immune responses via signaling pathways that resemble the Toll pathway. Further analogies can be drawn with the C system. As noted above, activation of the C system leads to the generation of several small fragments (termed C cytokines, or C-kine) that are involved in cell chemotaxis, and activation of phagocytes to promote killing of the pathogen. One role of the *Drosophila* Toll pathway is to activate the synthesis of an antimicrobial peptide (drosomycin) from the fat body and the toll ligand, Spaetzle. Spaetzle is a small cytokine-like protein that is generated via different proteolytic cascades during development and after infection. In mammals and flies, these molecules are involved in engaging and controlling the innate immune response and in orchestrating the transition, if necessary, to an acquired immune response.

Conclusion

The essence of innate immunity is the detection of molecules that are unique to infectious organisms and noxious substances, to induce clearance of the intruders and, in mammals, to educate the acquired immune response. C is now widely accepted to constitute the critical link between the innate immune response and acquired immunity. C is involved in the selective recognition and clearance of potentially noxious substances, whether they are derived from the host following injury (apoptotic cells, toxic cell debris) or following an infectious challenge (microbial agents). Recent evidence from whole-genome sequencing studies and fly genetics indicate that C is the archetypal component of innate immunity.

Acknowledgements

The work by Philippe Gasque and B. Paul Morgan was supported by the Medical Research Council, and the Wellcome Trust.

References

1. Medzhitov R, Janeway Jr CA. Innate immunity: The virtues of a nonclonal system of recognition. Cell 1997; 91:295-298.
2. Gordon S. The role of the macrophage in immune regulation. Res Immunol 1998; 149:685-688.
3. Savill J, Fadok V. Corpse clearance defines the meaning of cell death. Nature 2000; 407:784-788.
4. Platt NR, da Silva RP, Gordon S. Recognizing death: The phagocytosis of apoptotic cells. Trends Cell Biol 1998; 8:365-372.
5. Franc N, White CK, Ezekowitz RA. Phagocytosis and development: Back to the future. Curr Opin Immunol 1999; 11:47-52.
6. Gasque P, Dean YD, McGreal EP et al. Complement components of the innate immune system in health and disease in the CNS. Immunopharmacology 2000; 49:171-186.
7. Frank MM, Fries LF. The role of complement in inflammation and phagocytosis. Immunology Today 1991; 12:322-326.
8. Dodds AW, Law SK. The phylogeny and evolution of the thioester bond-containing proteins C3, C4 and alpha 2-macroglobulin. Immunol Rev 1998; 166:15-26.
9. Smith LC, Clow LA, Terwilliger DP. The ancestral complement system in sea urchins. Immunol Rev 2001; 180:16-34.
10. Sahu A, Lambris JD. Structure and biology of complement protein C3, a connecting link between innate and acquired immunity. Immunol Rev 2001; 180:35-48.
11. Lagueux M, Perrodou E, Levashina EA et al. Constitutive expression of a complement-like protein in toll and JAK gain-of-function mutants of Drosophila. Proc Natl Acad Sci USA 2000; 97:11427-11432.
12. Levashina EA, Moita LF, Blandin S et al. Conserved role of a complement-like protein in phagocytosis revealed by dsRNA knockout in cultured cells of the mosquito, Anopheles gambiae. Cell 2001; 104:709-718.
13. Petry FK, Reid KBM, Loos M. Gene expression of the a-chain and B-chain of mouse C1q in different tissues and the characterization of the recombinant a-chain. J Immunol 1991; 147:3988-3993.
14. Morgan BP, Gasque P. Extrahepatic complement biosynthesis: Where, when and why? Clin Exp Immunol 1997; 107:1-7.

15. Farries TC, Lachmann PJ, Harrison RA. Analysis of the interactions between properdin, the third component of complement (C3), and its physiological activation products. Biochem J 1988; 252:47-54.
16. Gewurz H, Ying SC, Jiang H et al. Nonimmune activation of the classical complement pathway. Behring Inst Mitt 1993; 93:138-147.
17. Kilpatrick JM, Volanakis JE. Molecular genetics, structure and function of C-reactive protein. Immunol Res 1991; 10:43-53.
18. Armstrong PB, Armstrong MT, Quigley JP. Involvement of alpha 2-macroglobulin and C-reactive protein in a complement-like hemolytic system in the arthropod, Limulus polyphemus. Mol Immunol 1993; 30:929-934.
19. Liu TY, Minetti CA, Fortes-Dias CL et al. C-reactive proteins, limunectin, lipopolysaccharide-binding protein, and coagulin. Molecules with lectin and agglutinin activities from Limulus polyphemus. Ann N Y Acad Sci 1994; 712:146-154.
20. Tharia HA, Shrive AK, Mills JD et al. Complete cDNA sequence of SAP-like pentraxin from Limulus polyphemus: Implications for pentraxin evolution. J Mol Biol 2002; 316:583-597.
21. Bharadwaj D, Stein MP, Volzer M et al. The major receptor for C-reactive protein on leukocytes is Fc gamma receptor II. J Exp Med 1999; 190:585-590.
22. Petersen SV, Thiel S, Jensenius JC. The mannan-binding lectin pathway of complement activation: Biology and disease association. Mol Immunol 2001; 38:133-149.
23. Jack DL, Klein NJ, Turner MW. Mannose-binding lectin: Targeting the microbial world for complement attack and opsonophagocytosis. Immunol Rev 2001; 180:86-99.
24. Stover C, Endo Y, Takahashi M et al. The human gene for mannan-binding lectin-associated serine protease-2 (MASP-2), the effector component of the lectin route of complement activation, is part of a tightly linked gene cluster on chromosome 1p36.2-3. Genes Immun 2001; 2:119-127.
25. Dahl MR, Thiel S, Matsushita M et al. MASP-3 and its association with distinct complexes of the mannan-binding lectin complement activation pathway. Immunity 2001; 15:127-135.
26. Rossi V, Cseh S, Bally I et al. Substrate specificities of recombinant mannan-binding lectin-associated serine proteases-1 and -2. J Biol Chem 2001; 276:40880-40887.
27. Morgan BP. Complement membrane attack on nucleated cells: Resistance, recovery and nonlethal effects. Biochem J 1989; 264:1-14.
28. Morgan BP, Gasque P. Expression of complement in the brain: Role in health and disease. Immunology Today 1996; 17:461-466.
29. Morgan BP. Complement regulatory molecules: Application to therapy and transplantation. Immunology Today 1995; 16:257-259.
30. Speth C, Dierich MP, Gasque P. Neuroinvasion by pathogens: A key role of the complement system. Mol Immunol 2002; 38:669-679.
31. Korb LC, Ahearn JM. C1q binds directly and specifically to surface blebs of apoptotic human keratinocytes: Complemement deficiency and systemic lupus erythematous revisited. J Immunol 1997; 158:4525-4528.
32. Mevorach D, Mascarenhas JO, Gershov D et al. Complement-dependent clearance of apoptotic cells by human macrophages. J Exp Med 1998; 188:2313-2320.
33. Taylor PR, Carugati A, Fadok VA et al. A hierarchical role for classical pathway complement proteins in the clearance of apoptotic cells in vivo. J Exp Med 2000; 192:359-366.
34. Ogden CA, de Cathelineau A, Hoffmann PR et al. C1q and mannose binding lectin engagement of cell surface calreticulin and CD91 initiates macropinocytosis and uptake of apoptotic cells. J Exp Med 2001; 194:781-795.
35. Botto M, Dell'Agnola C, Bygrave AE et al. Homozygous C1q deficiency causes glomerulonephritis associated with multiple apoptotic bodies. Nat Genet 1998; 19:56-59.
36. Mitchell DA, Taylor PR, Cook HT et al. Cutting edge: C1q protects against the development of glomerulonephritis independently of C3 activation. J Immunol 1999; 162:5676-5679.
37. Guan E, Robinson SL, Goodman EB et al. Cell-surface protein identified on phagocytic cells modulates the C1q-mediated enhancement of phagocytosis. J Immunol 1994; 152:4005-4016.
38. Nepomuceno RR, Henschen-Edman AH, Burgess WH et al. cDNA cloning and primary structure analysis of C1qR(p), the human C1q/Mbl/Spa receptor that mediates enhanced phagocytosis in vitro. Immunity 1997; 6:119-129.
39. Petrenko O, Beavis A, Klaine M et al. The molecular characterization of the fetal stem cell marker AA4. Immunity 1999; 10:691-700.
40. Dean YD, McGreal EP, Akatsu H et al. Molecular and cellular properties of the rat AA4 antigen, a C-type lectin-like receptor with structural homology to thrombomodulin. J Biol Chem 2000; 275:34382-34392.

41. Dean YD, McGreal EP, Gasque P. Endothelial cells, megakaryoblasts, platelets and alveolar epithelial cells express abundant levels of the mouse AA4 antigen, a C-type lectin-like receptor involved in homing activities and innate immune host defense. Eur J Immunol 2001; 31:1370-1381.
42. Lovik G, Vaage JT, Dissen E et al. Characterization and molecular cloning of rat C1qRp, a receptor on NK cells. Eur J Immunol 2000; 30:3355-3362.
43. Fonseca MI, Carpenter PM, Park M et al. C1qR(P), a myeloid cell receptor in blood, is predominantly expressed on endothelial cells in human tissue. J Leuk Biol 2001; 70:793-800.
44. McGreal EP, Ikewaki N, Akatsu H et al. Human C1qRp is identical with CD93 and the mNI-11 antigen but does not bind C1q. J Immunol 2002; 168:5222-5232.
45. Ikewaki N, Inoko H. Development and characterization of a novel monoclonal antibody (mNI-11) that induces cell adhesion of the LPS-stimulated human monocyte-like cell line U937. J Leuk Biol 1996; 59:697-708.
46. Ikewaki N, Tamauchi H, Yamada A et al. A unique monoclonal antibody mNI-11 rapidly enhances spread formation in human umbilical vein endothelial cells. J Clin Immunol 2000; 20:317-324.
47. Steinberger P, Szekeres A, Wille S et al. Identification of human CD93 as the phagocytic C1q receptor (C1qRp) by expression cloning. J Leuk Biol 2002; 71:133-140.
48. Krych-Goldberg M, Atkinson JP. Structurefunction relationships of complement receptor type 1. Immunol Rev 2001; 180:112-122.
49. Nicholson-Weller A, Klickstein LB. C1q-binding proteins and C1q receptors. Curr Opin Immunol 1999; 11:42-46.
50. Johnson S, Michalak M, Opas M et al. The ins and outs of calreticulin: From the ER lumen to the extracellular space. Trends Cell Biol 2001; 11:122-129.
51. Ghebrehiwet B, Lim BL, Kumar R et al. gC1q-R/p33, a member of a new class of multifunctional and multicompartmental cellular proteins, is involved in inflammation and infection. Immunol Rev 2001; 180:65-77.
52. Ehlers MR. CR3: A general purpose adhesion-recognition receptor essential for innate immunity. Microbes Infect 2000; 2:289-294.
53. Cabanas C, Sanchez-Madrid F. CD11c (leukocyte integrin CR4 alpha subunit). J Biol Regul Homeost Agents 1999; 13:134-136.
54. Myones BL, Dalzell JG, Hogg N et al. Neutrophil and monocyte cell surface p150,95 has iC3b-receptor (CR4) activity resembling CR3. J Clin Invest 1988; 82:640-651.
55. Miyazawa S, Azumi K, Nonaka M. Cloning and characterization of integrin alpha subunits from the solitary ascidian, Halocynthia roretzi. J Immunol 2001; 166:1710-1715.
56. Diamond MS, Garcia-Aguilar J, Bickford JK et al. The I domain is a major recognition site on the leukocyte integrin Mac-1 (CD11b/CD18) for four distinct adhesion ligands. J Cell Biol 1993; 120:1031-1043.
57. Hogg N, Stewart MP, Scarth SL et al. A novel leukocyte adhesion deficiency caused by expressed but nonfunctional beta2 integrins Mac-1 and LFA-1. J Clin Invest 1999; 103:97-106.
58. Aderem A, Underhill DM. Mechanisms of phagocytosis in macrophages. Annu Rev Immunol 1999; 17:593-623.
59. Ember JA, Jagels MA, Hugli TE. Characterization of complement anaphylatoxins and their biological responses. In: Volanakis JE, Frank MM, eds. The human complement system in health and disease. New York: Dekker M Inc., 1998:241-284.
60. Ames RS, Li Y, Sarau HM et al. Molecular cloning and characterization of the human anaphylatoxin C3a receptor. J Biol Chem 1996; 271:20231-20234.
61. Crass T, Raffetseder U, Martin U et al. Expression cloning of the human C3a anaphylatoxin receptor (C3aR) from differentiated U-937 cells. Eur J Immunol 1996; 26:1944-1950.
62. Zwirner J, Gotze O, Begemann G et al. Evaluation of C3a receptor expression on human leucocytes by the use of novel monoclonal antibodies. Immunology 1999; 97:166-172.
63. Martin U, Bock D, Arseniev L et al. The human C3a receptor is expressed on neutrophils and monocytes, but not on B or T lymphocytes. J Exp Med 1997; 186:199-207.
64. Daffern PJ, Pfeifer PH, Ember JA et al. C3a is a chemotaxin for human eosinophils but not for neutrophils. I. C3a stimulation of neutrophils is secondary to eosinophil activation. J Exp Med 1995; 181:2119-2127.
65. Gasque P, Singhrao SK, Neal JW et al. The receptor for complement anaphylatoxin C3a is expressed by myeloid cells and nonmyeloid cells in inflamed human central nervous system: Analysis in multiple sclerosis and bacterial meningitis. J Immunol 1998; 160:3543-3554.
66. Hack CE, Nuijens JH, Felt-Bersma RJ et al. Elevated plasma levels of the anaphylatoxins C3a and C4a are associated with a fatal outcome in sepsis. Am J Med 1989; 86:20-26.

67. Zilow G, Sturm JA, Rother U et al. Complement activation and the prognostic value of C3a in patients at risk of adult respiratory distress syndrome. Clin Exp Immunol 1990; 79:151-157.
68. Fischer WH, Jagels MA, Hugli TE. Regulation of IL-6 synthesis in human peripheral blood mononuclear cells by C3a and C3a(desArg). J Immunol 1999; 162:453-459.
69. Takabayashi T, Vannier E, Burke JF et al. Both C3a and C3a(desArg) regulate interleukin-6 synthesis in human peripheral blood mononuclear cells. J Infect Dis 1998; 177:1622-1628.
70. Kildsgaard J, Hollmann TJ, Matthews KW et al. Cutting edge: Targeted disruption of the C3a receptor gene demonstrates a novel protective anti-inflammatory role for C3a in endotoxin-shock. J Immunol 2000; 165:5406-5409.
71. Fearon DT, Locksley RM. The instructive role of innate immunity in the acquired immune response. Science 1996; 272:50-54.
72. Pepys MB. Role of complement in induction of antibody production in vivo. Effect of cobra factor and other C3-reactive agents on thymus-dependent and thymus-independent antibody responses. J Exp Med 1974; 140:126-145.
73. Dempsey PW, Allison ME, Akkaraju S et al. C3d of complement as a molecular adjuvant: Bridging innate and acquired immunity. Science 1996; 271:348-350.
74. Heyman B. Regulation of antibody responses via antibodies, complement, and Fc receptors. Annu Rev Immunol 2000; 18:709-737.
75. Sekine H, Kenjo A, Azumi K et al. An ancient lectin-dependent complement system in an ascidian: Novel lectin isolated from the plasma of the solitary ascidian, Halocynthia roretzi. J Immunol 2001; 167:4504-4510.
76. Hoffmann JA, Reichhart JM. Drosophila innate immunity: An evolutionary perspective. Nat Immunol 2002; 3:121-126.

Chapter 5

Carbohydrate Recognition Receptors on Antigen Presenting Cells

Philip R. Taylor, Gordon D. Brown, Luisa Martinez-Pomares and Siamon Gordon

Abstract

This chapter describes the receptors on the surface of antigen presenting cells that have been shown to bind to carbohydrates, including those found on the surface of microbes and viruses, and glycosylated endogenous molecules such as hormones and lysosomal enzymes. The same repertoire of receptors, often the same individual receptor, on the surface of the antigen presenting cell is involved in the recognition of these two very distinct types of antigen, i.e., both "nonself" and "self". These receptors play diverse roles in immunity and homeostasis, including enhancement of antigen capture for presentation, cellular activation in response to pathogen-associated motifs, the maintenance of serum glycoprotein levels and the regulation of cell migration.

Introduction

The way in which antigen presenting cells (APCs), that is, dendritic cells (DCs), macrophages (Mϕs) and B cells, can recognize carbohydrates is crucial to our understanding of how these structures can be internalized by APCs and how this affects their phenotype and function. This chapter will concentrate on the receptors present on the surface of APCs and their role in this recognition process. Some of the carbohydrate receptors discussed may show only limited expression on APCs, perhaps having more definite expression and function outside the realms of innate immunity, but are included here for completeness. It is not forgotten that many other soluble carbohydrate recognition systems, such as the collectins, pentraxins and complement, are able to interact with the surface of APCs, but these systems and proteins are dealt with in detail in other chapters in this book. The Siglec family, which includes sialoadhesin, a Mϕ expressed receptor for the recognition of terminal sialic acids on oligosaccharides, will not be included here.

Antigen presenting cells are endowed with an array of surface receptors for the recognition of repetitive structures, including a broad range of carbohydrates, on the surface of microbes and viral particles. These pattern recognition receptors (PRRs),[1] are able to recognize these pathogen associated molecular patterns (PAMPs). Recognition of PAMPs by nonopsonic PRRs on the APCs helps determine the mechanism of microbial uptake (phagocytosis/endocytosis), and influences the intracellular signalling events and cellular activation that follow. Thus these innate recognition receptors can promote the uptake of foreign antigens by APCs and aid in the development of an appropriate immune response.

In this chapter we will attempt to provide an up to date evaluation of the carbohydrate recognition receptors present on the surface of APCs (summarized in Table 1), concentrating

Immunobiology of Carbohydrates, edited by Simon Y.C. Wong and Gemma Arsequell.
©2003 Eurekah.com and Kluwer Academic / Plenum Publishers.

Table 1. PRRs with reported carbohydrate specificity present on APCs

Receptor	Main Cellular Expression	Ligand Specificity	References
Mannose receptor (CD206)	Mϕ and non-vascular endothelium	Terminal fucose/ mannose/N-acetylglucosamine	78, 79
Endo180	Fibroblast, endothelial cells and Mϕ	N-acetylglucosamine	4
Langerin (CD207)	Langerhans cells, DC in thymic medulla and sub populations of DC in lymph nodes and spleen	Mannan	34, 45
DC-SIGN (CD209)	DC	Mannose	32, 37, 40, 42
DC-SIGNR	DC	Mannose	33, 37
NKCL/dectin-2*	Mϕ, monocytes, multipotent precursors, neutrophils, DC and Langerhans cells	Mannose	35, 36, 46
CR3	Mϕ, DC, NK cells and granulocytes	β-D-glucan	54, 55
βGR/dectin-1	Mϕ, neutrophil and DC	β-(1,3)-D-glucan	55, 65, 66
Mϕ galactose receptor	Peritoneal and tumoricidal Mϕ	Terminal galactose/N-acetylgalactose	69, 71
Kupffer cell receptor	Kupffer cells	Terminal galactose/N-acetylgalactose/fucose	72, 74
Sialoadhesin	Mϕ	Terminal sialic acids	80

*Literature reports regarding both the cellular distribution and lectin activity of NKCL/dectin-2 remain controversial.

on those with recognized carbohydrate binding potential and assessing the evidence regarding their specificity and function in the immune system.

The Macrophage Mannose Receptor Family

The Mϕ mannose receptor (MR) represents the best characterized of a family of four related cell surface receptors. Each member of the family shares the same overall structure (the structure of the Mϕ MR is represented schematically in Fig. 1). These receptors are type I membrane molecules with a N-terminal cysteine-rich domain followed by a fibronectin type II repeat and a series of C-type lectin carbohydrate recognition domains (CRDs). Only two of the members of this family: the MR itself[2,3] and Endo180,[4] have been demonstrated to bind to carbohydrates in a Ca^{2+} dependent manner. The two additional members of the family, the phospholipase A2 receptor[5,6] and DEC-205,[7] do not have the conserved residues required for Ca^{2+}-dependent sugar recognition nor have they been shown to bind carbohydrates.

The Mϕ Mannose Receptor (CD206)

The Mϕ MR (CD206), which is widely known as a PRR with specificity for L-fucose, D-mannose and D-N-acetylglucosamine, was identified by its ability to bind to endogenous glycosylated self-ligands such as β-glucuronidase (for a review see ref. 8). In normal mice, the expression of the MR is predominantly restricted to populations of macrophages and endothelial cells (both hepatic sinusoidal and lymphatic), but its expression is notably absent from DCs

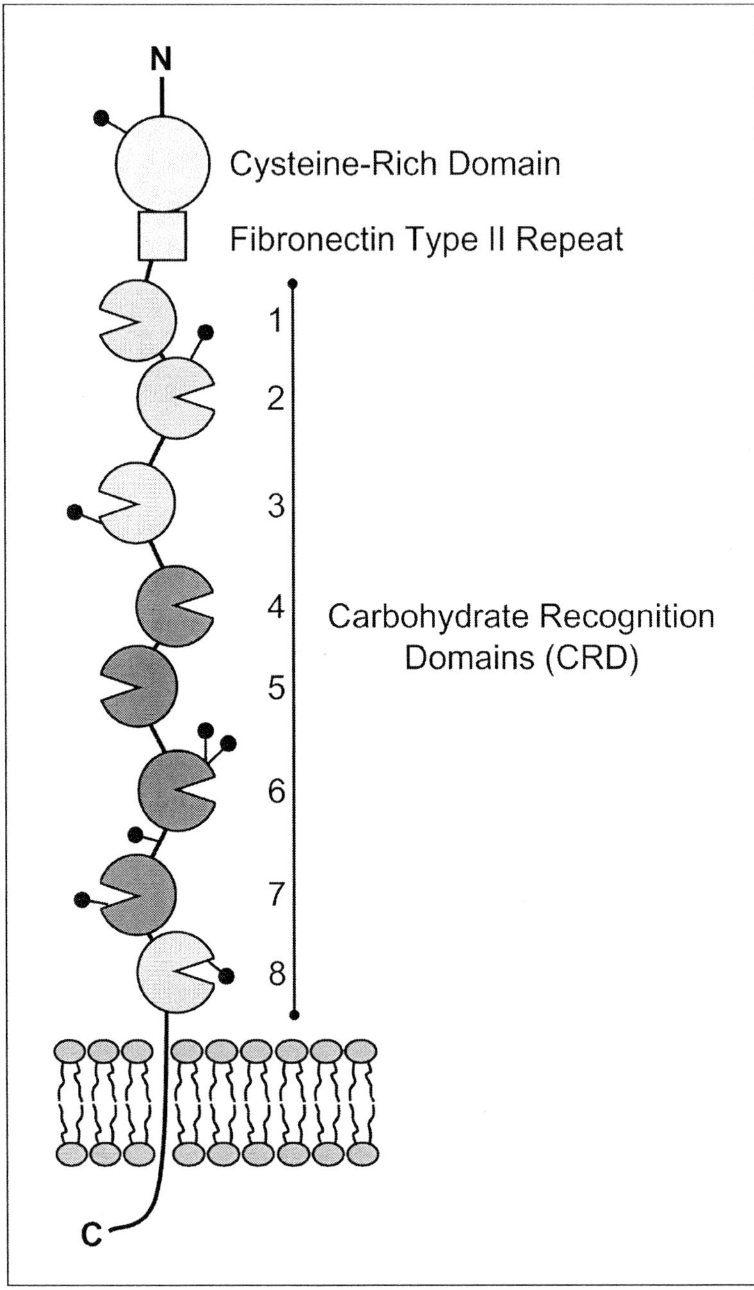

Figure 1. Schematic image of the Mφ mannose receptor as a representative member of the MR gene family. The MR has an N-terminal CR domain, adjacent to a fibronectin type II domain followed by 8 CRDs. CRDs 4-7 (dark grey) have been shown to be sufficient for efficient mannose recognition. N-glycosylation sites are as indicated (●).

in situ.[9] However, the MR has been detected on cultured DCs where it can concentrate antigen in MHCII-rich intracellular compartments for enhanced presentation.[10,11] The absence of MR expression on DCs in situ, in normal mice, but its presence on DCs in vitro indicates that its expression is regulated. This is consistent with a role for the MR in homeostasis under normal steady-state conditions but is suggestive that under inflammatory conditions it may be upregulated on professional APCs. This hypothesis has recently been supported by the observation of MR expression on epidermal DCs, but not Langerhans cells, in skin lesions taken from patients with atopic dermatitis and psoriasis.[12]

As well as being an endocytic receptor for the clearance of endogenous glycoproteins and foreign mannosylated-antigens the MR has also been ascribed a role in phagocytosis of numerous unopsonized particles, including yeasts,[13] *Candida albicans*,[14] zymosan,[15] *Leishmania*,[16,17] and *Pneumocystis carinii*.[18] Evidence for a direct role of the MR in the phagocytic process was obtained by conferring the ability to phagocytose yeast and *Pneumocystis carinii* on the normally poorly-phagocytic COS-1 cells after transfection with the MR cDNA.[3,18] The MR has also been implicated in the recognition of virus.[19] The MR thus appears to have a dual role in the living animal: (i) as a PRR for host defence via recognition of exposed carbohydrate structures on the surface of microbes and viral particles; and (ii) as a homeostatic regulator of selected plasma glycoproteins as has recently been confirmed with the development of a MR-deficient mouse.*20* The MR on lymphatic endothelium has also been suggested to be involved in lymphocyte migration via interaction with L-selectin.[21]

The MR also has a second lectin activity within the cysteine-rich (CR) domain for sulfated N-acetylgalactosamine.[22] Only endogenous ligands of the CR domain have so far been identified, including specific sulfated glycoforms of CD45 and sialoadhesin.[23] These ligands are found on discrete populations of macrophages: the marginal zone metallophilic Mϕs of the spleen and the subcapsular sinus Mϕs of the lymph nodes. It is speculated (since these cells do not themselves express MR) that these CR-domain ligands may be involved in the traffic of a soluble form of the MR (as a soluble form of this receptor has been found in mouse serum) or in cell-cell interactions.[23-25] The CR domain also binds to sulfated N-linked sugars found in highly glycosylated hormones such as lutropin, and is believed to contribute to the clearance of these types of molecules from the circulation via multiple interactions between the hormone and both the CR domain and CRDs.[22,26] The structure of both lectin domains has been determined.[27,28]

Endo180/uPARAP

Endo180, like the Mϕ MR, contains 8 CRDs, the second of which possesses the conserved amino acids required for Ca^{2+}-dependent sugar binding.[4] Endo180 was demonstrated to be able to bind to N-acetylglucosamine, but not mannose and is expressed predominantly by fibroblasts, endothelial cells and macrophages.[4] Endo180 has also been identified as uPARAP (urokinase-type plasminogen activator receptor associated protein). uPARAP forms a complex on the surface of the cell with the pro-form of the urokinase-type plasminogen activator and its receptor. It is speculated that Endo180/uPARAP may play a role in the internalization of this complex as, like other members of the family, it is a continuously recycling endocytic receptor. Additional ligands, including collagenase-3, have been proposed and a role for Endo180/uPARAP in the regulation of protease turnover and tissue degradation has been suggested (for a review, see ref. 29). However, recent studies aimed at addressing this activity have failed to support this hypothesis.[30]

DEC-205 (CD205)

A third member of the Mϕ mannose receptor family, DEC-205 (CD205), is able to target antigen efficiently for presentation to T cells.[7] The functional activity associated with DEC-205, and its fairly restricted expression to DCs, make it an interesting comparison to the Mϕ MR as it has not been reported to bind carbohydrates. Like the MR, DEC-205 recycles to the cell surface after endocytosis, but unlike the MR, it is able to traffic to late endosomes or lysosomes, which have high levels of MHCII, resulting in efficient antigen presentation.[31] To date, functional data regarding DEC-205 have been obtained either by targeting specific antibodies to the receptor or by constructing chimeric molecules with defined ligand binding potential. The natural physiological ligands of DEC-205 remain unknown and their identification will have important implications for our understanding of the types of antigens processed for presentation.

Other Receptors with Mannose-Specificity

There is significant evidence in the literature for the existence of alternative receptors with mannose binding specificity such as DC-SIGN,[32] DC-SIGNR,[33] Langerin[34] and, perhaps controversially, NKCL/dectin-2.[35,36] The recent release of draft genome sequences for both human and mouse has identified a number of related molecules so this area is destined to become still more complex.

DC-SIGN and DC-SIGNR

DC-SIGN[32] and a related genetically linked protein DC-SIGNR (DC-SIGN-related)[33] are both type II membrane proteins with C-terminal C-type CRDs which are able to bind mannose by a novel mechanism. This type of CRD has been structurally characterized.[37,38] Both molecules can mediate cell-cell contact within the immune system via interaction with ICAM-3[32,39] and both are able to bind to human immunodeficiency virus-1 via gp120[39-41] facilitating viral infectivity. DC-SIGN has also been suggested to support the tethering and rolling of dendritic cells on ICAM-2 under flow conditions[42] and can internalize antigen for presentation to T cells.[43]

Since the discovery of DC-SIGN, five homologues have been found in mouse.[44] Although the carbohydrate binding potential of these molecules has not been assessed it is possible that a large degree of redundancy and/or diversification exists amongst these carbohydrate recognition systems.

Langerin

Langerin is a specialized receptor, identified initially in Langerhans cells.[34] It is a type II membrane protein, with a Ca^{2+}-dependent lectin domain thought to be able to recognize mannan.[34] It is expressed predominantly by Langerhans cells but also by thymic medullary DCs and small subpopulations of DCs in the spleen and lymph nodes.[45] Transfection of Langerin into fibroblasts induces the formation of specialized organelles known as Birbeck granules. Antibody bound to Langerin is internalized into these organelles (but not into MHCII-rich compartments) where it remains for up to 2 hours after treatment.[34] It has been suggested that a novel antigen-processing pathway exists, where mannosylated-antigen is internalized into specialized organelles (the formation of which are induced by the receptor) and is retained intact—possibly until appropriate migrational or activation signals are received.[34]

NKCL/Dectin-2

Fernandes and colleagues originally identified NKCL as an overexpressed transcript in the spleen in a mouse model of chronic leukemia.[35] NKCL was found to be a member of the group V C-type lectins, a type II membrane molecule, that retained the second Ca^{2+} binding site which forms part of the carbohydrate recognition site and was able to bind mannose in a Ca^{2+}-dependent manner.[35] It was subsequently shown by Northern blot that NKCL was expressed by cells of the myeloid-lineages and by multipotent precursors.[35] Ariizumi and colleagues identified the same molecule, renamed dectin-2, by subtractive cDNA cloning of genes expressed in the murine DC cell line XS52, but not the murine Mϕ cell line J774.[36] NKCL has 100% identity with the α-isoform of dectin-2. Ariizumi and coworkers reported that dectin-2

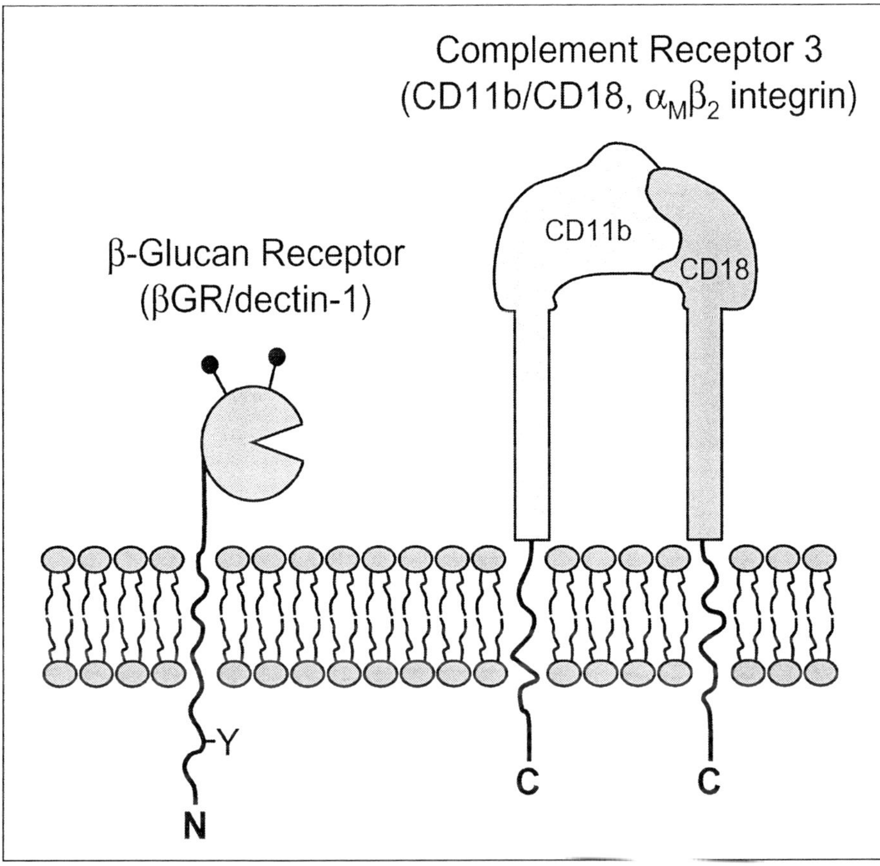

Figure 2. Representation of the two main candidate β-glucan recognition receptors, βGR/dectin-1 and CR3. The βGR/dectin-1 is a type II molecule with a single NK like C-type lectin domain with two N-linked glycosylation sites (●) on a short extracellular stalk and with an intracellular ITAM (Y) motif. CR3 is a heterodimer of CD11b and CD18. The numerous N-glycosylation sites of CR3 are omitted for simplicity.

did not bind to mannose. They also concluded that dectin-2 was restricted to DC-cell lines and primary epidermal cells that were depleted with anti-Ia mAb (implying that they were Langerhans cells).[36] Subsequent promoter analysis indicated that the dectin-2 5'-proximal promoter was strongly biased to expression in Langerhans cells.[46]

Although these studies differ in their assessment of precisely which cells express NKCL/dectin-2, perhaps the most critical difference is whether or not it can recognize mannose.[35,36] Interestingly, as mentioned above, Langerhans cells express high levels of this putative mannose binding receptor, as they do Langerin,[34,35] but they do not express the Mϕ MR.[9,12] Freshly isolated Langerhans cells were able to actively phagocytoze unopsonized zymosan by two mechanisms, one inhibited by mannan and one by β-glucan.[47] Inhibition by mannan implies the existence of a mannose receptor so both receptors, NKCL/dectin-2 and Langerin, could be considered candidates. Equally, alternative mannose binding receptors, either currently undescribed or not currently assigned to Langerhans cells may be responsible for this pattern recognition function.

β-Glucan Recognition by APCs

Glucans are polymers of glucose that are found in the cell walls of plants, fungi and bacteria and as components of the fungal cytosol and as polymers secreted by β-glucan producing microorganisms.[48,49] The presence of these glucans in the circulation of patients has become a diagnostic feature of invasive fungal infections.[50] Soluble and, importantly, pure β-glucans have significant activatory effects on macrophages and neutrophils, influencing the subsequent cytokine production and enhancing humoral and cellular immunity with few clinical side-effects (for reviews see refs. 48, 49).

Several candidate receptors for β-glucans have been identified including: lactosylceramide,[51] scavenger receptors,[52] complement receptor 3 (CR3)[53,54] and the β-glucan receptor (βGR) / dectin-1.[55] The main candidates are CR3, on which most attention has focused, and the recently identified βGR/dectin-1 (Fig. 2).

Complement Receptor 3

CR3 (CD11b/CD18, $\alpha_M\beta_2$integrin) was first recognized to possess lectin activity towards β-glucans approximately 15 years ago[53] and since then has been considered one of the major components of β-glucan recognition.[56] Internalization of opsonized targets by CR3 requires a second signal which mobilizes CR3 on the cell surface and endows the receptor with the capacity to trigger phagocytosis (for a review see ref. 57). Opsonization of a target cell, such as a tumor cell, with iC3b is insufficient for the induction of cytotoxicity/cytolysis.[58] More recent studies have suggested that the binding of soluble zymosan polysaccharides (SZP) to the lectin site of CR3 primed the receptor for cytotoxicity towards such iC3b opsonized targets.[59] When incorporated into animal tumor models, SZP conferred a significant protective effect, resulting in a marked reduction in tumor size.[60] This model required antibody- and C3-mediated opsonization of tumor cells in vivo and their recognition by SZP-activated CR3-bearing leukocytes.

The SZP preparations used, however, have varied dramatically in their relative richness of mannose or β-glucan[60,61] and their structural characterization has never been published. The lectin binding site, which has been mapped to a C-terminal domain distal to the binding site of the complement fragment iC3b[54,62] has also been shown to exhibit some promiscuity, recognizing polysaccharides containing mannose and N-acetyl-D-glucosamine, but not α-mannans.[54] Whilst CR3 does appear to possess a functional lectin site the exact carbohydrate-specificity of this site remains unclear and hence so does its importance in the nonopsonic recognition of β-glucans by APCs.

The β-Glucan Receptor

The existence of a leukocyte β-glucan receptor was first recognized nearly 20 years ago as a nonopsonic receptor for particulate activators of complement.[63,64] Dectin-1[65] was recently identified as a βGR by screening a retroviral cDNA expression library derived from a mouse Mϕ cell line, RAW264.7.[55] The βGR/Dectin-1 is an NK cell-like C-type lectin that is also believed to play a role in the binding and costimulation of T-cells via interaction with endogenous ligands.[55,65,66] It has been shown to bind *Saccharomyces cerevisiae*, heat-killed *Candida albicans* and the yeast-derived particle zymosan. Most significantly this binding is specifically inhibited by β-glucans containing β-(1,3)- and/or β-(1,6)-linkages.[55] Retroviral transduction of normally nonphagocytic NIH3T3 mouse fibroblast-like cell-line with the βGR conferred very efficient β-glucan-dependent phagocytosis of zymosan upon these cells.[55]

Using a monoclonal antibody directed against the murine βGR[67] we have been able to show that the surface expression of the βGR is broader than first indicated with highest levels on cells of the myeloid-lineages as well as on DCs (Taylor et al, submitted manuscript). This antibody almost totally blocked the nonopsonic recognition of zymosan by primary Mϕs, to a level equivalent to that obtained with pure structurally defined β-glucans.[67] In the same assay an antibody against CR3, which has been shown to block the lectin activity of CR3, was

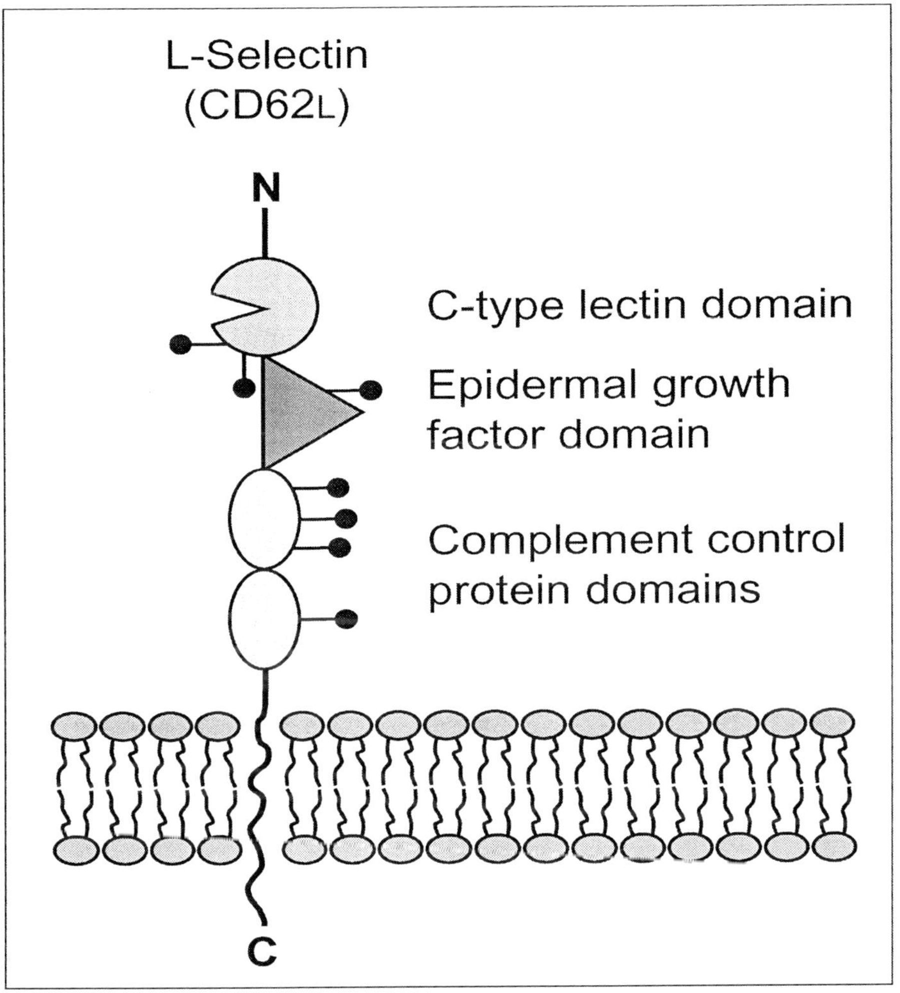

Figure 3. L-selectin is a type I membrane molecule with an N-terminal C-type lectin domain (mid-grey) adjacent to a single EGF super family domain (dark grey) and two complement control protein domains. N-glycosylation sites are as indicated (●).

unable to block the nonopsonic recognition of zymosan. Similarly, no significant defect in the nonopsonic recognition of zymosan by $CD11b^{-/-}$ Mϕ was observed.[67] Based on these observations we believe that the βGR/Dectin-1, not CR3, is the major receptor on APCs, for the nonopsonic recognition of β-glucans.

Further study is required to delineate the exact role of βGR/dectin-1 in the cellular and humoral responses induced by β-glucans. The βGR contains an immunoreceptor tyrosine based activation motif (ITAM).[65] The role of this ITAM in the internalization of soluble and particulate β-glucans and the subsequent cellular activation events associated with these carbohydrates is yet to be determined, but it is likely that some of the downstream signaling events will involve recruitment of Toll-like receptors, such as TLR-2, to the phagosome.[68]

Galactose Receptors

The Mϕ Galactose Receptor

The Mϕ galactose receptor was identified on peritoneal Mϕs and can recognize galactose and N-acetylgalactosamine.[69] It was shown to be expressed by tumoricidal Mϕs and it was believed to contribute to the binding of tumor cells to Mϕs.[70] Consistent with this, the carcinoma associated Tn antigen was found to be a naturally occurring ligand of the Mϕ galactose receptor.[71] It is compelling to think that the Mϕ galactose receptor may contribute to tumor immunity and it is hence also not unreasonable to suspect that it may also play a role in the recognition of microbes.

The Kupffer Cell Receptor

Kupffer cells, specialized Mϕs in the liver, express a receptor for galactose, N-acetyl galactose and, with lower affinity, fucose.[72,73] This receptor is interesting because recognition of exposed galactose on proteins was found to be dependent on the size of the protein, with large proteins/particles preferentially being cleared in vivo by the Kupffer cell receptor and smaller proteins being cleared by alternative receptors on hepatocytes.[74]

L-Selectin (CD62L)

L-selectin is expressed by lymphocytes, monocytes, granulocytes and some NK cells. L-selectin consists of a N-terminal C-type lectin domain adjacent to an epidermal growth factor domain and two "short consensus repeats" or complement control protein domains (Fig. 3). L-selectin has been shown to bind to sialylated Lewis x-related anionic polysaccharide sequences.[75] It can also bind to structurally distinct carbohydrates such as heparin sulfate. Specific glycoforms of CD34, GlyCAM-1 and MAdCAM-1 are ligands of L-selectin, with fucosylation, sialylation and sulfation important modifications required for binding.[76] As a consequence of this, L-selectin mediates the initial tethering and rolling of leukocytes on endothelial cells. It is important for the homing of naïve lymphocytes to the peripheral lymph nodes and Peyer's patches and the recruitment of peripheral blood leukocytes to sites of inflammation. Consistent with these proposed functions, leukocyte rolling and migration to inflammatory stimuli are significantly impaired in L-selectin-deficient mice.[77] Thus the role of L-selectin differs again from the other carbohydrate recognition receptors described in this chapter, its role being critical for the appropriate regulation of cellular influx to inflammatory lesions and in the homing of lymphocytes to peripheral lymphoid organs. Hence L-selectin is required for both mediation of the innate inflammatory response and the control of cell migration for the development of an antigen-specific acquired immune response.

Summary

One of the most striking features of several of these APC carbohydrate recognition receptors is that they can recognize both endogenous and exogenous ligands, sometimes independently of their ability to recognize carbohydrates, and can have different functions within the immune system or in the maintenance of homeostasis. For example, the Mϕ MR has well-defined dual roles in pathogen recognition and in the regulation of plasma glycoprotein levels. The βGR/dectin-1 appears to be the major leukocyte β-glucan receptor and yet binds T-cells in a β-glucan-independent costimulatory manner. DC-SIGN and DC-SIGNR are both involved in cell-cell interactions within the immune system via ICAM-3 but their ability to interact with the, evolutionarily very recent, human immunodeficiency virus-1 suggests that they may be able to serve as a PRR for other pathogens. As models of carbohydrate recognition systems on APCs, these molecules show that we have evolved complex and multifunctional mechanisms for the recognition of foreign carbohydrates and "self" antigens with quite diverse consequences. Coligation of additional surface receptors, such as those for complement and immunoglobulin, and the recruitment of potent signaling molecules such as the Toll-like receptors may provide the additional level of "decision-making" that ensures the appropriate cellular responses are achieved distinguishing "self" from "nonself".

References

1. Medzhitov R, Janeway Jr CA. Innate immunity: The virtues of a nonclonal system of recognition. Cell 1997; 91:295-298.
2. Taylor ME, Conary JT, Lennartz MR et al. Primary structure of the mannose receptor contains multiple motifs resembling carbohydrate-recognition domains. J Biol Chem 1990; 265:12156-12162.
3. Ezekowitz RA, Sastry K, Bailly P et al. Molecular characterization of the human macrophage mannose receptor: Demonstration of multiple carbohydrate recognition-like domains and phagocytosis of yeasts in Cos-1 cells. J Exp Med 1990; 172:1785-1794.
4. Sheikh H, Yarwood H, Ashworth A et al. Endo180, an endocytic recycling glycoprotein related to the macrophage mannose receptor is expressed on fibroblasts, endothelial cells and macrophages and functions as a lectin receptor. J Cell Sci 2000; 113:1021-1032.
5. Lambeau G, Ancian P, Barhanin J et al. Cloning and expression of a membrane receptor for secretory phospholipases A2. J Biol Chem 1994; 269:1575-1578.
6. Ishizaki J, Hanasaki K, Higashino K et al. Molecular cloning of pancreatic group I phospholipase A2 receptor. J Biol Chem 1994; 269:5897-5904.
7. Jiang W, Swiggard WJ, Heufler C et al. The receptor DEC-205 expressed by dendritic cells and thymic epithelial cells is involved in antigen processing. Nature 1995; 375:151-155.
8. Pontow SE, Kery V, Stahl PD. Mannose receptor. Int Rev Cytol 1992:221-244.
9. Linehan SA, Martinez-Pomares L, Stahl PD et al. Mannose receptor and its putative ligands in normal murine lymphoid and nonlymphoid organs: In situ expression of mannose receptor by selected macrophages, endothelial cells, perivascular microglia and mesangial cells, but not dendritic cells. J Exp Med 1999; 189:1961-1972.
10. Sallusto F, Cella M, Danieli C et al. Dendritic cells use macropinocytosis and the mannose receptor to concentrate macromolecules in the major histocompatibility complex class II compartment: Downregulation by cytokines and bacterial products. J Exp Med 1995; 182:389-400.
11. Engering AJ, Cella M, Fluitsma D et al. The mannose receptor functions as a high capacity and broad specificity antigen receptor in human dendritic cells. Eur J Immunol 1997; 27:2417-2425.
12. Wollenberg A, Mommaas M, Oppel T et al. Expression and function of the mannose receptor CD206 on epidermal dendritic cells in inflammatory skin diseases. J Invest Dermatol 2002; 118:327-334.
13. Warr GA A macrophage receptor for (mannose/glucosamine)-glycoproteins of potential importance in phagocytic activity. Biochem Biophys Res Commun 1980; 93:737-745.
14. Karbassi A, Becker JM, Foster JS et al. Enhanced killing of Candida albicans by murine macrophages treated with macrophage colony-stimulating factor: Evidence for augmented expression of mannose receptors. J Immunol 1987; 139:417-421.
15. Sung SS, Nelson RS, Silverstein SC. Yeast mannans inhibit binding and phagocytosis of zymosan by mouse peritoneal macrophages. J Cell Biol 1983; 96:160-166.
16. Blackwell JM, Ezekowitz RA, Roberts MB et al. Macrophage complement and lectin-like receptors bind Leishmania in the absence of serum. J Exp Med 1985; 162:324-331.
17. Wilson ME, Pearson RD. Roles of CR3 and mannose receptors in the attachment and ingestion of Leishmania donovani by human mononuclear phagocytes. Infect Immun 1988; 56:363-369.
18. Ezekowitz RA, Williams DJ, Koziel H et al. Uptake of Pneumocystis carinii mediated by the macrophage mannose receptor. Nature 1991; 351:155-158.
19. Reading PC, Miller JL, Anders EM. Involvement of the mannose receptor in infection of macrophages by influenza virus. J Virol 2000; 74:5190-5197.
20. Lee SJ, Evers S, Roeder D, Parlow AF et al. Mannose receptor-mediated regulation of serum glycoprotein homeostasis. Science 2002; 295:1898-1901.
21. Irjala H, Johansson EL, Grenman R et al. Mannose receptor is a novel ligand for L-selectin and mediates lymphocyte binding to lymphatic endothelium. J Exp Med 2001; 194:1033-1042.
22. Fiete DJ, Beranek MC, Baenziger JU. A cysteine-rich domain of the "mannose" receptor mediates GalNAc-4-SO4 binding. Proc Natl Acad Sci USA 1998; 95:2089-2093.
23. Martinez-Pomares L, Crocker PR, Da Silva R et al. Cell-specific glycoforms of sialoadhesin and CD45 are counter-receptors for the cysteine-rich domain of the mannose receptor. J Biol Chem 1999; 274:35211-35218.
24. Martinez-Pomares L, KoscoVilbois M, Darley E et al. Fc chimeric protein containing the cysteine-rich domain of the murine mannose receptor binds to macrophages from splenic marginal zone and lymph node subcapsular sinus and to germinal centers. J Exp Med 1996; 184:1927-1937.

25. Martinez-Pomares L, Mahoney JA, Kaposzta R et al. A functional soluble form of the murine mannose receptor is produced by macrophages in vitro and is present in mouse serum. J Biol Chem 1998; 273:23376-23380.
26. Simpson DZ, Hitchen PG, Elmhirst EL et al. Multiple interactions between pituitary hormones and the mannose receptor. Biochem J 1999; 343:403-411.
27. Feinberg H, Park-Snyder S, Kolatkar AR et al. Structure of a C-type carbohydrate recognition domain from the macrophage mannose receptor. J Biol Chem 2000; 275:21539-21548.
28. Liu Y, Chirino AJ, Misulovin Z et al. Crystal structure of the cysteine-rich domain of mannose receptor complexed with a sulfated carbohydrate ligand. J Exp Med 2000; 191:1105-1116.
29. Engelholm, LH, Nielsen BS, Dano K et al. The urokinase receptor associated protein (uPARAP/endo180): A novel internalization receptor connected to the plasminogen activation system. Trends Cardiovasc Med 2001; 11:7-13.
30. Bailey L, Wienke D, Howard M et al. Investigation of the role of Endo180/urokinase-type plasminogen activator receptor-associated protein as a collagenase 3 (matrix metalloproteinase 13) receptor. Biochem J 2002; 363:67-72.
31. Mahnke K, Guo M, Lee S et al. The dendritic cell receptor for endocytosis, DEC-205, can recycle and enhance antigen presentation via major histocompatibility complex class II-positive lysosomal compartments. J Cell Biol 2000; 151:673-684.
32. Geijtenbeek TB, Torensma R, van Vliet SJ et al. Identification of DC-SIGN, a novel dendritic cell-specific ICAM-3 receptor that supports primary immune responses. Cell 2000; 100:575-585.
33. Soilleux EJ, Barten R, Trowsdale J. DC-SIGN; A related gene, DC-SIGNR; and CD23 form a cluster on 19p13. J Immunol 2000; 165:2937-2942.
34. Valladeau J, Ravel O, Dezutter-Dambuyant C et al. Langerin, a novel C-type lectin specific to Langerhans cells, is an endocytic receptor that induces the formation of Birbeck granules. Immunity 2000; 12:71-81.
35. Fernandes MJ, Finnegan AA, Siracusa LD et al. Characterization of a novel receptor that maps near the natural killer gene complex: Demonstration of carbohydrate binding and expression in hematopoietic cells. Cancer Res 1999; 59:2709-2717.
36. Ariizumi K, Shen GL, Shikano S et al. Cloning of a second dendritic cell-associated C-type lectin (dectin-2) and its alternatively spliced isoforms. J Biol Chem 2000; 275:11957-11963.
37. Mitchell DA, Fadden AJ, Drickamer K. A novel mechanism of carbohydrate recognition by the C-type lectins DC-SIGN and DC-SIGNR. Subunit organization and binding to multivalent ligands. J Biol Chem 2001; 276:28939-28945.
38. Feinberg H, Mitchell DA, Drickamer K et al. Structural basis for selective recognition of oligosaccharides by DC-SIGN and DC-SIGNR. Science 2001; 294:2163-2166.
39. Bashirova AA, Geijtenbeek TB, van Duijnhoven GC et al. A dendritic cell-specific intercellular adhesion molecule 3-grabbing nonintegrin (DC-SIGN)-related protein is highly expressed on human liver sinusoidal endothelial cells and promotes HIV-1 infection. J Exp Med 2001; 193:671-678.
40. Geijtenbeek TB, Kwon DS, Torensma R et al. DC-SIGN, a dendritic cell-specific HIV-1-binding protein that enhances trans-infection of T cells. Cell 2000; 100:587-597.
41. Pohlmann S, Baribaud F, Lee B et al. DC-SIGN interactions with human immunodeficiency virus type 1 and 2 and simian immunodeficiency virus. J Virol 2001; 75:4664-4672.
42. Geijtenbeek TB, Krooshoop DJ, Bleijs DA et al. DC-SIGN-ICAM-2 interaction mediates dendritic cell trafficking. Nat Immunol 2000; 1:353-357.
43. Engering A, Geijtenbeek TB, van Vliet SJ et al. The dendritic cell-specific adhesion receptor DC-SIGN internalizes antigen for presentation to T cells. J Immunol 2002; 168:2118-2126.
44. Park CG, Takahara K, Umemoto E et al. Five mouse homologues of the human dendritic cell C-type lectin, DC-SIGN. Int Immunol 2001; 13:1283-1290.
45. Valladeau J, Clair-Moninot V, Dezutter-Dambuyant C et al. Identification of mouse langerin/CD207 in Langerhans cells and some dendritic cells of lymphoid tissues. J Immunol 2002; 168:782-792.
46. Bonkobara M, Zukas PK, Shikano S et al. Epidermal Langerhans cell-targeted gene expression by a dectin-2 promoter. J Immunol 2001; 167:6893-6900.
47. Reis e Sousa C, Stahl PD, Austyn JM. Phagocytosis of antigens by Langerhans cells in vitro. J Exp Med 1993; 178:509-519.
48. Williams DL, Mueller A, Browder W. Glucan-based macrophage stimulators. Clin Immunother 1996; 5:392-399.
49. Williams DL. Overview of (1,3)-beta-D-glucan immunobiology. Mediators of Inflammation 1997; 6:247-250.
50. Reiss E, Obayashi T, Orle K et al. Nonculture based diagnostic tests for mycotic infections. Med Mycol 2000; 38:147-159.

51. Zimmerman JW, Lindermuth J, Fish PA et al. A novel carbohydrate-glycosphingolipid interaction between a beta-(1-3)-glucan immunomodulator, PGG-glucan, and lactosylceramide of human leukocytes. J Biol Chem 1998; 273:22014-22020.
52. Pearson A, Lux A, Krieger M. Expression cloning of dSR-CI, a class C macrophage-specific scavenger receptor from Drosophila melanogaster. Proc Natl Acad Sci USA 1995; 92:4056-4060.
53. Ross GD, Cain JA, Lachmann PJ. Membrane complement receptor type three (CR3) has lectin-like properties analogous to bovine conglutinin as functions as a receptor for zymosan and rabbit erythrocytes as well as a receptor for iC3b. J Immunol 1985; 134:3307-3315.
54. Thornton BP, Vetvicka V, Pitman M et al. Analysis of the sugar specificity and molecular location of the beta-glucan-binding lectin site of complement receptor type 3 (CD11b/CD18). J Immunol 1996; 156:1235-1246.
55. Brown GD, Gordon S. Immune recognition. A new receptor for beta-glucans. Nature 2001; 413:36-37.
56. Ross GD. Regulation of the adhesion versus cytotoxic functions of the Mac-1/CR3/alphaMbeta2-integrin glycoprotein. Crit Rev Immunol 2000; 20:197-222.
57. Underhill DM, Ozinsky A. Phagocytosis of microbes: Complexity in action. Annu Rev Immunol 2002; 20:825-852.
58. Perlmann P, Perlmann H, Muller-Eberhard HJ. Cytolytic lymphocytic cells with complement receptor in human blood. Induction of cytolysis by IgG antibody but not by target cell-bound C3. J Exp Med 1975; 141:287-296.
59. Vetvicka V, Thornton BP, Ross GD. Soluble beta-glucan polysaccharide binding to the lectin site of neutrophil or natural killer cell complement receptor type 3 (CD11b/CD18) generates a primed state of the receptor capable of mediating cytotoxicity of iC3b-opsonized target cells. J Clin Invest 1996; 98:50-61.
60. Yan J, Vetvicka V, Xia Y et al. Beta-glucan, a "specific" biologic response modifier that uses antibodies to target tumors for cytotoxic recognition by leukocyte complement receptor type 3 (CD11b/CD18). J Immunol 1999; 163:3045-3052.
61. Xia Y, Vetvicka V, Yan J et al. The beta-glucan-binding lectin site of mouse CR3 (CD11b/CD18) and its function in generating a primed state of the receptor that mediates cytotoxic activation in response to iC3b-opsonized target cells. J Immunol 1999; 162:2281-2290.
62. Xia Y, Ross GD. Generation of recombinant fragments of CD11b expressing the functional beta-glucan-binding lectin site of CR3 (CD11b/CD18). J Immunol 1999; 162:7285-7293.
63. Kadish JL, Choi CC, Czop JK. Phagocytosis of unopsonized zymosan particles by trypsin-sensitive and beta-glucan-inhibitable receptors on bone marrow-derived murine macrophages. Immunol Res 1986; 5:129-138.
64. Czop JK, Fearon DT, Austen KF. Opsonin-independent phagocytosis of activators of the alternative complement pathway by human monocytes. J Immunol 1978; 120:1132-1138.
65. Ariizumi K, Shen GL, Shikano S et al. Identification of a novel, dendritic cell-associated molecule, dectin-1, by subtractive cDNA cloning. J Biol Chem 2000; 275:20157-20167.
66. Willment JA, Gordon S, Brown GD. Characterisation of the human beta-glucan receptor and its alternatively spliced isoforms. J Biol Chem 2001; 276:43818-43823.
67. Brown GD, Taylor PR, Reid DM et al. Dectin-1 is a major beta-glucan receptor on macrophages. J Exp Med 2002; 196:407-412.
68. Underhill DM, Ozinsky A, Hajjar AM et al. The Toll-like receptor 2 is recruited to macrophage phagosomes and discriminates between pathogens. Nature 1999; 401:811-815.
69. Kelm S, Schauer R. The galactose-recognizing system of rat peritoneal macrophages; Identification and characterization of the receptor molecule. Biol Chem Hoppe Seyler 1988; 369:693-704.
70. Sato M, Kawakami K, Osawa T et al. Molecular cloning and expression of cDNA encoding a galactose/N-acetylgalactosamine-specific lectin on mouse tumoricidal macrophages. J Biochem (Tokyo) 1992; 111:331-336.
71. Suzuki N, Yamamoto K, Toyoshima S et al. Molecular cloning and expression of cDNA encoding human macrophage C-type lectin. Its unique carbohydrate binding specificity for Tn antigen. J Immunol 1996; 156:128-135.
72. Lehrman MA, Hill RL. The binding of fucose-containing glycoproteins by hepatic lectins. Purification of a fucose-binding lectin from rat liver. J Biol Chem 1986; 261:7419-7425.
73. Hoyle GW, Hill RL. Structure of the gene for a carbohydrate-binding receptor unique to rat kupffer cells. J Biol Chem 1991; 266:1850-1857.
74. Biessen, EA, Bakkeren HF, Beuting DM et al. Ligand size is a major determinant of high-affinity binding of fucose- and galactose-exposing (lipo)proteins by the hepatic fucose receptor. Biochem J 1994; 299:291-296.
75. Varki A. Selectin ligands. Proc Natl Acad Sci USA 1994; 91:7390-7397.

76. Rosen SD, Bertozzi CR. Two selectins converge on sulfate. Leukocyte adhesion. Curr Biol 1996; 6:261-264.
77. Arbones ML, Ord DC, Ley K et al. Lymphocyte homing and leukocyte rolling and migration are impaired in L-selectin-deficient mice. Immunity 1994; 1:247-260.
78. Schlesinger PH, Doebber TW, Mandell BF et al. Plasma clearance of glycoproteins with terminal mannose and N-acetylglucosamine by liver nonparenchymal cells. Studies with beta-glucuronidase, N-acetyl-beta-D-glucosaminidase, ribonuclease B and agalacto-orosomucoid. Biochem J 1978; 176:103-109.
79. Stahl PD, Rodman JS, Miller MJ et al. Evidence for receptor-mediated binding of glycoproteins, glycoconjugates, and lysosomal glycosidases by alveolar macrophages. Proc Natl Acad Sci USA 1978; 75:1399-1403.
80. Crocker PR, Kelm SC, Dubois C et al. Purification and properties of sialoadhesin, a sialic acid-binding receptor of murine tissue macrophages. EMBO J 1991; 10:1661-1669.

Chapter 6

Toll-Like Receptor:
Specificity and Signaling

Osamu Takeuchi, Tsuneyasu Kaisho and Shizuo Akira

Abstract

Toll-like receptors (TLRs) are type I transmembrane proteins consisting of extracellular leucine-rich repeats and a cytoplasmic domain resembling that of the Interleukin 1 receptor. Different TLRs recognize distinct pathogen-associated molecular patterns, such as lipopolysaccharide, lipoprotein and bacterial DNA. Ligand recognition by the respective TLRs can activate not only a common signaling pathway leading to NF-κB activation but also TLR-specific pathways. Mouse models lacking TLRs and their signaling molecules have greatly contributed to the clarification of the innate immune system. In this article, we review recent progress in the study of how TLRs establish immune responses against various pathogenic stimuli.

Introduction

The invasion of pathogens evokes immune responses which can discriminate self from nonself. The mammalian immune system is divided into two classes: innate and acquired. Acquired immunity can defend against pathogens the host has never experienced, by detecting them with rearranged antigen receptors. The system is advantageous in that the receptor can recognize a specific pathogen with high affinity and that acquired immune cells can persist as memory cells in the host. In contrast, the innate immune system is characterized by the usage of germline-encoded pattern recognition receptors to detect deleterious pathogens.[1] In mammals, the innate immune recognition system can further be divided into two subclasses. One recognizes pathogens in the blood stream. This is represented by a lectin pathway, which includes mannose-binding lectin (MBL) that recognizes carbohydrates such as mannose and N-acetylglucosamine on the surface of pathogens.[2] Upon binding of a pathogen to MBL, an MBL-associated serine protease (MASP), which complexes with MBL, mediates the activation of the complement system. The reactions cause the lysis of microorganisms without any gene induction in host cells. The other system is mediated through transmembrane receptors, termed Toll-like receptors (TLRs). Different TLRs are responsible for the recognition of a variety of pathogen-associated molecular patterns (PAMPs). TLRs are mainly expressed on the surface of various cells such as monocytes/macrophages, dendritic cells (DCs), vascular endothelial cells and epithelial cells. Stimulation with pathogen components initiates the activation of intracellular signaling pathways via TLRs, leading to cytokine production, and augmenting surface expression of costimulatory molecules, thereby activating and modulating acquired immunity. In this regard, the TLR recognition system is a bridge between innate and acquired immunity. This chapter focuses on the specificity and chemical structure of TLR ligands and intracellular signaling pathways of TLRs.

Immunobiology of Carbohydrates, edited by Simon Y.C. Wong and Gemma Arsequell.
©2003 Eurekah.com and Kluwer Academic / Plenum Publishers.

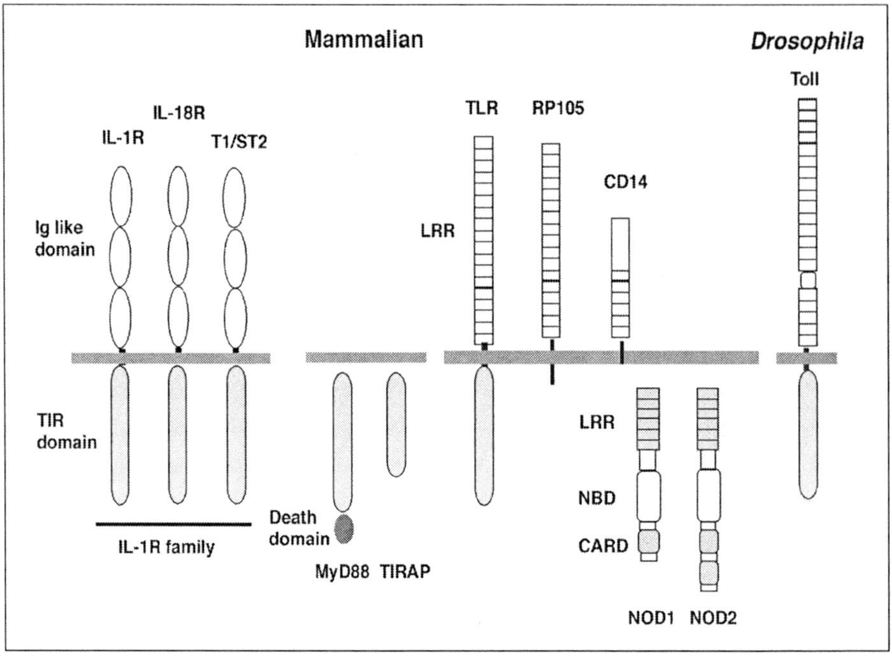

Figure 1. Proteins characterized by a TLR domain and LRR. A TIR domain is shared by IL-1R family members, TLRs, MyD88 and TIRAP. Ig-like modules in the extracellular region distinguish IL-1R family from TLR. In addition to TLRs, RP105, CD14, and NOD family members feature LRR motifs.

Toll-Like Receptor Family and Relatives

Toll was originally identified as a *Drosophila* gene essential for dorsoventral patterning in early embryogenesis.[3] Subsequent studies disclosed critical roles for Toll protein in anti-fungal and anti-Gram-positive bacterial host defense.[4] Toll is a type I transmembrane protein that possesses an extracellular leucine-rich repeat (LRR) domain and a cytoplasmic domain homologous to that of the mammalian Interleukin-1 receptor (IL-1R). In *Drosophila*, fungi and Gram-positive bacteria can activate protease cascades which mediate the processing of an endogenous protein, Spaetzle, from its precursor.[5] Subsequent ligation of Spaetzle to Toll initiates the activation of an intracellular signaling cascade leading to the induction of genes encoding anti-fungal or anti-bacterial peptides.

Ten mammalian counterparts of Toll have been discovered and designated TLR1-10.[6-11] They are structurally related to Toll, indicating the evolutionally conserved roles of TLRs and Toll (Fig. 1). IL-1R family members also possess a Toll/IL-1R (TIR) domain, although they contain an immunoglobulin-like domain in their extracellular portion. The family includes IL-1R accessory protein, IL-18R and T1/ST2.[12,13] In addition to transmembrane proteins, several cytoplasmic proteins, myeloid differentiation factor 88 (MyD88) and TIR domain-containing adaptor protein (TIRAP), also contain a TIR domain.[14,15] These proteins function as adaptors in TLR/IL-1R signaling.

A diverse group of proteins also contains LRR without a TIR domain. For instance, CD14 carries LRR and is embedded into the cell membrane through a glycosylphosphatidylinositol (GPI) anchor.[16,17] In addition, RP105 is a transmembrane protein expressed on the surface of B cells and consists of an extracellular LRR domain and a short cytoplasmic tail.[18] Furthermore, nucleotide-binding oligomerization domain (NOD) proteins contain N-terminal caspase-recruitment domains (CARDs), a nucleotide-binding domain, and LRR.[19] Several

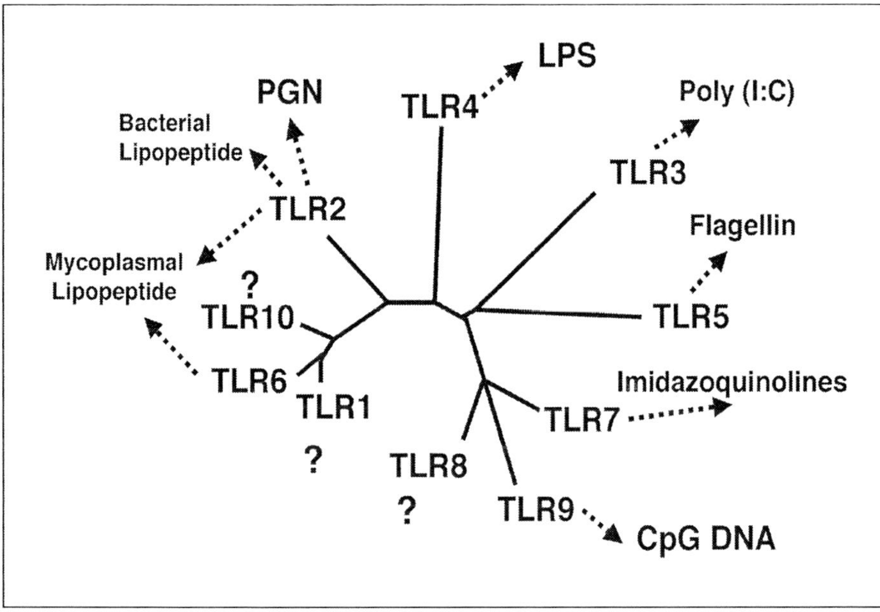

Figure 2. Phylogenetic analysis of TLR family members and their ligands. Protein sequences of human TLR1 to TLR10 were aligned and a nonrooted phylogenic tree was constructed by the neighbor-joining method. Branch lengths were proportional to evolutionary distances. The arrows indicate ligands for TLRs. TLR2 and TLR6 are required for MALP-2 signaling. PolyI:C, LPS, flagellin, imidazoquinolines and CpG-DNA are recognized by TLR3, 4, 5, 7, and 9, respectively.

reports have shown that a frameshift mutation and two missense mutations of NOD2 are associated with Crohn's disease, a chronic inflammatory bowel disease.[20,21] These proteins with LRR are also involved in immune responses (see below).

Pathogen-Associated Molecular Patterns: Targets of Sentinels

The vast number of microorganisms that pose potential threats to human health can be classified into several groups, such as Gram-positive bacteria, Gram-negative bacteria, mycobacteria, fungi and viruses. Each of the subgroup members possesses several common structural features, called PAMPs, which are not found in the host.[1] These include lipopolysaccharide (LPS) from Gram-negative bacteria, lipoteichoic acid (LTA) from Gram-positive bacteria, mycobacterial lipoarabinomannan (LAM), peptidoglycan (PGN), lipoprotein, flagellin, unmethylated DNA with a CpG motif (CpG-DNA) and viral double stranded RNA (dsRNA). Host immune cells target such common nonself structures to sense the invasion of pathogens promptly by a limited number of innate immune receptors. PAMP recognition by distinct TLRs leads to the initiation of intracellular signaling cascades which ultimately results in the production of a set of proinflammatory cytokines, such as tumor necrosis factor α (TNFα), IL-1, IL-6 and IL-12. TLR signaling also induces maturation of DCs and makes them competent for supporting mainly T helper 1 (Th1) cell development. So far, ligands for TLR2, TLR3, TLR4, TLR5, TLR6, TLR7 and TLR9 have been reported (Fig. 2). In the following sections, we review the ligand specificities of each TLR.

TLR2

Bacterial lipoproteins, one of most important PAMPs, are produced by many different species of bacterial pathogens and are known to induce the production of cytokines in macrophages.[22] All lipoproteins contain a lipoylated N-terminal amino acid residue. Lipoproteins prepared from a variety of pathogens were shown to stimulate cells through TLR2. These include *Mycobacterium tuberculosis*, *Borrelia burgdorferi*, *Treponema pallidum* and *Mycoplasma fermentans*. Furthermore, synthetic lipopeptides mimicking the N-terminal end of bacterial lipoproteins also stimulate TLR2-expressing cells.[23-26] Moreover, TLR2-deficient macrophages lack cytokine production induced by these lipoprotein/lipopeptides.[27,28] These results indicate that TLR2 intrinsically recognizes chemical structures consisting of fatty acid and several amino acids.

PGN, another PAMP, is a polymer of alternating N-acetylglucosamine and N-acetylmuramic acid cross-linked by short peptides. It is a major component for the cell wall of most bacteria, although it is more abundant in Gram-positive bacteria.[29] PGN prepared from Gram-positive bacteria also activates TLR2-expressing cells.[30,31] TLR2-deficient macrophages have severely impaired responses to PGN prepared from *S. aureus*. This indicates the essential role of TLR2 in PGN recognition.[32] However, PGN preparations used in these studies were isolated from bacteria, and there is no report showing that synthetic PGN activates cells via TLR2. Therefore, it is still possible that some unknown structure(s) contaminating these PGN preparations, and not the simple GlcNAc and MurNAc polymers themselves, stimulates cells via TLR2. Further investigation will be needed to identify the exact chemical structure of PGN recognized by TLR2.

Although TLR2 was implicated in LPS recognition, this has been attributed to "endotoxin-protein" contaminant(s) of LPS preparations.[33] However, some reports have shown that the involvement of TLR2 in the recognition of LPSs purified from *Porphyromonas gingivalis* and *Leptospira interrogans*. Notably, these LPSs have a different structure from enterobacteria-derived LPS.[25,34] Not only lipids but also proteins purified from bacteria are reported to stimulate cells via TLR2. These include neisserial porins and a *P. gingivalis* fimbrial protein.[35,36] Furthermore, TLR2 recognizes LAM from *M. tuberculosis* and GPI anchors purified from *Trypanosoma cruzi*, a protozoan parasite which is a causative agent of Chagas' disease.[37,38]

Several reports have also shown that TLR2 is involved in anti-bacterial host defense. For example, TLR2-deficient mice are highly susceptible to *S. aureus* infection.[39] Moreover, a 19kD-lipoprotein promotes the killing of *M. tuberculosis* via TLR2, indicating that TLR2-mediated signaling leads to the augmentation of bactericidal activity.[40] In summary, TLR2 recognizes a variety of PAMPs from a wide range of pathogens and TLR2 signaling has so far been shown to be critically important in the activation of host defense responses against Gram-positive bacterial and mycobacterial infections.

Possible Role of TLR3 in Double Stranded RNA Recognition

Double-stranded RNA (dsRNA) is produced during viral infection and can trigger anti-viral responses within the host cell.[41] Polyinosine-polycytidylic acid (polyI:C) is a synthetic RNA molecule that can mimic the action of viral dsRNA. The administration of polyI:C into cells induces a strong anti-viral response at least partly through the production of type I interferons (IFN).[42,43] Alexopoulou et al reported that polyI:C is possibly recognized by TLR3.[44] They showed that the overexpression of TLR3 on transfected 293 cells results in their ability to induce NF-κB activation in response to polyI:C. In addition, polyI:C-induced cytokine production was partially impaired in TLR3-deficient macrophages. However, TLR3-deficient mice can still respond to polyI:C to some extent, which is in contrast to the abolished responses of other TLR-deficient mice to their specific ligands. For example, responses to LPS and CpG-DNA are completely abrogated in TLR4- and TLR9-deficient mice, respectively (as described below). Intracellular recognition of dsRNA leading to IFN production is thought to be mediated, at least in part, through dsRNA-dependent protein kinase, PKR, based on studies using an immortalized fibroblast cell line from PKR-deficient mice.[45-47] However, there has been one report indicating that polyI:C-induced activation of NF-κB occurred normally in embryonic fibroblasts from PKR-deficient mice, while the phosphorylation of eIF-2α was abrogated.[48] Further research will be needed to clarify how host cells recognize polyI:C.

TLR4

LPS, a major component of the outer membrane of Gram-negative bacteria, is one of the best known PAMPs.[49] Numerous studies have described its critical roles in Gram-negative septic shock. Recently, several studies have clarified that TLR4 is essential for LPS recognition. First, a positional cloning strategy has identified a gene responsible for LPS recognition in C3H/HeJ and C57BL/10ScCr mice, both of which are known to be hyporesponsive or unresponsive to LPS. In C3H/HeJ mice, a missense point mutation resulted in the replacement of the highly conserved proline residue with histidine in the cytoplasmic domain of the TLR4 protein, converting it into a dominant negative form.[50] In C57BL/10ScCr mice, the *Tlr4* locus was entirely deleted.[51] Furthermore, TLR4-deficient mice were resistant to LPS-induced shock.[52] In TLR4-deficient cells, LPS-induced cellular responses as well as the activation of signaling molecules was completely abrogated.

On the cell surface, LPS is recognized by a complex mechanism. LPS is first captured by LPS-binding protein (LBP) in the blood, and then the LPS-LBP complex binds CD14, a GPI-anchored protein. LBP and CD14 function as amplifiers for LPS recognition. However, they are not essential for LPS-mediated effects, because mutant mice lacking LBP or CD14 retain LPS responsiveness.[53-55] Subsequently, the presence of LPS is detected on the target cell surface by a complex of TLR4 and a TLR4-binding extracellular protein, MD2.[56] A protein crosslinking experiment has revealed that LPS binds directly to TLR4 and MD2 in the presence of CD14.[57] Coexpression of TLR4 and MD2 greatly enhances the LPS-mediated NF-κB activation.[56] Moreover, mutant CHO cell lines having a point mutation in MD-2 were not activated by LPS, indicating a critical role of MD-2 in LPS recognition.[58]

There are several other candidates for molecules involved in LPS recognition. Ogata et al reported that targeted disruption of RP105 resulted in an impaired proliferative response of B cells to LPS and speculated that RP105 is a component of the LPS receptor complex in B cells.[59] Inohara et al also showed that overexpression of NOD1 or NOD2 in human embryonic kidney 293 cells conferred responsiveness to LPS.[21,60] They described that LPS-binding activity could be specifically coimmunoprecipitated with NOD1 from cytosolic extracts. It was also reported that a dominant-negative form of NOD1 blocked the activation of NF-κB induced by microinjection of LPS as well as infection by *Shigella flexneri*, implying that cytoplasmic LPS might be detected by NOD proteins.[61] Furthermore, a fluorescence resonance energy transfer study also revealed that several proteins including HSP70 and 90 form an activation cluster mediating the response to LPS.[62]

Various PAMPs other than LPS have been reported to activate cells via TLR4. A plant-derived anti-cancer drug, Taxol, mimics the response of LPS and can function as a mouse TLR4 ligand.[63] In addition, the F protein of respiratory syncytial virus (RSV) can also induce an immune response via a CD14- and TLR4-dependent pathway.[64] Macrophages from C3H/HeJ mice and C57BL10/ScCr mice displayed a defect in IL-6 production in response to RSV F protein, and RSV persisted longer in C57BL10/ScCr mice than control mice. In addition to PAMPs, TLR4 also recognizes endogenous products. Under inflammatory conditions, components of extracellular matrix are degraded or processed. These substances including fragments of fibronectin and degraded products of hyaluronic acid can stimulate cells though TLR4.[65,66] Furthermore, several studies also demonstrated that both human and chlamydial HSP60 are putative ligands for TLR2 and TLR4.[67,68]

TLR5

Recent studies demonstrated that bacterial flagellin is also recognized by epithelial cells as one of the PAMPs. Exposure of the basolateral surface of intestinal epithelial cells to purified flagellin activates NF-κB and leads to the production of cytokines such as IL-8.[69] Expression of

TLR5 in CHO cells resulted in a flagellin-induced activation of NF-κB, suggesting that TLR5 recognizes flagellin.[70] This is also underscored by the fact that mutant *S. typhimurium* strains with defective flagellin genes cannot activate cells expressing TLR5. Interestingly, TLR5 is expressed on the basolateral, but not apical, membrane of the intestinal epithelium.[71] This polarized expression of TLR5 indicates that intestinal bacteria can induce immune responses via TLR5, after they breach the epithelial barrier and reach the basolateral surface.

TLR6

Bacterial lipoproteins could be divided into two subgroups based on the difference in the pattern of lipoylation. A Gram-negative bacterial lipoprotein (BLP) has N-acyl S-diacyl cysteine at the N-terminal cysteinyl residue. In contrast, mycoplasmal MALP-2 from *M. fermentans* has only a diacylated cysteine, since mycoplasma lacks N-acyl-transferase responsible for the addition of a third fatty acid chain.[72,73] Analysis of TLR6-deficient mice revealed that TLR6 can discriminate di-acylated from tri-acylated lipoproteins.[28] TLR6-deficient macrophages did not produce a detectable level of cytokine in response to MALP-2, whereas the cells responded normally to tri-acylated lipopeptide. In contrast, TLR2-deficient cells did not respond to either MALP-2 or BLP, indicating that TLR2 does not distinguish between these two classes of lipopeptide. In addition, the expression of chimeric constructs consisting of extracellular TLR2 and cytoplasmic TLR6 (TLR2/TLR6 chimera) and the reciprocal (TLR6/TLR2) chimera in TLR2/TLR6 double mutant fibroblasts conferred MALP-2-induced IL-6 production. This demonstrates that both extracellular and cytoplasmic regions of TLR2 and TLR6 are required for the recognition and signaling of MALP-2. These results demonstrate that TLR6 and TLR2 act together to recognize MALP-2, and TLR6 is responsible for recognizing the subtle structural difference between MALP-2 and BLP. Ozinsky et al also showed the cooperation between TLR2 and TLR6 recognizing bacterial ligand by expressing dominant-negative versions of TLR6 and TLR2 in RAW cells.[74] They showed that TLR2 associates with TLR6 in the absence of ligand, suggesting that a heterodimer formation may be crucial for ligand recognition. Recently, a heat-labile soluble factor released by Group B *Streptococcus* was also reported to activate cells via TLR6 and TLR2.[75] Further study will clarify the chemical structures recognized by TLR6/TLR2 heterodimers.

TLR7

TLR7 is structurally related to TLR9 and its gene locus is mapped to the X chromosome.[9] A recent study implicated TLR7 in antiviral host defense. Imidazoquinolines are small chemical compounds possessing immunostimulating properties.[76] They are known to have antiviral activity and one of them, Imiquimod is now clinically used for the treatment of genital warts caused by human papilloma virus.[77] A derivative of Imiquimod, R-848, is more potent than Imiquimod and is expected to be more effective as an antiviral drug.[78] The antiviral activity of Imiquimod in mice is ascribed to a unique ability to induce IFNα production. Imidazoquinolines induce the production of proinflammatory cytokines in macrophages, activate B cell blastogenesis and maturate DCs. When injected into mice, they increase serum levels of cytokines including IFNα. It has been shown that all of these biological activities are abolished in TLR7-deficient mice as well as MyD88-deficient mice.[79] In addition, expression of human TLR7 in 293 cells confers R-848-induced NF-κB activation, indicating that human and mouse TLR7 recognize the same ligand. It remains unknown the biological significance, if any, of TLR7 recognizing chemically synthesized compounds. It is likely that these compounds mimic biological TLR7 ligands which may be virus- or host-derived products that are generated upon viral infections.

TLR9

Prokaryotic DNA is different from that of the host in terms of the frequency of CpG motifs and their degree of methylation.[80] Unmethylated CpG motifs, are present at much higher frequency in bacterial DNA than in mammalian DNA. Bacterial DNA triggers an immune

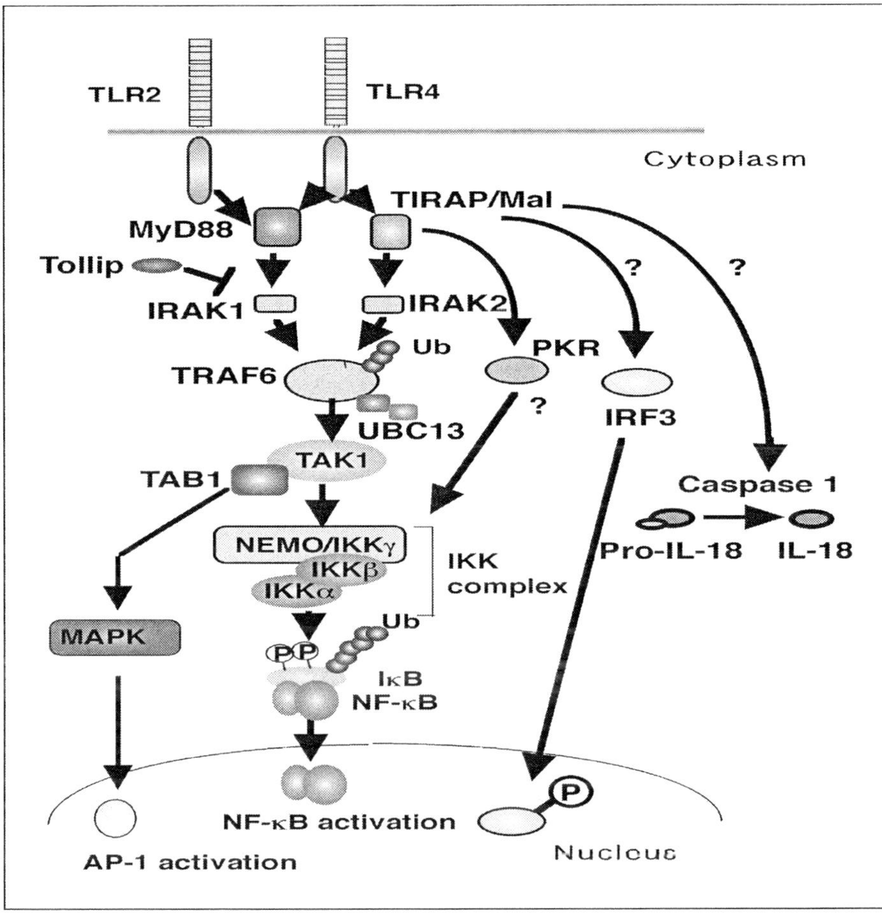

Figure 3. Schematic representation of TLR signaling pathway. TLR2 ligand-induced NF-κB activation is fully MyD88-dependent, whereas TLR4 activates both MyD88-dependent and -independent signaling pathways. In addition, TLR4 signaling induces the activation of IRF3 as well as the processing of pro-IL-18.

response represented by cytokine production and Th1 cell differentiation. Synthetic oligonucleotides containing unmethylated CpG motifs mimic the immunostimulatory effect of bacterial DNA, whereas methylated ones do not exhibit such activity. TLR9-deficient mice lack all CpG-DNA-induced activities, such as the induction of proinflammatory cytokines and B cell proliferative responses, and the activation of NF-κB. This demonstrates that TLR9 is the sole receptor for CpG-DNA.[81] Human and mouse cells respond differentially to distinct CpG-DNAs. Genetic complementation studies using human and mouse TLR9 genes clarified that TLR9 is involved in species-specific responses to CpG-DNAs.[82]

Taken together, accumulating evidence indicates that PAMPs from viruses, bacteria, parasites and endogenous proteins activate immune cells via distinct TLRs. However, the ligands for TLR1, 8 and 10 are still unknown. Further study should identify these ligands.

TLR Signaling Pathway

The TIR domain containing type I receptors shares at least one common signaling pathway involving MyD88, the nuclear translocation of NF-κB and the activation of MAP kinases such as c-Jun N-terminal kinase (JNK) and p38 (Fig. 3). Upon stimulation by a ligand, recruitment of MyD88 to the IL-1R complex or TLR occurs, which then triggers autophosphorylation of IL-1R associated kinase (IRAK).[83-86] Prior to stimulation with a ligand, IRAK is associated with a recently identified molecule referred to as Tollip. The IRAK-Tollip complex is recruited to the activated receptor complex, where the dissociation of IRAK from Tollip occurs.[87] Overexpression of Tollip suppresses IL-1β, TLR2 and TLR4-mediated NF-κB activation, suggesting that Tollip acts as a regulator for several TIR-domain containing receptors.[87,88] Activated IRAK is then released from the receptor, binds and activates TNFR-associated factor 6 (TRAF6), which stimulates the IκB kinase (IKK) complex and MAP kinase. Experiments with TRAF6-deficient mice demonstrated that TRAF6 is essential for cellular responses to IL-1, LPS, CD40 ligand and RANK.[89,90] TRAF6 activates two proteins involved in the ubiquitination process, Ubc13 and Uev1A.[91] Interestingly, although polyubiquitination usually leads to proteasome-mediated protein degradation, ubiquitination via this Ubc complex does not. Ubiqutinated TRAF6 activates a complex of TGFβ activating kinase 1 (TAK1) and associated proteins, TAB1 and TAB2, to phosphorylate the IKK complex directly.[92] The activated IKK complex phosphorylates IκB, leading to proteasome-mediated degradation of IκB, and then NF-κB translocates from the cytosol to the nucleus to mediate the expression of proinflammatory cytokines.[93] Activated TAK1 also phosphorylates MKK6, leading to the activation of JNK and p38 kinases, suggesting that the signaling pathway leading to activation of MAP kinase diverges from that of NF- κB at the TAK1 level.[92]

In addition, atypical protein kinase C (ζPKC) and its interacting protein, p62, were also involved in IKK activation in the IL-1 signaling pathway by interacting with TRAF6.[94] Recent studies in knockout animals disclosed a requirement of PKC family members for the activation of NF-κB-dependent gene expression. Leitges et al generated ζPKC-deficient mice and showed that ζPKC is critical for κB-dependent transcriptional activity as well as for the phosphorylation of NF-κB p65.[95] Notably, TNFα- and IL-1-induced IKK activation was attenuated in ζPKC–deficient lung, whereas it was not impaired in embryonic fibroblasts. Castrillo et al established PKCε-deficient mice and demonstrated that macrophages from these mice exhibited severely reduced proinflammatory cytokine production in response to LPS.[96] Activation of the IKK complex as well as nuclear translocation of NF-κB was also impaired in PKCε-deficient macrophages. Increasing number of molecules are reported to participate in TLR/IL-1R signaling coupled with the activation of NF-κB and MAP kinases. The next attractive issue is the analysis of cell- and stimulation-specific signaling pathways.

Essential Role of MyD88 in TLR/IL-1 Signaling

MyD88 consists of a N-terminal death domain and C-terminal TIR domain.[97] MyD88 associates with TLR/IL-1R by a homophilic interaction through a TIR domain. Then a death domain of MyD88 recruits IRAK to activate the signaling cascade.

Analysis of MyD88-deficient mice revealed an essential role for MyD88 in the signaling through various TLR/IL-1R family members. MyD88-deficient macrophages did not produce any cytokines in response to IL-1R/TLR ligands including IL-1, IL-18, TLR2 ligands such as PGN and MALP-2, a TLR4 ligand LPS, TLR7 ligand, imidazoquinolines and a TLR9 ligand CpG-DNA.[27,79,98-101] Therefore, we concluded that MyD88 is an essential mediator for the signaling of TLR2, TLR4, TLR7, TLR9 and the IL-1R family. Treatment of MyD88-deficient mice with a TLR5 ligand, flagellin, also failed to produce IL-6.[70] A recent report showed that MyD88-deficient DCs failed to produce IL-12 and IL-6 in response to polyI:C.[44] However, polyI:C-induced IFN_{β} mRNA expression was detected in MyD88-deficient fibroblasts at levels comparable to that in wild-type cells (our unpublished observation). Further investigation and cautious interpretation of the results will be required to elucidate the role of the TLR-MyD88 pathway in the recognition of polyI:C. In addition to known TLR ligands, TNFα production in response to heat-killed Group B streptococcus was also abrogated in MyD88-deficient macrophages, but TLR2-, TLR4-, TLR6- and TLR9-deficient macrophages produced

TNFα normally.[75] Furthermore, Adachi et al showed that the liver injury and IL-12 production caused by infection with *Plasmodium berghei* was impaired in MyD88-deficient mice, suggesting that the TLR-MyD88 signaling pathway is critical for the recognition of this malaria-causing parasite.[102] However, TLR2, TLR4 and TLR6-deficient mice showed normal responses to *P. berghei* infection. These observations suggest that the PAMPs on Group B streptococcus and *P. berghei* might be recognized by TLR(s) other than TLR2, 4, 6 and 9.

Furthermore, it was also shown that MyD88 plays a central role in anti-bacterial host defense. MyD88-deficient macrophages did not produce TNFα and IL-6 in response to heat-killed whole *S. aureus*. Accordingly, MyD88-deficient mice were highly susceptible to *S. aureus* infection.[39] After i.v. infection with 1 x 10^7 *S. aureus*, all MyD88-deficient mice died within 4 days, whereas all wild-type mice survived. Although TLR2-deficient mice were also susceptible to *S. aureus* infection, the phenotype of MyD88-deficient mice was much more severe than that of TLR2-deficient mice. The prominent susceptibility of MyD88-deficient mice to the inoculation of *S. aureus* suggests that MyD88 is critical for the elimination of invading bacteria in the early stage of infection, and prior to the development of acquired immunity.

To investigate the role of MyD88 in antigen-specific T cell responses, MyD88-deficient mice were immunized with ovalubumin (OVA) mixed with complete Freund's adjuvant (CFA). T cells obtained from the draining lymph nodes of MyD88-deficient mice did not proliferate or produce IFNγ in response to stimulation with OVA, indicating that the adjuvant-induced Th1 response is mediated through a MyD88-dependent signaling pathway.[103] Therefore, MyD88-dependent signaling is critical not only for innate immunity but also for an optimal acquired response.

TLR4-Specific Signaling Pathway

In vitro experiments suggest that MyD88 is an adaptor protein that mediates the association between TLR and IRAK. Consistent with this notion, MALP-2- or CpG-DNA-stimulated MyD88-deficient macrophages failed to activate signaling molecules such as IRAK, NF-κB and JNK.[27,101] However, further investigation revealed that LPS-induced activation of NF-κB or JNK was not abrogated, but delayed, in MyD88-deficient macrophages.[99] Since TLR4-deficient cells failed to activate both NF-κB and JNK in response to LPS, the MyD88-independent signaling pathway seems to originate from TLR4.

The MyD88-independent signaling pathway was further investigated using subtractive hybridization screening to identify LPS-inducible genes in MyD88-deficient macrophages. Prominent of these are the interferon (IFN)-inducible genes including genes for IP-10, GARG16 and IRG1.[104] We found that LPS activates IFN regulatory factor (IRF) 3 leading to the induction of the genes containing an IFN-stimulated regulatory element (ISRE) (eg. the IP-10 gene) in MyD88-deficient as well as in wild-type cells. In contrast, TLR2 ligand did not activate IRF3 or increase the expression of IFN-inducible genes. Thus, TLR4 signaling specifically induces a functional activation of IRF3 in a MyD88-independent manner.

Cytokines and bacterial components are known to induce the maturation of DCs whose hallmark is the ability to activate T cells. When MyD88-deficient bone marrow-derived DCs were cultured with LPS, they increased the amount of surface costimulatory molecules such as CD40, CD80 and CD86, and activated T cells to proliferate.[105] In contrast, DCs from TLR4-deficient or C3H/HeJ mice failed to maturate in response to LPS. Moreover, CpG-DNA did not cause MyD88-deficient DCs to mature, indicating that the MyD88-independent pathway originating from TLR4 is responsible for LPS-induced DC maturation. MyD88-independent LPS responses were also shown in other cells. Kupffer cells are known to store a IL-18 precursor protein in their cytosol and stimulation with LPS induced the maturation of IL-18 via cleavage with caspase-1.[106] MyD88-deficient cells still secreted IL-18, but not IL-1 and IL-12, upon stimulation with LPS, indicating that LPS activates caspase-1 in a MyD88-independent fashion.

Which molecules are involved in the MyD88-independent pathway? Another cytoplasmic protein containing a TIR domain, designated TIRAP/MAL, has been identified. Medzhitov and coworkers, who named this protein TIR domain containing adaptor protein (TIRAP), showed that its overexpression in 293 cells led to the activation of NF-κB and TIRAP associated with TLR4 and PKR.[15] Furthermore, the expression of a dominant negative version of TIRAP inhibited LPS-induced NF-κB activation and DC maturation, whereas it did not affect CpG-DNA signaling, suggesting that TIRAP is involved in a MyD88-independent pathway specific for TLR4. O'Neill and coworkers named the same protein MyD88-adaptor-like (MAL).[107] They showed that MAL activates NF-κB via IRAK2, but not IRAK1, by forming heterodimers with MyD88. Although both reports suggested the participation of TIRAP/MAL in MyD88-independent signaling, further investigation such as an analysis of adaptor-deficient mice is needed to clarify the functional role of this protein in the MyD88-independent pathway.

Perspectives

Among the ten identified TLRs, specific ligands for seven have been discovered. The ligands are diverse and include nucleotides, proteins and glycolipids. However, direct association of PAMPs with respective TLRs has been reported only in the case of LPS, and this is based on a ligand-receptor crosslinking experiment and species specificity. The next key issue to resolve is how LRR proteins recognize verified structures of PAMPs. In *Drosophila*, pattern recognition receptors circulating in the blood detect pathogens and subsequently activate a protease cascade to process an endogenous Toll ligand, Spatezle.[5] Although a human counterpart of Spaetzle has not been discovered, the existence of an intermediate protease cascade could still be hypothesized.

Future investigation is to clarify TLR-specific signaling pathways. We have discussed TLR4-specific signaling pathways leading to the induction of IFN-inducible genes where MyD88-independent activation of NF-κB and IRF3 is responsible. Although none of the TLR2, TLR7 and TLR9 ligands activated NF-κB and MAP kinases in MyD88-deficient mice, signaling of these TLRs results in the induction of a set of genes that are unique to a particular TLR. For example, TLR7 signaling results in the production of type I IFN in addition to proinflammatory cytokines. Similarly, TLR9 signaling has the potential to make macrophages secrete significantly more IL-12 p40 than other TLR ligands such as MALP-2 and LPS, although the production of TNFα and IL-6 is comparable. These observations imply the existence of common signaling pathways shared by all TLRs, and also currently unknown signaling pathways that are specifically induced by each TLR. Future studies may be able to identify the precise differences between each TLR signaling pathway. Finally, inflammatory diseases and severe infections are still a serious health problem. The manipulation of the TLR signaling pathways should be a powerful tool to control aberrant inflammation as well as infection.

Note Added in Proof

Analysis on TIRAP/MAL-deficient mice has clarified that TIRAP/MAL is involved in the MyD88-dependent pathway in TLR2 and TLR4 signaling.[108,109] Notably, LPS can induce NF-κB activation, IFN-inducible gene expression and DC maturation not only in TIRAP/MAL-deficient but also in MyD88 and TIRAP/MAL-double deficient mice.[109] Another adapter, TIR domain-containing adapter inducing IFN-β (TRIF), is a candidate molecule involved in the MyD88-independent pathway.[110]

Acknowledgement

We thank E. Horita for secretarial assistance.

References

1. Medzhitov R, Janeway Jr CA. Innate immunity: The virtues of a nonclonal system of recognition. Cell 1997; 91:295-298.
2. Matsushita M, Endo Y, Nonaka M et al. Complement-related serine proteases in tunicates and vertebrates. Curr Opin Immunol 1998; 10:29-35.
3. Hashimoto C, Hudson KL, Anderson KV. The toll gene of *Drosophila*, required for dorsal-ventral embryonic polarity, appears to encode a transmembrane protein. Cell 1988; 52:269-279.
4. Lemaitre B, Nicolas E, Michaut L et al. The dorsoventral regulatory gene cassette spatzle/toll/cactus controls the potent antifungal response in *Drosophila* adults. Cell 1996; 86:973-983.
5. Hoffmann JA, Reichhart JM. Drosophila innate immunity: An evolutionary perspective. Nat Immunol 2002; 3:121-126.
6. Medzhitov R, Preston-Hurlburt P, Janeway Jr CA. A human homologue of the *Drosophila* toll protein signals activation of adaptive immunity. Nature 1997; 388:394-397.
7. Rock FL, Hardiman G, Timans JC et al. A family of human receptors structurally related to *Drosophila* toll. Proc Natl Acad Sci USA 1998; 95:588-593.
8. Takeuchi O, Kawai T, Sanjo H et al. TLR6: A novel member of an expanding toll-like receptor family. Gene 1999; 231:59-65.
9. Chuang TH, Ulevitch RJ. Cloning and characterization of a sub-family of human toll-like receptors: hTLR7, hTLR8 and hTLR9. Eur Cytokine Netw 2000; 11:372-378.
10. Du X, Poltorak A, Wei Y et al. Three novel mammalian toll-like receptors: Gene structure, expression, and evolution. Eur Cytokine Netw 2000; 11:362-371.
11. Chuang T, Ulevitch RJ. Identification of hTLR10: A novel human toll-like receptor preferentially expressed in immune cells. Biochim Biophys Acta 2001; 1518:157-161.
12. Parnet P, Garka KE, Bonnert TP et al. IL-1Rrp is a novel receptor-like molecule similar to the type I interleukin-1 receptor and its homologues T1/ST2 and IL-1R AcP. J Biol Chem 1996; 271:3967-3970.
13. Mitcham JL, Parnet P, Bonnert TP et al. T1/ST2 signaling establishes it as a member of an expanding interleukin- 1 receptor family. J Biol Chem 1996; 271:5777-5783.
14. Hultmark D. Macrophage differentiation marker MyD88 is a member of the Toll/IL-1 receptor family. Biochem Biophys Res Commun 1994; 199:144-146.
15. Horng T, Barton GM, Medzhitov R. TIRAP: An adapter molecule in the toll signaling pathway. Nat Immunol 2001; 2:835-841.
16. Ferrero E, Goyert SM. Nucleotide sequence of the gene encoding the monocyte differentiation antigen, CD14. Nucleic Acids Res 1988; 16:4173.
17. Haziot A, Chen S, Ferrero E et al. The monocyte differentiation antigen, CD14, is anchored to the cell membrane by a phosphatidylinositol linkage. J Immunol 1988; 141:547-552.
18. Miyake K, Yamashita Y, Ogata M et al. RP105, a novel B cell surface molecule implicated in B cell activation, is a member of the leucine-rich repeat protein family. J Immunol 1995; 154:3333-3340.
19. Inohara N, Koseki T, del Peso L et al. Nod1, an Apaf-1-like activator of caspase-9 and nuclear factor-kappaB. J Biol Chem 1999; 274:14560-14567.
20. Hugot JP, Chamaillard M, Zouali H et al. Association of NOD2 leucine-rich repeat variants with susceptibility to Crohn's disease. Nature 2001; 411:599-603.
21. Ogura Y, Bonen DK, Inohara N et al. A frameshift mutation in NOD2 associated with susceptibility to Crohn's disease. Nature 2001; 411:603-606.
22. Henderson B, Poole S, Wilson M. Bacterial modulins: A novel class of virulence factors which cause host tissue pathology by inducing cytokine synthesis. Microbiol Rev 1996; 60:316-341.
23. Brightbill HD, Libraty DH, Krutzik SR et al. Host defense mechanisms triggered by microbial lipoproteins through toll-like receptors. Science 1999; 285:732-736.
24. Aliprantis AO, Yang RB, Mark MR et al. Cell activation and apoptosis by bacterial lipoproteins through toll- like receptor-2. Science 1999; 285:736-739.
25. Hirschfeld M, Kirschning CJ, Schwandner R et al. Cutting edge: Inflammatory signaling by borrelia burgdorferi lipoproteins is mediated by toll-like receptor 2. J Immunol 1999; 163:2382-2386.
26. Lien E, Sellati TJ, Yoshimura A et al. Toll-like receptor 2 functions as a pattern recognition receptor for diverse bacterial products. J Biol Chem 1999; 274:33419-33425.
27. Takeuchi O, Kaufmann A, Grote K et al. Cutting edge: Preferentially the R-stereoisomer of the mycoplasmal lipopeptide macrophage-activating lipopeptide-2 activates immune cells through a toll-like receptor 2- and MyD88-dependent signaling pathway. J Immunol 2000; 164:554-557.
28. Takeuchi O, Kawai T, Muhlradt PF et al. Discrimination of bacterial lipoproteins by toll-like receptor 6. Int Immunol 2001; 13:933-940.

29. Strominger JL, Ghuysen JM. Mechanisms of enzymatic bacteriaolysis. Cell walls of bacteri are solubilized by action of either specific carbohydrases or specific peptidases. Science 1967; 156:213-221.
30. Yoshimura A, Lien E, Ingalls RR et al. Cutting edge: Recognition of Gram-positive bacterial cell wall components by the innate immune system occurs via toll-like receptor 2. J Immunol 1999; 163:1-5.
31. Schwandner R, Dziarski R, Wesche H et al. Peptidoglycan- and lipoteichoic acid-induced cell activation is mediated by toll-like receptor 2. J Biol Chem 1999; 274:17406-17409.
32. Takeuchi O, Hoshino K, Kawai T et al. Differential roles of TLR2 and TLR4 in recognition of gram-negative and gram-positive bacterial cell wall components. Immunity 1999; 11:443-451.
33. Hirschfeld M, Ma Y, Weis JH et al. Cutting Edge: Repurification of lipopolysaccharide eliminates signaling through both human and murine toll-like receptor 2. J Immunol 2000; 165:618-622.
34. Werts C, Tapping RI, Mathison JC et al. Leptospiral lipopolysaccharide activates cells through a TLR2-dependent mechanism. Nat Immunol 2001; 2:346-352.
35. Massari P, Henneke P, Ho Y et al. Cutting edge: Immune stimulation by neisserial porins is toll-like receptor 2 and MyD88 dependent. J Immunol 2002; 168:1533-1537.
36. Asai Y, Ohyama Y, Gen K et al. Bacterial fimbriae and their peptides activate human gingival epithelial cells through toll-like receptor 2. Infect Immun 2001; 69:7387-7395.
37. Means TK, Lien E, Yoshimura A et al. The CD14 ligands lipoarabinomannan and lipopolysaccharide differ in their requirement for toll-like receptors. J Immunol 1999; 163:6748-6755.
38. Campos MA, Almeida IC, Takeuchi O et al. Activation of toll-like receptor-2 by glycosylphosphatidylinositol anchors from a protozoan parasite. J Immunol 2001; 167:416-423.
39. Takeuchi O, Hoshino K, Akira S. Cutting edge: TLR2-deficient and MyD88-deficient mice are highly susceptible to Staphylococcus aureus infection. J Immunol 2000; 165:5392-5396.
40. Thoma-Uszynski S, Stenger S, Takeuchi O et al. Induction of direct antimicrobial activity through mammalian toll-like receptors. Science 2001; 291:1544-1547.
41. Jacobs BL, Langland JO. When two strands are better than one: The mediators and modulators of the cellular responses to double-stranded RNA. Virology 1996; 219:339-349.
42. Akiyama Y, Stevenson GW, Schlick E et al. Differential ability of human blood monocyte subsets to release various cytokines. J Leukoc Biol 1985; 37:519-530.
43. Manetti R, Annunziato F, Tomasevic L et al. Polyinosinic acid: Polycytidylic acid promotes T helper type 1-specific immune responses by stimulating macrophage production of interferon-alpha and interleukin-12. Eur J Immunol 1995; 25:2656-2660.
44. Alexopoulou L, Holt AC, Medzhitov R et al. Recognition of double-stranded RNA and activation of NF-kappaB by toll- like receptor 3. Nature 2001; 413:732-738.
45. Yang YL, Reis LF, Pavlovic J et al. Deficient signaling in mice devoid of double-stranded RNA-dependent protein kinase. Embo J 1995; 14:6095-6106.
46. Uetani K, Der SD, Zamanian-Daryoush M et al. Central role of double-stranded RNA-activated protein kinase in microbial induction of nitric oxide synthase. J Immunol 2000; 165:988-996.
47. Maggi Jr LB, Heitmeier MR, Scheuner D et al. Potential role of PKR in double-stranded RNA-induced macrophage activation. Embo J 2000; 19:3630-3638.
48. Iordanov MS, Wong J, Bell JC et al. Activation of NF-kappaB by double-stranded RNA (dsRNA) in the absence of protein kinase R and RNase L demonstrates the existence of two separate dsRNA-triggered antiviral programs. Mol Cell Biol 2001; 21:61-72.
49. Ulevitch RJ, Tobias PS. Receptor-dependent mechanisms of cell stimulation by bacterial endotoxin. Annu Rev Immunol 1995; 13:437-457.
50. Poltorak A, He X, Smirnova I et al. Defective LPS signaling in C3H/HeJ and C57BL/10ScCr mice: Mutations in Tlr4 gene. Science 1998; 282:2085-2088.
51. Poltorak A, Smirnova I, Clisch R et al. Limits of a deletion spanning Tlr4 in C57BL/10ScCr mice. J Endotoxin Res 2000; 6:51-56.
52. Hoshino K, Takeuchi O, Kawai T et al. Cutting edge: Toll-like receptor 4 (TLR4)-deficient mice are hyporesponsive to lipopolysaccharide: evidence for TLR4 as the Lps gene product. J Immunol 1999; 162:3749-3752.
53. Jack RS, Fan X, Bernheiden M et al. Lipopolysaccharide-binding protein is required to combat a murine gram-negative bacterial infection. Nature 1997; 389:742-745.
54. Wurfel MM, Monks BG, Ingalls RR et al. Targeted deletion of the lipopolysaccharide (LPS)-binding protein gene leads to profound suppression of LPS responses ex vivo, whereas in vivo responses remain intact. J Exp Med 1997; 186:2051-2056.
55. Haziot A, Ferrero E, Kontgen F et al. Resistance to endotoxin shock and reduced dissemination of gram-negative bacteria in CD14-deficient mice. Immunity 1996; 4:407-414.

56. Shimazu R, Akashi S, Ogata H et al. MD-2, a molecule that confers lipopolysaccharide responsiveness on toll- like receptor 4. J Exp Med 1999; 189:1777-1782.
57. da Silva Correia J, Soldau K, Christen U et al. Lipopolysaccharide is in close proximity to each of the proteins in its membrane receptor complex. transfer from CD14 to TLR4 and MD-2. J Biol Chem 2001; 276:21129-21135.
58. Schromm AB, Lien E, Henneke P et al. Molecular genetic analysis of an endotoxin nonresponder mutant cell line: A point mutation in a conserved region of MD-2 abolishes endotoxin-induced signaling. J Exp Med 2001; 194:79-88.
59. Ogata H, Su I, Miyake K et al. The toll-like receptor protein RP105 regulates lipopolysaccharide signaling in B cells. J Exp Med 2000; 192:23-29.
60. Inohara N, Ogura Y, Chen FF et al. Human Nod1 confers responsiveness to bacterial lipopolysaccharides. J Biol Chem 2001; 276:2551-2554.
61. Girardin SE, Tournebize R, Mavris M et al. CARD4/Nod1 mediates NF-kappaB and JNK activation by invasive Shigella flexneri. EMBO Rep 2001; 2:736-742.
62. Triantafilou K, Triantafilou M, Dedrick RL. A CD14-independent LPS receptor cluster. Nat Immunol 2001; 2:338-345.
63. Kawasaki K, Akashi S, Shimazu R et al. Mouse toll-like receptor 4.MD-2 complex mediates lipopolysaccharide- mimetic signal transduction by Taxol. J Biol Chem 2000; 275:2251-2254.
64. Kurt-Jones EA, Popova L, Kwinn L et al. Pattern recognition receptors TLR4 and CD14 mediate response to respiratory syncytial virus. Nat Immunol 2000; 1:398-401.
65. Okamura Y, Watari M, Jerud ES et al. The extra domain A of fibronectin activates toll-like receptor 4. J Biol Chem 2001; 276:10229-10233.
66. Termeer C, Benedix F, Sleeman J et al. Oligosaccharides of Hyaluronan activate dendritic cells via toll-like receptor 4. J Exp Med 2002; 195:99-111.
67. Vabulas RM, Ahmad-Nejad P, da Costa C et al. Endocytosed HSP60s use toll-like receptor 2 (TLR2) and TLR4 to activate the toll/interleukin-1 receptor signaling pathway in innate immune cells. J Biol Chem 2001; 276:31332-31339.
68. Ohashi K, Burkart V, Flohe S et al. Cutting edge: Heat shock protein 60 is a putative endogenous ligand of the toll-like receptor-4 complex. J Immunol 2000; 164:558-561.
69. Gewirtz AT, Simon Jr PO, Schmitt CK et al. Salmonella typhimurium translocates flagellin across intestinal epithelia, inducing a proinflammatory response. J Clin Invest 2001; 107:99-109.
70. Hayashi F, Smith KD, Ozinsky A et al. The innate immune response to bacterial flagellin is mediated by toll- like receptor 5. Nature 2001; 410:1099-1103.
71. Gewirtz AT, Navas TA, Lyons S et al. Cutting edge: Bacterial flagellin activates basolaterally expressed TLR5 to induce epithelial proinflammatory gene expression. J Immunol 2001; 167:1882-1885.
72. Muhlradt PF, Kiess M, Meyer H et al. Isolation, structure elucidation, and synthesis of a macrophage stimulatory lipopeptide from *Mycoplasma fermentans* acting at picomolar concentration. J Exp Med 1997; 185:1951-1958.
73. Chambaud I, Wroblewski H, Blanchard A. Interactions between mycoplasma lipoproteins and the host immune system. Trends Microbiol 1999; 7:493-499.
74. Ozinsky A, Underhill DM, Fontenot JD et al. The repertoire for pattern recognition of pathogens by the innate immune system is defined by cooperation between toll-like receptors. Proc Natl Acad Sci USA 2000; 97:13766-13771.
75. Henneke P, Takeuchi O, van Strijp JA et al. Novel engagement of CD14 and multiple toll-like receptors by group B streptococci. J Immunol 2001; 167:7069-7076.
76. Miller RL, Gerster JF, Owens ML et al. Imiquimod applied topically: A novel immune response modifier and new class of drug. Int J Immunopharmacol 1999; 21:1-14.
77. von Krogh G, Lacey CJ, Gross G et al. European course on HPV associated pathology: Guidelines for primary care physicians for the diagnosis and management of anogenital warts. Sex Transm Infect 2000; 76:162-168.
78. Stephenson J. New therapy promising for genital herpes. Jama 2001; 285:2182-2183.
79. Hemmi H, Kaisho T, Takeuchi O et al. Small anti-viral compounds activate immune cells via the TLR7 MyD88-dependent signaling pathway. Nat Immunol 2002; 3:196-200.
80. Lipford GB, Heeg K, Wagner H. Bacterial DNA as immune cell activator. Trends Microbiol 1998; 6:496-500.
81. Hemmi H, Takeuchi O, Kawai T et al. A toll-like receptor recognizes bacterial DNA. Nature 2000; 408:740-745.
82. Bauer S, Kirschning CJ, Hacker H et al. Human TLR9 confers responsiveness to bacterial DNA via species-specific CpG motif recognition. Proc Natl Acad Sci USA 2001; 98:9237-9242.
83. Muzio M, Ni J, Feng P et al. IRAK (Pelle) family member IRAK-2 and MyD88 as proximal mediators of IL- 1 signaling. Science 1997; 278:1612-1615.

84. Wesche H, Henzel WJ, Shillinglaw W et al. MyD88: An adapter that recruits IRAK to the IL-1 receptor complex. Immunity 1997; 7:837-847.
85. Burns K, Martinon F, Esslinger C et al. MyD88, an adapter protein involved in interleukin-1 signaling. J Biol Chem 1998; 273:12203-12209.
86. Medzhitov R, Preston-Hurlburt P, Kopp E et al. MyD88 is an adaptor protein in the hToll/IL-1 receptor family signaling pathways. Mol Cell 1998; 2:253-258.
87. Burns K, Clatworthy J, Martin L et al. Tollip, a new component of the IL-1RI pathway, links IRAK to the IL-1 receptor. Nat Cell Biol 2000; 2:346-351.
88. Bulut Y, Faure E, Thomas L et al. Cooperation of toll-like receptor 2 and 6 for cellular activation by soluble tuberculosis factor and Borrelia burgdorferi outer surface protein A lipoprotein: Role of toll-interacting protein and IL-1 receptor signaling molecules in toll-like receptor 2 signaling. J Immunol 2001; 167:987-994.
89. Lomaga MA, Yeh WC, Sarosi I et al. TRAF6 deficiency results in osteopetrosis and defective interleukin-1, CD40, and LPS signaling. Genes Dev 1999; 13:1015-1024.
90. Naito A, Azuma S, Tanaka S et al. Severe osteopetrosis, defective interleukin-1 signalling and lymph node organogenesis in TRAF6-deficient mice. Genes Cells 1999; 4:353-362.
91. Deng L, Wang C, Spencer E et al. Activation of the IkappaB kinase complex by TRAF6 requires a dimeric ubiquitin-conjugating enzyme complex and a unique polyubiquitin chain. Cell 2000; 103:351-361.
92. Wang C, Deng L, Hong M et al. TAK1 is a ubiquitin-dependent kinase of MKK and IKK. Nature 2001; 412:346-351.
93. Karin M, Ben-Neriah Y. Phosphorylation meets ubiquitination: The control of NF-[kappa]B activity. Annu Rev Immunol 2000; 18:621-663.
94. Sanz L, Diaz-Meco MT, Nakano H et al. The atypical PKC-interacting protein p62 channels NF-kappaB activation by the IL-1-TRAF6 pathway. Embo J 2000; 19:1576-1586.
95. Leitges M, Sanz L, Martin P et al. Targeted Disruption of the zetaPKC gene results in the impairment of the NF-kappaB pathway. Mol Cell 2001; 8:771-780.
96. Castrillo A, Pennington DJ, Otto F et al. Protein kinase cepsilon is required for macrophage activation and defense against bacterial infection. J Exp Med 2001; 194:1231-1242.
97. Hardiman G, Rock FL, Balasubramanian S et al. Molecular characterization and modular analysis of human MyD88. Oncogene 1996; 13:2467-2475.
98. Adachi O, Kawai T, Takeda K et al. Targeted disruption of the MyD88 gene results in loss of IL-1- and IL- 18-mediated function. Immunity 1998; 9:143-150.
99. Kawai T, Adachi O, Ogawa T et al. Unresponsiveness of MyD88-deficient mice to endotoxin. Immunity 1999; 11:115-122.
100. Takeuchi O, Takeda K, Hoshino K et al. Cellular responses to bacterial cell wall components are mediated through MyD88-dependent signaling cascades. Int Immunol 2000; 12:113-117.
101. Hacker H, Vabulas RM, Takeuchi O et al. Immune cell activation by bacterial CpG-DNA through myeloid differentiation marker 88 and tumor necrosis factor receptor-associated factor (TRAF)6. J Exp Med 2000; 192:595-600.
102. Adachi K, Tsutsui H, Kashiwamura S et al. Plasmodium berghei infection in mice induces liver injury by an IL-12- and toll-like receptor/myeloid differentiation factor 88-dependent mechanism. J Immunol 2001; 167:5928-5934.
103. Schnare M, Barton GM, Holt AC et al. Toll-like receptors control activation of adaptive immune responses. Nat Immunol 2001; 2:947-950.
104. Kawai T, Takeuchi O, Fujita T et al. Lipopolysaccharide stimulates the MyD88-independent pathway and results in activation of IFN-regulatory factor 3 and the expression of a subset of lipopolysaccharide-inducible genes. J Immunol 2001; 167:5887-5894.
105. Kaisho T, Takeuchi O, Kawai T et al. Endotoxin-induced maturation of MyD88-deficient dendritic cells. J Immunol 2001; 166:5688-5694.
106. Seki E, Tsutsui H, Nakano H et al. Lipopolysaccharide-induced IL-18 secretion from murine Kupffer cells independently of myeloid differentiation factor 88 that is critically involved in induction of production of IL-12 and IL-1beta. J Immunol 2001; 166:2651-2657.
107. Fitzgerald KA, Palsson-McDermott EM, Bowie AG et al. Mal (MyD88-adapter-like) is required for toll-like receptor-4 signal transduction. Nature 2001; 413:78-83.
108. Horng T, Barton GM, Flavell RA et al. The adaptor molecule TIRAP provides signalling specificity for Toll-like receptors. Nature 2002; 420:329-333.
109. Yamamoto M, Sato S, Hemmi H et al. Essential role for TIRAP in activation of the signalling cascade shared by TLR2 and TLR4. Nature 2002; 420:334-339.
110. Yamamoto M, Sato S, Mori K et al. Cutting edge: A novel Toll/IL-1 receptor domain-containing adapter that preferentially activates the IFN-beta promoter in the Toll-like receptor signalling. J Immunol 2002; 169:6668-6672.

CHAPTER 7

C-Type Lectin and Lectin-Like Receptors in the Immune System

Sally Rogers and Simon Y.C. Wong

Abstract

Protein-carbohydrate interactions have been shown to mediate a variety of biological activities such as homeostasis and immune responses. Immune activities include pathogen recognition and neutralization, leukocyte trafficking, phagocytosis, antigen uptake and processing, and apoptosis. Most of the carbohydrate-binding proteins (lectins) involved in these immunological processes belong to the C-type animal lectin superfamily. Elegant studies have provided precise molecular details of calcium-dependent binding to carbohydrates by the carbohydrate recognition domain (CRD) of serum mannose binding protein (MBP) and the selectins, whose carbohydrate ligands have been characterized extensively. However, many C-type animal lectin superfamily members do not have in their CRDs the critical amino acid residues required for MBP to bind carbohydrates. This chapter summarizes the known and predicted structures and immune functions of several groups of the C-type lectin superfamily that have at least one lectin or lectin-like domain.

Introduction

Carbohydrate recognition by animal lectins is associated with discrete protein domains, termed carbohydrate recognition domains (CRDs). CRDs can be classified into families based on amino acid sequence motifs and functional properties. The C-type animal lectins were originally classified as a distinct family of structurally related CRDs which bind carbohydrate in a calcium dependent fashion.[1] These protein-carbohydrate interactions have been shown to mediate a wide array of biological events, such as cell-cell adhesion, serum glycoprotein turnover and pathogen neutralization. Furthermore, there is increasing evidence that C-type lectins are important not only in many aspects of innate immunity, but also in directing the adaptive immune response. Comparison of the amino acid sequences of the CRDs from the C-type lectins revealed a highly conserved pattern of 32 amino acids spaced over a stretch of approximately 120 amino acids.[2] This motif is essential for correct folding of the structural domain and has been used to identify putative CRDs in the primary sequence of many proteins. However, it remains to be determined if all of these C-type CRDs bind carbohydrate, since an increasing number of proteins containing the C type CRD structural motif, but missing the critical residues for calcium binding, have been identified.

The group of receptors lacking the critical residues for calcium binding was initially comprised of receptors expressed predominantly by natural killer cells, hence this structural domain is often referred to as a NK domain (NKD). However, the expression of many of the recently discovered receptors belonging to this group are not restricted to NK cells, but found on a wide range of leukocytes. In order to reflect the cellular distribution, this type of structural domain is now more appropriately referred to as a C-type lectin-like domain (CTLD). This

Immunobiology of Carbohydrates, edited by Simon Y.C. Wong and Gemma Arsequell.
©2003 Eurekah.com and Kluwer Academic / Plenum Publishers.

term is also used to encompass both the NKDs and CRDs. The ligands for the majority of these newly identified receptors have yet to be determined, and the absence of the critical residues for calcium binding would suggest that they might not be able to bind carbohydrate. This has been shown to be true for particular receptors including type II antifreeze glycoprotein, lithostathine and NKG2, with ligands ranging from inorganic molecules to proteins.[3-6] The wide range of ligands for CTLDs reflects the diversity of the CTLDs themselves, both in terms of amino acid sequence within the domain, and also the overall domain organization of CTLD-containing proteins. These characteristics, together with gene organization information, have proved useful in classifying these proteins into fourteen groups (website at http:// ctld.glycob.ox.ac.uk).[1] Only five groups will be discussed in detail as several receptors in these groups are known to participate in immune processes including self and nonself discrimination, leukocyte recruitment, and antigen capture and processing (**Fig.** 1).

CTLDs in the Immune Response—Ligand Binding, Specificity and Function

The importance of CTLDs in the immune response has been appreciated for a long time, particularly in terms of pathogen recognition and cell-cell interactions. However, the ever-increasing amount of evidence available is beginning to reveal the extent to which CTLDs are utilized by the immune system. The identification and characterization of CTLDs expressed by antigen presenting cells (APCs) has highlighted the importance of C-type lectins in the initiation of the adaptive immune response. Studies of the Group V CTLDs encoded in the NK complex are also demonstrating the critical role that these receptors play in viral and tumour immunity. In order to make sense the diverse array of functions of and ligands for CTLDs, it is perhaps easiest to focus on certain aspects. In particular, the determination of the crystal structures of 11 CTLDs has revealed the remarkable conservation of overall structure of the domains. The greatest diversity between CTLDs is seen in the loop regions, and the importance of amino acid sequences in these regions for determining ligand specificity is becoming clear. Finally, dimer or oligomer formation is another critical factor when discussing the role of CTLDs in the immune response. Five of the seven major groups of CTLDs are predominantly associated with immune function, and the recent findings for each of these groups will be discussed, and compared with the best characterized CTLD, the CRD of MBP.

Group II of CTLD-Containing Proteins in the Immune System—Endocytic Type II Transmembrane Proteins

The Group II CTLDs are anchored to the cell membrane by N-terminal hydrophobic sequences, giving the receptors a characteristic type II orientation. Analysis of the amino acid sequences has revealed that the critical residues for calcium binding are present in the Group II CTLDs, which are separated from the transmembrane region by a series of short sequence repeats. The prototype for this group is the asialoglycoprotein receptor, a hepatic cell surface protein that directs turnover of serum glycoproteins.[7] Other receptors in this group include the low affinity IgE receptor CD23,[8] and a recently identified dendritic cell (DC)-specific receptor DC-SIGN,[9] both of which also mediate endocytosis.

The asialoglycoprotein receptor, which mediates clearance of galactose or N-acetylgalactosamine terminated glycoproteins from the circulation. Following binding of the ligand at the cell surface, the asialoglycoprotein receptor is endocytosed via clathrin-coated pits and directed to the endosomes. The bound carbohydrate is released here in a pH dependent manner, and targeted for degradation in lysosomes. The empty asialoglycoprotein receptor is recycled back to the cell surface. Site directed mutagenesis of the major subunit of the asialoglycoprotein receptor has been used to identify the residues which are essential for the pH dependent release of carbohydrate ligands at endosomal pH.[10] In particular, residues His256, Asp266 and Arg270 have been shown to have an important role. The proximity of these three residues to the ligand binding sites at Ca^{2+} binding site 2 of the domain suggests that they form a pH-sensitive switch

Group	Example	Ligands	Functions	Architecture
II	DC-SIGN	Internal mannose	Mediates initial T cell/DC interaction Internalises soluble ligand for presentation Binds HIV gp120	
	Asialoglycoprotein Receptor	GalNAC, terminal galactose	Directs turnover of serum glycoproteins	
	CD23	IgE	Binds and regulates synthesis of IgE	
	Kupffer Cell Receptor	GalNAC, terminal galactose	DC recruitment to the liver	
III	Mannose Binding Protein	GlcNAc, terminal mannose	Binds to bacterial and fungal pathogens, leads to opsonization and initiation of complement cascade	
IV	E-Selectin	sialyl Lewis X, PSGL-1, ESL-1 VIM-2	Leukocyte trafficking	
	P-Selectin	PSGL-1	Lymphocyte "homing"	
	L-Selectin	sulfo, sialyl Lewis X (GlyCAM-1, CD34, MAdCAM-1, podocalyxin)		
V	Ly49 Multigene Family	MHC class I	Activatory/inhibitory signalling molecules Ly49H critical for resistence to mCMV	
	NKG2D	ULBPs, Rae, Hsp60, MICA & B	Activatory signalling receptor Binds tumour antigens	
	CD69	Unknown	Early activation antigen	
VI	Macrophage Mannose Receptor	Mannose	Scavenger receptor, binds and internalises pathogenic microorganisms	
	Endo-180	GlcNAc	Internalisation of glycosylated ligands	
	DEC205	Unknown	Antigen uptake	
	PLA$_2$ Receptor	Phospholipase A$_2$	Endocytic receptor	

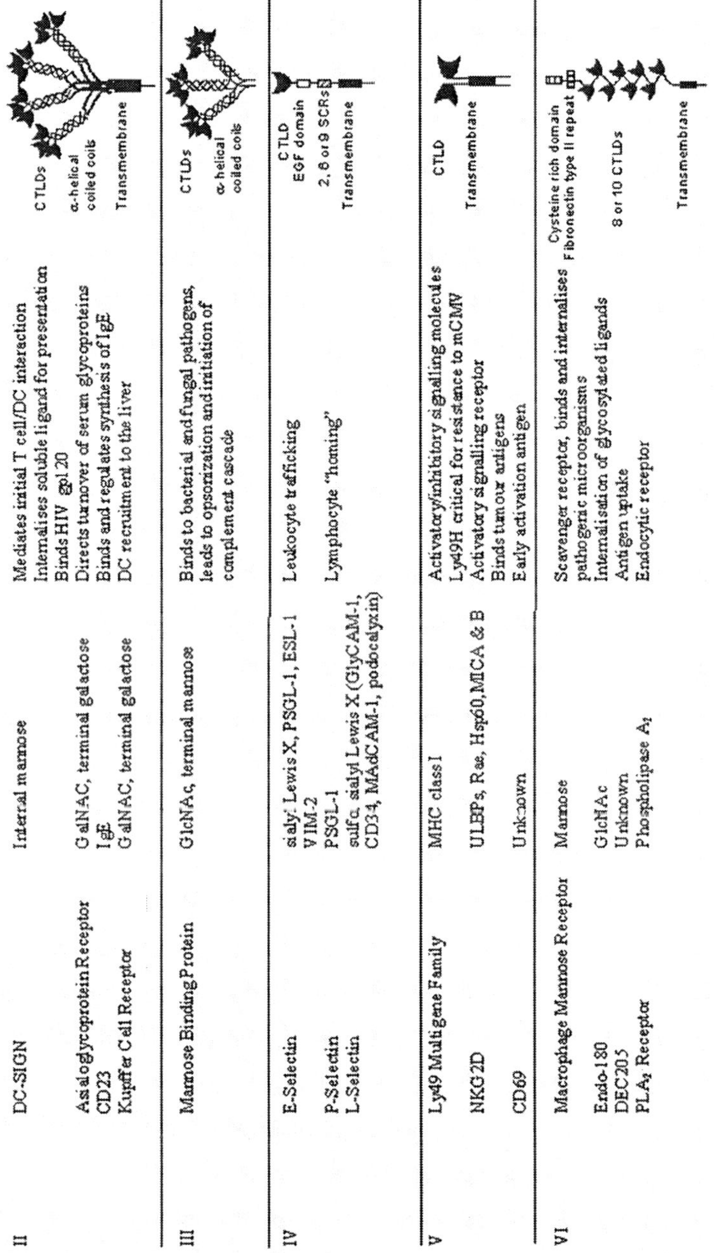

Figure 1. Summary of the functions of five groups of proteins containing a CTLD in the immune response.

for Ca^{2+} and ligand binding. Interestingly, in other nonendocytic CTLDs the residues corresponding to His256, Asp266 and Arg270 differ, providing more evidence that these residues may be useful in determining the function of a CTLD.

Recently, two novel isoforms of the asialoglycoprotein receptor have been identified.[11] Unlike the hepatic asialoglycoprotein receptor, these new isoforms are expressed predominantly by immature DCs with an interstitial phenotype, and therefore have been termed DC-asialoglycoprotein receptor (DC-ASGPR). Furthermore, they have been shown to function as endocytic receptors, delivering bound carbohydrates to the early endosomes. The natural ligand for DC-ASGPR has not been identified, but analysis of the amino acid sequence of the CTLD of both isoforms revealed the presence of the QPD peptide motif in the putative binding loop, suggesting that these novel receptors could also bind galactosylated glycoproteins. An immune function has also been suggested recently for another liver specific Group II CTLD. The Kupffer receptor is expressed only by rat Kupffer cells, and has high affinity for oligosaccharides with terminal galactose, N-acetylgalactosamine and fucose, but its function has not been well characterized. Uwatoku and colleagues have shown that DC recruitment to the liver is dependent on the presence of Kupffer cells, and that initial binding is mediated by N-acetylgalactosamine specific, calcium dependent receptors.[12] This suggests a possible role for the Kupffer receptor in recruitment of DCs to the liver.

Perhaps one of the most exciting developments in this research area in the last few years has been the identification and characterization of DC-SIGN (dendritic cell specific ICAM-3 grabbing nonintegrin). The salient features of this molecule will be reviewed here for comparison with other CTLDs. DC-SIGN was originally identified in 1992 through its ability to bind the HIV-1 envelope in the absence of CD4,[13] but its expression on DCs was only characterized in 2000, when Geijtenbeek and coworkers demonstrated the critical role of the DC-SIGN/ICAM-3 interaction in the initial contact between DC and T cells.[9] Furthermore, DC-SIGN displays high affinity for ICAM-2, thereby supporting tethering and rolling of DC, a prerequisite for emigration from the blood.[14] DC-SIGN has also been shown to function as an endocytic receptor that is rapidly internalized upon binding of soluble ligands and targeted to late endosomes/lysosomes.[15] Moreover, ligands internalized by DC-SIGN can subsequently be efficiently presented to $CD4^+$ T cells. Finally, DC-SIGN binds the HIV-1 envelope glycoprotein gp120 at mucosal sites of initial infection, allowing transport of the virus to secondary lymphoid organs and highly efficient infection of $CD4^+$ T cells.

Determination of the crystal structure of DC-SIGN bound to a pentasaccharide has revealed that the CTLD adopts the typical C-type lectin fold.[16] In contrast to the crystal structure of MBP as described later, an extra disulfide bond is observed in DC-SIGN, which links strand $\beta 1$ to an additional β strand, $\beta 0$. This additional disulfide bridge is observed in the crystal structures of some other CTLDs, all of which are referred to as long-form CTLDs. The study also revealed that DC-SIGN interacts with internal saccharide residues, unlike the MBP protein that interacts with terminal sugar residues. Another recent report has highlighted the role of oligomerization of DC-SIGN in ligand recognition, showing that the extracellular domain of the molecule forms a tetramer stabilized by an α-helical neck region.[17] This particular conformation is unusual among CTLDs, and suggests that DC-SIGN distinguishes its natural ligands by recognizing oligosaccharides spaced at appropriate distances on the surface of a limited number of glycoproteins, including ICAM-3 and gp120.

Together, these recent reports suggest that DC-SIGN recognizes carbohydrate differently from MBP, although both receptors are specific for mannose. Each CTLD of MBP binds to a single terminal mannose residue and high avidity is achieved by the clustering of CTLDs in trimeric units. The broad spacing of individual MBP CTLDs is critical for determining the specificity of MBP for widely spaced oligosaccharides such as those found on pathogenic surfaces, a defining property for the role of MBP in the innate immune response. In contrast, DC-SIGN may recognize multiple self-oligosaccharides in specific arrangements on the surfaces of select endogenous glycoproteins, a necessary ability for its function in several aspects of the initiation of the adaptive immune response. Unfortunately, it would appear that HIV has exploited this exquisite mode of recognition to promote its infection of T cells. Future studies will almost certainly be focused on various ways of disrupting the interaction between DC-SIGN and gp120 of HIV. Whether carbohydrate ligands could lower the frequency of or inhibit totally T cell infection by HIV remains to be investigated.

Group III of CTLD-Containing Proteins in the Immune System—Pathogen Recognition

The Group III lectins, the collectins, are arguably the best characterized group of CTLDs, and include the prototypic C-type CRD of MBP.[2] The collectins are characterized by the presence of amino terminal collagen-like domains, and carboxy terminal CRDs. The collectins are important innate immune receptors, distinguishing "self" and "nonself" by recognizing terminal monosaccharides characteristic of bacterial and fungal surfaces. Binding of serum MBP to surfaces of microbes leads to opsonization and initiation of the complement cascade. Other collectins, such as the surfactant proteins SP-A and SP-D, are secreted in the lung as part of the surfactant that lines the alveoli, providing a first line defence against airborne pathogens.

The structure of the CRD of MBP was the first to be determined, and revealed a core fold consisting of five β-strands and two α-helices connected by loop regions (Fig. 2).[18] The β-strands 1 and 5 form an anti-parallel β-sheet, placing the beginning and end of the CRD next to each other in opposite orientations. This positioning explains the observation that a CTLD can be located anywhere in a polypeptide chain. Two disulfide bonds are formed by four invariant cysteines, the first linking α-helix 1 to β-strand 5, and the second linking the beginning of β-strand 3 to the loop following β-strand 4. MBP contains two Ca^{2+} binding sites (designated site 1 and site 2) which are important for maintaining the structural fold as well as for carbohydrate binding.[2]

MBP exhibits a broad carbohydrate binding specificity, including N-acetylglucosamine, mannose and fucose- a reflection of the variety of pathogenic surfaces to which it binds. All these monosaccharides contain equatorial 3- and 4-hydroxyl groups (which are involved in direct coordination of Ca^{2+} at site 2) and noncovalent and hydrogen bonds with the MBP itself, creating an intimately associated complex of protein, carbohydrate and calcium. N-acetylglucosamine, mannose and fucose are not common in terminal positions on mammalian oligosaccharides, but are frequently found on the surface of microorganisms, providing a mechanism for self versus nonself discrimination. The oligomeric structure of MBP is another principal factor in preventing an immune response to self. MBP is organized as a "bouquet" containing three trimers of the MBP monomer. This monomer contains four discrete regions of primary structure, starting with a short cysteine-rich domain. This is followed by 18-20 short sequence repeats forming an α-helical coiled coil, then a stalk region of around 30 amino acids.

The CRD is located at the carboxy terminus. A hydrophobic interface between the neck region and the CRD maintains a fixed spatial relationship between the binding sites of the trimer, such that they are 53 Å and 45 Å apart in rat and human MBP oligomers, respectively. Studies have shown that the terminal mannose residues in vertebrate oligosaccharides are about 20-30 Å apart, therefore the binding sites in the MBP trimer are too far apart to interact multivalently with self oligosaccharides. The interactions between MBP and monovalent ligands are extremely weak, with dissociation constants in the mM range. Highly avid multivalent interactions can only be achieved on binding to the high-mannose structures on pathogenic surfaces, which present dense, repetitive arrays of ligands that can span the distance between the binding sites on the MBP trimer. The critical amino acid residues responsible for the fine ligand specificity of MBP has been revealed by sequence comparison, and mutational analysis. For example, the Glu and Asn residues at positions 185 and 187 of MBP are present in other family members specific for mannose-type ligands, whereas C-type lectins that recognise galactose type ligands have Gln and Asp at these positions. Furthermore, replacement of Glu 185 and Asn 187 of MBP with Gln and Asp conferred preferential binding of mutant MBP for galactosides.[10]

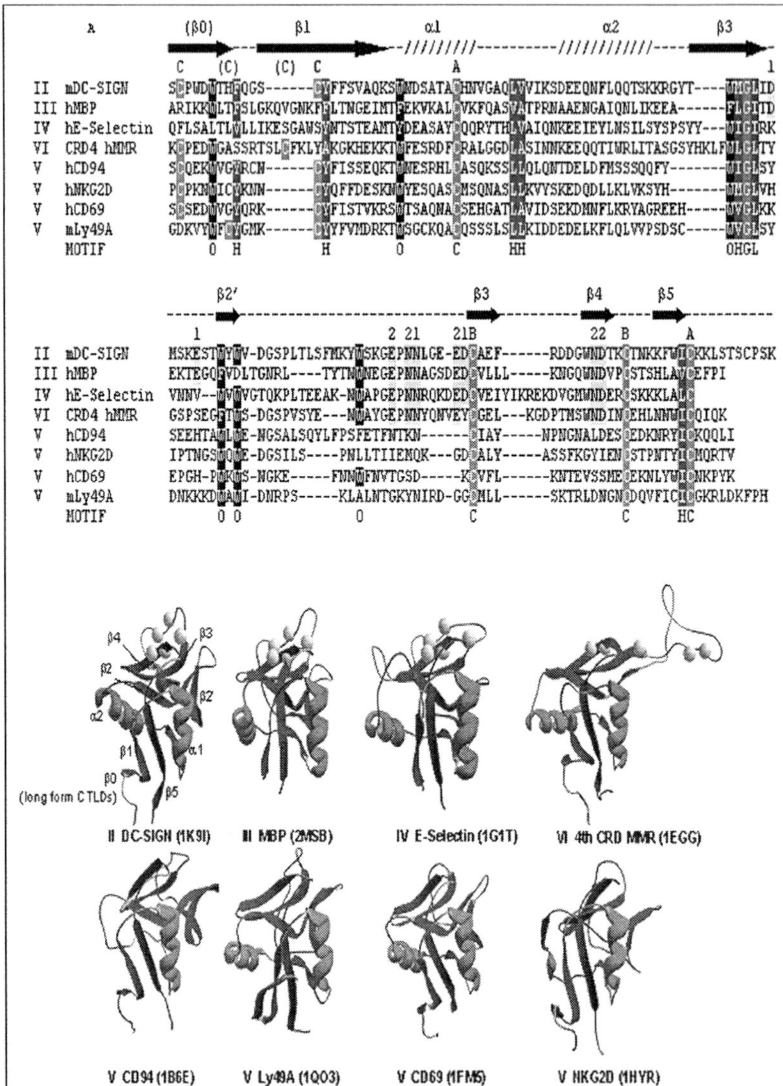

Figure 2. Amino acid and structure comparison of CTLDs. A) Alignment of the amino acid sequences of eight CTLDs, showing the variation observed between the conserved motif residues. Sequences start at the β strand found in the long-form CTLDs, βO. Dashes indicated gaps introduced for optimum alignment. Key residues that define the CTLD motif are highlighted, and designated below the alignment as follows: O denotes aromatic, H denotes hydrophobic and G denotes glycine. Conserved cysteines are designated C, and upper case letters above the alignment indicate disulphide-linked pairs. Residues that coordinate Ca^{2+} in the carbohydrate binding site are designated 1 and 2 above the alignment to indicate site 1 and site$_1$ as described in the text. Secondary structure elements are shown above the alignment; arrows represent β strands, slashes represent α helices and dashed lines represent the looped region. B) Ribbon diagrams of the crystal structures of eight CTLDs involved in the immune response, showing that the CTLD fold is well conserved. Group$_1$ receptor name and PDB accession number are indicated beneath each structure. α helices are shown as red coils and β strands are blue arrows. The position of residues responsible for Ca^{2+} coordination at binding site 2 in the carbohydrate binding CTLDs are shown as yellow spheres. Ribbon diagrams were prepared using Swiss-PDB Viewer.

Group IV of CTLD-Containing Proteins in the Immune System—Leukocyte Adhesion Events

Group IV lectins are the selectins. The selectins are expressed by platelets (P-selectin), vascular endothelium (E- and P-selectin) and leukocytes (L-selectin). Although they consist of only three members, the selectins have been among the most intensively investigated C-type lectin families. They play a critical role in the recruitment of leukocytes to sites of inflammation and emigration of leukocytes from the blood to lymphoid tissues. The structure and function of selectins have been studied by a large number of groups since the determination of their gene sequences in 1989.[19-23] These research efforts have resulted in a large body of evidence on the role of selectins and their carbohydrate ligands (on counter-receptors) in leukocyte-endothelial cell interactions. We will summarize representative studies that led to the identification of the selectins as carbohydrate-binding proteins, and the counter-receptors that display carbohydrate ligands for the selectins.[2,24-32] The story of the selectins illustrates some of the major challenges facing immunologists in determining the precise role of carbohydrate-protein interactions in the immune system.

Leukocyte trafficking to sites of infection and injury is one of the key events in the generation of immune and inflammatory responses. Studies investigating the movement of leukocytes from the blood vasculature to peripheral lymph nodes (PLNs) have demonstrated that specific adhesive events are involved in this "homing" process. These events enable the leukocytes initially to roll along the vascular surface and then to adhere firmly to the endothelial cells prior to the extravasation process. The binding of lymphocytes to the high endothelial venules (HEVs) of PLNs as determined by in vitro cell adhesion assays was found to be inhibitable by carbohydrates. Inhibitors included simple negatively charged sugars such as mannose-6-phosphate, or polymers of phosphomannan ester (PPME) from yeast cell wall and fucose-4-sulfate (fucoidan) from seaweed. Concentrations of monomeric sugars required for inhibition were in the millimolar range: too high to make this interaction physiologically relevant. However, for PPME and fucoidan, they were in the nanomolar range. Carbohydrate polymers with repetitive structural units would thus be better than monosaccharides as mimics of complex carbohydrate structures on cell surfaces. In addition to inhibition assays, the role of carbohydrates could also be deduced by determining the effect on this type of adhesive event following modification of cell surface glycans. For example, removal of sialic acids by sialidase treatment of HEVs abolished the ability of lymphocytes to bind to them. These studies provided strong evidence that carbohydrate-protein interactions are involved in the binding of lymphocytes to PLN, although the molecular nature of the receptor(s) and its endogenous carbohydrate ligand(s) were not defined.[26]

The breakthrough in the search for the putative "lymphocyte homing" receptor occurred following the demonstration that a monoclonal antibody (MEL-14) could inhibit the binding of PPME to lymphocytes in vitro and, perhaps more significantly, the homing of lymphocytes in vivo. Molecular cloning of the MEL-14 antigen revealed a novel cell adhesion molecule. It consists of three different protein domains: one C-type lectin domain at its amino terminus, followed by an epidermal growth factor (EGF) domain and two tandem short consensus repeats (SCRs) that are found in complement binding domains. Two other adhesion molecules having similar domain organization were discovered around the same time. One was termed the endothelial leukocyte adhesion molecule 1 (ELAM-1) and it appeared to mediate specific adhesive interactions between neutrophils/monocytes and inflamed endothelium. The other was referred to as platelet activation-dependent granule external membrane protein (PADGEM) or gmp140 and it appeared to mediate adhesive interactions between neutrophils and thrombin-activated platelets or endothelial cells. Similar to MEL-14, ELAM-1 and PADGEM have one C-type lectin domain at their amino terminus followed by one EGF domain. However, they have six and nine, instead of two, tandem SCRs respectively. These proteins were renamed as L-selectin (MEL-14), E-selectin (ELAM-1) and P-selectin (PADGEM) to standardize the nomenclature and more importantly to reflect their similarity in structural organization and function.

The presence of a C-type lectin domain in each of the three selectins has generated immense interest in determining whether or not selectins bind carbohydrates. A few members of the C-type family of animal lectins (eg. asialoglycoprotein receptor and mannose binding proteins) were already known to recognize carbohydrate structures via the carbohydrate recognition domain (CRD) in a calcium-dependent manner.[33] However, even for these well-characterized members, the carbohydrate ligands for them were loosely defined (mannose-containing glycoconjugates or polysaccharides). The identification of the selectin family, now more than a decade ago, has initiated the search for their counter-receptors as well as studies into the potential role of carbohydrate recognition in leukocyte trafficking, inflammation and immunity.

Do selectins bind carbohydrate structures? Several approaches have confirmed that all three selectins bind carbohydrates. The simplest approach was to inhibit selectin-mediated adhesion events by the use of antibodies with known carbohydrate specificity or liposomes composed of defined glycolipids. A myeloid cell-surface lactosaminoglycan carbohydrate antigen, sialyl Lewis x (sLe^x), was identified as the minimal structure recognized by the selectins. This tetrasaccharide (N-acetylneuraminic acid α2-3galactose β1-4[fucoseα1-3] N-acetyl glucosamine) has also been shown to be critical in selectin binding by other approaches including the expression of glycosyltransferases in transfected cells to create the sLe^x structure on their cell surface glycoconjugates for binding to selectins. A related structure termed sialyl Lewis a (sLe^a) was also found to support selectin binding in vitro. Unfortunately, a variety of carbohydrate structures which have very little resemblance to sLe^x and sLe^a were also reported to be capable of binding to the selectins.[25] How can the high specificity of the selectin-mediated interactions occur in vivo with such a bewildering number of possible ligands, each of which bind the selectins with low affinity? Careful analyses of the available data have revealed the limitations of the binding assays used in various in vitro studies. These reviews have also outlined steps to resolve the controversies regarding the identities of the physiologically relevant carbohydrate ligands for the selectins.[27,28]

What are the counter-receptors for the selectins and what carbohydrate structures do they display (see reference 34 and references therein)? The identities of the counter-receptors are also not necessarily restricted to a single entity. GlyCAM-1, CD34, MAdCAM-1 and podocalyxin have been reported to bind to L-selectin. The common characteristic among these structurally diverse glycoproteins is that they are mucin-type proteins, containing multiple O-linked glycans that serve as ligands for L-selectin. Sulfation of the galactose and N-acetylglucosamine residues of sLe^x on O-linked glycans provides the most effective ligands for L-selectin. The importance of sulfate modification of L-selectin activity in HEV of PLN has been inferred by the use of an inhibitor of sulfate incorporation and analysis of mice lacking a specific sulfo-transferase. For E-selectin, the counter-receptor candidates are VIM-2 (a glycolipid), ESL-1, PSGL-1 and several glycoproteins on monocytes, neutrophils, eosinophils and subpopulations of B and T lymphocytes. There is evidence to support that these counter-receptors present N- or O-linked glycans containing sLe^x or its isomers that are required for E-selectin binding. In contrast to L- and E-selectin, all of the counter-receptor activities for P-selectin can be attributed to PSGL-1 on neutrophils as indicated by studies using PSGL-1 null mice. The glycans on PSGL-1 have not been determined, but O-linked glycans with sLe^x or sLe^x-like structures have been implicated to be critical in P-selectin binding. As is the case for L-selectin, sulfation increases the affinity of P-selectin binding to sLe^x. Interestingly the sulfation occurs not on sLe^x but on one or more tyrosine residues near the N-terminus of PSGL-1. Extensive biochemical analyses of the selectins and their potential carbohydrate ligands have shown definitively that they recognize sialylated, fucosylated carbohydrate ligands related to sLe^x, although the exact structures are still not entirely clear. While all three selectins bind sLe^x with low affinity, ligand specificity for each could further be defined. The interaction of sLe^x with E-selectin appears to be the strongest (affinity estimated to be approximately 100 μM). In vivo, selectin-carbohydrate interactions may be strengthened by the presence of a cluster of sLe^x (E-selectin) or by sulfate modification either on sLe^x (L-selectin) or on tyrosine residues in the vicinity of sLe^x (P-selectin). The molecular basis for the differences between E- and P-selectin has recently been illustrated by the X-ray structures of the CRD-EGF domains of these two selectins with bound ligands, sLe^x and PSGL-1 peptide.[35] Similar X-ray structures of L-selectin with bound ligands are not yet available.

While a number of lectins and potential carbohydrate-binding proteins in the immune system have now been identified, the precise carbohydrate ligands for the majority of these have not been defined. The story of the selectins has revealed the key challenges associated with studies on protein-carbohydrate interactions in the immune system. One of the major challenges is to deal with the complexity of the glycosylation process and various cell- or tissue-specific controls of glycosylation (eg. glycosyltransferase expression). Heterogeneity of carbohydrate structures on a given protein or glycoforms[36] has made analysis of potential carbohydrate ligands very difficult. There is also insufficient information on the carbohydrate structures on host- or pathogen-derived glycoconjugates. Although some of the technical challenges have been overcome by recent advances in sugar analysis and sequencing, a large gap remains in our knowledge of the precise sugar structures on glycoconjugates that are available in minute quantities from natural sources (eg. blood, cells and tissues). Other challenges are the inherent low-affinity binding of monovalent carbohydrate structures to lectins, and the limited availability of physiologically relevant carbohydrate ligands and assays that mimic the physiological conditions under which lectins interact with their carbohydrate ligands. In the case of the selectins, a number of approaches (eg. use of inhibitory antibodies, biochemical and genetic alteration of cell surface glycosylation phenotypes and X-ray crystallography) were required to probe the details of their specificities and mechanisms of carbohydrate-recognition.

A few concepts have emerged from the selectin studies.[34] Although not counter intuitive to our knowledge of other animal lectin-carbohydrate interactions and known glycosylation pathways, they are nevertheless worth reviewing. (1) Potential carbohydrate ligands would not necessarily be restricted to a single structure. (2) The related structures would most probably be present on more than one counter-receptor. Redundancy of carbohydrate ligands and counter-receptors would be advantageous in preserving important lectin-mediated functions. This has been clearly demonstrated in mice lacking certain counter-receptors or glycosyltransferases. The L-selectin mediated lymphocyte homing process is essentially normal in mice lacking either the counter-receptor GlyCAM-1 or CD34. Mice lacking GST-3, a main transferase for the sulfation of L-selectin ligands, also retain significant L-selectin binding activities. (3) The cell type-specific glycosylation pattern and other post-translational modifications of a protein may determine whether or not it serves as a counter-receptor. For example, CD34 only supports L-selectin binding when expressed on L-selectin ligand-positive HEV. (4) Carbohydrate ligand specificity may be defined not only by the glycan structures on the counter-receptors but also the modifications (eg. sulfation) on either the glycan or the protein part of the counter-receptor(s). Further investigations into the selectins and their counter-receptors would provide important insights into the role of carbohydrate-protein interactions in leukocyte trafficking in health and disease. These studies would also provide a structural basis for the development of highly effective inhibitors of selectin-mediated pathophysiology.

Group V CTLD-Containing Proteins in the Immune System—NKC Encoded Receptors

The Group V CTLDs are the most intriguing group of C-type lectins. The overall architecture of receptors in this group is similar to the Group II lectins, with the extracellular CTLD situated at the C-terminus of the protein, which is anchored to the cell membrane by a hydrophobic transmembrane region at the N-terminus of the protein. The CTLD is separated from the cell membrane by a flexible neck region, but unlike the Group II receptors, there are no short sequence repeats in the neck regions of the group V receptors. Analysis of the amino acid sequences of the Group V CTLDs has revealed that the critical residues necessary for calcium binding have been lost, although the determination of several crystal structures of CTLDs from this group has revealed that the overall structure is well conserved. Some of the Group V lectins, such as Ly49A[37] and NKG2A,[5,6] have been shown to bind proteins encoded by the major histocompatibility complex (MHC) proteins as opposed to oligosaccharides. The ligands for other Group V lectins have largely not been determined. Studies have indicated that a few of the Group V receptors may interact with carbohydrates, but incontravertible evidence has been lacking. In addition, the reported carbohydrate-binding properties of rat NKR-P1 and mouse CD69 have been questioned recently. Thus the role of carbohydrates, if any, in the functions of Group V CTLD-containing proteins remains unclear.

Interestingly, all of the Group V CTLD genes are clustered on chromosome 12 in humans, in a region termed the NK complex or NKC.[38] Many of the genes in the NKC are involved in the development and regulation of NK cells, which are large granular lymphocytes of a similar lineage to T cells. However, unlike T cells, NK cells recognize their targets as having down-regulated levels of self-MHC class I, a common consequence of viral infection or transformation.[39] NK cells represent a critical component of innate immunity, particularly in the first stages of pathogen invasion. They are able to detect and lyse transformed cells, and those infected by viral and other microbial pathogens, and can direct the ensuing adaptive immune response via secretion of soluble factors such as IFNγ, IL-1, IL-13, GM-CSF and TGFβ.

Several of the Group V C-type lectin-like molecules expressed by NK cells, including most of the Ly49 receptors and NKG2A, bind to MHC class I molecules on a potential target cell. Upon ligation, these C-type lectin-like receptors deliver an inhibitory signal to the NK cell, preventing cytolysis of healthy cells expressing normal levels of self-class I MHC. In contrast, some of the other C-type lectin-like receptors expressed by NK cells associate with activating signaling adaptor molecules, such as DAP12. The ligands for some of these activating C-type lectin-like receptors have only recently been identified, and evidence shows that they bind to molecules expressed predominantly by "stressed" cells. For example, the human NKG2D homodimer binds to the polymorphic nonclassical MHC molecules MICA and MICB, which are frequently expressed on epithelial tumors and virally infected cells.[40] Thus a fine balance of inhibitory and activatory signals are delivered to NK cells via C-type lectin-like receptors, providing a mechanism for distinguishing infected or transformed cells from healthy cells.

Figure 3 shows the organization of the group V CTLDs encoded in the human and murine NKC, which are generally well conserved between the two species. The Ly49 and NKG2 genes are situated at one end of the NKC, and encode receptors expressed predominantly on NK cells. Interestingly, the only Ly49 gene in humans is a pseudogene.[41] Instead, another structurally distinct group of receptors, the killer immunoglobulin receptors (KIRs) have evolved to perform many of the functions in humans that the Ly49 receptors do in mice. Lox-1, DECTIN-1, CLEC-1 and CLEC–2 form a subgroup of C-type lectin-like receptors in the NKC that are expressed predominantly on myeloid, dendritic and endothelial cells.[42] Telomeric of this group is another group of single copy genes expressed on a range of lymphocytes, including the early activation antigen CD69 and the activation induced C-type lectin (AICL). At this end of the NKC is another group of receptors expressed predominantly on NK cells, including the NKR-P1 family in mice, and the KLRF1 receptor in humans. Interestingly, like Ly49, there is only a single NKR-P1 gene in humans, although it is believed to be functional. The ligands of many of these receptors have been identified, although the role of carbohydrate in the interactions remains controversial. The focus of the following discussion will be on the function of several of the best-characterized group V CTLD receptors in the immune response, and the role that carbohydrate has in the receptor/ligand interactions. For this purpose, we will concentrate on the Ly49 and NKG2D receptors, and review the data on carbohydrate binding by NKR-P1 and CD69.

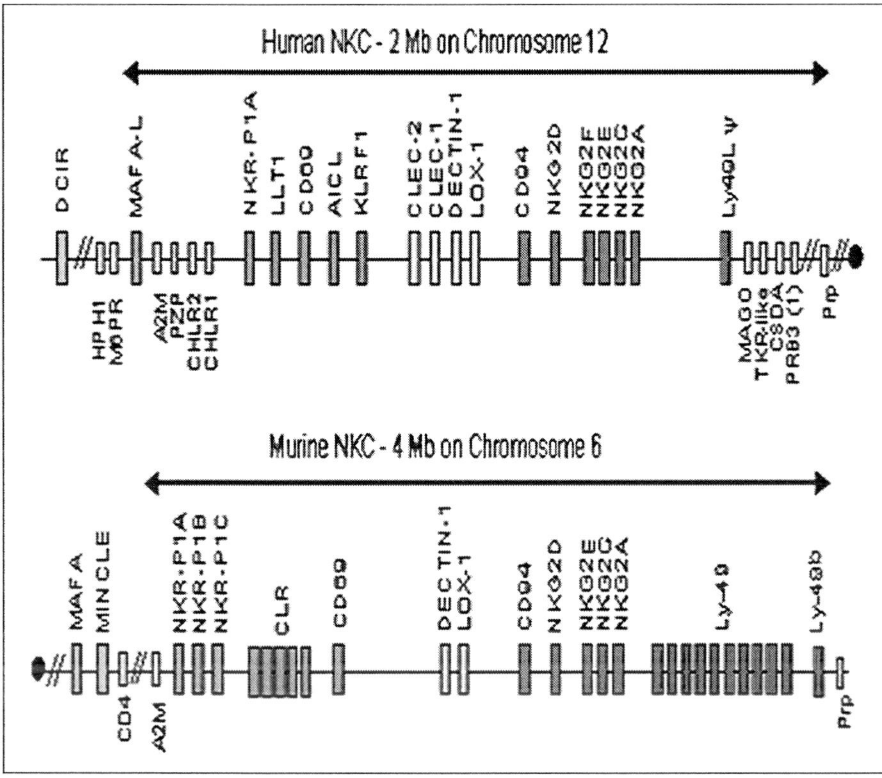

Figure 3. Organization of the extended NKC in humans and mice. The Human NKC on chromosome 12, with the syntenic mouse NKC on chromosome 6. Genes encoding a Group V CTLD are indicated with large boxes, and are labelled above. Small, unfilled boxes represent non-lectin-like genes, and are labelled below. Ψ indicates a pseudogene, and forward slashes indicate gaps. Centromere is indicated by filled circle. The large Ly49 family that is present in the mouse NKC is absent from the human NKC, with only one pseudogene remaining.

Ly49 Receptor Family

The Ly49 receptors were among the first C-type lectin-like NK receptors to be identified, and are encoded by a large number of genes in the NKC.[43,44] There are at least 14 Ly49 genes (designated A to N) in the NKC of C57BL/6 mice, however the number of genes expressed varies enormously between different mouse strains. All of the Ly49 receptors have a virtually identical organization, consisting of a Group V CTLD at the C-terminus, which is separated from the transmembrane region by a short neck. All but Ly49D and Ly49H have an immunomodulatory tyrosine-based inhibitory motif (ITIM) in the cytoplasmic tail, and function as inhibitory receptors. By contrast, Ly49D and Ly49H lack the ITIM but possess a charged residue in the transmembrane region for association with DAP12, an activating adaptor signaling molecule.[45]

Ly49 molecules are expressed as homodimers on NK cells, NKT cells, γδ T cells and also a population of CD8+ T cells with a memory phenotype, although T cells do not express the activating receptors Ly49D and Ly49H. Individual NK cells express a diverse repertoire of Ly49 genes, with as many as five different Ly49 molecules identified on the cell surface, although every NK cell expresses at least one inhibitory receptor. Two papers have recently shown the critical role that Ly49 receptors play in the immune response.[46,47] Both demonstrate that

Ly49H expression is essential for resistance to murine cytomegalovirus (CMV) in vivo, and suggest that the activating Ly49H receptor may bind to gp m144, a CMV encoded MHC class I homologue expressed on the surface of infected cells.[48] More recently, Arase et al have reported that Ly49H directly recognizes a class I MHC-like homologue encoded by murine CMV.[49]

The role of carbohydrate in the interaction between the Ly49 receptors and their natural ligands, the classical class I MHC molecules is debatable. Analysis of the amino acid sequences of the Ly49 receptors shows that they have lost the critical motifs for calcium binding. Furthermore, nonglycosylated class I MHC tetramers bind to inhibitory Ly49 receptors, clearly demonstrating that carbohydrates are not essential for the interaction.[50,51] However, several studies have shown that Ly49A and Ly49C can bind carbohydrates, particularly sulfated polysaccharides,[52,53] and a functional role for carbohydrate binding has been demonstrated. Ly49A-mediated inhibition of NK cytolysis is partially abrogated by soluble carbohydrates, and also by treatment of the target cells with tunicamycin, an inhibitor of N-glycosylation.[52] Moreover, mutation of the glycosylation site at position 176, but not at position 86, of H-2D^d significantly reduced binding of Ly49A and Ly49C.[54] Interestingly, the N-glycosylation site at position 186 is highly conserved among murine class I MHCs, but is not found among those of other species. This study provides support that carbohydrate may be involved in the interaction between Ly49 and class I MHC molecules, and suggests that the Ly49 multigene family may have evolved in concert with the class I MHC in mice.

Finally, the crystal structure of Ly49A bound to its natural ligand H-2D^d was recently determined (Fig. 2), providing possibly a structural basis for carbohydrate-protein interaction.[37] This structure revealed that the Ly49A homodimer interacts with H-2D^d at two distinct interfaces. At site 1, a single Ly49A subunit binds at one side of the MHC peptide-binding platform, spanning the N-terminus of helix $\alpha 1$ and the C-terminus of helix $\alpha 2$. The interaction does not involve the peptide antigen, but is located just above the conserved N-glycosylation site at position 176. The binding site on Ly49A corresponds to the carbohydrate-binding site of MBP, and contains an open cavity-facing residue 176 of H-2D^d. Modeling of carbohydrates attached to residue 176 suggests that this pocket could easily accommodate an N-acetylglucosamine residue, and also a fucose residue, which is a common modification of N-linked oligosaccharides on MHC class I molecules. At site 2, the Ly49A dimer is bound to a cavity formed by the $\alpha 2$ and $\alpha 3$ domains of H-2D^d and $\beta 2$ microglobulin, which lies underneath the peptide binding platform and partially overlaps the CD8 binding site. The same binding site on the Ly49A CTLD is involved as in the site 1 interaction, and the arrangement locates the ligand binding face of Ly49A in the proximity of the glycosylation site at position 86 on the H-2D^d molecule.

Collectively, these findings support a role for ligand glycosylation in the interaction with Ly49 molecules, although it is clearly not essential for binding. The current opinion is that glycosylation of class I MHC molecules may influence the affinity or duration of interactions with Ly49A and Ly49C, thereby fine-tuning the signals that are delivered to NK cells. The absence of bound calcium observed in the interaction may reflect the evolutionary progression of Ly49A from a sugar binding protein to a primarily protein-binding molecule that retains a sugar-binding capacity.

NKG2 Receptor Family

The NKG2 receptors are also encoded in the NKC, but unlike the Ly49 molecules, they are well conserved between rodents and humans. The overall protein architecture is typical of Group V, with the CTLD encoded at the extracellular C-terminus, which is joined to a transmembrane region by a short neck region. There are four NKG2 genes in mice, NKG2A, NKG2C, NKG2D and NKG2E, and an extra fifth gene in humans, NKG2F. All the genes except NKG2D show a high degree of sequence identity to each other, and form a heterodimer on the cell surface with another NKC encoded C-type lectin-like molecule CD94. The NKG2/CD94 dimers are specific for the nonclassical class I MHC molecule HLA-E in humans[6] or Qa-1 in mice.[5] These nonclassical MHC molecules bind nonamer peptides that are derived from the leader sequences of classical class I molecules, and bound peptide is absolutely required for the stable cell surface expression of HLA-E. Thus HLA-E is a reliable indication of intact antigen processing machinery and may be used as another level of detection of down-regulated classical class I MHC expression. However, there is evidence that CMV infection can actually up-regulate expression of HLA-E via the viral protein gpUL40, which mimics class I MHC leader peptide sequences.[55] Both NKG2A/CD94 and NKG2C/CD94 complexes bind to HLA-E, and upon ligation, NKG2A delivers an inhibitory signal to the NK cell via an ITIM in the cytoplasmic domain. In contrast, ligation of NKG2C results in an activation of the NK cell via DAP12.[56] There is evidence to suggest that the inhibitory signal may dominate when delivered in conjunction with an activating signal, as NKG2A/CD94 has a higher affinity for HLA-E.[57]

In contrast, NKG2D is only distantly related to the other NKG2 genes. It does not interact with CD94, but instead forms a homodimer that associates with the activating signaling adaptor protein DAP10 on the surface of all NK cells, $CD8^+$ $\alpha\beta TCR^+$ T cells, $\gamma\delta$ T cells and activated macrophages.[40,58,59] This wide cellular distribution suggests a broader role in the immune response than for the other isoforms, which are generally restricted to NK or T cells. In humans, the NKG2D receptor binds to the nonclassical MHC molecules MICA and MICB.[40] NKG2D ligation by MIC molecules stimulates effector responses from NK and $\gamma\delta$ T cells and may positively modulate $CD8\alpha\beta$ T cell responses.[40,60] Ligation of NKG2D on macrophages triggers $TNF\alpha$ production, and release of nitric oxide.

Recently another group of ligands for NKG2D have been identified.[61] UL16 is a CMV encoded protein that binds to a family of human proteins called UL16 binding proteins (ULBPs). Three GPI anchored ULBPs have been identified so far that share homology with the $\alpha 1$ and $\alpha 2$ domains of class I MHC. Ligation of NKG2D by ULBPs delivers an activating signal to the NK cell and it has been proposed that UL16 may help hCMV avoid the immune response by serving as decoy ligands for NKG2D. MICA and MICB homologues have been found in primates, but not in rodents. In mice, the ligands for NKG2D are encoded by the retinoic acid early inducible (RAE-1) and H60 minor histocompatibility antigens.[58,62] Although the expression of these proteins is low on normal adult tissues, they are constitutively expressed by some tumors. Furthermore, it has recently been shown that tumors expressing RAE-1 molecules can be recognized by NK cells and are rejected.[63]

The crystal structure of an NKG2D dimer bound to MICA has been determined in the last year, and reveals an interaction that is similar to the T cell receptor-MHC class I complexes.[64] Each NKG2D monomer predominantly contacts either the $\alpha 1$ or $\alpha 2$ domains of MICA. Unlike the Ly49A/H-2Dd interface, none of the eight potential N- glycosylation sites on MICA are near the interface, suggesting that the binding is mediated solely by protein/protein interactions. The crystal structure of murine NKG2D dimer bound to RAE-1β has also been determined[65] and resembles the human NKG2D/MICA complex. All of the potential glycosylation sites identified on RAE-1β are distant from the binding interface, also suggesting that carbohydrates are not involved in binding. This is consistent with previous findings by O'Callaghan.[66] Moreover, murine CD94/NKG2A heterodimers bind Qa-1b tetramers, which are devoid of N-glycosylation, providing further evidence that carbohydrate is not necessary for NKG2/ligand interactions.[5] It appears that NKG2 molecules may not be lectins at all, forming protein-protein complexes in the absence of sugars.

CD69 and NKR-P1 Receptors

The crystal structure of another Group V CTLD encoded in the NKC has also been published[67] and reveals intriguing insights into the nature of its ligand. CD69 is one of the earliest cell surface molecules to appear following lymphocyte activation. It is a disulfide linked homodimer expressed on the surface of a variety of hematopoietic cell surfaces, including T, B, and NK cells, monocytes, neutrophils and platelets.[68,69] Crosslinking of CD69 on lymphocytes stimulates proliferation, cytokine secretion, Ca^{2+} flux and cytolytic responses, and on neutrophils, monocytes and platelets leads to nitric oxide release and degranulation. Despite extensive efforts, the physiological ligand for CD69 remains unknown. Early reports demonstrated binding of soluble recombinant CD69 to N-acetylglucosamine,[70] but were not supported by a subsequent report. No detectable binding was observed to the monosaccharides glucose, galactose, mannose, fucose or N-acetylglucosamine, and only weak binding to the polysaccharide fucoidan by the soluble recombinant extracellular domain of CD69 was observed. The crystal structure of CD69 shows that the overall CTLD fold is well conserved, containing three disulfide bonds and the extra β strand, β0, that defines the CTLD of CD69 as the long form. Like other group V CTLDs, CD69 probably lacks any Ca^{2+} binding activity as only one of the four residues coordinated to the first Ca^{2+} in MBP is conserved in CD69, and no bound Ca^{2+} is observed in the crystal structure. Despite this, additional electron density with characteristics of a puckered six-membered ring was observed in the crystal structure of CD69. Residues Asp171 and Glu185 of CD69 appear to contact the extra electron density. Both these residues are located in the putative ligand binding site of CD69, which corresponds to the carbohydrate binding site of MBP, and also the region of Ly49A implicated in potential carbohydrate recognition. The molecule represented by the extra electron density is proposed to have bound to the recombinant CD69 after refolding, and may be part of either a mimetic or natural carbohydrate ligand for CD69.

Finally, the NKR-P1 family of C-type lectin-like receptors encoded in the NKC was among the first to be identified. There are six NKR-P1 genes in mice.[71,72] NKR-P1A and NKR-P1C are activatory receptors, both containing a charged residue in the transmembrane region. It has been shown that this charged residue is responsible for association of the prototype mouse NK cell marker, the NK1.1 antigen (NKR-P1C), with the activatory signalling adaptor molecule FcεRIγ.[73] In contrast murine NKR-P1B contains an ITIM in the cytoplasmic tail, predicting an inhibitory function. There is only a single NKR-P1 gene in the human NKC, NKR-P1A, which unlike its murine counterpart appears to deliver inhibitory signals via an undefined mechanism.[74] The protein is expressed as a homodimer on the surface of NK cells, and also on $CD4^+$ T cells, resting monocytes and DCs. Furthermore, hNKR-P1A was shown to mediate upregulation of IL-1 and IL-12 in monocytes and DCs, and was suggested to be involved in transendothelial migration of resting $CD4^+$ T cells.

The ligand for NKR-P1 is unknown, and early results from a study which indicated that cell surface glycolipids are physiological ligands for rat NKR-P1A was later retracted.[75,76] However, there have been two further reports by the same author that provide evidence for carbohydrate ligands for rat NKR-P1A, in particular N-acetylglucosamine and the chitobiose core of incompletely glycosylated N-linked glycans.[77,78] The observation that soluble recombinant NKR-P1A CTLD can refold in the absence of Ca^{2+},[79] together with the absence of Ca^{2+} binding amino acid residues, suggests that NKR-P1 may bind ligands in a calcium independent manner.

In summary, the Group V CTLDs perform a wide array of functions in the immune response, with important roles in tumour immunity and resistance to viral infection. Their importance is demonstrated by the observation that CMV appears to have evolved mechanisms specifically to avoid detection by C-type lectin-like receptors expressed on NK cells. All Group V CTLDs form dimers at the cell surface, and are encoded by genes located in the NKC. However, many of the receptors encoded in this region are expressed on a wide range of lymphoid and myeloid cells, including DC, macrophages, T cells and B cells. The determination of crystal structures has revealed that the Group V CTLDs have conserved the overall C-type lectin protein fold, despite having lost the critical residues for Ca^{2+} binding. Furthermore, there is evidence that this group of CTLDs may represent an evolutionary distant relative of other C-type lectins. The divergence of these CTLDs appears to result in a switch from a carbohydrate-binding motif to a domain that mediate primarily protein-protein interactions. For some, such as Ly49A, additional recognition of carbohydrate may function to mediate the affinity of the interaction, but for others, such as NKG2D, it appears that carbohydrates are not involved in the interaction at all. Future challenges will be to conclusively elucidate the role of oligosacccharides in the Ly49A interaction with class I MHC, and also to identify the ligands for receptors such as NKR-P1A and CD69. The therapeutic potential of Group V CTLDs in both neoplasia and pathogenic infections warrants further investigation.

Group VI CTLD-Containing Proteins in the Immune System—Phagocytosis and Antigen Uptake and Processing

The group VI CTLDs include the macrophage mannose receptor (MR), Endo-180, DEC-205 and the Phospholipase-A2 (PLA_2) receptor, which all share the same overall domain organization consisting of an extracellular N-terminal cysteine-rich domain, a fibronectin type II repeat and then eight or 10 CTLDs. The mannose receptor and Endo-180 have been shown to bind carbohydrates in a calcium dependent fashion,[80] but DEC-205 and PLA_2 do not contain the calcium binding residues in any of their CTLDs. The mannose receptor is known to have both phagocytic and endocytic functions.[2] There is evidence that other members of this group function as endocytic receptors.

The mannose receptor is the best characterized member of the group, and is known to function as a scavenger receptor by binding and internalizing pathogenic micro-organisms.[2] It is expressed by macrophages and liver endothelial cells, and binds carbohydrates in a calcium dependent fashion. The mannose receptor is important in the immune response against pathogens such as *Mycobacterium tuberculosis* and *Pneumocystis carnii*, and also plays a role in the clearance of the pituitary hormones lutropin and thyrotropin from the circulation by binding to terminal sulfated N-acetylgalactosamine residues.[81,82] It is likely that the presence of multiple CTLDs within the single polypeptide chain of the mannose receptor is important for determining specificity, as it has been shown that high affinity binding requires multivalent interactions between oligosaccharides and several CTLDs, in particular CTLDs 4 to 8. It follows that the spatial arrangement of the CTLDs must influence which ligands are able to interact in a multivalent manner, similar to the situation in MBP, in which the orientation of the three CTLDs is fixed to ensure that only nonself oligosaccharides are recognized. Like MBP, the mannose receptor is specific for mannose, N-acetylglucosamine and fucose residues. The crystal structure of the fourth CRD of the mannose receptor has been determined, and reveals that the basic C-type lectin fold is well conserved.[83] A recent study has shown a role for the mannose receptor in mediating lymphocyte exit from lymphoid tissues.[84] The mannose receptor was shown to be expressed on human lymphatic endothelium, and binds to another C-type lectin-like receptor, L-selectin expressed on lymphocytes.

Conclusions

The C-type lectin-like superfamily members have many functions in the immune response. They are critical in cell-cell interactions, not only in mediating exit of lymphocytes from the blood at sites of inflammation (the selectins), but also in the exit of cells from lymphoid tissues (mannose receptor and L-selectin) and in recruitment of DCs to organs such as the liver (Kupffer receptor). The role of the canonical C-type lectin MBP as an innate pattern recognition receptor has been well-characterized, but the recent analysis of a number of C-type lectin-like receptors expressed on APCs (as discussed in other chapters in this book) has highlighted their role in the induction of the adaptive immune response. Furthermore, the association of DC-SIGN and Ly49H with HIV and CMV infections respectively has emphasized that C-type lectin-like receptors have a crucial function in viral immunity. Finally, the elucidation of ligands expressed predominantly on transformed cells for the C-type lectin-like NK receptor NKG2D has exposed the mechanisms by which NK cells are activated to reject tumors. However, it remains to be determined whether or not carbohydrates contribute or responsible for their activities.

References

1. Drickamer K. Evolution of Ca(2+)-dependent animal lectins. Prog Nucleic Acid Res Mol Biol 1993; 45:207-232.
2. Weis WI, Taylor ME, Drickamer K. The C-type lectin superfamily in the immune system. Immunol Rev 1998; 163:19-34.
3. Loewen MC, Gronwald W, Sonnichsen FD et al. The ice-binding site of sea raven antifreeze protein is distinct from the carbohydrate-binding site of the homologous C-type lectin. Biochemistry 1998; 37:17745-17753.
4. Bertrand JA, Pignol D, Bernard JP et al. Crystal structure of human lithostathine, the pancreatic inhibitor of stone formation. EMBO J 1996; 15:2678-2684.
5. Vance RE, Kraft JR, Altman JD et al. Mouse CD94/NKG2A is a natural killer cell receptor for the nonclassical major histocompatibility complex (MHC) class I molecule Qa-1(b). J Exp Med 1998; 188:1841-1848.
6. Braud VM, Allan DS, O'Callaghan CA et al. HLA-E binds to natural killer cell receptors CD94/NKG2A, B and C. Nature 1998; 391:795-799.
7. Weigel PH. Endocytosis and function of the hepatic asialoglycoprotein receptor. Subcell Biochem 1993; 19:125-161.
8. Novak N, Kraft S, Bieber T. IgE receptors. Curr Opin Immunol 2001; 13:721-726.
9. Geijtenbeek TB, Torensma R, van Vliet SJ et al. Identification of DC-SIGN, a novel dendritic cell-specific ICAM-3 receptor that supports primary immune responses. Cell 2000; 100:575-585.
10. Drickamer K. Engineering galactose-binding activity into a C-type mannose-binding protein. Nature 1992; 360:183-186.
11. Valladeau J, Duvert-Frances V, Pin JJ et al. Immature human dendritic cells express asialoglycoprotein receptor isoforms for efficient receptor-mediated endocytosis. J Immunol 2001; 167:5767-5774.
12. Uwatoku R, Suematsu M, Ezaki T et al. Kupffer cell-mediated recruitment of rat dendritic cells to the liver: roles of N-acetylgalactosamine-specific sugar receptors. Gastroenterology 2001; 121:1460-1472.
13. Curtis BM, Scharnowske S, Watson AJ. Sequence and expression of a membrane-associated C-type lectin that exhibits CD4-independent binding of human immunodeficiency virus envelope glycoprotein gp120. Proc Natl Acad Sci USA 1992; 89:8356-8360.
14. Geijtenbeek TB, Krooshoop DJ, Bleijs DA et al. DC-SIGN-ICAM-2 interaction mediates dendritic cell trafficking. Nat Immunol 2000; 1:353-357.
15. Engering A, Geijtenbeek TB, van Vliet SJ et al. The dendritic cell-specific adhesion receptor DC-SIGN internalizes antigen for presentation to T cells. J Immunol 2002; 168:2118-2126.
16. Feinberg H, Mitchell DA, Drickamer K et al. Structural basis for selective recognition of oligosaccharides by DC-SIGN and DC-SIGNR. Science 2001; 294:2163-2166.
17. Mitchell DA, Fadden AJ, Drickamer K. A novel mechanism of carbohydrate recognition by the C-type lectins DC-SIGN and DC-SIGNR. Subunit organization and binding to multivalent ligands. J Biol Chem 2001; 276:28939-28945.
18. Weis WI, Kahn R, Fourme R et al. Structure of the calcium-dependent lectin domain from a rat mannose-binding protein determined by MAD phasing. Science 1991; 254:1608-1615.
19. Siegelman MH, van der Rijn M, Weissman IL. Mouse lymph node homing receptor cDNA clone encodes a glycoprotein revealing tandem interaction domains. Science 1989; 243:1165-1172.
20. Lasky LA, Singer MS, Yednock TA et al. Cloning of a lymphocyte homing receptor reveals a lectin domain. Cell 1989; 56:1045-1055.
21. Tedder TF, Isaacs CM, Ernst TJ et al. Isolation and chromosomal localization of cDNAs encoding a novel human lymphocyte cell surface molecule, LAM-1. Homology with the mouse lymphocyte homing receptor and other human adhesion proteins. J Exp Med 1989; 170:123-133.
22. Bevilacqua MP, Stengelin S, Gimbrone MA et al. Endothelial leukocyte adhesion molecule 1: An inducible receptor for neutrophils related to complement regulatory proteins and lectins. Science 1989; 243:1160-1165.
23. Johnston GI, Cook RG, McEver RP. Cloning of GMP-140, a granule membrane protein of platelets and endothelium: sequence similarity to proteins involved in cell adhesion and inflammation. Cell 1989; 56:1033-1044.
24. Bevilacqua MP. Endothelial-leukocyte adhesion molecules. Annu Rev Immunol 1993; 11:767-804.
25. Varki A. Selectin ligands. Proc Natl Acad Sci USA 1994; 91:7390-7397.
26. Lasky LA. Selectin-carbohydrate interactions and the initiation of the inflammatory response. Annu Rev Biochem 1995; 64:113-139.
27. Kansas GS. Selectins and their ligands: Current concepts and controversies. Blood 1996; 88:3259-3287.

28. Varki A. Selectin ligands: Will the real ones please stand up? J Clin Invest 1997; 99:158-162.
29. Crockett-Torabi E. Selectins and mechanisms of signla transduction. J Leuk Biol 1998; 63:1-14.
30. Gonzalez-Amaro R, Sanchez-Madrid F. Cell adhesion molecules: Selectins and integrins. Crit Rev Immunol 1999; 19:389-429.
31. Vestweber D, Blanks JE. Mechanisms that regulate the function of the selectins and their ligands. Physiol Rev 1999; 79:181-213.
32. Ley K. Functions of selectins. Results Probl Cell Differ 2001; 33:177-200.
33. Drickamer K. Two distinct classes of carbohydrate-recognition domains in animal lectins. J Biol Chem 1998; 263:9557-9560.
34. Lowe JB. Glycosylation in the control of selectin counter-receptor structure and function. Immunol Rev 2002; 186:19-36.
35. Somers WS, Tang J, Shaw GD et al. Insights into the molecular basis of leukocyte tethering and rolling revealed by structures of P- and E-selectin bound to Slex and PSGL-1. Cell 2000; 103:467-479.
36. Rademacher TW, Parekh RB, Dwek RA. Glycobiology. Annu Rev Biochem 1988; 57:785-838.
37. Tormo J, Natarajan K, Margulies DH et al. Crystal structure of a lectin-like natural killer cell receptor bound to its MHC class I ligand. Nature 1999; 402:623-631.
38. Yokoyama WM, Seaman WE. The Ly-49 and NKR-P1 gene families encoding lectin-like receptors on natural killer cells: the NK gene complex. Annu Rev Immunol 1993; 11:613-635.
39. Ljunggren HG, Karre K. In search of the 'missing self': MHC molecules and NK cell recognition. Immunol Today 1990; 11:237-244.
40. Bauer S, Groh V, Wu J et al. Activation of NK cells and T cells by NKG2D, a receptor for stress-inducible MICA. Science 1999;285:727-729.
41. Westgaard IH, Berg SF, Orstavik S et al. Identification of a human member of the Ly-49 multigene family. Eur J Immunol 1998; 28:1839-1846.
42. Sobanov Y, Bernreiter A, Derdak S et al. A novel cluster of lectin-like receptor genes expressed in monocytic, dendritic and endothelial cells maps close to the NK receptor genes in the human NK gene complex. Eur J Immunol 2001; 31:3493-3503.
43. Takei F, McQueen KL, Maeda M et al. Ly49 and CD94/NKG2: developmentally regulated expression and evolution. Immunol Rev 2001; 181:90-103.
44. Anderson SK, Ortaldo JR, McVicar DW. The ever-expanding Ly49 gene family: Repertoire and signaling. Immunol Rev 2001; 181:79-89.
45. Smith KM, Wu J, Bakker AB et al. Ly-49D and Ly-49H associate with mouse DAP12 and form activating receptors. J Immunol 1998; 161:7-10.
46. Lee SH, Girard S, Macina D et al. Susceptibility to mouse cytomegalovirus is associated with deletion of an activating natural killer cell receptor of the C-type lectin superfamily. Nat Genet 2001; 28:42-45.
47. Brown MG, Dokun AO, Heusel JW et al. Vital involvement of a natural killer cell activation receptor in resistance to viral infection. Science 2001; 292:934-937.
48. Farrell H, Degli-Esposti M, Densley E et al. Cytomegalovirus MHC class I homologues and natural killer cells: an overview. Microbes Infect 2000; 2:521-532.
49. Arase H, Mocarski ES, Campbell AE et al. Direct recognition of cytomegalovirus by activating and inhibitory NK cell receptors. Science 2002; 296:1323-1326.
50. Hanke T, Takizawa H, McMahon CW et al. Direct assessment of MHC class I binding by seven Ly49 inhibitory NK cell receptors. Immunity 1999; 11:67-77.
51. Michaelsson J, Achour A, Salcedo M et al. Visualization of inhibitory Ly49 receptor specificity with soluble major histocompatibility complex class I tetramers. Eur J Immunol 2000; 30:300-307.
52. Daniels BF, Nakamura MC, Rosen SD et al. Ly-49A, a receptor for H-2Dd, has a functional carbohydrate recognition domain. Immunity 1994; 1:785-792.
53. Brennan J, Takei F, Wong S et al. Carbohydrate recognition by a natural killer cell receptor, Ly-49C. J Biol Chem 1995; 270:9691-9694.
54. Lian RH, Freeman JD, Mager DL et al. Role of conserved glycosylation site unique to murine class I MHC in recognition by Ly-49 NK cell receptor. J Immunol 1998; 161:2301-2306.
55. Tomasec P, Braud VM, Rickards C et al. Surface expression of HLA-E, an inhibitor of natural killer cells, enhanced by human cytomegalovirus gpUL40. Science 2000; 287:1031.
56. Lanier LL, Corliss B, Wu J et al. Association of DAP12 with activating CD94/NKG2C NK cell receptors. Immunity 1998; 8:693-701.
57. Vales-Gomez M, Reyburn HT, Erskine RA et al. Kinetics and peptide dependency of the binding of the inhibitory NK receptor CD94/NKG2-A and the activating receptor CD94/NKG2-C to HLA-E. EMBO J 1999; 18:4250-4260.

58. Diefenbach A, Jamieson AM, Liu SD et al. Ligands for the murine NKG2D receptor: Expression by tumor cells and activation of NK cells and macrophages. Nat Immunol 2000; 1:119-126.
59. Wu J, Song Y, Bakker AB et al. An activating immunoreceptor complex formed by NKG2D and DAP10. Science 1999; 285:730-732.
60. Groh V, Steinle A, Bauer S et al. Recognition of stress-induced MHC molecules by intestinal epithelial gamma delta T cells. Science 1998; 279:1737-1740.
61. Cosman D, Mullberg J, Sutherland CL et al. ULBPs, novel MHC class I-related molecules, bind to CMV glycoprotein UL16 and stimulate NK cytotoxicity through the NKG2D receptor. Immunity 2001; 14:123-133.
62. Cerwenka A, Bakker AB, McClanahan T et al. Retinoic acid early inducible genes define a ligand family for the activating NKG2D receptor in mice. Immunity 2000; 12:721-727.
63. Diefenbach A, Jensen ER, Jamieson AM et al. Rae1 and H60 ligands of the NKG2D receptor stimulate tumour immunity. Nature 2001; 413:165-171.
64. Li P, Morris DL, Willcox BE et al. Complex structure of the activating immunoreceptor NKG2D and its MHC class I-like ligand MICA. Nat Immunol 2001; 2:443-451.
65. Li P, McDermott G, Strong RK. Crystal structures of RAE-1beta and its complex with the activating immunoreceptor NKG2D. Immunity 2002; 16:77-86.
66. O'Callaghan CA, Cerwenka A, Willcox BE et al. Molecular competition for NKG2D: H60 and RAE1 compete unequally for NKG2D with dominance of H60. Immunity 2001; 15:201-211.
67. Natarajan K, Sawicki MW, Margulies DH et al. Crystal structure of human CD69: A C-type lectin-like activation marker of hematopoietic cells. Biochemistry 2000; 39:14779-14786.
68. Testi R, D'Ambrosio D, De Maria R et al. The CD69 receptor: a multipurpose cell-surface trigger for hematopoietic cells. Immunol Today 1994; 15:479-483.
69. Marzio R, Mauel J, Betz-Corradin S. CD69 and regulation of the immune function. Immunopharmacol Immunotoxicol 1999; 21:565-582.
70. Bezouska K, Nepovim A, Horvath O et al. CD 69 antigen of human lymphocytes is a calcium-dependent carbohydrate- binding protein. Biochem Biophys Res Commun 1995; 208:68-74.
71. Giorda R, Trucco M. Mouse NKR-P1. A family of genes selectively coexpressed in adherent lymphokine-activated killer cells. J Immunol 1991; 147:1701-1708.
72. Plougastel B, Matsumoto K, Dubbelde C et al. Analysis of a 1-Mb BAC contig overlapping the mouse Nkrp1 cluster of genes: Cloning of three new Nkrp1 members, Nkrp1d, Nkrp1e, and Nkrp1f. Immunogenetics 2001; 53:592-598.
73. Arase N, Arase H, Park SY et al. Association with FcRgamma is essential for activation signal through NKR-P1 (CD161) in natural killer (NK) cells and NK1.1+ T cells. J Exp Med 1997; 186:1957-1963.
74. Lanier LL, Chang C, Phillips JH. Human NKR-P1A. A disulfide-linked homodimer of the C-type lectin superfamily expressed by a subset of NK and T lymphocytes. J Immunol 1994; 153:2417-2428.
75. Bezouska K, Yuen CT, O'Brien J et al. Oligosaccharide ligands for NKR-P1 protein activate NK cells and cytotoxicity. Nature 1994; 372:150-157.
76. Bezouska K, Yuen CT, Jacqui O'Brien et al. Nature 1996; 372:150-157 (retraction of reference 75).
77. Bezouska K, Kren V, Kieburg C et al. GlcNAc-terminated glycodendrimers form defined precipitates with the soluble dimeric receptor of rat natural killer cells, sNKR-P1A. FEBS Lett 1998; 426:243-247.
78. Bezouska K, Sklenar J, Dvorakova J et al. NKR-P1A protein, an activating receptor of rat natural killer cells, binds to the chitobiose core of uncompletely glycosylated N-linked glycans, and to linear chitooligomers. Biochem Biophys Res Commun 1997; 238:149-153.
79. Kogelberg H, Lawson AM, Muskett FW et al. Expression in Escherichia coli, folding in vitro, and characterization of the carbohydrate recognition domain of the natural killer cell receptor NKR-P1A. Protein Expr Purif 2000; 20:10-20.
80. Sheikh H, Yarwood H, Ashworth A et al. Endo180, an endocytic recycling glycoprotein related to the macrophage mannose receptor is expressed on fibroblasts, endothelial cells and macrophages and functions as a lectin receptor. J Cell Sci 2000; 113:1021-1032.
81. Fiete D, Baenziger JU. Isolation of the SO^4-4-GalNAc beta1,4GlcNAc beta1,2Man alpha-specific receptor from rat liver. J Biol Chem 1997; 272:14629-14637.
82. Fiete D, Beranek MC, Baenziger JU. The macrophage/endothelial cell mannose receptor cDNA encodes a protein that binds oligosaccharides terminating with SO^4-4- GalNAc beta1,4GlcNAc beta or Man at independent sites. Proc Natl Acad Sci USA 1997; 94:11256-11261.
83. Feinberg H, Park-Snyder S, Kolatkar AR et al. Structure of a C-type carbohydrate recognition domain from the macrophage mannose receptor. J Biol Chem 2000; 275:21539-21548.
84. Irjala H, Johansson EL, Grenman R et al. Mannose receptor is a novel ligand for L-selectin and mediates lymphocyte binding to lymphatic endothelium. J Exp Med 2001; 194:1033-1042.

Chapter 8

The Sialic Acid-Binding Siglec Family

Lars Nitschke and Paul R. Crocker

Abstract

Siglecs are a group of adhesion receptors with tyrosine-based inhibitory signaling motifs. They are mainly expressed by cells of the hematopoietic system. Recently new members of the Siglec family were identified both in the human and the mouse. These comprise a CD33-related subgroup. All Siglecs specifically recognize sialic acids that are broadly expressed on many cell surfaces. These properties suggest that Siglecs may be involved in cell-cell interactions resulting in intracellular inhibitory signals.

Introduction to the Siglec Family

The sialic acid (Sia)-binding Ig-like lectins (Siglecs) are adhesion and signaling receptors that interact specifically with sialic acids. Sia is abundantly expressed on cell surfaces and on secreted glycoproteins. More than 40 forms exist in nature and they can be attached in a variety of linkages to other sugars thereby generating a considerable degree of molecular diversity and specificity.[1] On the cellular surface, sialic acids are a major carrier of negative charge and can reduce nonspecific cell adhesion through a charge repulsion effect. However, the Siglecs have the potential to promote cellular interactions via Sia recognition. The characterization of Siglecs as Sia-binding proteins began with independent studies on sialoadhesin (Sn), which is a receptor on macrophages,[2] and CD22 on B cells.[3] When both cDNAs were cloned, a significant degree of sequence similarity between these two members of the immunoglobulin (Ig) superfamily was observed.[4] Two other known mammalian members of the Ig superfamily, namely myelin associated glycoprotein (MAG) and CD33, were found to share sequence similarity with Sn and were subsequently shown to be Sia-binding proteins.[5,6] Most recently, several new members of the Siglec family have been discovered, most of them by sequencing of randomly selected human cDNAs. Apart from MAG (Siglec-4a), which is found exclusively in the nervous system, all Siglecs are expressed within the hemopoietic and the immune systems. Most of the Siglecs have a quite restricted expression pattern, present only on some cell types of the immune system. Siglecs often carry one or several tyrosine-based inhibitory motifs (ITIMs) in their intracellular tails (Fig. 1) ITIM motifs are known to be involved in negative regulation of cellular signaling. These properties suggest that most Siglec family members may be involved in cell-cell interactions that result in intracellular inhibitory signals.

Sialoadhesin (Sn, Siglec-1, CD169) is expressed on macrophages in a regulated fashion, with high levels on distinct sets of macrophages, such as those present in the perifollicular zones of lymphoid tissues.[7] Sn has a preference for Sia in α2,3 linkage (eg NeuNAc-α2,3-Gal-β1,3GalNAc) expressed, for example, on the surface of myeloid cells and in the extracellular matrix. Certain pathogenic microorganisms also express sialylated glycoconjugates on their surface. Therefore, it is thought that Sn may function in host defense as well as acting as an accessory molecule to promote interactions of macrophages with the host microenvironment. CD22 (Siglec-2) is expressed specifically on B-cells. It has a high degree of specificity for α2,6 linked Sia (2,6Sia)

Immunobiology of Carbohydrates, edited by Simon Y.C. Wong and Gemma Arsequell.
©2003 Eurekah.com and Kluwer Academic / Plenum Publishers.

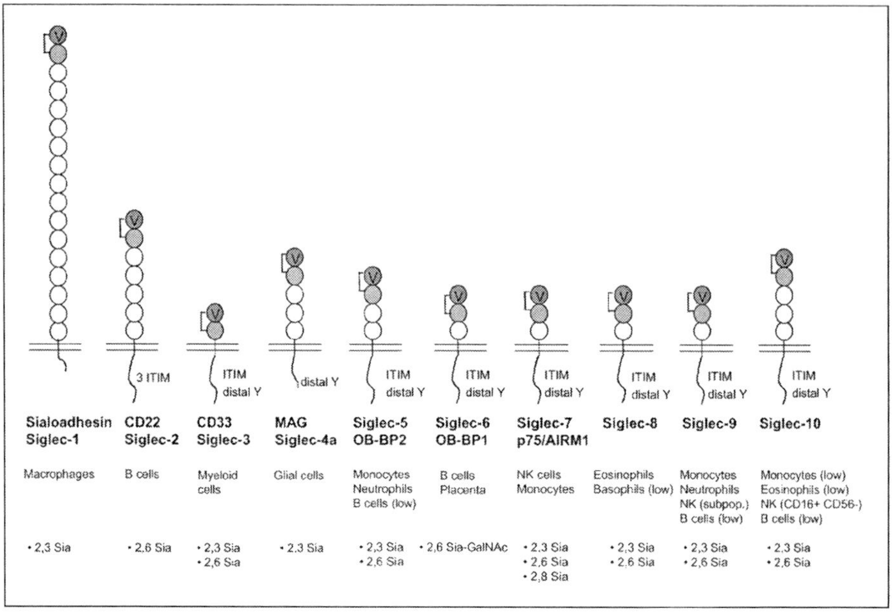

Figure 1. The human Siglec family. The diagram shows the domain organization and presence of ITIM motifs or the membrane-distal SLAM-like Tyr motifs (distal Y) on the cytoplasmic tails of human Siglecs. Greatest sequence similarity between Siglecs occurs in the V-set domain (red) and the adjacent 1 to 2 N-terminal domains (orange). The sialic acid binding site is contained in the V-set domain. Names of the various members, their expression pattern and sialic acid specificities are shown.

and carries three ITIM motifs on its intracellular tail. Potentially, CD22 can bind to target cells such as lymphocytes or cytokine-activated endothelial cells that express high levels of α2,6 Sia on the surface.[5,8,9] The role of CD22 in modulating B-cell signaling is quite well understood, due to the characterization of CD22-deficient mice, as will be discussed below. CD33 (Siglec-3) was originally shown to have specificity for α2,3-linked Sia (6), although recent observations suggest that CD33 may prefer the α2,6 linkage.[10] CD33 has been extensively used as a marker of myeloid cell progenitors that is absent from pluripotential hemopoietic stem cells. During myelopoiesis, CD33 is lost from neutrophils but retained on monocytes and some tissue macrophages.[6] MAG (Siglec 4a) and its close homologue SMP (Siglec 4b), which has only been identified in birds, are expressed on neuronal cells. MAG had been postulated as being important for the process of myelin formation. However, MAG-deficient mice develop essentially normally and only show some defects in myelin-axon interactions at older age (Fig. 1).[11,12]

New CD33-Related Siglecs

Identification and cloning of new human members of the Siglec family in the last 3 years demonstrate that the Siglecs comprise a large gene family. All of these new members (hSiglec-5 to Siglec-11) are most strongly related to CD33 (with 50 to 80 % amino acid identity), thus comprise a subfamily (Fig. 1).[13-22] An unusual CD33-related Siglec-like molecule (Siglec-L1) with two tandem V-set domains, each lacking a critical arginine residue that ligates Sia, has also been characterized by two laboratories.[23,24] One group described weak Sia-dependent binding whereas the other laboratory could not demonstrate this unless one of the key arginine residues was reintroduced.[23,24] Interestingly, the primate orthologue of this Siglec-like protein contains a canonical arginine residue and is capable of avid Sia-dependent binding, suggesting that a mutation leading to the loss of Sia binding occurred relatively recently during evolution.

The genes encoding the CD33-related Siglecs are clustered on human chromosome 19q13.3-13.4. Little is currently known concerning the functional properties of these new Siglecs, but they have been characterized in terms of tissue distribution, sugar-binding specificity and in initial signaling studies. By using specific monoclonal antibodies (mAbs), the expression pattern of each of the new human Siglecs on blood leukocytes was analysed.[13-22] The new CD33-related Siglecs are expressed mostly in the myeloid lineage. Each new Siglec exhibits a very specific and often very restricted expression pattern among hematopoietic cells suggesting specific functions. Several Siglecs can be present on the same cell type (e.g., monocytes express CD33 and Siglecs-5,-7,-9 and -10) which suggests some functional redundancy.

All CD33-related Siglecs carry a membrane-proximal ITIM-like motif. In addition to ITIM, most CD33-like Siglecs have a well-conserved membrane-distal tyrosine based motif, TEYSE(I/V), that is similar to a motif in SLAM (signaling lymphocytes activation molecule) known to be important for recruitment of SAP (SLAM-associated protein).[25] SAP has been shown to prevent interactions with the SHP-2 protein tyrosine phosphatase as well as promoting additional signaling pathways that affect both activatory and inhibitory responses.[26] Following treatment of cells with pervanadate to inhibit tyrosine phosphatases, hCD33 or Siglec-7 become phosphorylated.[17,27-29] CD33 could also be tyrosine-phosphorylated following crosslinking with anti-CD33 mAb; this could be inhibited by the src kinase inhibitor, PP2.[27] Siglec-10 could be tyrosine phosphorylated in vitro by different kinases such as STAT and src-related kinases.[30] Under these conditions, Siglec-7 recruited only SHP-1, whereas CD33 and Siglec-10 bound both SHP-1 and SHP-2.[27-30] Functional evidence that CD33-related Siglecs can mediate inhibitory signals has been obtained by cocrosslinking CD33 with mAb to an activating receptor, the FcγR1. This coligation resulted in reduced Ca^{2+} flux.[28,29] Similarly, antibody-induced crosslinking of Siglec-7 on NK cells led to reduced killing of P815 cells in a redirected killing assay.[17] The physiological relevance of these interesting findings is so far unknown.

Recently, a comparison between human and mouse genomes has identified a region on mouse chromosome 7 which is syntenic to the CD33-related gene cluster on human chromosome 19.[22] A remarkable difference between the mouse and human Siglec gene cluster was found. While the human gene cluster of CD33-related genes contains 7 Siglec genes and 13 pseudogenes, the corresponding region in the mouse only contains 4 Siglec genes and 2 pseudogenes. Because of this divergence it is difficult to assign human and mouse orthologues in a one to one fashion. The reported orthologue of hCD33 ('mCD33')[31] was identified in this region but in contrast to the human protein, the mouse protein lacks ITIM-like motifs in the cytoplasmic tail. The other 3 murine Siglecs were given a temporary letter code: mSiglecE is most-related to hSiglec-7,-8 and -9, mSiglecF is the closest relative of hSiglec-5 and -6 while mSiglecG corresponds to hSiglec10.[22] An additional CD33-like mouse Siglec gene (mSiglec-H) has been identified which also lacks ITIM-like motifs and for which there is no equivalent gene in humans.[22] Sia-binding specificities and tissue distribution of the new mouse orthologues are just beginning to be examined. mSiglecE or MIS (mouse inhibitory Siglec) was shown to have a high mRNA expression level in spleen cells and to recruit SHP-1 and SHP-2 when expressed in a human promonocytic cell line treated with pervanadate.[32,33] In these cells, mSiglecE could inhibit FcγR1-induced Ca^{2+} mobilization when both molecules were crosslinked together with antibodies. mSiglecF is expressed on immature cells of the myelomonocytic lineage and a subset of CD11b (Mac-1)-positive cells.[22]

Mode of Sialic Acid Recognition by Siglecs

Mammalian lectins that mediate Sia-dependent interactions have only been identified relatively recently. An important class of such lectins are the selectins, which mediate lymphocyte homing and leukocyte migration into tissues during inflammation. Like the Siglecs, physiological interactions mediated by the selectins are Sia dependent. The ligands for selectins contain sialylated, fucosylated and sulfated components (sialyl-Lewis^x derivatives). In contrast

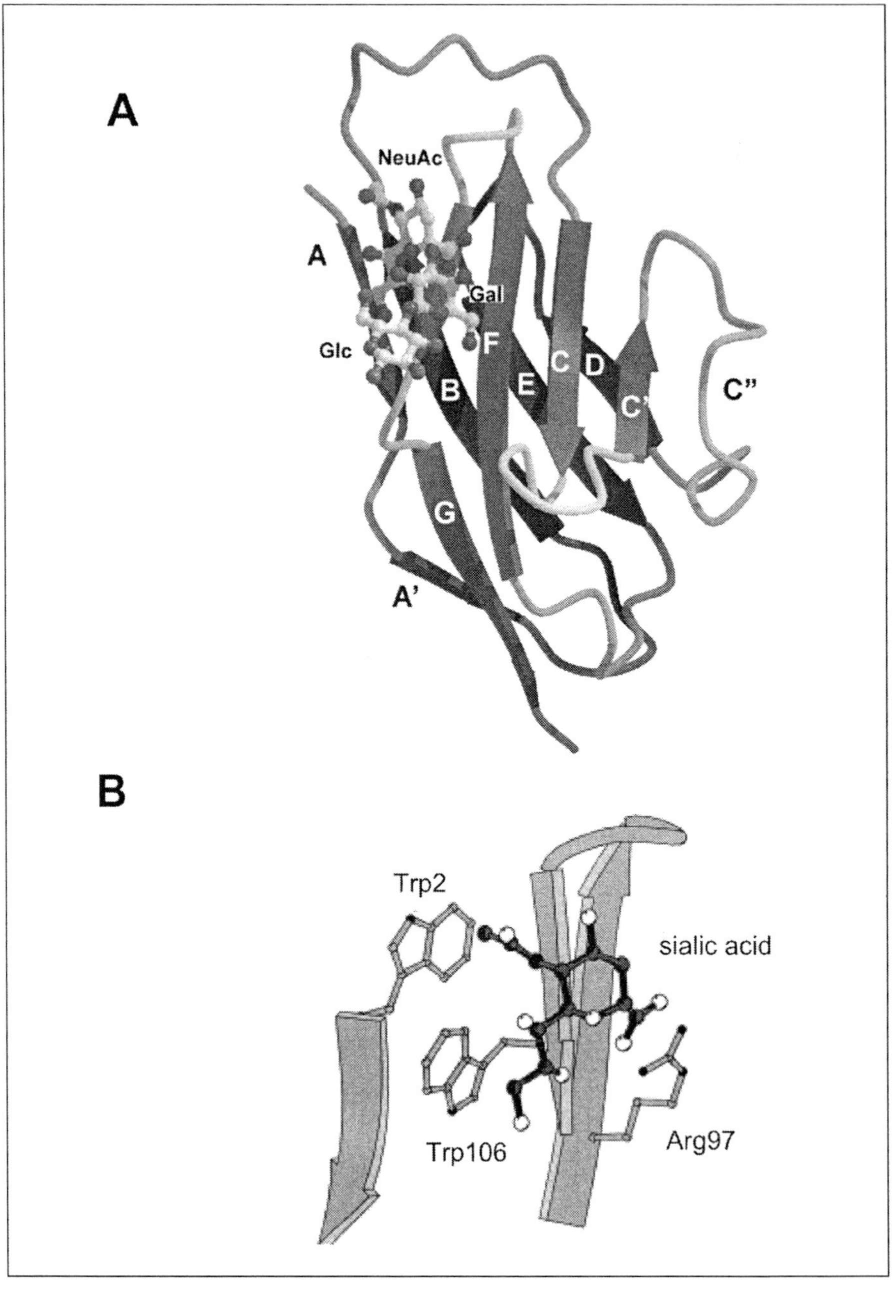

Figure 2. Structure of the N-terminal domain of sialoadhesin.
a) The V-set domain of Sn in complex with 3' sialyllactose is shown. The 3' sialyllactose makes interactions with residues positioned on the A, G and F strands.
b) Structure of the sialic acid binding site in Sn. Positions of key residues near the sialic acid residue in the crystal structure of the N-terminal domain of Sn complexed with 3' sialylactose are shown. For details see reference 34.

to selectins, that primarily ligate fucose, Siglecs are Sia binding proteins which form direct molecular contacts with the chemical substituents of Sia, namely the carboxylate group, the glycerol side chain and the N-acetyl group.[34]

The extracellular region of each Siglec is made up of an N-terminal V-set domain, followed by varying numbers of C2-set domains (Fig. 1). The two N-terminal domains are predicted to be linked by an unusual disulfide bond that is likely to have a major influence on their orientation. However, the significance of this disulfide bond for ligand recognition is unclear because truncation deletion studies, site-directed mutagenesis experiments, X-ray crystallography and NMR analysis have shown that the Sia binding site is contained exclusively within the N-terminal V-set domain.[34-37] The recent structural determination by X-ray crystallography of the isolated V-set domain of Sn in a complex with α2,3 sialyllactose has directly demonstrated the interaction of Sn to its ligand[34] (Fig. 2a). This structure has provided a template for Sia recognition that is likely to be applicable to other Siglecs.

By solving the crystal structure of Sn complexed to α2,3 sialyllactose it became clear that most of the molecular contacts occur with the Sia, rather than the attached sugar units. As shown in (Fig. 2b), an arginine residue (Arg97 in Sn) which is highly conserved in the Siglec family forms a salt bridge with the carboxylate group of sialic acid. Two well-conserved aromatic groups (both tryptophan for Sn) are involved in hydrophobic interactions with the N-acetyl and glycerol side chains of Sia.[34] Results of site directed mutagenesis studies carried out with both Sn and CD22 are in good agreement with the crystal structure. In particular, the crucial importance of the conserved arginine has been demonstrated by the finding that even a conservative substitution with lysine leads to about 10 fold reduction in binding of Sn and undetctable binding in the case of CD22.[35,36] Recent studies of Sn-sialoside interactions by NMR have provided evidence that the molecular interactions characterized in crystals also occur in solution and have demonstrated that the affinity of Sn for sialosides is very low, around 10^{-3} M.[37] The very low affinity of Sn and other Siglecs could be important in their ability to mediate cell-cell interactions in plasma, since despite being abundant, poorly clustered glycans on plasma proteins would not be expected to compete efficiently with the highly clustered Sia present on cell surfaces.

The Influence of CIS Inhibition on Adhesion to Other Cells in Trans

With the exception of Siglec-6, which shows a restricted specificity for the sialyl Tn antigen (Neu5Ac α2-6GalNAc α), all known Siglecs recognize forms and linkages of Sia that are found at the cell surface and on extracellular plasma proteins. Since Sia are abundantly expressed on cell surfaces, it could be expected that Siglecs would be involved in multiple cellular interactions. However, while all of the Siglecs studied to date can mediate Sia-dependent binding to cells as purified proteins, there are striking differences in their ability to do this when expressed on cell surfaces. This was first noted for CD33, which when expressed transiently on COS cells, required sialidase treatment to promote binding to cells carrying sialylated ligands.[6] When CD22 was expressed in COS or CHO cells, high levels of cellular binding were observed.[3,5] However these cells lack ST6GalI, the sialyltransferase that creates ligands recognized by CD22. When ST6GalI was coexpressed with CD22 in both CHO or COS cells, all CD22-dependent binding activity was lost. The binding could be restored by sialidase treatment of the transfected cells.[8,38]

Since Siglecs recognize forms of Sia that are commonly found at the cell surface, the Siglec binding site can consequently be masked by cis-interactions with sialylated ligands on the same plasma membrane. A recent study, using multivalent glycoprobes, showed that unmasked forms of Siglecs on human blood leukocytes could not be detected without prior sialidase treatment to remove inhibitory Sia. Unmasking of blood mononuclear cells also occurred after cellular activation with phorbol ester plus ionomycin treatment. These results are similar to those reported previously for CD22, which was shown to be masked on most human or mouse B-cells,[40,41] but it can become unmasked after cellular activation. This unmasking effect may

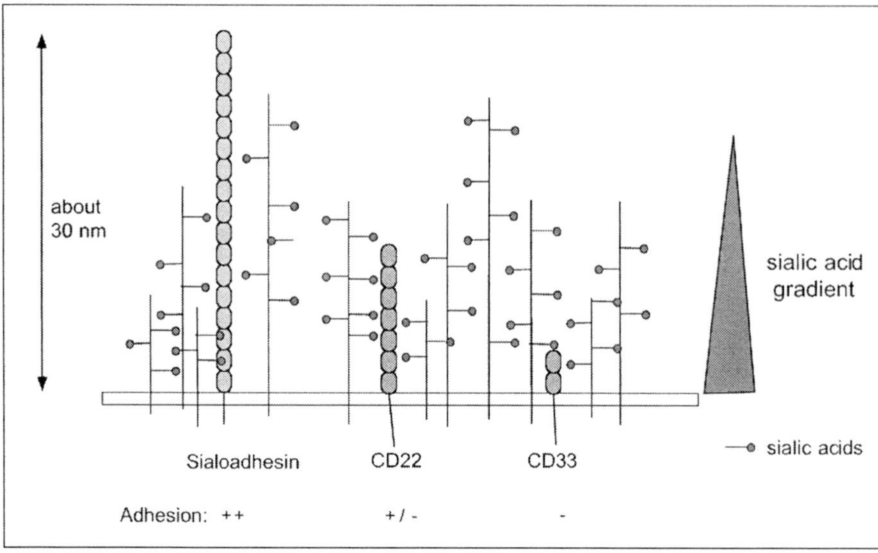

Figure 3. Potential cis-inhibition of Siglec binding sites. Diagram illustrating how extension of the sialic acid binding site (N-terminal V-set domain) of Siglecs away from the plasma membrane could be important in regulation of cell-cell interactions. According to this model, Siglecs like Sn that extend the siliac acid binding site a long distnace from the plasma membrane are likely to be less prone to cis-inhibition by sialic acids present on the plasma membrane. In comparison, the binding site of CD33 is believed to be completely masked due to its relatively small size.

be due to reduced cell surface expression of Sia resulting from either increased activity of a sialidase or decreased expression of a sialyltransferase. Cis-interactions to Sia on the same cellular surface could potentially regulate Siglec functions, e.g., by preventing cell-cell contacts until required or by affecting intracellular signaling functions. The findings with CD33, CD22 and other Siglecs provide a possible explanation why Sn has evolved the large number (i.e., 17) of Ig domains. This extended structure may position the Sia binding site away from the plasma membrane, in order to avoid contact with cis inhibitory Sia present on the same cellular surface (Fig. 3).

CD22: An Example of a Siglec on B Cells with Inhibitory and Adhesion Functions

Most knowledge of biological function for Siglecs has been collected for CD22 (Siglec-2). CD22 is partly associated with the B-cell antigen receptor (BCR) on the B-cell plasma membrane and inhibits BCR signaling by recruiting and activating SHP-1.[42] This role was clearly demonstrated by the generation of CD22-deficient mice. The strongest phenotype of these mice was the greatly enhanced Ca^{2+} mobilization in their B-cells after BCR stimulation.[43-46] Accordingly, B cells of CD22-deficient mice showed a mildly activated phenotype in vivo. When CD22-deficient mice become older they often produce autoantibodies.[47] CD22 seems to be important for controlling the signaling threshold, particularly at early B-cell activation. However, when T-cell help is provided, $CD22^{-/-}$ B cells respond normally, as shown by immune responses and germinal center formation. While the SHP-1-mediated inhibitory function of CD22 was evident in CD22-deficient mice, the role of other signaling molecules binding to the phosphorylated tail of CD22, such as Syk, PLCγ2, Grb-2 or Shc, is still unknown.[48,49]

So far it has been unclear by which mechanism CD22 associates with the BCR and whether the ligand binding of CD22 also influences its inhibitory signaling function. Cis-binding to

2,6Sia on the B-cell surface could potentially control CD22 association to the BCR or tyrosine phosphorylation of this Siglec. By pretreating B cells with anti-CD22 mAb coupled to beads, a higher B-cell response by anti-IgM mAb was triggered than without pretreatment.[50] This result was interpreted as sequestering CD22 away from the BCR and increased signaling resulting from the release of inhibition. Recently, the role of the Sia-binding CD22 domain in regulating signaling was addressed by identification of a highly specific synthetic sialic acid analogue that bound hCD22 with 200-fold increased affinity. Pretreatment of human B cells with this synthetic sialoside led to increased Ca^{2+} mobilization after addition of anti-IgM. In the presence of the sialoside inhibitor, CD22 was less tyrosine-phosphorylated and recruited less SHP-1 than in control treatments.[51] These results show for the first time that ligand-binding, most likely a cis interaction, is required for efficient tyrosine-phosphorylation and subsequent inhibitory function of a Siglec. Similar results were obtained by another group using a different approach.[52] It seems that CD22 interaction with a specific sialylated ligand on the B-cell surface has to be postulated in order to explain these results. It will be a challenging task to find this specific ligand on the B-cell surface. Also, it will be interesting to see whether such a mechanism applies to other Siglecs.

The Sia-binding domain of CD22 is also involved in trans interactions. It was found that 2,6Sia is expressed on endothelium of bone marrow sinusoids.[53] This specific expression of 2,6Sia on bone marrow endothelium, but not on endothelium of other organs, suggested a specific function. We could show by injecting CD22-Fc protein into mice, that this fusion protein bound to bone marrow endothelium and blocked recirculation of mature B-cells back to the bone marrow. This finding, together with the phenotype of CD22-deficient mice which showed a strongly reduced population of mature, recirculating B cells in the bone marrow, suggested that CD22 is a bone marrow homing receptor for recirculating B cells.[53] Similar to selectins which induce lymphocyte homing to the lymph nodes by binding to specific ligands expressed on high endothelial venules, CD22 may direct mature B-cells to the parenchyme of the bone marrow. Further studies are needed to reveal the mechanistic details of this important process.

Conclusions

The Siglec family is an unusual and interesting group of adhesion and signaling receptors. Although the molecular basis for Sia recognition and specificities are largely known, for most members little is known about the functional significance of these interactions. The highly conserved tyrosine-based motifs and association with SHP phosphatases suggest inhibitory functions, but very few data concerning the functional relevance are available. CD22 is a Siglec where these questions have been addressed recently and novel results have been obtained on the role of Sia-binding in regulation of inhibitory function and in cell-cell contacts. Especially for the Siglecs expressed on cells of the innate immune system it is tempting to speculate that Sia residues could act as broadly expressed "self" ligands in a mechanism to distinguish foreign pathogens from self. This theory is based on the fact that Sia rarely occurs in lower organisms, although some pathogens have acquired the capacity to synthesize Sia as a form of host molecular mimicry. By binding to Sia on host cells Siglecs could control activation thresholds via their inhibitory domains. Such a mechanism has recently been demonstrated for CD22 when B cells are stimulated by membrane-bound antigens.[54]

References

1. Kelm S, Schauer R. Sialic acids in molecular and cellular interactions. Int Rev Cytol 1997; 175:137-240.
2. Crocker PR, Gordon S. Properties and distribution of a lectin-like hemagglutinin differentially expressed by murine stromal tissue macrophages. J Exp Med 1986; 164:1862-1875.
3. Stamenkovic I, Sgroi D, Aruffo A et al. The B lymphocyte adhesion molecule CD22 interacts with leukocyte common antigen CD45RO on T cells and alpha 2-6 sialyltransferase, CD75, on B cells. Cell 1991; 66:1133-1144.
4. Crocker PR, Mucklow S, Bouckson V et al. Sialoadhesin, a macrophage sialic acid binding receptor for haemopoietic cells with 17 immunoglobulin-like domains. EMBO J 1994; 13:4490-4503.

5. Kelm S, Pelz A, Schauer R et al. Sialoadhesin, myelin-associated glycoprotein and CD22 define a new family of sialic acid-dependent adhesion molecules of the immunoglobulin superfamily. Curr Biol 1994; 4:965-972.
6. Freeman SD, Kelm S, Barber EK et al. Characterization of CD33 as a new member of the sialoadhesin family of cellular interaction molecules. Blood 1995; 85:2005-2012.
7. Crocker PR, Gordon S. Mouse macrophage hemagglutinin (sheep erythrocyte receptor) with specificity for sialylated glycoconjugates characterized by a monoclonal antibody. J Exp Med 1989; 169:1333-1346.
8. Hanasaki K, Varki A, Powell LD. CD22-mediated cell adhesion to cytokine-activated human endothelial cells. Positive and negative regulation by alpha 2-6-sialylation of cellular glycoproteins. J Biol Chem 1995; 270:7533-7542.
9. Engel P, Nojima Y, Rothstein D et al. The same epitope on CD22 of B lymphocytes mediates the adhesion of erythrocytes, T and B lymphocytes, neutrophils, and monocytes. J Immunol 1993; 150:4719-4732.
10. Brinkman-Van der Linden EC, Varki A. New aspects of siglec binding specificities, including the significance of fucosylation and of the sialyl-Tn epitope. Sialic acid-binding immunoglobulin superfamily lectins. J Biol Chem 2000; 275:8625-8632.
11. Li C, Tropak MB, Gerlai R et al. Myelination in the absence of myelin-associated glycoprotein. Nature 1994; 369:747-750.
12. Fruttiger M, Montag D, Schachner M et al. Crucial role for the myelin-associated glycoprotein in the maintenance of axon-myelin integrity. Eur J Neurosci 1995; 7:511-515.
13. Cornish AL, Freeman S, Forbes G et al. Characterization of siglec-5, a novel glycoprotein expressed on myeloid cells related to CD33. Blood 1998; 92:2123-2132.
14. Patel N, Brinkman-Van der Linden EC, Altmann SW et al. OB-BP1/Siglec-6. A leptin- and sialic acid-binding protein of the immunoglobulin superfamily [published erratum appears in J Biol Chem 1999 Sep 24;274(39):28058]. J Biol Chem 1999; 274:22729-22738.
15. Nicoll G, Ni J, Liu D et al. Identification and characterization of a novel siglec, siglec-7, expressed by human natural killer cells and monocytes. J Biol Chem 1999; 274:34089-34095.
16. Angata T, Varki A. Siglec-7: A sialic acid-binding lectin of the immunoglobulin superfamily. Glycobiology 2000; 10:431-438.
17. Falco M, Biassoni R, Bottino C et al. Identification and molecular cloning of p75/AIRM1, a novel member of the sialoadhesin family that functions as an inhibitory receptor in human natural killer cells. J Exp Med 1999; 190:793-802.
18. Floyd H, Ni J, Cornish AL et al. Siglec-8. A novel eosinophil-specific member of the immunoglobulin superfamily. J Biol Chem 2000; 275:861-866.
19. Zhang JQ, Nicoll G, Jones C et al. Siglec-9, a novel sialic acid binding member of the immunoglobulin superfamily expressed broadly on human blood leukocytes. J Biol Chem 2000; 275:22121-22126.
20. Angata T, Varki A. Cloning, characterization, and phylogenetic analysis of Siglec-9, a new member of the CD33-related Group of Siglecs. Evidence for coevolution with sialic acid synthesis pathways. J Biol Chem 2000; 275:22127-22135.
21. Munday J, Kerr S, Ni J et al. Identification, characterization and leucocyte expression of Siglec-10, a novel human sialic acid-binding receptor. Biochem J 2001; 355:489-497.
22. Angata T, Hingorani R, Varki NM et al. Cloning and characterization of a novel mouse Siglec, mSiglec-F: Differential evolution of the mouse and human (CD33) Siglec-3-related gene clusters. J Biol Chem 2001; 276:45128-45136.
23. Angata T, Varki NM, Varki A. A second uniquely human mutation affecting sialic acid biology. J Biol Chem 2001; 276:40282-40287.
24. Yu Z, Lai CM, Maoui M et al. Identification and characterization of S2V, a novel putative siglec that contains two V set Ig-like domains and recruits protein-tyrosine phosphatases SHPs. J Biol Chem 2001; 276:23816-23824.
25. Sayos J, Wu C, Morra M et al. The X-linked lymphoproliferative-disease gene product SAP regulates signals induced through the coreceptor SLAM [see comments]. Nature 1998; 395:462-469.
26. Veillette A. The SAP family: A new class of adaptor-like molecules that regulates immune cell functions. Sci STKE 2002:E8.
27. Taylor VC, Buckley CD, Douglas M et al. The myeloid-specific sialic acid-binding receptor, CD33, associates with the protein-tyrosine phosphatases, SHP-1 and SHP-2. J Biol Chem 1999; 274:11505-11512.
28. Ulyanova T, Blasioli J, Woodford-Thomas TA et al. The sialoadhesin CD33 is a myeloid-specific inhibitory receptor. Eur J Immunol 1999; 29:3440-3449.

29. Paul SP, Taylor LS, Stansbury EK et al. Myeloid specific human CD33 is an inhibitory receptor with differential ITIM function in recruiting the phosphatases SHP-1 and SHP-2. Blood 2000; 96:483-490.
30. Whitney G, Wang S, Chang H et al. A new siglec family member, siglec-10, is expressed in cells of the immune system and has signaling properties similar to CD33. Eur J Biochem 2001; 268:6083-6096.
31. Tchilian EZ, Beverley PC, Young BD et al. Molecular cloning of two isoforms of the murine homolog of the myeloid CD33 antigen. Blood 1994; 83:3188-3198.
32. Yu Z, Maoui M, Wu L et al. MSiglec-E, a novel mouse CD33-related siglec (sialic acid-binding immunoglobulin-like lectin) that recruits Src homology 2 (SH2)-domain- containing protein tyrosine phosphatases SHP-1 and SHP-2. Biochem J 2001; 353:483-492.
33. Ulyanova T, Shah DD, Thomas ML. Molecular cloning of MIS, a myeloid inhibitory siglec, that binds protein-tyrosine phosphatases SHP-1 and SHP-2. J Biol Chem 2001; 276:14451-14458.
34. May AP, Robinson RC, Vinson M et al. Crystal structure of the N-terminal domain of sialoadhesin in complex with 3' sialyllactose at 1.85 A resolution. Mol Cell 1998; 1:719-728.
35. Vinson M, van der Merwe PA, Kelm S et al. Characterization of the sialic acid-binding site in sialoadhesin by site-directed mutagenesis. J Biol Chem 1996; 271:9267-9272.
36. van der Merwe PA, Crocker PR, Vinson M et al. Localization of the putative sialic acid-binding site on the immunoglobulin superfamily cell-surface molecule CD22. J Biol Chem 1996; 271:9273-9280.
37. Crocker PR, Vinson M, Kelm S et al. Molecular analysis of sialoside binding to sialoadhesin by NMR and site-directed mutagenesis. Biochem J 1999; 341:355-361.
38. Braesch-Andersen S, Stamenkovic I. Sialylation of the B lymphocyte molecule CD22 by alpha 2,6-sialyltransferase is implicated in the regulation of CD22-mediated adhesion. J Biol Chem 1994; 269:11783-11786.
39. Razi N, Varki A. Cryptic sialic acid binding lectins on human blood leukocytes can be unmasked by sialidase treatment or cellular activation. Glycobiology 1999; 9:1225-1234.
40. Razi N, Varki A. Masking and unmasking of the sialic acid-binding lectin activity of CD22 (Siglec-2) on B lymphocytes. Proc Natl Acad Sci USA 1998; 95:7469-7474.
41. Floyd H, Nitschke L, Crocker PR. A novel subset of murine B cells that expresses unmasked forms of CD22 is enriched in the bone marrow: Implications for B-cell homing to the bone marrow. Immunology 2000; 101:342-347.
42. Doody GM, Justement LB, Delibrias CC et al. A role in B cell activation for CD22 and the protein tyrosine phosphatase SHP. Science 1995; 269:242-244.
43. Nitschke L, Carsetti R, Ocker B et al. CD22 is a negative regulator of B-cell receptor signalling. Curr Biol 1997; 7:133-143.
44. O'Keefe TL, Williams GT, Davies SL et al. Hyperresponsive B cells in CD22-deficient mice. Science 1996; 274:798-801.
45. Otipoby KL, Andersson KB, Draves KE et al. CD22 regulates thymus-independent responses and the lifespan of B cells. Nature 1996; 384:634-637.
46. Sato S, Miller AS, Inaoki M et al. CD22 is both a positive and negative regulator of B lymphocyte antigen receptor signal transduction: Altered signaling in CD22-deficient mice. Immunity 1996; 5:551-562.
47. O'Keefe TL, Williams GT, Batista FD et al. Deficiency in CD22, a B cell-specific inhibitory receptor, is sufficient to predispose to development of high affinity autoantibodies. J Exp Med 1999; 189:1307-1313.
48. Cyster JG, Goodnow CC. Tuning antigen receptor signaling by CD22: Integrating cues from antigens and the microenvironment. Immunity 1997; 6:509-517.
49. Nitschke L, Floyd H, Crocker PR. New functions for the sialic acid-binding adhesion molecule CD22, a member of the growing family of Siglecs. Scand J Immunol 2001; 53:227-234.
50. Tooze RM, Doody GM, Fearon DT. Counterregulation by the coreceptors CD19 and CD22 of MAP kinase activation by membrane immunoglobulin. Immunity 1997; 7:59-67.
51. Kelm S, Gerlach J, Brossmer R et al. The ligand-binding domain of CD22 is needed for inhibition of the BCR signal, as demonstrated by a novel human CD22-specific inhibitor compound. J Exp Med 2000; in press.
52. Jin L, McLean PA, Neel BG et al. Sialic acid binding domains of CD22 are required for negative regulation of B cell receptor signaling. J Exp Med 2002; 195:1199-1205.
53. Nitschke L, Floyd H, Ferguson DJ et al. Identification of CD22 ligands on bone marrow sinusoidal endothelium implicated in CD22-dependent homing of recirculating B cells. J Exp Med 1999; 189:1513-1518.
54. Lanoue A, Batista FD, Stewart M et al. Interaction of CD22 with alpha2,6-linked sialoglycoconjugates: Innate recognition of self to dampen B cell autoreactivity? Eur J Immunol 2002; 32:348-355.

CHAPTER 9

Antibody Responses to Polysaccharides

Carola G. Vinuesa and Ian C.M. MacLennan

Abstract

The polysaccharide capsules of *Haemophilus influenzae b,* pneumococci and meningococci protect these bacteria from innate immune mechanisms. Consequently, antibody responses to these encapsulated organisms are crucial for host defence. These responses are different from those stimulated by conventional protein antigens because the requirement for T cell help is bypassed. The clinical importance of antibody responses to the polysaccharides that coat encapsulated bacteria is underlined by the high rate of morbidity and mortality from these organisms during infancy, when the immature immune system cannot make T-independent responses. The reason for the delay in developing this crucial defence is not known and an evolutionary advantage is not obvious. Our understanding of immune responses to encapsulated bacteria has increased substantially over the last few years. Specialized subsets of B cells—marginal zone B cells and peritoneal B1 cells— are responsible for the protective extrafollicular antibody response. Recent studies show that rapid B cell activation and fast antibody production, which are crucial to prevent dissemination of encapsulated pathogens, depend on interactions between the innate and adaptive immune systems.

Introduction

The innate immune system has considerable powers that protect individuals with antibody deficiency from bacterial infection. Nevertheless, those bacteria that have a well-organized polysaccharide capsule present a particular hazard to these patients. These bacteria are the pneumococci, meningococci and *Haemophilus influenzae* b (Hib). While they can exist in encapsulated and nonencapsulated states, they grow mainly within the body in their encapsulated form. For convenience they will be referred to as encapsulated bacteria in this chapter, although many other bacteria have coats of varying sophistication outside their cell wall.

Polysaccharide capsules confer virulence, in part because they enable bacteria to evade adaptive and specific immune defence mechanisms.[1] Capsules inhibit phagocytosis, resist the lytic action of complement; and obscure choline residues in the cell wall from recognition by C-reactive protein and natural antibody. Thus, the capacity to generate opsonizing antibody targeted towards capsular polysaccharides is critical for the early host defence against these organisms.

Paradoxically, the adaptive immune system is slow to develop the ability to produce protective antibody against purified capsular polysaccharides.[2-4] This delay is associated with a peak of susceptibility to infection between 6 months of age, when the protection afforded by maternally-derived antibody is lost, and 5 years, when a full capacity to respond to capsular polysaccharides is acquired. During the first few years of life the ability to respond to polysaccharides is acquired gradually and responsiveness to some serotypes develops later than to others.[3] Worldwide, this late development of a capacity to produce an adaptive response to capsular polysaccharides contributes to an enormous number of deaths in infants from infection with encapsulated bacteria.[5] The effectiveness of antibody in this age group is illustrated by the

Immunobiology of Carbohydrates, edited by Simon Y.C. Wong and Gemma Arsequell.
©2003 Eurekah.com and Kluwer Academic / Plenum Publishers.

dramatic protection afforded by the Hib conjugate vaccines.[6] The protein associated with these conjugates enables them to elicit T cell help and it is this property that results in a protective immune response against Hib from birth.[7] At birth the human immune system is already able to mount specific antibody responses against viruses as well as bacterial exotoxins and endotoxins.

There have been considerable advances in our understanding of the cellular and molecular basis of antibody responses to polysaccharide antigens. Despite this, the selective advantage for the late emergence of responsiveness to many of these antigens in the face of the massive disadvantage associated with lack of antibody against encapsulated bacteria in infancy remains unclear.

B Cell Activation and Signaling Pathways Triggered by Polysaccharide Antigens

In conventional or thymus-dependent (TD) immune responses to protein antigens, B cell activation and antibody production only occurs after B cells have received two signals: one from antigen binding to the B cell receptor for antigen (BCR) and the other from T cells in the form of cell-cell contact and cytokine signaling. In order for B cells to elicit T cell help they must make cognate interaction with rare primed T cells expressing T cell receptors that recognize antigen-derived peptides presented by the class II major histocompatibility complexes on the B cell. Antibody responses to many pathogens depend on dual recognition by T and B cells. This requirement for two kinds of receptors and signals is a key mechanism that enables immune cells to distinguish foreign from self-antigens.[8] This constraint greatly reduces the risk of autoantibody production, as both B cells and T cells must simultaneously escape tolerance to the same self-antigen. For several important classes of microorganisms, the second signal can come from specific components of pathogens themselves, enabling antibody production in the absence of T cells. These antigens are called thymus-independent (TI) antigens. They can elicit antibodies in athymic (nude) mice and other strains of T cell-deficient mice.

In contrast to proteins, capsular polysaccharides express regularly spaced repeating antigenic determinants (epitopes) that induce multivalent BCR cross-linking.[9] This, together with their high molecular weight and their poor internalization by B cells, results in potent and persistent signaling, which obviates the requirement for specific T cell help. In any case T help is unavailable because the B cell does not process and present epitopes from polysaccharides.[10,11] The polysaccharide component also appears to play a role, since high molecular weight, repetitive polypeptides do not elicit TI responses. Nevertheless, highly repetitive antigenic epitopes displayed on the surface of some virions[12] and polymerized bacterial flagellin[13] elicit TI antibody production. The protein content of these antigens also results in a TD response.

TI antigens are sub-divided into type 1 (TI-1), exemplified by lipopolysaccharide (LPS, or endotoxin) and type-2 (TI-2), which mainly consists of pure polysaccharides. This classification originates from the observation that polysaccharides fail to elicit B cell responses in CBA/N (X-linked immunodeficiency) mice, even though they respond normally to both TD and TI-1 antigens.[14] The cause of this murine immunodeficiency has been identified as a point mutation in the gene encoding Bruton's tyrosine kinase (Btk).[15] Btk participates in several signaling pathways downstream from the B cell receptor (BCR), which are critical for survival and differentiation of activated B cells in response to polysaccharide antigens. The BCR consists of membrane Ig, which contains a short intracytoplasmic tail noncovalently linked to CD79a (Igα and (CD79b) Igβ.[16] Signaling after BCR ligation is initiated by these molecules. A coreceptor complex consisting of CD19, CD21, TAPA-1 and Leu13, regulates the threshold for B cell activation.[17] Importantly, CD21 (or complement receptor 2, CR2) binds the complement degradation fragment C3d. It has been shown that antigens that have bound C3d can deliver signals by cross-linking CD21 and the BCR. This lowers by 10-100 fold the threshold for B cell activation.[18] Many TI-2 antigens directly activate the complement system without the need for specific Ig.[19] This generates C3 convertases that break down C3 to C3a and C3b. C3b binds to the bacteria and breaks down further to the stable C3d fragment that binds CD21.

The BCR complex phosphorylates protein tyrosine kinases (Lyn, Fyn, and Syk), which then activate phosphoinositol-3 kinase. The phosphoinositol-3 kinase generates inositol phosphate metabolites including PIP_3, which bind to the pleckstrin homology domain of Btk and localizes it to the cell membrane. Activated Syk phosphorylates the adaptor molecule BLNK, which enables Btk to be phosphorylated by Syk. Btk in turn activates phospholipase C-γ2. Downstream targets of this activation pathway include diacylglycerol, inositol triphosphate, and protein kinase C-β, which lead to induction of calcium fluxes. Activated Btk plays a crucial role in BCR-dependent B cell survival through activation of NF-κB,[20] which is known to control the expression of the anti-apoptotic molecules Bcl-2 and Bcl-xL.[21] Mouse studies have shown that selective TI-2 unresponsiveness does not only occur in mice deficient in Btk or its adaptor BLNK, but can also be found after deletion of phosphoinositol-3 kinase, protein kinase C-β and phospholipase C-γ2 genes.

Point mutations of Btk in humans cause X-linked agammaglobulinaemia (XLA), which is characterized by B cell deficiency and low or absent serum immunoglobulin. Exceptionally Btk-deficient patients have B cells. In this situation they make TD responses but have a selective deficiency in their ability to respond to TI-2 antigens.[22,23] This is consistent with evidence that the bulk of morbidity in XLA is due to respiratory tract infections with encapsulated bacteria. Recent studies have identified patients presenting as XLA phenocopies but who harbour non-Btk gene defects. For example, patients with mutations in the gene for the linker protein BLNK, present with clinical features typical of XLA.[24,25] Other patients with selective susceptibility to encapsulated bacteria may be found with similar signaling defects as those identified in mouse studies (i.e., phosphoinositol-3 kinase, protein kinase C-β, phosphlipase C-γ2 etc).

The Three Main Sources of Antibodies

Antibody is produced during three main processes. First a significant proportion of serum IgM and the IgA secreted in the gut arises from the spontaneous or nonspecifically-induced maturation of B1 cells to produce antibody. The antibodies produced by this process are termed natural antibodies.

Second, antibody is induced by antigen-driven direct differentiation of B cells to proliferate as plasmablasts. This occurs in extra-follicular foci in the spleen, or the medullary cords in lymph nodes. The plasmablasts differentiate to become plasma cells, which are generally short-lived. Both TD and TI antigens induce these extra-follicular antibody responses.

Finally plasma cells are also derived from germinal center B cells. This is the source of persistent, high affinity antibody produced in TD antibody responses. All three types of antibody production have to be considered in relation to responses to encapsulated bacteria.

Natural Antibody and B1 B Cells

The natural serum antibody is mainly IgM,[26] but B1 cells also contribute substantially to IgA secreted from the gut.[27] This antibody is produced by spontaneous maturation of a sub-group of B cells known as B1 cells. The antigen-independence of at least some natural antibody production is shown by its presence at normal levels in germ-free rodents that have been fed on a chemically defined diet and housed in a hypoallergenic environment.[26,28,29] Many natural antibodies appear to have broad specificity: natural antibodies bind to a wide range of bacterial antigens and many have weak affinity for self-antigens. Natural antibody with specificity for phosphorylcholine provides a relevant example in the context of the present chapter. Phosphorylcholine has been identified as a component of the capsules of 6 of the more than 90 pneumococcal serotypes.[30] This is typically a component of bacterial cell wall C-polysaccharide and is a target for protective antibody against many bacteria. Unfortunately this is inaccessible in most,[31] but not all[32] encapsulated bacteria. The importance of phosphorylcholine as a target is suggested by it being bound by other innate immune molecules such as C reactive protein and platelet activating factor receptor.[33]

While it is reasonable to attribute natural antibody production to B1 cells, available evidence suggests this is not their only function and that they also produce antibody in responses to TI and perhaps TD antigens that engage their BCR.[34] In addition it appears that B1 cells can differentiate to become marginal zone (MZ) B cells, which are typically the responsive cellular subset to polysaccharide antigens.[35-39] Importantly this is only one of the sources of MZ B cell development. As we shall see later they also develop from the recirculating pool and from germinal center B cells. The main characteristics of B1 cells are summarized in (Table 1).

It was initially suggested that B1 and conventional follicular B cells (B2) belong to separate lineages. There is strong evidence that B1 cells are produced and maintained by processes that differ from those associated with recirculating B cells and their progeny. B1 cells in mice appear to be produced almost exclusively before 7 weeks of age by a process that does not require IL-7. By contrast, recirculating B cells are produced throughout life by IL-7-dependent B lymphopoiesis.[40] Once produced, the B1 cell subset appears capable of self-renewal and, unlike other mature B cell subsets, does not require the tumour necrosis factor family ligand BAFF (B cell activating factor) for continued survival.[41,42] Recirculating B cells, by contrast, have limited capacity for self-renewal and require survival signals from BAFF to survive. B1 cells are also generally more resistant to apoptosis and this seems to be related with their lower expression of Fas. Nevertheless, they show heightened apoptosis in response to anti-IgM.[43] Another significant difference is the constitutive expression of STAT-3 by B1 cells compared to B2 cells.[44]

B1 cells appear to undergo a selection process involving BCR ligation. There is a close correlation between BCR activation threshold and production of B1 cells. A decrease in the BCR activation threshold leads to an expansion of B1 cells, while mutations that increase the threshold for B cell activation abolish the B1 population.[45-47] Thus, B1 cells are selectively absent from both Btk-deficient and $CD19^{-/-}$ mice, as well as from mice bearing other mutations that affect the Btk signaling pathway. Conversely, the population is increased in $CD22^{-/-}$, $CD72^{-/-}$, or $Lyn^{-/-}$mice, where the BCR signaling threshold is decreased. These data are also consistent with the finding that treatment with Cyclosporin A, which interferes with signaling downstream of the BCR, appears to block selection into the B1 repertoire.[48] Certain BCR specificities appear to guide the development of B1 or B2 cells.[49] Indeed, strong self-antigen binding appears to be a prerequisite for B1 cell selection.[50]

There are a number of features associated with B1 cells, which are not necessarily absolute criteria for classifying a B cell as a B1 cell:

(A). The Ig Heavy (H) chain rearrangements during B1 lymphopoiesis typically occur without the action of terminal deoxynucleotidyl transferase (Tdt).[51] The flanking sequences on the downstream side of each Ig V segment gene and the upstream side of each D segment gene vary. Consequently, not all V segments will recombine with all DJ rearrangements when the sequences are germ-line. This restricts the V gene usage in the absence of Tdt and provides an explanation for the restricted pattern of B1 cells' V segment usage.[52] These features of B1 lymphopoiesis mean that the B1 repertoire is far more closely defined by the genome than the B2 repertoire, which has more V, D, and J segment usage and extensive junctional diversity introduced by Tdt. Thus, the B1 repertoire of natural antibodies represents an innate selection of antibodies which are likely to have been selected for their protective capacity against common life-threatening infection.

(B). B1 cells typically are located in the mouse peritoneal and pleural cavities, but not all B cells in these cavities are B1 cells.[53] Recently the importance of the chemokine CXCL-13 produced by peritoneal macrophages in inducing B1 cells to home to the body cavities has been defined. Mice deficient in this chemokine fail to produce natural antibodies against phosphorylcholine or even produce an antibody response to phosphorylcholine-containing streptococcal C-polysaccharide.[54] This implies that the peritoneal microenvironment is critical for the production of at least some natural antibodies.

(C). Many of these are $CD5^+$ and these are termed B1a cells, while those that are $CD5^-$ have been termed B1b cells. CD5 is not an exclusive marker of B1a cells and can be found in B2 cells after activation. Consequently, the definition of B1a and B1b repertoires by peritoneal IgM^{high}, IgD^-, $CD5^+$ and IgM^{high}, IgD^-, $CD5^-$ may be misleading.[55] Other phenotypic differences that help distinguish peritoneal B1 from B2 cells are the absent expression of CD23 and high expression of CD11b (Mac-1) by peritoneal B1 cells.

It is difficult to reconcile human and mouse data in relation with B1 cell populations and most of the characteristics described above apply mainly to mouse B1 cells, which have been studied in more detail. An apparent difference is that human B1 cells have been found with somatic mutations in their Ig V-region genes, whereas mouse B1 cells appear not to participate in germinal centre responses. It is difficult to be certain that these are real differences because of the limitations of defining human B1 cells from phenotype alone. In humans, at birth almost all B cells are B1 cells as defined by CD5 and CD1c expression, and the dominance of this phenotype is only lost in the second year of life. Apparently both $CD5^+$ and $CD5^-$ B cells are capable of producing IgM antibodies spontaneously, while this property is not seen in adult $CD5^-$ B cells and is only a property of a proportion of $CD5^+$ adult B cells.[56]

Extra-Follicular Antibody Responses

These responses are initiated when B cells bind their specific antigen. The sequence of events that takes place during B cell activation and differentiation into extra-follicular antibody-secreting plasma cells is illustrated in (Fig. 1). The B cells rapidly migrate to the T zones of secondary lymphoid tissues on binding antigen.[57,58] This is associated with altered responsiveness to chemokines, which also renders the B cells exquisitely efficient at interacting with primed T cells.[59] For responses to protein-based antigens, cognate interaction with T cells and the associated help is essential for further progress towards antibody production. This may occur without T cells in response to capsular polysaccharides or antigens based on bacterial cell wall lipopolysaccharide. Nevertheless protein or other T cell-recognized molecules associated with these antigens may allow T cell help to augment polysaccharide-specific antibody responses. In addition nonspecific T cell help may be delivered through cytokine release or interactions through nonantigen-specific surface molecules.[60]

During the next stage of the extra-follicular response B cells differentiate into plasmablasts, which move to sites where they proliferate. This extra-follicular pathway of B cell differentiation is regulated by the expression of the transcriptional repressor Blimp-1.[61,62] Plasmablasts are the immediate precursors of antibody-secreting plasma cells. They are in cell cycle and have upregulated distinctive markers of commitment to the plasma cell lineage such as syndecan-1 (CD138) and J chain. They already have large amounts of intracytoplasmic immunoglobulin. Within the spleen, plasmablasts typically proliferate in foci where the red pulp joins the white pulp (Fig. 1).[63] In lymph nodes, proliferation occurs in the medullary cords.[64] In both sites the plasmablasts proliferate in close contact with $CD11c^{high}$ dendritic cells, which appear to be required for plasmablast survival and differentiation into mature plasma cells.[65] Recent evidence suggests that the molecular basis of this survival is likely to be the secretion of BAFF by dendritic cells. This secretion may be triggered by innate contact between the dendritic cell precursors and bacteria,[66] or indirectly by type-1 interferons secreted by plasmacytoid dendritic cells.[67] There are no obvious extra-follicular foci of plasmablast growth in the secondary lymphoid tissues at the body surfaces, i.e., the tonsils, Peyer's patches and appendix. It may be that plasmablast precursors induced in these organs migrate to the medullary cords of downstream lymph nodes, but there appear to be no studies investigating this possibility.

Plasmablast growth in the spleen and lymph nodes continues for around 3 days after which those cells that receive appropriate survival and differentiation signals become mature nonproliferating plasma cells. Most of these plasma cells have a relatively brief lifespan,[68,69] but some survive for much longer and localize close to blood vessels and collagen fibres.[69] It appears that long-term survival is secured by the plasma cells locating in niches where they can

Table 1. Characteristics of recirculating, B1 and MZ B cells in mice

	Recirculating B Cells	B1 Cells	Marginal Zone B Cells
Phenotype	IgD^{high}, IgM^{low}, $CD21^{int}$, $CD23^{high}$	IgM^{high}, IgD^{-}, $CD11b^{high}$ (Mac-1), $CD23^{-/low}$, can be $CD5^{+}$ (B1a) or $CD5^{-}$ (B1b)	IgM^{high}, $IgD^{low/-}$, $CD21^{high}$, $CD23^{low/-}$
Origin	Production throughout life, dependent on IL-7 Recruited from recent bone marrow emigrants. Selection requires signalling through the BCR	Produced early in life independently of IL-7 Selected after BCR ligation, guided by certain antigen-specificities: expanded if B cell activation threshold is decreased (as in $CD22^{-/-}$, $CD72^{-/-}$ & $Lyn^{-/-}$mice), and reduced if threshold is raised ($Btk^{-/-}$, $CD19^{-/-}$mice)	Absent at birth and not fully mature until 2 years of age Diverse origins: 1. Immature marrow B cells 2. Recirculating follicular B cells 3. Memory B cells 4. B1 cells At least a subset is ligand selected and requires signalling through Btk and CD19
Localization / homing	Follicles of all secondary lymphoid tissues. Homing requires expression of CXCR5 by B cells.	Peritoneal and pleural cavities, dependent on macrophage-derived CXCL-13. A small proportion can be selected into the splenic marginal zone.	Mainly limited to the marginal zone of the spleen. Compartmentalization requires expression of Pyk-2 and the integrins LFA-1 and $\alpha4\beta1$ by B cells

Continued on next page

Table 1. Characteristics of recirculating, B1 and MZ B cells in mice (continued)

	Recirculating B Cells	B1 Cells	Marginal Zone B Cells
Antibody produced	Broad repertoire Lack self reactive specificities (in the presence of a normal repertoire) Participate in T-dependent responses (to protein antigens.)	Natural serum IgM and gut IgA, broadly reactive specificities Restricted V segment usage. Lack junctional diversity introduced by tdt Contribute to specific polysaccharide antigen extrafollicular responses	Key producers of antibody against blood borne bacteria. Contains partially self-reactive specificities. Specialised in T-independent responses to polysaccharide antigens and bacterial cell wall lipopolysaccharide.
Survival / lifespan	Disappear after 4 months in the absence of bone marrow influx. Require signals through BCR and BAFF receptor for survival.	Unlimited lifespan due to self-renewal Do not require BAFF signals for survival	Long lifespan. Numbers remain constant for at least one year in the absence of bone marrow influx Require BAFF signals for survival.
Activation / signalling	Resting status, low level expression of co-stimulatory molecules.	Higher resistance to Fas-induced apoptosis. Constitutive STAT-3 expression	Partially activated status, express higher levels of co-stimulatory molecules. Heightened proliferative responses in response to Ag

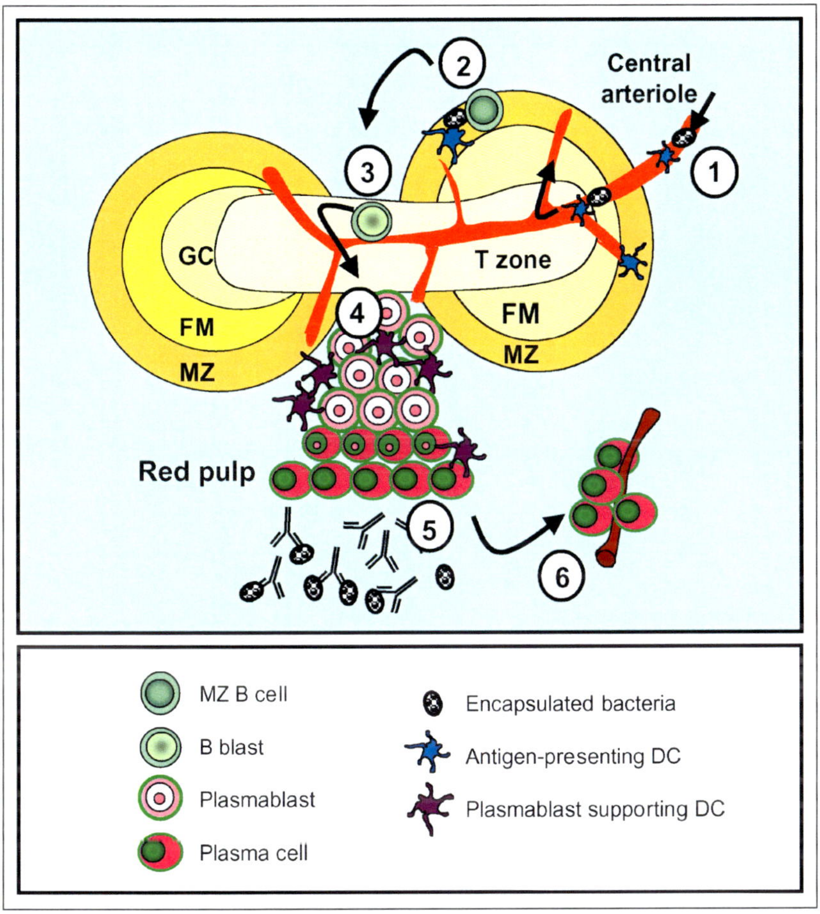

Figure 1. Sequence of events during splenic responses to polysaccharides. 1) Antigen enters the spleen from the blood stream via the central arteriole, which drains into the marginal zone sinusoids. Intact bacteria can be transported in the blood by dendritic cell precursors (CD11c^{low}), which are induced to enter the spleen upon uptake of antigen. 2) Marginal zone (MZ) B cells bind bacterial polysaccharide antigens through their B cell receptors. This induces them to enter cell cycle and change integrin expression and chemokine responsiveness, which enables them to leave the MZ. 3) Activated B cells migrate to the outer T zone of the spleen. Interaction with T cells is not essential for their subsequent differentiation but can augment antibody production. 4) B blasts differentiate into plasmablasts, which are still cycling but have large amounts of intracytoplasmic Ig. Plasmablasts move to the bridging channels situated at the junctions between the T zone and the red pulp. At this location they undergo exponential growth in close contact with CD11c^{high} dendritic cells. Their survival seems to be dependent on dendritic cell-derived TNF-family ligands BAFF and APRIL. 5) Plasmablasts come out of cell cycle and differentiate into fully mature plasma cells that secrete large amounts of polysaccharide-specific low affinity antibody. 6) Fully differentiated plasma cells leave the bridging channels and compete for niches close to splenic blood vessels for long-term survival.

receive signals that avoid activation of the apoptotic program; the nature for these signals is obscure.

A proportion of the antibody-producing cells in extra-follicular responses have undergone immunoglobulin class switch recombination. In TD responses this is triggered by signals received from T cells at the time of cognate interaction.[58] With TI responses to polysaccharide

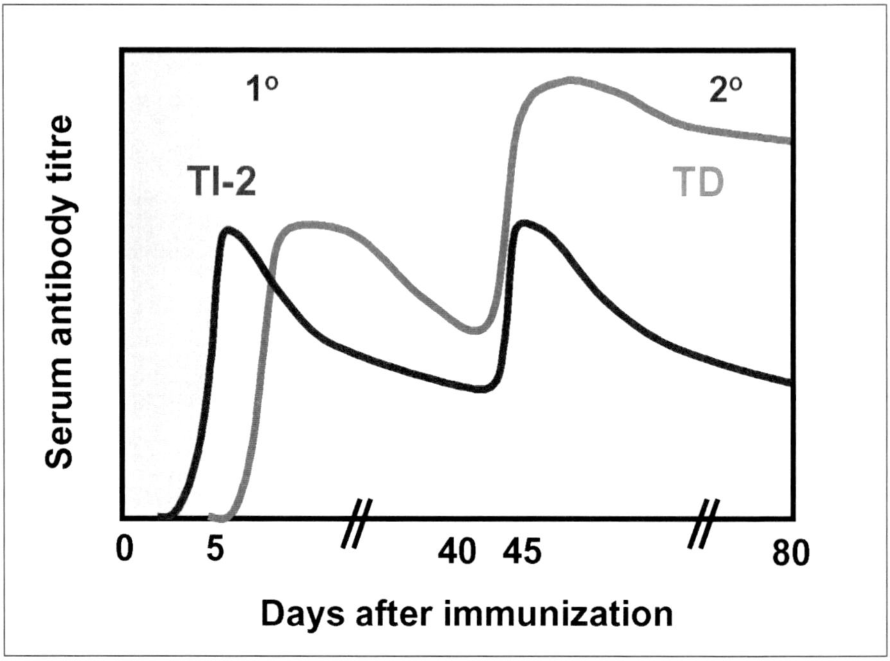

Figure 2. Comparison between the kinetics of TI-2 responses (anti-polysaccharide) and TD responses (against protein antigens). Primary immunization with a polysaccharide antigen (TI-2) results in faster antibody production compared with responses to protein antigens (TD). The delay in the latter response is mainly due to the time required for T cells to be primed on the surface of dendritic cells. However, the kinetics of a secondary challenge are similar for both types of immune responses, since primed T cells are readily available to provide help to B cells that have bound protein antigens. The magnitude of the secondary TD response is greater than that of the TI-2 response, since memory B cells are only generated in the former, which means a greater frequency of antigen-specific B cells can be synchronously activated upon reencounter with the same antigen.

the switching is mainly to IgG3 in mice with a little IgG1 and IgG2b; in humans characteristically IgG2 and IgG1 are produced with a greater IgG1 component in children, which is gradually reversed with age.[2] Innate immune signals can induce class switching in TI responses and this is exemplified by switching induced by type 1 interferons.[67] These are produced by blood dendritic cell precursors known as plasmacytoid dendritic cells, classified as pDC2.[70] This interferon in turn is capable of inducing another subset of blood derived dendritic cell precursors, pDC1, to secrete BAFF, which is capable of inducing B cells to switch Ig class.[71] Studies of strong interferon-α induction by viral antigens or bacterial DNA containing the CpG motif results in switching to all IgG isotypes. This has been observed both during an in vivo immune response in mice[67] and during in vitro culture of human blood-derived B and dendritic cells.[71] Nevertheless, switching is often far more selective in the majority of both TD and TI responses. This difference could reflect the accumulation of signals delivered both at the time of initial B cell activation as well as those received by plasmablasts as they are proliferating.

The advantage of the extra-follicular antibody response is its relative speed compared with that of antibody production resulting from germinal centre formation. TI antigens can induce antibody production by about 48 hours. This generally takes one to three days longer in TD responses when there is a lag due to T cell priming (Fig. 2). Even the TI response is too slow in the face of a bacteraemia in log phase of growth. Innate immunity, natural antibody, maternally

Table 2. Differences between T-dependent and T-independent type II responses

Responses to Polysaccharides (TI-2)	Responses to Protein Antigens (TD)
Absent during the first 2 years of life, reduced until 5 years of age	Present from birth
Main responsive B cell subset:	Main responsive B cell subset:
Marginal zone and B1 cells	Follicular and Marginal zone
T cell help not essential but can augment antibody production.	T cell help essential for B cell activation.
Dendritic cells present antigen and deliver innate signals to augment antibody synthesis	Dendritic cells prime T cells and may enhance B cell activation and antibody secretion
Fast extra-follicular antibody production	Delayed follicular and extra-follicular antibody production
Non-productive follicular response:	Productive follicular (germinal center) reaction:
No memory	Memory forms (increased secondary response)
No affinity maturation	Affinity maturation
Predominant Ig isotypes: IgG3 in mice and IgG2 in humans. IgA in mucosal surfaces.	Ig switching to most isotypes: IgE, IgG1, IgG2a (mice), IgE, IgG1, IgG3 and IgG4 (humans)

transferred antibody and prophylactic immunization provide more efficient protection although each can have loopholes resulting in clinical disease.

Follicular Antibody Responses

Germinal center formation often has not been considered in relation to antibody responses to bacterial polysaccharides. The pure polysaccharides exceptionally may induce germinal centers, but these are sterile since the polysaccharide cannot evoke the T cell help required to select cells in germinal centers.[72] Productive germinal centers require protein-linked antigen and therefore, productive follicular antibody responses are critically dependent on T cell help (Table 2). This becomes highly relevant to responses to conjugate vaccines; where bacterial carbohydrate epitopes are linked to carrier protein, or where there are natural conjugates formed with fragments of bacterial protein and polysaccharide.

The initial stages of TD germinal center activation are not obviously different from those associated with extra-follicular responses. B cells are activated via their receptors for antigen; they migrate to the T zone and there make cognate interaction with T cells. It is not known how the signals that induce growth in germinal centers differ from those resulting in growth in extra-follicular foci as plasmablasts. Nevertheless, once signals for divergent differentiation are registered, the commitment to one or the other pathway of differentiation is reinforced by key transcriptional regulators. B cells committed to germinal center formation upregulate Bcl-6 and this suppresses the plasmablast program.[73] Conversely, those cells that proliferate as plasmablasts upregulate Blimp-1[61,62] and this represses the expression of genes involved in B cell receptor signaling and proliferation, while allowing expression of further key molecules associated with plasma cell development, such as XBP-1.[62,74]

Germinal centers are oligoclonal, with on average only some three B cells founding each germinal center.[57,75] The sequence of their development has been reviewed.[76] In brief, once in the follicle the founding B cells grow exponentially for 3-4 days with a cell cycle time of about 6 hours. By the end of this period the blasts have activated a hypermutation process directed against their immunoglobulin variable region genes. These cells that have undergone hypermutation next start to undergo a process of selection. Critical to this is antigen held on follicular dendritic cells. Positive selection depends on B cells first binding antigen. They then process this and present it to local T cells. The B cells that make successful cognate interaction and receive T cell help, differentiate in one of three possible directions. Some differentiate to become memory B cells;[77] others differentiate to become plasma cells, which mainly colonize the bone marrow[68,78] or gut.[79,80] The rest of the selected cells remain within the germinal center to undergo further rounds of Ig variable region mutation and selection. Those cells within the germinal center that are not selected die in situ by apoptosis.[81] The cells proliferating in the germinal center at the selection phase continue to have a cell cycle time of about 6 hours, but the balance between death, emigration, and proliferation maintains the germinal center at a relatively uniform size. The size of germinal centers declines with time depending on antigen supply. Thus, following a single immunization with inert antigen a germinal center may last 3-4 weeks.[57] Conversely during a persistent virus infection germinal centers may last for months.[82] Even when the duration of a germinal center reaction is relatively short the antibody produced after a secondary immunization may last for months or years (Fig. 2).[83] The longevity of the response appears to result in part from long lived plasma cells[84,85] or plasma cell precursors[86] and partially on continued activation of memory B cells.[87]

In the present context there is evidence for memory B cell formation in responses to Hib-conjugate vaccines,[88] and secondary type antibody responses to meningococcal-conjugate vaccines in children as long as 5 years after primary immunization.[89]

MZ B Cells and Their Specialized Response to Polysaccharide Antigens

In contrast to TD responses, which occur in all secondary lymphoid organs, the spleen has a far greater capacity to respond to TI-2 antigens than lymph nodes.[90] Consequently, splenectomy affects TI-2 antibody responses more than TD responses.[91,92] Some studies have suggested that specialized splenic macrophages are crucial for removing antibody-coated encapsulated organisms from the blood.[93] However, TI-2 responsiveness is maintained even after depletion of these cells from mice. The unique role of the spleen is also attributed to the importance of MZ B cells in responding to TI-2 antigens. The MZ in infants is immature and is characterized by low or absent expression of CD21 on the surface of MZ B cells.[94] This developmental immaturity correlates very closely and could be at the cause of the unresponsiveness to encapsulated bacteria during the first years of life.

The main phenotypic and functional features that characterize MZ B cells have been summarized in (Table 1). MZ B cells comprise about 30% of human splenic B cells, are rare in lymph nodes, other than mesenteric lymph nodes,[95] and similar cells have been identified deep to the dome epithelium in Peyer's patches[96] and in the tonsilar crypt epithelium.[97] In the spleen they are perfused by blood sinusoids, which in mice are separated from the follicles by an endothelial structure called the marginal sinus. Although nonactivated MZ B cells are sessile and nonrecirculating,[98] when their B cell receptors (BCR) bind antigen they move rapidly to the outer T zones of the spleen.[99,100] MZ B cells are morphologically and phenotypically distinct: they are larger, have less condensed chromatin, and express higher levels of costimulatory molecules (CD80 and CD86) than recirculating follicular B cells.[101] These features suggest that MZ B cells are in a state of partial activation, and are consistent with evidence that they can undergo rapid differentiation to plasma cells after ligation with antigen.[36] Other phenotypic features that distinguish this population are high levels of IgM and low levels of IgD, as well as high levels of CD21 expression and low levels of CD23. Conventional recirculating B cells, which populate B cell areas of lymph nodes, are not capable of responding to TI-2 antigens in vivo.[35] Accumulating evidence suggests that B cells responsible for TI-2 responses are selected into the MZ. The MZ B cell subset is enriched in B cells with specificities for bacterial pathogens such as phosphorylcholine.[36] These properties have suggested that MZ B cells represent the "first-line of defence" against rapidly-dividing microorganisms. In the absence of MZ B cells, B1 cells can produce adaptive responses to TI-2 antigen. Tanigaki et al[102] found that mice deficient in MZ B cells that have normal numbers of B1 cells make antibody responses to haptenated Ficoll. The increased susceptibility of these mice to blood-borne bacterial infection may reflect in part the limitations of the B1 repertoire as well as an advantage to adaptive responses mounted by MZ B cells. The advantages might include a faster response, enhanced IgG production, and higher antibody titres.

Ontogeny and Selection of MZ B Cells

MZ B cells are heterogeneous and their ontogeny remains to be established with certainty. There is evidence that MZ B cells are recruited from different sources. In rats, recirculating follicular B cells can replenish a depleted MZ.[35,103,104] Memory B cells from germinal centres also colonize the MZ.[99,105,106] Recent analysis of gene expression by MZ B cells confirms that this population includes cells carrying both germline and somatically mutated Ig V_H regions.[107,108] B1 cells and possibly immature B cells can also become MZ cells.[34,109] The gradual emergence of protective immunity against encapsulated bacteria may reflect the diverse sources from which the MZ repertoire is recruited.

In contrast to follicular B cell selection, MZ B cell selection may be inversely related with the strength of BCR signaling, because the MZ is small in CD19$^{-/-}$ and Aiolos$^{-/-}$ mice, where the BCR signaling threshold is increased.[109,110] On the other hand, two strains of mice expressing BCR transgenes have been identified in which B cells are preferentially selected to enter the MZ based on Ig heavy chain expression, raising the possibility of positive selection.[69,109] Furthermore, selection of these B cells is lost in the absence of Btk. These apparently conflicting results may be related with the heterogeneity of MZ B cells.

Maintenance and Survival of MZ B Cells

While recirculating follicular B cells numbers decline slowly over months in the absence of a continuous influx from the bone marrow, the number of B1 and MZ B cells can remain unchanged for a long time.[111] This probably reflects the ability of cells from the B1 cell pool to differentiate into MZ B cells. Although MZ B cell numbers are maintained in this situation the MZ repertoire is likely to be restricted as MZ B cell recruitment from the recirculating pool diminishes. Cytokines, chemokines, and adhesion molecules play a role in establishing and maintaining the MZ. The TNF family member BAFF, with its receptors on B cells (BAFF-R, TACI and BCMA) is important for maintaining both MZ and recirculating B cell numbers. These decline rapidly when BAFF signaling is blocked by soluble BAFF receptors—soluble BCMA,[42] or TACI-Ig.[41] This decline is too rapid to be attributable to failure of new B cell production. Nevertheless BAFF does support the survival of transitional B cells through increasing the levels of expression of Bcl-2 and Bcl-x_L, which are critical players in the maturation of immature B cells.[42,112] Mice transgenic for BAFF have increased numbers of transitional B cells and MZ B cells.

BAFF is expressed by monocytes/macrophages, dendritic cells and some activated T cells. The presence of unique macrophage populations (MZ macrophages and metallophillic macrophages) within the MZ raises the possibility that they may provide the stromal support required for MZ B cell survival. MZ metallophils are absent in early life and their appearance coincides with formation of the MZ.[113] Furthermore, in mice in which LTβR ligation is disrupted, there is no identifiable MZ. Analysis of splenocytes from TNF/LTα$^{-/-}$ mice by flow cytometry has shown that B cells bearing the phenotypic hallmarks of MZ cells are absent.[114] Both MZ macrophages and metallophillic macrophages are also absent from these mice.

Mice lacking the tyrosine kinase Pyk-2 have a selective MZ deficiency.[38] Adoptive transfer studies indicate that Pyk-2 must be expressed in the B cells rather than the stromal components of the spleen in order for MZ to develop.[38] Pyk-2 participates in signaling pathways downstream of chemokine and growth factor receptors but the putative chemokine has not been identified. Administration of pertussis toxin, which blocks signaling through chemokine receptors,[115] has also been shown to deplete MZ B cells.[38] Taken together, these findings imply that MZ B cell organization, like other B cell compartments, depends on secretion of a chemokine by mesenchymal/stromal elements.

An important role for integrins and their receptors in the long-term maintenance of MZ B cells has recently being suggested. MZ B cells express higher levels of the $\alpha 1\beta 2$ heterodimer LFA-1 and $\alpha 4\beta 1$ integrins.[116] Their ligands ICAM-1 and VCAM-1 are expressed in the MZ and their expression is dependent on the presence of lymphotoxin $\alpha 1\beta 2$. Mice deficient in this cytokine lack MZ B cells, and brief treatment of mice with lymphotoxin $\alpha 1\beta 2$ antagonist (lymphotoxin-βR-Ig) causes a dramatic reduction in the number of MZ B cells.[116] It seems therefore that lymphotoxin $\alpha 1\beta 2$ directly induces ICAM-1 and VCAM-1 expression on MZ stromal cells and this is required for maintenance of B cells within the MZ. It has also been shown that LPS-induced activation decreases MZ integrin activity and this allows migration of antigen-specific B cells to the sites of antibody formation.[116]

Autoreactive B Cells in the MZ and Implications for Antibody Responses to Polysaccharides

There are several lines of evidence showing that a proportion of the B cells that accumulate in the MZ have self-reactive specificities. B cells bearing a transgene for a germline rearranged heavy chain V_H81X that generate multi-reactive IgM antibody preferentially accumulate in the MZ.[117] In mice bearing transgenic B cell receptors reactive with self-antigen, the copies of the transgene are subject to the process of receptor editing and this results in B cells expressing receptors with more than one specificity. These partially autoreactive B cells that express dual receptors (one of them anti-self) have been shown to accumulate in the MZ.[118] It has also been shown that the MZ population in rats has been selected based on the V_H repertoire, which differs from the follicular B cell repertoire.[107] The third hyper-variable or complementarity determining region (CDR3) of immunoglobulin variable region genes spans the D and J segments and their junction with the V segment. The action of Tdt introduces new base pairs at the junctions between these gene segments. When this occurs it extends the length of CDR3. A striking finding is that segments from MZ B cells contain significantly shorter CDR3 regions than those from follicular B cells.[107] This, together with the finding that somatically mutated memory B cells tend to have short CDR3 regions, suggests MZ B cells may be selected based on their heavy chain structure, which may or may not reflect antigen-binding.[119] There seems to be an association between short CDR3 regions and multi-reactivity, and short CDR3 regions seem to bind polysaccharides more easily, suggesting a positive link between having a multi-reactive MZ repertoire and mounting optimal responses to encapsulated bacteria. The restriction of these multi-reactive cells in this location may help to prevent them from posing a serious threat for development of autoimmune disease. This may reflect a failure of MZ B cells activated by TI antigens to form productive germinal centers that produce long-lived plasma cells and memory B cells.

There may be advantages in having a weakly broad and self-reactive population in the MZ. The repeating epitopes of capsular polysaccharides allow stronger cross-linking of BCR than self-carbohydrate carrying the same determinant at lower epitope frequency. This is likely to offer a therapeutic index where the MZ B cell responds to capsular polysaccharide, but is not activated by the cross-reacting self-antigen. The same valency difference is likely to operate allowing effector mechanism activation by antibody bound to the bacteria but not that bound to self-antigen.

Despite the above considerations there is evidence that there is negative selection of B cells that react both with self-antigens and capsular polysaccharide antigens. There are several reports showing that bacterial capsular polysaccharides cross-react with polysialic acid on neural tissues in early life.[120] During fetal life, a high proportion of sialic acid in the brain is in the form of polysialic acid. This decreases during development,[121,122] to about 10% of fetal levels in adult life. It may be that the decline in auto-antigen with age is directly associated with loss of tolerance to these carbohydrate epitopes. An extreme example of a bacterial polysaccharide that is identical to a self-antigen is found in the capsule of meningococcus B. Anti-*N. meningitidis* B capsular polysaccharide antibodies have been shown to cross-react with fetal brain and to a lesser degree with adult human brain.[120,123] Thus, at least for this capsular polysaccharide, tolerance appears to be responsible for the lack of responsiveness to pure polysaccharide vaccines against *N. meningitidis* B. Synthesis of a modified form of this capsular polysaccharide by substituting an acetyl group by a propionyl group has been shown to overcome tolerance, and render the polysaccharide immunogenic.[124]

Other TI Responses: Critical Costimulation by Toll-Like Receptors

As mentioned above, during TD responses B lymphocytes integrate two signals from (i) the antigen and (ii) helper T cells, before they make antibody. By contrast, during responses to TI antigens the second signal is not absolutely required in the presence of strong BCR crosslinking, or can come from specific components of the bacterial pathogens themselves, resulting in antibody production in the absence of T cells. One of the best understood of pathogen-derived costimulatory signals is the one delivered by lipopolysaccharide (LPS). LPS has been classified as a TI-1 antigen because it can elicit an antibody response in both nude and Xid (Btk-deficient) mice. LPS is an essential component of the outer cell wall of gram-negative bacteria and is a powerful stimulant of both the innate and adaptive arms of the immune system. The effects of LPS on the innate immune system dominate the clinical picture of septic shock. The unique lipid moiety of LPS represents a highly conserved structural motif of this class of pathogens which is recognized through a surface receptor present on all B cells, Toll-like receptor 4 (TLR4).[125] TLR4 signals through a highly conserved and phylogenetically ancient pathogen-sensing pathway involving MyD88 and activation of the transcription factor NFκB. High concentrations of LPS are mitogenic for all B cells. At lower concentrations, simultaneous recognition of LPS by membrane Ig and by TLR4 selectively triggers antibody production by bacteria-specific B cells. This is achieved through potent and synergistic signaling for B cell proliferation between the two receptors.[126-128]

Upon LPS-induced activation, B cells migrate to the outer T zone, and then move to the red pulp as they differentiate into plasma cells. Antibody responses to LPS do not require T cells but their presence can influence the response.

LPS is not alone in being capable of eliciting TI responses in Btk-deficient mice. Bacterial DNA has also been recently shown to be a potent adjuvant of antibody responses and is capable of providing a sufficient B cell costimulus to bypass the requirement for T cell help. Bacterial DNA, which is relatively rich in unmethylated CpG dinucleotides compared to mammalian DNA, is recognized as a conserved pathogen-specific motif by TLR9 (Krieg et al, 1995), which can trigger direct B cell activation, also signaling through the MyD88 pathway to activate NFκB.[129]

Finally, TI 2 responses have recently also been shown to be enhanced by signals delivered through TLR. It has been described before how capsular polysaccharides can stimulate B cells without the need of a costimulatory signal. This is because epitopes on polysaccharide antigens are arranged in a repetitive fashion enabling them to cross-link multiple B cell receptors and thereby provide a strong stimulus capable of inducing B cells to enter cell cycle. The response to *Neisseria meningitidis* is an example of how costimulation through TLR at the time of antigen encounter can enhance anti-polysaccharide responses. Neisserial porins bind TLR-2 on B cells and this signal has been shown to increase substantially the amount of anti-polysaccharide antibodies produced.[130,131]

When the source of the costimulatory signal derives from a pathogen rather than from T cells, the risk of activating self-reactive B cells emerges. A clear example of how this can occur and may contribute to the production of autoantibodies has recently been shown in a mouse model of systemic autoimmune disease bearing transgenic BCR specific for self-IgG. Simultaneous recognition of self-IgG bound to DNA by membrane Ig and TLR9 leads to rheumatoid factor production.[132] This finding cautions against the potential dangers of the clinical use of TLR-ligands as adjuvants to enhance polysaccharide and other types of immune responses.

Conclusions

The polysaccharide capsules of pneumococci, meningococci and Hib protect these bacteria from innate immune mechanisms. Consequently, antibody responses to these encapsulated organisms are crucial for host defence. These responses are different from those stimulated by conventional protein antigens because the requirement for T cell help is bypassed. The clinical importance of antibody responses to the polysaccharides that coat encapsulated bacteria is underlined by the high rate of morbidity and mortality from these organisms during infancy, when the immature immune system cannot make TI-2 responses. The reason for the delay in developing this crucial defence is not known and an evolutionary advantage is not obvious.

Our understanding of immune responses to encapsulated bacteria has increased substantially over the last few years. Specialized subsets of B cells—marginal zone B cells and peritoneal B1 cells—are responsible for the protective extra-follicular antibody response. Furthermore, recent studies have revealed that rapid B cell activation and fast antibody production, which are crucial to prevent dissemination of encapsulated pathogens, depend on interactions between the innate and adaptive immune systems. Many encapsulated bacteria induce the production of the complement degradation fragment C3d, which binds to their surface. C3d signals through CD21, which is expressed at high levels on marginal zone B cells. Simultaneous cross-linking of CD21 and the antigen specific receptor on B cells lowers the threshold for B cell activation. A subset of blood dendritic cells can take up encapsulated bacteria and enter the spleen to present intact antigen to MZ B cells. Dendritic cells also deliver signals that are critical for survival of antibody-secreting cells. Other innate players in T-independent antibody responses are the highly conserved Toll family of pattern recognition receptors that bind specific microbial motifs and deliver costimulatory signals to B cells.

A better understanding of immune responses to polysaccharides has had a tangible effect on the approach to immunization against encapsulated pathogens. Pure polysaccharide-based vaccines are inadequate for infant immunization because they cannot elicit a response during the first years of life and do not generate memory B cells. By contrast, protein-polysaccharide conjugate vaccines are taken up by polysaccharide-specific B cells, which process the protein component and by presenting the resulting peptides to T cells can elicit cognate help. T-dependent immune responses are intact from birth and this strategy has seen the virtual elimination of *Haemophilus infuenzae* b meningitis in countries that have fully implemented conjugate vaccination. For pneumococci, conjugating sufficient polysaccharide serotypes within the same vaccine to provide comprehensive protection remains technically challenging. Further characterization of the ways the innate and adaptive immune systems cooperate to enhance antibody responses against polysaccharides may provide clues for the generation of vaccines capable of protecting against multiple capsular serotypes simultaneously.

References

1. Tuomanen E. Molecular and cellular biology of pneumococcal infection. Curr Op Microbiol 1999; 2:35-39.
2. Hazlewood M, Nusrat R, Kumararatne DS et al. The acquisition of anti-pneumococcal capsular polysaccharide Haemophilus influenzae type b and tetanus toxoid antibodies with age in the UK. Clin Exp Immunol 1993; 93:157-164.
3. Leinonen M, Sakkinen A, Kalliokoski R et al. Antibody response to 14-valent pneumococcal capsular polysaccharide vaccine in preschool age children. Pediatr Infect Dis 1986; 5:39-44.
4. Peltola H, Kayhty H, Sivonen A et al. Haemophilus influenzae type b capsular polysaccharide vaccine in children: A double-blind field study of 100,000 vaccinees 3 months to 5 years of age in Finland. Pediatrics 1977; 60:730-737.

5. Murray CJ, Lopez AD. Global mortality, disability, and the contribution of risk factors: Global burden of disease study. Lancet 1997; 349:1436-1442.
6. Eskola J, Kayhty H, Takala AK et al. A randomized, prospective field trial of a conjugate vaccine in the protection of infants and young children against invasive Haemophilus influenzae type b disease. N Engl J Med 1990; 323:1381-1387.
7. Anderson P. Antibody responses to Haemophilus influenzae type b and diphtheria toxin induced by conjugates of oligosaccharides of the type b capsule with the nontoxic protein CRM197. Infect Immun 1983; 39:233-238.
8. Bretscher P, Cohn M. A theory of self nonself discrimination. Science 1970; 169:1042.
9. Mond JJ, Lees A, Snapper CM. T cell-independent antigens type 2. Annu Rev Immunol 1995; 13:655-692.
10. Brunswick M, Finkelman FD, Highet PF et al. Picogram quantities of anti-Ig antibobies coupled to dextran induce B cell proliferation. J Immunol 1988; 140:3364-3374.
11. Fehr T, Bachmann MF, Bluethmann H et al. T-independent activation of B cells by vesicular stomatitis virus: No evidence for the need of a second signal. Chem Immunol 1996; 168:184-192.
12. Fehr T, Naim NY, Bachmann MF et al. T-cell independent IgM and enduring protective IgG antibodies induced by chimeric measles viruses. Nature Med 1998; 4:945-948.
13. Feldmann M. Induction of immunity and tolerance in vitro by hapten protein conjugates. II. Carrier independence of the response to dinitrophenylated polymerized flagellin. Eur J Immunol 1972; 2:130-137.
14. Scher I. The CBA/N mouse strain: An experimental model illustrating the influence of the X-chromosome on immunity. Adv Immunol 1982; 33:1-71.
15. Thomas JD, Sideras P, Smith CI et al. Colocalization of X-linked agammaglobulinaemia and X-linked immunodeficiency genes. Science 1993; 261:355-358.
16. Reth M. Antigen receptor tail clue. Nature 1989; 338:383-384.
17. Fearon DT, Carter RH. The CD19/CR2/TAPA-1 complex of B lymphocytes: Linking natural to acquired immunity. Annu Rev Immunol 1995; 13:127-149.
18. Dempsey PW, Allison MED, Akkaraju S et al. C3d of complement as a molecular adjuvant bridging innate and acquired immunity. Science 1996; 271:348-350.
19. Peset Llopis MJ, Harms G, Hardonk MJ et al. Complement-mediated localization preferentially on CD21-positive marginal zone B cells and follicular dendritic cells. J Allergy Clin Immunol 1996; 97:1015-1024.
20. Bajpai UD, Zhang K, Teutsch M et al. Bruton's tyrosine kinase links the B cell receptor to nuclear factor kappaB activation. J Exp Med 2000; 191:1735-1744.
21. Karin M, Lin A. NF-kappaB at the crossroads of life and death. Nat Immunol 2002; 3:221-227.
22. Nonoyama S, Tsukada S, Yamadori T et al. Functional analysis of peripheral blood B cells in patients with X-linked agammaglobulinemia. J Immunol 1998; 161:3925-3929.
23. Wood PM, Mayne A, Joyce H et al. A mutation in Bruton's tyrosine kinase as a cause of selective anti-polysaccharide antibody deficiency. J Pediatr 2001; 139:148-145.
24. Minegishi Y, Rohrer J, Conley ME. Recent progress in the diagnosis and treatment of patients with defects in early B cell development. Curr Opin Pediatr 1999a; 11:528-532.
25. Minegishi Y, Rohrer J, Coustan-Smith E et al. An essential role for BLNK in human B cell development. Science 1999b; 286:1954-1957.
26. Bos NA, Kimura H, Meeuwsen CG et al. Serum immunoglobulin levels and naturally occurring antibodies against carbohydrate antigens in germ-free BALB/c mice fed chemically defined ultrafiltered diet. Eur J Immunol 1989a; 19:2335-2339.
27. Kroese FG, Ammerlaan WA, Deenen GJ et al. A dual origin for IgA plasma cells in the murine small intestine. Adv Exp Med Biol 1995; 371A:435-440.
28. Bos NA, Meeuwsen CG, Hooijkaas H et al. Early development of Ig-secreting cells in young of germ-free BALB/c mice fed a chemically defined ultrafiltered diet. Cell Immunol 1987; 105:235-245.
29. Bos NA, Meeuwsen CG, Van Wijngaarden P et al. B cell repertoire in adult antigen-free and conventional neonatal BALB/c mice. II. Analysis of antigen-binding capacities in relation to VH gene usage. Eur J Immunol 1989b; 19:1817-1822.
30. Sorensen UB, Agger R, Bennedsen J et al. Phosphorylcholine determinants in six pneumococcal capsular polysaccharides detected by monoclonal antibody. Infect Immun 1984; 43:873-878.
31. Nielsen SV, Sorensen UB, Henrichsen J. Antibodies against pneumococcal C-polysaccharide are not protective. Microb Pathog 1993; 14:299-305.
32. Briles DE, Claflin JL, Schroer K et al. Mouse IgG3 antibodies are highly protective against infection with Streptococcus pneumoniae. Nature 1981; 294:88-90.

33. Gould JM, Weiser JN. The inhibitory effect of C-reactive protein on bacterial phosphorylcholine platelet-activating factor receptor-mediated adherence is blocked by surfactant. J Infect Dis 2002; 186:361-371.
34. Martin F, Kearney JF. B1 cells: Similarities and differences with other B cell subsets. Curr Opin Immunol 2001; 13:195-201.
35. Lane PJ, Gray D, Oldfield S et al. Differences in the recruitment of virgin B cells into antibody responses to thymus-dependent and thymus-independent type-2 antigens. Eur J Immunol 1986; 16:1569-1575.
36. Oliver AM, Martin F, Gartland GL et al. Marginal zone B cells exhibit unique activation, proliferative and immunoglobulin secretory responses. Eur J Immunol 1997; 27:2366-2374.
37. Fagarasan S, Honjo T. T-Independent immune response: New aspects of B cell biology. Science 2000; 290:89-92.
38. Guinamard R, Okigaki M, Schlessinger J et al. Absence of marginal zone B cells in Pyk-2-deficient mice defines their role in the humoral response. Nat Immunol 2000; 1:31-36.
39. Martin F, Oliver AM, Kearney JF. Marginal zone and B1 B cells unite in the early response against T- independent blood-borne particulate antigens. Immunity 2001; 14:617-629.
40. Carvalho TL, Mota-Santos T, Cumano A et al. Arrested B lymphopoiesis and persistence of activated B cells in adult interleukin 7(-/-) mice. J Exp Med 2001; 194:1141-1150.
41. Schneider P, Takatsuka H, Wilson A et al. Maturation of marginal zone and follicular B cells requires B cell activating factor of the tumor necrosis factor family and is independent of B cell maturation antigen. J Exp Med 2001; 194:1691-1697.
42. Thompson JS, Schneider P, Kalled SL et al. BAFF binds to the tumor necrosis factor receptor-like molecule B cell maturation antigen and is important for maintaining the peripheral B cell population. J Exp Med 2000; 192:129-135.
43. Wang J, Taniuchi I, Maekawa Y et al. Expression and function of Fas antigen on activated murine B cells. Eur J Immunol 1996; 26:92-96.
44. Karras JG, Wang Z, Huo L et al. Signal transducer and activator of transcription-3 (STAT3) is constitutively activated in normal, self-renewing B-1 cells but only inducibly expressed in conventional B lymphocytes. J Exp Med 1997; 185:1035-1042.
45. Clarke SH, McCray SK. VH CDR3-dependent positive selection of murine VH12-expressing B cells in the neonate. Eur J Immunol 1993; 23:3327-3334.
46. Cyster JG, Goodnow CC. Protein tyrosine phosphatase 1C negatively regulates antigen receptor signaling in B lymphocytes and determines thresholds for negative selection. Immunity 1995b; 2:13-24.
47. Rickert RC, Rajewsky K, Roes J. Impairment of T-cell-dependent B-cell responses and B-1 cell development in CD19-deficient mice. Nature 1995; 376:352-355.
48. Arnold LW, McCray SK, Tatu C et al. Identification of a precursor to phosphatidyl choline-specific B-1 cells suggesting that B-1 cells differentiate from splenic conventional B cells in vivo: Cyclosporin A blocks differentiation to B-1. J Immunol 2000; 164:2924-2930.
49. Lam KP, Rajewsky K. B cell antigen receptor specificity and surface density together determine B-1 versus B-2 cell development. J Exp Med 1999; 190:471-477.
50. Hayakawa K, Asano M, Shinton SA et al. Positive selection of natural autoreactive B cells. Science 1999; 285:113-116.
51. Benedict CL, Gilfillan S, Thai TH et al. Terminal deoxynucleotidyl transferase and repertoire development. Immunol Rev 2000; 175:150-157.
52. Feeney AJ, Tang A, Ogwaro KM. B-cell repertoire formation: Role of the recombination signal sequence in nonrandom V segment utilization. Immunol Rev 2000; 175:59-69.
53. Herzenberg LA. B-1 cells: The lineage question revisited. Immunol Rev 2000; 175:9-22.
54. Ansel KM, Harris RB, Cyster JG. CXCL13 is required for B1 cell homing, natural antibody production, and body cavity immunity. Immunity 2002; 16:67-76.
55. Kantor AB, Merrill CE, Herzenberg LA et al. An unbiased analysis of V(H)-D-J(H) sequences from B-1a, B-1b, and conventional B cells. J Immunol 1997; 158:1175-1186.
56. Bhat NM, Kantor AB, Bieber MM et al. The ontogeny and functional characteristics of human B-1 (CD5+ B) cells. Int Immunol 1992; 4:243-252.
57. Liu YJ, Zhang J, Lane PJ et al. Sites of specific B cell activation in primary and secondary responses to T cell-dependent and T cell-independent antigens. Eur J Immunol 1991; 21:2951-2962.
58. Toellner KM, Gulbranson-Judge A, Taylor DR et al. Immunoglobulin switch transcript production in vivo related to the site and time of antigen-specific B cell activation. J Exp Med 1996; 183:2303-2312.
59. Casamayor-Palleja M, Mondiere P, Verschelde C et al. BCR ligation reprograms B cells for migration to the T zone and B-cell follicle sequentially. Blood 2002; 99:1913-1921.

60. Wu Z-Q, Khan AQ, Chu C-L et al. The mechanism underlying T cell help for induction of an antigen-specific in vivo humoral immune response to intact Streptococcus pneumoniae is dependent on the type of antigen. J Immunol 2002; 168:5551-5557.
61. Mock BA, Liu L, LePaslier D et al. The B-lymphocyte maturation promoting transcription factor BLIMP1/PRDI-BF1 maps to D6S447 on human chromosome 6q21-q22.1 and the syntenic region of mouse chromosome 10. Genomics 1996; 37:24-28.
62. Shaffer AL, Lin KI, Kuo TC et al. Blimp-1 orchestrates plasma cell differentiation by extinguishing the mature B cell gene expression program. Immunity 2002; 17:51-62.
63. Jacob J, Kassir R, Kelsoe G. In situ studies of the primary immune response to (4-hydroxy-3-nitrophenyl)acetyl. I. The architecture and dynamics of responding cell populations. J Exp Med 1991; 173:1165-1175.
64. Luther SA, Gulbranson-Judge A, Acha-Orbea H et al. Viral superantigen drives extrafollicular and follicular B cell differentiation leading to virus-specific antibody production. J Exp Med 1997; 185:551-562.
65. Vinuesa CG, Gulbranson-Judge A, Khan M et al. Dendritic cells associated with plasmablast survival. Eur J Immunol 1999; 29:3712-3721.
66. Balazs M, Martin F, Zhou T et al. Blood dendritic cells interact with splenic marginal zone B cells to initiate T-independent immune responses. Immunity 2002; 17:341-352.
67. Le Bon A, Schiavoni G, D'Agostino G et al. Type I interferons potently enhance humoral immunity and can promote isotype switching by stimulating dendritic cells in vivo. Immunity 2001; 14:461-470.
68. Smith KG, Hewitson TD, Nossal GJ et al. The phenotype and fate of the antibody-forming cells of the splenic foci. Eur J Immunol 1996; 26:444-448.
69. Sze DM, Toellner KM, Garcia de Vinuesa C et al. Intrinsic constraint on plasmablast growth and extrinsic limits of plasma cell survival. J Exp Med 2000; 192:813-821.
70. Shortman K, Liu YJ. Mouse and human dendritic cell subtypes. Nat Rev Immunol 2002; 2:151-161.
71. Litinskiy MB, Nardelli B, Hilbert DM et al. DCs induce CD40-independent immunoglobulin class switching through BLyS and APRIL. Nat Immunol 2002; 3:822-829.
72. Vinuesa CG, Cook MC, Ball J et al. Germinal centers without T cells. J Exp Med 2000; 191:485-494.
73. Fearon DT, Manders P, Wagner SD. Arrested differentiation, the self-renewing memory lymphocyte and vaccination. Science 2001; 293:248-250.
74. Reimold AM, Iwakoshi NN, Manis J et al. Plasma cell differentiation requires the transcription factor XBP-1. Nature 2001; 412:300-307.
75. Kroese FG, Wubbena AS, Seijen HG et al. Germinal centers develop oligoclonally. Eur J Immunol 1987; 17:1069-1072.
76. MacLennan IC. Germinal centers. Annu Rev Immunol 1994; 12:117-139.
77. Coico RF, Bhogal BS, Thorbecke GJ. Relationship of germinal centers in lymphoid tissue to immunologic memory. VI. Transfer of B cell memory with lymph node cells fractionated according to their receptors for peanut agglutinin. J Immunol 1983; 131:2254-2257.
78. Benner R, Hijmans W, Haaijman JJ. The bone marrow: The major source of serum immunoglobulins, but still a neglected site of antibody formation. Clin Exp Immunol 1981; 46:1-8.
79. Hall JG, Smith ME. Homing of lymph-borne immunoblasts to the gut. Nature 1970; 226:262-263.
80. Husband AJ, Monie HJ, Gowans JL. The natural history of the cells producing IgA in the gut. Ciba Found Symp 1977; 29-54.
81. Liu YJ, Joshua DE, Williams GT et al. Mechanism of antigen-driven selection in germinal centers. Nature 1989; 342:929-931.
82. MacLennan IC, Gulbranson-Judge A, Toellner KM et al. The changing preference of T and B cells for partners as T-dependent antibody responses develop. Immunol Rev 1997; 156:53-66.
83. MacLennan IC, Liu YJ, Oldfield S et al. The evolution of B-cell clones. Curr Top Microbiol Immunol 1990; 159:37-63.
84. Manz RA, Cassese G. Thiel A and Radbruch A. Long-lived plasma cells survive independent of antigen. Curr Top Microbiol Immunol 1999; 246: 71-74.
85. Slifka MK, Ahmed R. Long-lived plasma cells: A mechanism for maintaining persistent antibody production. Curr Opin Immunol 1998; 10:252-258.
86. O'Connor BP, Cascalho M, Noelle RJ. Short-lived and long-lived bone marrow plasma cells are derived from a novel precursor population. J Exp Med 2002; 195:737-745.
87. Zinkernagel RM, Bachmann MF, Kundig TM et al. On immunological memory. Annu Rev Immunol 1996; 14:333-367.
88. Granoff DM, Holmes SJ, Osterholm MT et al. Induction of immunologic memory in infants primed with Haemophilus influenzae type b conjugate vaccines. J Infect Dis 1993; 168:663-671.

89. MacLennan J, Obaro S, Deeks J et al. Immunologic memory 5 years after meningococcal A/C conjugate vaccination in infancy. J Infect Dis 2001; 183:97-104.
90. Claassen E, Kors N, Dijkstra CD et al. Marginal zone of the spleen and the development and localisation of specific antibody-forming cells against thymus-independent type-2 antigens. Immunol 1986; 57:399-403.
91. Amlot PL, Hayes AE. Impaired human antibody response to the thymus independent antigen, DNP-Ficoll, after splenectomy. Implications for post-splenectomy infections. Lancet 1985; 1:1008-1011.
92. Gray D, Chassoux D, MacLennan ICM et al. Selective depression of thymus-independent anti-DNP antibody responses induced by adult but not neonatal splenectomy. Clin Exp Immunol 1985; 60:78-86.
93. Kraal G, Ter Hart H, Meelhuizen C et al. Marginal zone macrophages and their role in the immune response against T-independent type 2 antigens: Modulation of the cells with specific antibody. Eur J Immunol 1989; 19:675-680.
94. Timens W, Boes A, Rozeboomuiterwijk T et al. Immaturity of the human splenic marginal zone in infancy - possible contribution to the deficient infant immune-response. J Immunol 1989; 143:3200-3206.
95. Stein H, Bonk A, Tolksdorf G et al. Immunohistologic analysis of the organization of normal lymphoid tissue and nonHodgkin's lymphomas. J Histochem Cytochem 1980; 28:746-760.
96. Spencer J, Finn T, Pulford KAF et al. The human gut contains a novel population of B lymphocytes which resemble marginal zone cells. Clin Exp Immunol 1985; 62:607-612.
97. Liu YJ, Barthelemy C, de Bouteiller O et al. Memory B cells from human tonsils colonize mucosal epithelium and directly present antigen to T cells by rapid up-regulation of B7-1 and B7-2.PG. Immunity 1995; 3:239-248.
98. Gray D, MacLennan IC, Bazin H et al. Migrant mu+ delta+ and static mu+ delta- B lymphocyte subsets. Eur J Immunol 1982; 12:564-569.
99. Liu YJ, Oldfield S, MacLennan IC. Memory B cells in T cell-dependent antibody responses colonize the splenic marginal zones. Eur J Immunol 1988; 18:355-362.
100. Vinuesa CG, Sunners Y, Pongracz J et al. Tracking the response of Xid B cells in vivo: TI-2 antigen induces migration and proliferation but Btk is essential for terminal differentiation. Eur J Immunol 2001; 31:1340-1350.
101. Spencer J, Perry ME, Dunn-Walters DK. Human marginal zone B cells. Immunol Today 1998; 19:421-426.
102. Tanigaki K, Han H, Yamamoto N et al. Notch-RBP-J signaling is involved in cell fate determination of marginal zone B cells. Nat Immunol 2002; 3:443-450.
103. Kumararatne DS, MacLennan IC. Cells of the marginal zone of the spleen are lymphocytes derived from recirculating precursors. Eur J Immunol 1981; 11:865-869.
104. Dammers PM, de Boer NK, Deenen GJ et al. The origin of marginal zone B cells in the rat. Eur J Immunol 1999; 29:1522-1531.
105. Dunn-Walters DK, Isaacson PG, Spencer J. Analysis of mutations in immunoglobulin heavy chain variable region genes of microdissected marginal zone (MGZ) B cells suggests that the MGZ of human spleen is a reservoir of memory B cells. J Exp Med 1995; 182:559-566.
106. Shih TA, Meffre E, Roederer M et al. Role of BCR affinity in T cell-dependent antibody responses in vivo. Nat Immunol 2002; 3:570-575.
107. Dammers PM, Visser A, Popa ER et al. Most marginal zone B cells in rat express germline encoded Ig V(H) genes and are ligand selected. J Immunol 2000; 165:6156-6169.
108. Tierens A, Delabie J, Michiels L et al. Marginal-zone B cells in the human lymph node and spleen show somatic hypermutations and display clonal expansion. Blood 1999; 93:226-234.
109. Martin F, Kearney JF. Positive selection from newly formed to marginal zone B cells depends on the rate of clonal production, CD19, and btk. Immunity 2000; 12:39-49.
110. Wang JH, Avitahl N, Cariappa A et al. Aiolos regulates B cell activation and maturation to effector state. Immunity 1998; 9:543-553.
111. Hao Z, Rajewsky K. Homeostasis of peripheral B cells in the absence of B cell influx from the bone marrow. J Exp Med 2001; 194:1151-1164.
112. Batten M, Groom J, Cachero TG et al. BAFF mediates survival of peripheral immature B lymphocytes. J Exp Med 2000; 192:1453-1466.
113. Morris L, Crocker PR, Hill M et al. Developmental regulation of sialoadhesin (sheep erythrocyte receptor), a macrophage-cell interaction molecule expressed in lymphohemopoietic tissues. Dev Immunol 1992; 2:7-17.

114. Korner H, Winkler TH, Sedgwick JD et al. Recirculating and marginal zone B cell populations can be established and maintained independently of primary and secondary follicles. Immunol Cell Biol 2001; 79:54-61.
115. Cyster JG, Goodnow CC. Pertussis toxin inhibits migration of B and T lymphocytes into splenic white pulp cords. J Exp Med 1995a; 182:581-586.
116. Lu TT, Cyster JG. Integrin-mediated long-term B cell retention in the splenic marginal zone. Science 2002; 297:409-412.
117. Chen X, Martin F, Forbush KA et al. Evidence for selection of a population of multi-reactive B cells into the splenic marginal zone. Int Immunol 1997; 9:27-41.
118. Li Y, Li H, Weigert M. Autoreactive B cells in the marginal zone that express dual receptors. J Exp Med 2002; 195:181-188.
119. Brezinschek HP, Foster SJ, Dorner T et al. Pairing of variable heavy and variable kappa chains in individual naive and memory B cells. J Immunol 1998; 160:4762-4767.
120. Finne J, Leinonen M, Makela PH. Antigenic similarities between brain components and bacteria causing meningitis. Implications for vaccine development and pathogenesis. Lancet 1983; 2:355-357.
121. Krog L, Block E. Glycosylation of neural cell adhesion molecules of the immunoglobulin superfamily. APMIS Suppl 1992; 27:53-70.
122. Troy F. Polysialylation: From bacteria to brains. Glycobiol 1992; 2:5-23.
123. Finne J, Bitter-Suermann D, Goridis C et al. An IgG monoclonal antibody to group B meningococci cross-reacts with developmentally regulated polysialic acid units of glycoproteins in neural and extraneural tissues. J Immunol 1987; 138:4402-4407.
124. Fusco PC, Michon F, Tai JY et al. Preclinical evaluation of a novel group B meningococcal conjugate vaccine that elicits bactericidal activity in both mice and non human primates. J Infect Dis 1997; 175:364-372.
125. Poltorak A, He X, Smirmova I et al. Defective LPS signaling in C3H/HeJ and C57BL/10ScCr mice:Mutations in Tlr4 gene. Science1998; 282:2085-2088.
126. Cooke M, Heath A, Shokat K et al. Immunoglobulin signal transduction guides the specificity of B cell-T cell interactions and is blocked in tolerant self-reactive B cells. J Exp Med 1994; 179:425-434.
127. Coutinho A, Moller G. Immune activation of B cells: Evidence for one nonspecific triggering signal not delivered by the Ig receptors. Scand J Immunol 1974; 3:133.
128. DeFranco AL, Kung JT, Paul WE. Regulation of growth and proliferation in B cell subpopulations. Immunological Rev 1982; 64:161-182.
129. Krieg AM, Yi AK, Matson S et al. CpG motifs in bacterial DNA trigger direct B-cell activation. Nature 1995; 374:546-549.
130. Masari P, Henneke P, Ho Y et al. Immune stimulation by Neisserial porins is Toll-like receptor 2 and MyD88 dependent. J Immunol 2002; 168:1533-1537.
131. Snapper CM, Rosas FR, Kehry MR et al. Neisserial porins may provide critical second signals to polysaccharide-activated murine B cells for induction of immunoglobulin secretion. Infect Immun 1997; 65:3203-3208.
132. Leadbetter EA, Rifkin IR, Hohlbaum AM et al. Chromatin/IgG complexes activate autoreactive B cells by dual engagement of sIgM and Toll-like receptors. Nature 2002; 416:603-607.

Chapter 10

CD1-Restricted T Cell Responses against Microbial Glycolipids

Steven A. Porcelli, Lynn G. Dover and Gurdyal S. Besra

Introduction

The identification of the Major Histocompatibility Complex (MHC) class I and class II molecules as presenting elements for the recognition of peptide antigens by T cells is one of the most fundamental principles of the adaptive immune response. While this mechanism clearly lies at the heart of most adaptive immunity, it has also become clear in recent years that specific T cell recognition of other types of antigens through mechanisms that are not dependent on the MHC also exist. One of the best-characterized of the MHC-unrestricted pathways for specific T cell responses involves the CD1 system, a family of nonpolymorphic antigen presenting molecules that controls the recognition of lipid and glycolipid antigens by T cells. In the current review, we provide a general overview of the CD1 system with an emphasis on its potential role in adaptive immunity to microbial infection.

CD1 Genes and the Evolution of the CD1 Family

The human CD1 molecules are encoded on chromosome 1 (1q22-23) by five closely linked genes, and is thus unlinked to the MHC which is on chromosome 6 in humans.[1] Interestingly, the region spanning human chromosome 1q21-q25, which contains the CD1 locus and multiple other immunologically relevant genes, has been proposed to be a premier MHC paralogous region.[2] This suggests that these two loci were created by large scale genomic duplications around the time of vertebrate emergence, and may thus explain the distant but still striking relationship between CD1 and MHC genes and the proteins they encode. Thus, all CD1 genes share a similar intron-exon structure as MHC class I genes, and encode polypeptides with significant homology to both MHC class I and II proteins.[3,4]

The products of all five human CD1 genes (CD1A, CD1B, CD1C, CD1D and CD1E) have been identified in humans and encode distinct CD1 isoforms, designated CD1a, CD1b, CD1c, CD1d and CD1e, respectively.[5,6] Initial studies of CD1 proteins indicated little or no polymorphism of CD1 molecules expressed by different individuals, suggesting that CD1 is not subject to the same selective forces that support the extensive allelic polymorphism of classical MHC I and II gene loci.[7,8] This has been confirmed by recent studies on human CD1 alleles showing few variations between individuals from different ethnic backgrounds.[9,10] A separation of CD1 genes of humans and other mammals into two separate groups was first proposed by Calabi et al based on analysis of extensive sequence divergence between the CD1d-type proteins of humans, mice and rabbits and the other CD1 isoforms in these species.[11] This classification, which seems to have been supported by subsequent studies of tissue expression and function, designates the CD1a, -b and -c proteins as group 1 and the CD1d proteins as group 2.[5]

Immunobiology of Carbohydrates, edited by Simon Y.C. Wong and Gemma Arsequell.
©2003 Eurekah.com and Kluwer Academic / Plenum Publishers.

Table 1. Size and complexity of CD1 gene families among different mammals

Species	Total CD1 Genes	Number of Genes of Indicated Isoform				
		CD1A	CD1B	CD1C	CD1D	CD1E
Human	5	1	1	1	1	1
Mouse	2	0	0	0	2	0
Rat	1	0	0	0	1	0
Guinea pig	~10	?	4	3	?	1
Rabbit	~8	≥2	≥1	?	≥1	≥1
Sheep	~7	?	≥3	?	≥1	≥1
Pig	~4	≥1	?	?	?	?

Besides humans, several other mammalian species have been studied for their expression of CD1. Interestingly, CD1 genes and proteins have been found in all mammals so far examined, although there are remarkable differences in size and complexity of the CD1 families in different species (Table 1).[12-16] The presence of CD1 genes in different mammalian species that diverged approximately 70 million years ago demonstrates the existence of such genes prior to that event, and thus supports the hypothesis that the CD1 system is relatively ancient in evolutionary terms. Consistent with this view, CD1 proteins show comparable levels of similarity and divergence at the amino acid and nucleic acid sequence levels to both MHC class I and class II.[5,17] This similar level of homology suggests that all three of these families of antigen presenting molecules may have evolved from a single ancestral molecule, possibly a primordial antigen binding or presenting element, that existed before the divergence of the MHC class I, II and CD1 families. Because MHC class I and II gene families are both known to exist in even the most ancient vertebrates with jaws (e.g., cartilaginous fish), this idea suggests that CD1 genes could be present in most or all vertebrates. This hypothesis has not yet been examined, as no studies have been reported on the presence or absence of CD1 genes in vertebrates that predate the evolution of mammalian lineage.

Another important issue for understanding the evolution of CD1 molecules is the relationship between the different CD1 isoforms. Sequence comparison of CD1 genes and proteins between nonhuman species and humans consistently reveal that each nonhuman CD1 is clearly an orthologue (i.e., having a homologous structure between different species) of one of the human isoforms.[5] For example, several species have been shown to have CD1 sequences that can be identified as orthologues of human CD1b (e.g., guinea pig, sheep and rabbit), and the 60 to 70% homology between CD1a, -b and -e and their orthologues in the rabbit is closer than the homology that exists between these different human CD1 isoforms.[16] It is evident from this that CD1 isoforms are highly conserved throughout evolution, more so than the classical MHC class I and II molecules that generally do not show clear orthologues between different species. This suggests that the different CD1 isoforms may have become specialized early in mammalian evolution, and have subsequently preserved their function and structure during the evolutionary divergence of mammalian species because of strong selective forces.

A striking difference between the CD1 gene loci of rodents and humans is the absence of group 1 CD1 in mice and rats. The genomes of all mouse strains examined lack group 1 CD1 genes, but maintain two very similar (90-95% sequence identity) group 2 CD1 genes (CD1D1 and CD1D2) that are likely the result of a relatively recent gene duplication event.[18,19] Functional studies suggest that although the CD1D2 gene is functional and probably expressed in most strains of mice, the majority of the currently known in vivo functions of CD1 in the mouse appear to map to the CD1D1 gene.[20-22] With regard to the complete absence of group

1 CD1 in mice and rats, this appears most likely to have resulted from the deletion of the group 1 genes in the early ancestors of these rodents. This view is favored by the presence of group 1 CD1 genes in close evolutionary relatives of the muroid rodents, such as rabbits and guinea pigs.[12,16] In addition, the location of murine CD1 genes defines the conserved boundary point of an area of chromosomal synteny between mice and humans, which most likely represents the breakpoint of an ancient translocation that may have been associated with deletion of a portion of the CD1 locus that contained the rodent group 1 genes.[23,24]

Whether the loss of group 1 CD1 genes is associated with any signifying immunologic defect in mice is unclear, but most likely mice have evolved mechanisms to compensate for the absence of these molecules. It is important to note that the absence of group 1 CD1 disqualifies the mouse animal model for many studies on the CD1 system and makes it difficult to extrapolate findings in this model to humans. Other mammalian species (e.g., guinea pigs, rabbits or nonhuman primates) may thus provide more suitable animal model systems for investigations on the functions of group 1 CD1 molecules in vivo.[12,16]

CD1 Protein Structure

From their deduced amino acid sequences, CD1 polypeptides have a predicted molecular mass of approximately 36 to 37 kDa, but the addition of three or more N-linked glycans raises their observed masses into a range between 41 and 55 kDa.[25,26] All CD1 proteins studied to date are expressed as type I transmembrane proteins and associate noncovalently with β2-microglobulin (β2m). Because of the homology of CD1 with MHC class I and the common association with β2m, it was predicted that the folding of CD1 and MHC class I heavy chains would be similar.[5] This prediction was proven correct by the three-dimensional structure of murine CD1d1 elucidated by X-ray crystallography, which revealed a remarkable similarity between CD1 and MHC class I in their overall configurations (Fig. 1).[27] The membrane distal α1 and α2 domains of CD1d1 have the typical antigen-binding structure also found in MHC class I and II domains, consisting of two antiparallel α-helices overlying a β-pleated sheet. As in MHC class I heavy chains, the α3 domain has an immunoglobulin-like fold and makes the majority of the contacts with the β2m subunit. The presence of a putative ligand binding groove between the two α-helices is similar to MHC class I and II, and forms the opening into a cavity within the α1 and α2 domains. However, the cavity in CD1d1 is deeper than the peptide binding grooves of MHC molecules, and the groove appears to be closed at both ends and covered over much of its length so that it is only accessible to the exterior of the molecule near its center. Furthermore, it does not contain the six to nine small pockets that accommodate the side chains of amino acids of bound peptide ligands in MHC class I and II molecules, but rather two large pockets, designated A' and F'. These pockets and the interior of the groove are mainly lined by hydrophobic amino acid residues, and thus comprise a hydrophobic ligand binding site, such as seen in various lipid-binding proteins.[27] Structures of other CD1 molecules are not yet available, but based on amino acid sequence comparisons and molecular modeling, it is believed that they will present the same features as those observed in murine CD1d1.

Cellular Expression and Tissue Distribution of CD1 Proteins

Group 1 CD1 molecules were first identified as differentiation markers expressed on immature cortical thymocytes.[28] They are also expressed on a variety of specialized antigen-presenting cells that can be found in both lymphatic and nonlymphatic tissues.[29] Group 1 CD1 molecules are inducible in vitro on human peripheral blood monocytes by incubation with granulocyte-macrophage colony stimulating factor (GM-CSF) and interleukin 4 (IL-4), which under appropriate culture conditions transforms these cells into immature dendritic cells.[30] This suggests that group 1 CD1 molecules might be up-regulated on monocyte-derived cells in many inflammatory lesions, as well as on subsets of resident tissue dendritic cells. A subset of human B cells expresses CD1c, and this appears to be especially prominent among B cells localizing to the mantle zones of lymphoid follicles and germinal centers.[31,32]

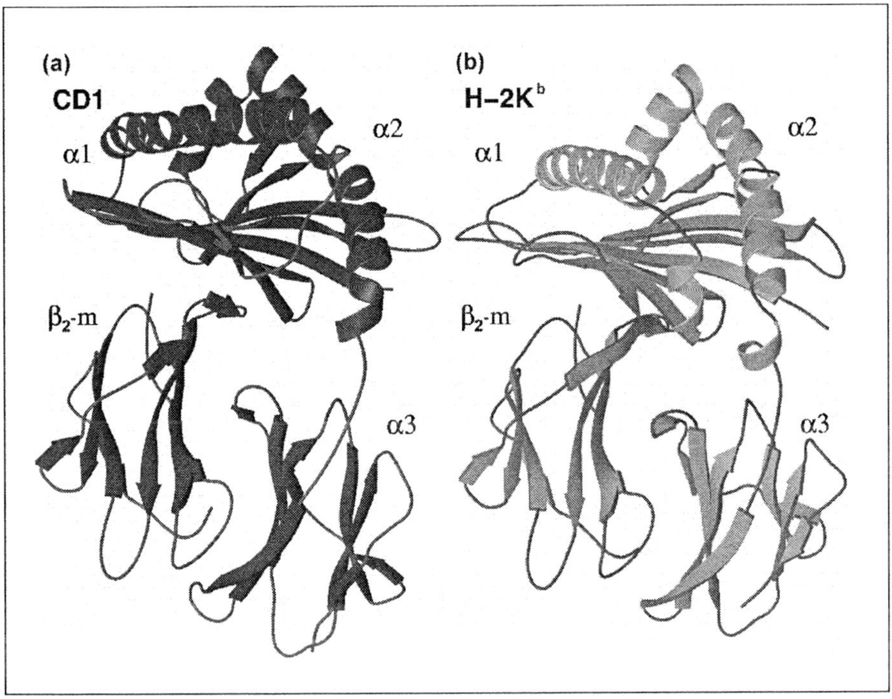

Figure1. Crystal structure of CD1 compared to MHC class I. The structure of the mouse CD1d1 protein as described by Zeng et al[27] is shown as a ribbon diagram in (a). Note the remarkable overall similarity in shape with the typical mouse MHC class I molecule H-2K^b shown in (b). Figure adapted with permission from Porcelli et al.[169]

Group 2 CD1 proteins (i.e., CD1d and its orthologues) appear to have a pattern of expression somewhat different that of group 1 CD1. Several studies indicate that CD1d may be expressed by gastrointestinal epithelial cells and also by hepatocytes, cell types on which group 1 CD1 seems to be absent.[33,34] Group 2 CD1 molecules are also expressed on hematopoietic cells in both humans and mice, with abundant expression on a majority of thymocytes.[35-37] During the thymocyte maturation process, CD1d seems to be down-regulated, but in mice the down-regulation does not appear to be complete and CD1d1 can still be detected on mature T cells.[35,38] Murine CD1d1 is also constitutively expressed on many dendritic cells and most B cells, and is highly up-regulated on splenic marginal zone B cells.[39] Cell surface levels of human CD1d do not appear to be up-regulated by GM-CSF treatment of human peripheral monocytes in vitro, and to date, no cytokines or other factors have been clearly shown to regulate expression of group 2 CD1 molecules on hematopoietic cells. These differences in cellular expression and regulation of group1 and group 2 CD1 molecules support the classification of CD1 molecules into these two categories.

T Cell Recognition of CD1 and CD1-Presented Antigens

The MHC class I-like structure and strong expression of CD1 on antigen-presenting cells provided the first suggestion that these proteins could function as antigen-presenting molecules and be recognized by specific T cells. The first direct evidence of a possible role of CD1 in T cell activation and function was provided by the isolation of CD4$^-$CD8$^-$ T cell clones expressing either αβ or γδ T cell receptors (TCRs) that lysed tumor cells expressing specific isoforms of human CD1 (i.e., CD1a or CD1c).[40] Similar findings were subsequently reported for circulating human γδ T cells and for human intestinal intraepithelial T lymphocyte lines.[41-43] After these first results in humans, CD1-reactive T cells were also described in mice. In normal mice, these cells were isolated from the NK1$^+$ T cell fraction, and in MHC class II-deficient mice they were contained in the small residual CD4$^+$ T cell population.[20,44] These CD1-reactive T cells recognized cells expressing CD1 molecules without addition of a foreign antigen, suggesting that they might represent a form of T cell autoreactivity that is present within the lymphocyte pool of normal animals.

Subsequent studies revealed examples of CD1-restricted T cells that were specific for foreign antigens. For example, T cells that recognized antigens from *Mycobacterium tuberculosis* or *M. leprae* presented by CD1b molecules could be isolated that were MHC class I and class II independent in their antigen specific responses.[45-47] T cells that were specifically reactive to mycobacterial antigens and restricted by CD1c or CD1a were also identified, thus establishing a role in microbial antigen presentation for each of the human group 1 CD1 proteins.[48,49] The role of the clonotypic TCR in recognition of microbial antigens by CD1-restricted T cells was proven by TCR gene transfer studies, in which transfection of cDNAs encoding the TCR α and β chains of CD1-restricted, mycobacterial antigen specific T cell clones into TCR deficient Jurkat cells transferred both CD1-restriction and antigen specificity to the resulting transfectants.[50] Although many of the earliest CD1-restricted T cell lines isolated were CD4$^-$CD8$^-$ (or double negative, DN), more recent studies have shown that CD1-restricted T cells can also be found among T cells that are CD8$^+$ or CD4$^+$.[49,51] However, the role of CD8 or CD4 molecules as coreceptors in the recognition of CD1-presented antigens remains unclear, although several studies suggest an interaction between murine CD1 and CD8 molecules that may contribute to TCR signaling.[52,53]

Studies on CD1-dependent T cell recognition in the mouse have thus far focused largely on a unique subset of lymphocytes known as natural killer (NK) T cells. NK T cells express cell surface proteins that are usually characteristic for the NK cell lineage, such as NKR-P1A, CD94 and CD69 in humans or Ly49, NK1.1 (NKR-P1C) and CD69 in mice. All of these molecules are encoded by the so-called NK locus.[53] The TCRs of the major population of CD1-restricted NK T cells are unusual in that they express an invariant TCRα chain (Vα14-Jα281 without N region additions or deletions) that is strongly homologous between humans and mice, and have a limited TCRβ chain repertoire (predominantly Vβ11 in humans and Vβ8, 7 and 2 in mice).[54] NK T cells secrete large amounts of IL-4 and other cytokines after their activation through the TCR, and appear to modulate inflammatory reactions and immune-mediated tissue injury. In fact, NK T cells have been implicated as important regulatory T cells for the control and prevention of several different autoimmune diseases, most notably autoimmune type I diabetes mellitus.[55] Like NK cells, NK T cells are also able to kill target cells, and they have been strongly implicated in tumor rejection in certain mouse models of malignancy.[56-58] In addition to these well described NK T cells bearing antigen receptors with the canonical invariant TCRα chain sequences, a population of CD1d-reactive murine T cells that expresses diverse TCRs has also been identified.[59] The phenotype, distribution and functions of these diverse TCR-bearing CD1d-reactive T cells are currently undefined.

It appears that most or possibly all of the cardinal features of murine NK T cells are highly conserved by an analogous human T cell subpopulation. Thus, studies have identified human NK T cells expressing the invariant Vα24-JαQ TCRα chain that is highly homologous to the TCRα chain expressed by murine NK T cells.[60-62] These human NK T cells are directly reactive (albeit weakly in most cases) with CD1d in vitro, and can typically produce high levels of both IL-4 and interferon γ (IFN-γ) upon TCR-mediated activation.[60] Their specific in vivo functions are still largely unknown, but recent studies link decreased numbers of NK T cells and alterations in NK T cell cytokine patterns to the progression of a variety of autoimmune diseases in humans.[63-65]

CD1-Presented Antigens

The first direct purification of an antigen recognized by a human CD1-restricted T cell line reactive with *M. tuberculosis* revealed the surprising finding that the specific antigen recognized was a lipid, in this particular case an abundant α-branched β-hydroxyl fatty acid of the mycobacterial cell wall known as mycolic acid.[66] Subsequently, multiple reports have identified specific foreign or self antigens presented by various CD1 proteins, and nearly all of the specific antigens identified have been found to be lipids. The only exceptions to this general rule have been two reports of CD1d-restricted T cell recognition of hydrophobic peptide sequences in the mouse.[67,68] Thus, while the presentation of variety of different types of hydrophobic ligands by CD1 molecules remains a formal possibility that needs to be subjected to further investigation, at this point the available evidence strongly favors the view that the overwhelming majority of molecules that function as CD1-presented ligands are lipids or glycolipids. Although all of the well characterized bacterial lipid antigens presented by CD1 have been derived from mycobacteria, one study has demonstrated T cell lines specific for the lipid fraction of *Haemophilus influenzae*, suggesting that other bacteria besides mycobacteria are likely to harbor CD1-presented lipid antigens.[69] The following sections briefly summarize the several currently known classes of lipids which thus far have been found to function as CD1-presented targets for T cell responses.

Glycosphingolipids

Both mouse and human CD1d-restricted NK T cells have been shown to respond to α-galactosyl ceramide (αGalCer), a naturally occurring glycosphingolipid originally isolated from a marine sponge.[70,71] This glycolipid structure is unusual in comparison with typical mammalian glycosphingolipids in three ways (**Fig.** 2). First and perhaps most unique is the α-linkage of the saccharide to the ceramide lipid. This bond is usually a β-linkage in the glycosphingolipids of most animals and plants, and this so far has been found universally to be the case in cells and tissues of vertebrate origin. Second, the sphingosine base of αGalCer is actually phytosphingosine, and thus has a saturated 3,4-dihydroxy structure as opposed to the usual single 3-hydroxy group and unsaturation that is found in the common sphingosine structure. Finally, the acyl chain of αGalCer is significantly longer at C26 compared to most ceramides in higher animals.

It remains unclear whether the strong recognition of αGalCer by a large fraction of CD1d-restricted NK T cells represents a purely fortuitous pharmacological effect or a cross-reaction with some as yet unknown structurally similar compound. Candidate compounds could occur naturally in infectious agents or as a result of alterations in cellular homeostasis. A substantial body of evidence has shown that αGalCer binds directly to CD1d, and site-directed mutagenesis implicates hydrophobic residues in both the A' and F' pockets of the CD1d groove in the anchoring of this glycolipid ligand.[72,73] Extensive studies of the structural requirements for αGalCer recognition by CD1d-restricted T cells show that some but not all of the unusual structural features of this glycolipid are necessary for its presentation and recognition by T cells. For example, the α-glycosidic linkage is an absolute requirement for recognition, whereas the 3,4-hydroxy phytosphingosine structure is not necessary. In addition, certain alterations of the hexose can be tolerated without loss of activity, such as substitution of glucose for galactose or the addition of additional monosaccharides to the C2 or C6 positions.[70,71]

Recent evidence from studies of human T cells also shows that group 1 CD1 molecules can present certain glycosphingolipids.[74,75] T cells responding to several different gangliosides, including GM1 and asialo-GM1 (Fig. 2), were isolated from the circulation of normal humans or from subjects with multiple sclerosis. These T cells were found to be dependent on CD1b for their recognition of gangliosides. More recently, it has been shown that soluble recombinant CD1b molecules can be loaded with gangliosides, and that these cell-free complexes when attached to a plastic surface can actually stimulate the ganglioside specific CD1b-restricted T

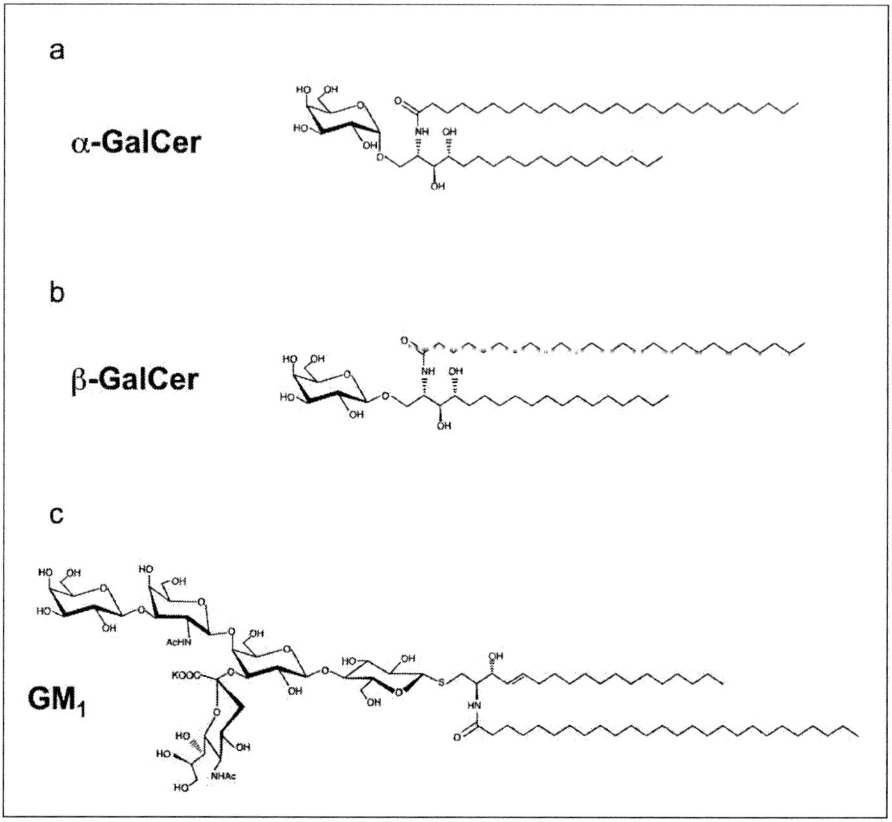

Figure 2. CD1-presented glycosphingolipid antigens. α-Galactosyl ceramide (a) binds to CD1d and acts as a strong agonist for human and mouse CD1d-restricted NK T cells. In contrast, β-galactosyl ceramide (b) is not recognized by these T cells. The ganglioside GM1 (c) is a mammalian glycosphingolipid that has been shown to be presented by human CD1b to T cells.

cells.[75,76] This provides strong evidence for the ability of at least one human group 1 CD1 molecule to bind these glycolipids in a way that gives rise to structures that can be specifically recognized by T cells. Since gangliosides such as GM1 represent abundant glycolipids in many normal cells and tissues, the T cells recognizing such ligands clearly have the potential to be autoreactive and thus potentially could be linked to certain forms of autoimmunity. How such reactivity to gangliosides is normally regulated, especially given the fact that CD1b-restricted ganglioside reactive cells can be found in normal subjects, remains to be resolved.

Mycolates and Glycosylated Mycolates

Mycolic acids are a homologous series of C_{60}-C_{90} α-alkyl, β-hydroxy fatty acids produced by all mycobacteria. Similar but shorter length mycolic acids are found in related taxa, for example in corynebacteria and nocardia. In mycobacteria, mycolic acids are present mainly as glycolipids, extractable by organic solvents mainly in the form of trehalose 6,6'-dimycolate (TDM, also known as 'cord factor'), glucose monomycolate (GMM) or as bound esters of arabinogalactan (AG) (Fig. 3).[77,78]

The prefix 'α' has also been attributed to the mycolates initially eluted from a column of adsorption chromatography. In some mycobacterial strains, shorter mycolates are present (around C_{60} instead of C_{90}) and have been designated as α'-mycolic acids.[77] The first α'-mycolic acids

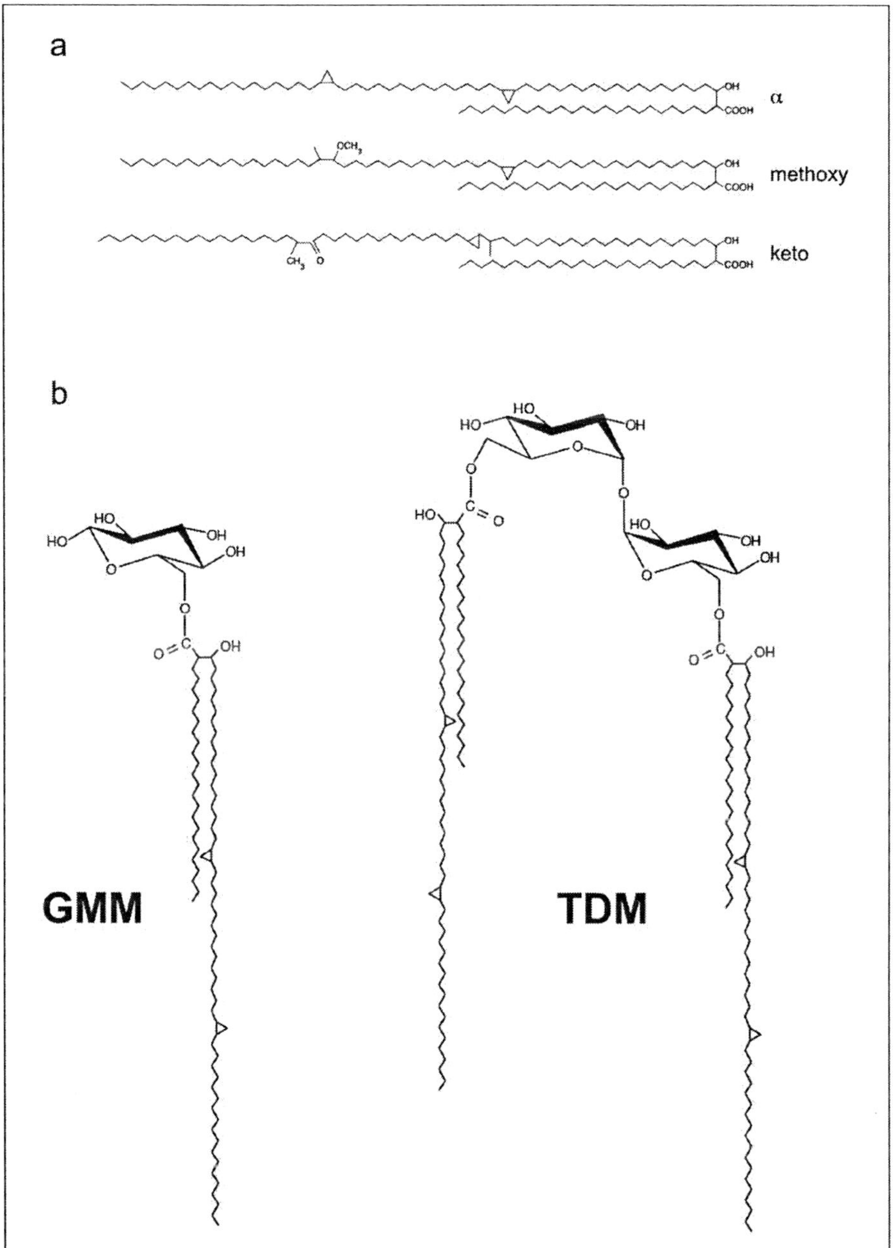

Figure 3. Mycolic acids and glycosylated mycolates. General structures of the three subclasses of mycolic acids found in *Mycobacterium tuberculosis* are illustrated (a). The longer meromycolate chain of these branched fatty acids can vary in length, whereas the shorter α-branch is fixed at a length of C24 in *M. tuberculosis*. The structures shown conform to average sized mycolates for *M. tuberculosis* of approximately C80. Note the presence of cyclopropyl groups and the oxygen containing R-groups on the meromycolate chains. Mycolic acids generally exist in the mycobacterial cell wall as esters of simple or complex carbohydrates (see, b), such as trehalose dimycolate (TDM) and glucose monomycolate (GMM).

were isolated from *Mycobacterium smegmatis* (C_{62}-C_{64}).[79] Although the basic structure is well conserved, there is considerable variation in functional groups within the long meromycolic branch of these acids. Other mycolic acids, eluted after α-mycolates, generally contain oxygen functions in the meromycolic acid branch and are referred to as keto-, methoxy-, epoxy- and wax ester-mycolates. These structural modifications occur at two positions in the meromycolic acid chain, referred to as distal and proximal (closest to the ß-hydroxy unit). In general, oxygenated modifications are restricted to the distal position, whereas nonoxygenated modifications occur at both the distal and the proximal positions and include cis or trans double bonds, cis or trans cyclopropane rings and adjacent methyl groups.

Several studies have shown that in any given mycolate class, there is considerable heterogeneity. It was shown that mycobacterial species produce several different α-chain lengths, ranging from 20 to 26 carbon atoms.[80] *M. smegmatis* was found to contain α branches containing 22 and 24 carbons and *M. tuberculosis* 24 and 26 carbons, with negligible amounts of C_{22}.[81] In addition to the α branch, variations also occur in the meromycolic acid chain. Kaneda et al demonstrated that α-mycolic acids from *M. smegmatis* range from 74 to 81 carbon atoms, and *M. tuberculosis* from 76 to 86 carbon atoms.[82] In *M. tuberculosis*, it was also found that oxygenated mycolates are longer than α-mycolates. Indeed, Yuan et al examined recently the carbon content of each mycolate class in *M. tuberculosis* and reported that α-mycolates contained 76-82 carbons, whereas methoxy and keto mycolic acids possess 83-90 and 84-89 carbons, respectively.[83] Microheterogeneity has also been linked to the internal position of the functional groups. All these variations in chain length and position of the functional groups contribute to the total number of molecules defining each class of mycolic acids. Danielson and Gray reported for example that in *M. smegmatis*, about 100 structural isomers are present in the mixture of α-mycolates.[84]

Both free mycolic acids and the monoglucosylated form of these lipids (glucose-6-monomycolate, GMM) have been identified as specific antigens presented to T cells by human CD1b.[66,85] The recognition of these antigens by CD1b-restricted T cells has been demonstrated to be highly specific for the polar cap structure of the lipid or glycolipid, and is mediated by specific interactions involving clonotypic TCRs on the responding T cells.[50,85-88] Although the specific role played by GMM in the structure and function of the mycobacterial cell wall is currently not known, the finding that T cells isolated from a human subject immune to *M. leprae* respond to GMM as a CD1b-presented antigen suggest that it is produced during growth of mycobacteria in tissues in vivo.[85] Interestingly, in vitro studies have recently shown that GMM is only synthesized at measurable levels when mycobacteria are provided with an exogenous source of free glucose.[87] This requirement for free glucose for the synthesis of GMM is seen for both pathogenic species of mycobacteria, such as *M. tuberculosis* and *M. avium*, and also for nonpathogenic saprophytes, such as *M. smegmatis* and *M. phlei*. In the case of *M. avium*, it was found that GMM was produced at levels that could be recognized by CD1b-restricted T cells specific for this glycolipid when the bacteria were grown in media supplemented with glucose at concentrations comparable to those normally found in mammalian tissues and body fluids (e.g., 100 mg/dl). Furthermore, glycolipids with chromatographic properties identical to GMM were shown to be present in *M. leprae* bacilli purified from animal tissue, and these were stimulatory for T cells specific for this antigen.[87] Together, these results demonstrate that GMM is likely to be formed by mycobacteria that invade tissues and replicate in vivo where free glucose is abundant, but is not likely to be formed by mycobacteria growing outside of an animal host. Thus, the recognition of GMM by CD1b-restricted T cells in humans has been proposed as a mechanism by which pathogenic mycobacteria that have productively infected tissues can be distinguished from ubiquitous nonpathogenic mycobacteria.

Glycosylated Phosphoisoprenoids

Recent work has also called attention to isoprenoid lipids as potential targets of the CD1-restricted adaptive immune response to mycobacteria. In a study of T cell responses to phospholipids of mycobacteria presented by the human CD1c protein, two novel fully

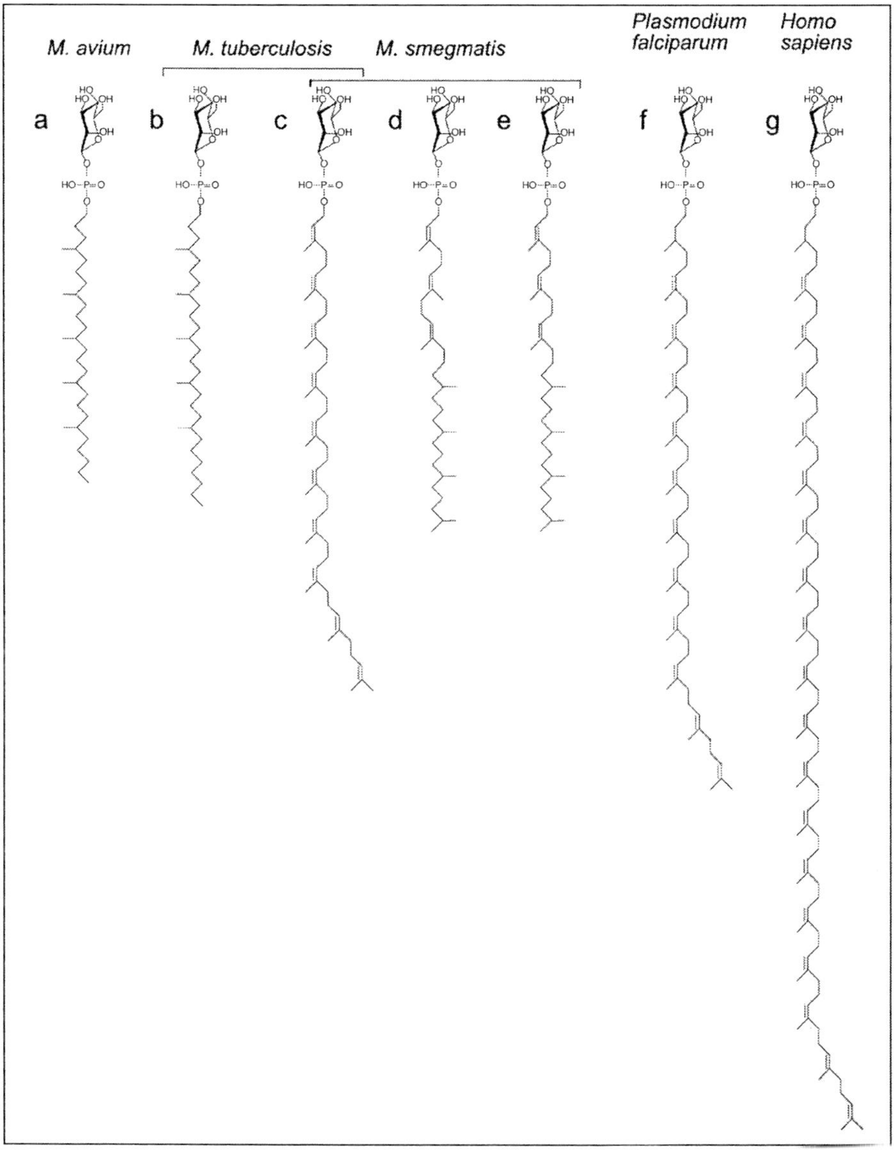

Figure 4. Isoprenoid phosphoglycolipids. The two structures at the far left (a and b) represent fully saturated β1-mannosyl phosphodolichols, which were identified as CD1c-presented isoprenoid type phosphoglycolipid antigens that stimulate human T cell responses. The other structures show the range and variations of mannosyl phosphoisoprenoid lipids that occur in mycobacteria and in lower and higher eukaryotes: (c) mycobacterial β1-mannosyl (mono-E, poly-Z) decaprenyl phosphate, (d) mycobacterial β1-mannosyl (di-E, mono-Z) heptaprenyl phosphate, (e) β1-mannosyl (tri-E, mono-Z) octaprenyl phosphate, (f) plasmodium β1-mannosyl (mono-E, poly-Z) C60 dolichol phosphate, and (g) human β1-mannosyl (mono-E, poly-Z) C95 dolichol phosphate.

saturated β1-mannosyl-phosphoisoprenoid lipids from *M. tuberculosis* and *M. avium* were isolated and structurally characterized (Fig. 4).[89] Analysis of *M. tuberculosis* infected human subjects indicated that T cells responding to these lipids expand in response to infection, and persist in the circulation as part of the memory T cell population specific for *M. tuberculosis*. The contribution of these CD1c-restricted T cells to the successful control of *M. tuberculosis* by the immune system is not yet known, but their existence provides novel targets for the T cell response that may prove relevant to vaccination or immunotherapy for mycobacterial infection.

Several isoprenoid lipids have been identified in the mycobacterial cell wall, where they are most likely involved in the synthesis of peptidoglycan and other components of the cell wall.[90-92] It is well established that polyisoprenoid phosphates are important intermediates involved in the biosynthesis of bacterial cell walls and various prokaryotic and eukaryotic glycosylation reactions, such as glycosylphosphatidylinositol (GPI) biosynthesis, N-linked, O-linked and C-linked glycosylation processes.[93-96] For instance, mannosyl phosphoryl-dolichol (polyprenols that are saturated in the α-isoprene unit) function as substrates, and transfer mannose residues to dolichyl pyrophosphorylglycans, which are then subsequently transferred to asparagine units during eukaryotic N-linked glycosylation processes. Similarly, prokaryotes use α—unsaturated monophospho- and pyrophosphoglycan polyprenols to translocate either single sugar residues such as mannose, or complex carbohydrates such as arabinogalactan in the case of mycobacteria, across the cytoplasmic membrane to synthesize bacterial cell wall glycoconjungates.[97,98]

Dolichol phosphates are synthesized in two stages. Firstly, 5-carbon units of isopentenyl diphosphate (IPP) and dimethyl allyl diphosphate (DMAPP) are synthesized. Initially, condensation of acetyl-CoA leads to formation of β-hydroxy-β-methyl-glutaryl-CoA, which is reduced to mevalonate and decarboxylated to form IPP, which is then isomerized to DMAPP.[99] Secondly, IPP is added to DMAPP to form geranyl diphosphate (GPP) and farnesyl diphosphate (FPP). FPP then serves as a primer for further extension through a 1-4 head to tail condensation with IPP, affording geranyl geranyl diphosphate (GGPP).[100] Chain elongation continues through a series of C_5 IPP additions catalyzed by prenyl diphosphate synthases until the appropriate chain length (di-E, poly-Z isoprenol diphosphate) is obtained. Finally, dephosphorylation results in the formation of the polyprenol phosphate. The synthesis of the dolichol unit results from enzymatic modification of the α-prenyl unit by a polyprenol reductase.[101] Among eukaryotic organisms, dolichols vary considerably in length. Humans produce dolichols that range in length from C_{90} to C_{100}, whereas fungi produce typically shorter C_{75}-C_{85} dolichols.[102,103] Protozoa, such as leishmania and trypanosomes, produce even shorter dolichols (C_{50}-C_{70}) (Fig. 4).[104]

Recent studies have shown that bacteria which synthesize α-unsaturated polyprenols use an alternate nonmevalonate pathway for the synthesis of IPP. This involves the condensation of pyruvate with glyceraldehye-3-phosphate (GAP) to form 1-deoxyxylulose-5-phosphate, which is then converted to 2-C-methyl-D-erythritol-4-phosphate and finally IPP.[105-110] The nonmevalonate pathway has been shown in a number of bacteria, including *Zymomonas mobilis*, *Methylobacterium fujisawaense*, *Escherichia coli*, *Alicyclobacillus acidoterrestris*, *Corynebacterium ammoniagenes*, and *Mycobacterium phlei*.[108,110] Recently, it was demonstrated that deoxyxylulose was utilized by *M. smegmatis* for the synthesis of a compound that behaved chromatographically similarly to menaquinone, confirming that mycobacteria utilize the nonmevalonate pathway for the synthesis of isoprenoid lipids.[111] However, not all bacteria use the nonmevalonate pathway to synthesize IPP. For instance, *Myxococcus fulvus*, *Staphylococcus carnosus*, *Lactobacillus planarum* and *Halobacterium cutirubrum* utilize the eukaryotic IPP pathway.[106]

The most common polyprenol phosphate structures are divided into four groups based on the stereochemistry of their carbon-carbon double bonds. These are 1) all E-prenol; 2) di-E-, poly-Z-prenol; 3) tri-E, poly-Z-prenol; and 4) all Z-prenol. Bacteria generally contain a single polyprenol phosphate consisting of 11 isoprene units, typically with the di-E, poly-Z-configuration. However, certain bacteria produce unusual isoprenoid glycolipids that may or may not be derived from IPP. For instance, certain archaebacteria produce terpenoid lipids that are composed of isoprenoid-like C_5 units but, are fully saturated.[112] Mycobacteria contain two unusual forms of polyprenol phosphate, a decaprenyl phosphate and a heptaprenyl phosphate (Fig. 4).[92] The decaprenyl phosphate contains one Ω-, one E-, and eight Z-isoprene units (mono-E, poly-Z).[113] The heptaprenyl phosphate consists of four saturated isoprene units at the Ω-end, and two E- and one Z-isoprene units or four saturated and three Z-isoprene units.[90,114] *M. tuberculosis* was initially found to possess only a decaprenyl phosphate, and the precise stereochemistry of the individual isoprene units was not determined.[115] The mycobacterial polyprenol phosphates serve as lipid carriers for glycosylation reactions involving mannose, arabinose and ribose in the synthesis of lipoarabinomannan, arabinogalactan and riban.[97,98,116] Interestingly, the heptaprenyl phosphate was further mycolated, termed Myc-PL, and was shown to be involved in the final deposition of mycolic acids to the mycobacterial cell wall arabinogalactan.[90]

The identification of CD1c-presented glycolipid antigens as fully saturated β1 -mannosyl-phosphoisoprenoid lipids from *M. tuberculosis* and *M. avium* extends the family of polyprenols found in mycobacteria.[89] Although much shorter than the mannosylated versions of decaprenyl, these nevertheless appear to be related in structure to the heptaprenyl phosphate described above. The pathway leading to the synthesis of these related lipids is currently unknown, and the possibility exists that they may be synthesized through an entirely novel pathway. For example, the *M. tuberculosis* genome encodes a number of polyketide synthases which can generate branched-chain lipids of great diversity, similar to polyprenols, by varying both substrates and the enzymatic sub-unit primers for extension.[117] The possibility that polyketide synthases may contribute to the production of the novel fully saturated isoprenoid type glycolipids presented by CD1c is thus an interesting possibility that remains to be explored.

Mycobacterial Lipoglycans and Other Phosphatidylinositols

Similar to T cell responses against mycolic acids or glycosylated mycolates, T cells specific for lipoarabinomannan (LAM) recognize this antigen following its presentation by the CD1b molecule.[46] These large and complex molecules, which are ubiquitously found in the envelopes of all mycobacterial species, are one of the dominant immunoreactive substances of the mycobacterial outer cell wall of these organisms. LAMs are heterogeneous but share a tripartite amphipathic structure consisting of a phosphatidyl-myo-inositol anchor, a mannan core with a branching arabinan polymer, and the cap motifs that decorate the termini of the branched arabinan (Fig. 5).[118,119]

The identification of LAM followed from early studies on the immunogenic properties of mycobacterial cell walls, which pointed towards the presence of a serologically active polysaccharide containing arabinose and mannose.[120-122] Misaki et al and Azuma et al established that the polysaccharide possessed an α(1→6)-linked D-Manp backbone to which were attached short side chains of α(1→2)-linked D-Manp and α(1→5)-linked D-Araf residues.[123,124] Further studies by Tsumita et al and Ohashi established two classes of arabinomannan (AM); these are an acylated form possessing palmitic and tuberculostearic acyl functions, which is now termed LAM, and a neutral or nonacylated AM.[125,126] Independent studies by Weber and Gray resulted in isolation and partial purification of an acidic AM from *M. smegmatis* which they found to contain approximately 56 arabinosyl residues, 11 mannosyl residues, 2 phosphates (acid and alkali labile), 6 monoesterified succinates and 4 ether linked lactate groups.[127] The acidic AM was reactive with antisera from rabbits immunized with cell walls of *M. smegmatis,* and it was concluded that both AM and arabinogalactan share a common immunodominant epitope, i.e., α(1→5)-linked D-Araf residues.

The structural features of LAM that determine its recognition by CD1-restricted T cells are only partially understood. Initial studies have shown that the acyl chains of the lipid portion are absolutely required for recognition, whereas a substantial portion of the carbohydrate making up the arabinan portion of the glycan seemed not to be necessary.[46] However, the ability of at least some CD1b-restricted T cells to discriminate between LAMs from different mycobacterial species suggested that subtle variations in the structure of LAM are likely to be important in its ability to stimulate this type of host response. The following sections review in detail the current understanding of the detailed structure and structural variations that occur in LAM.

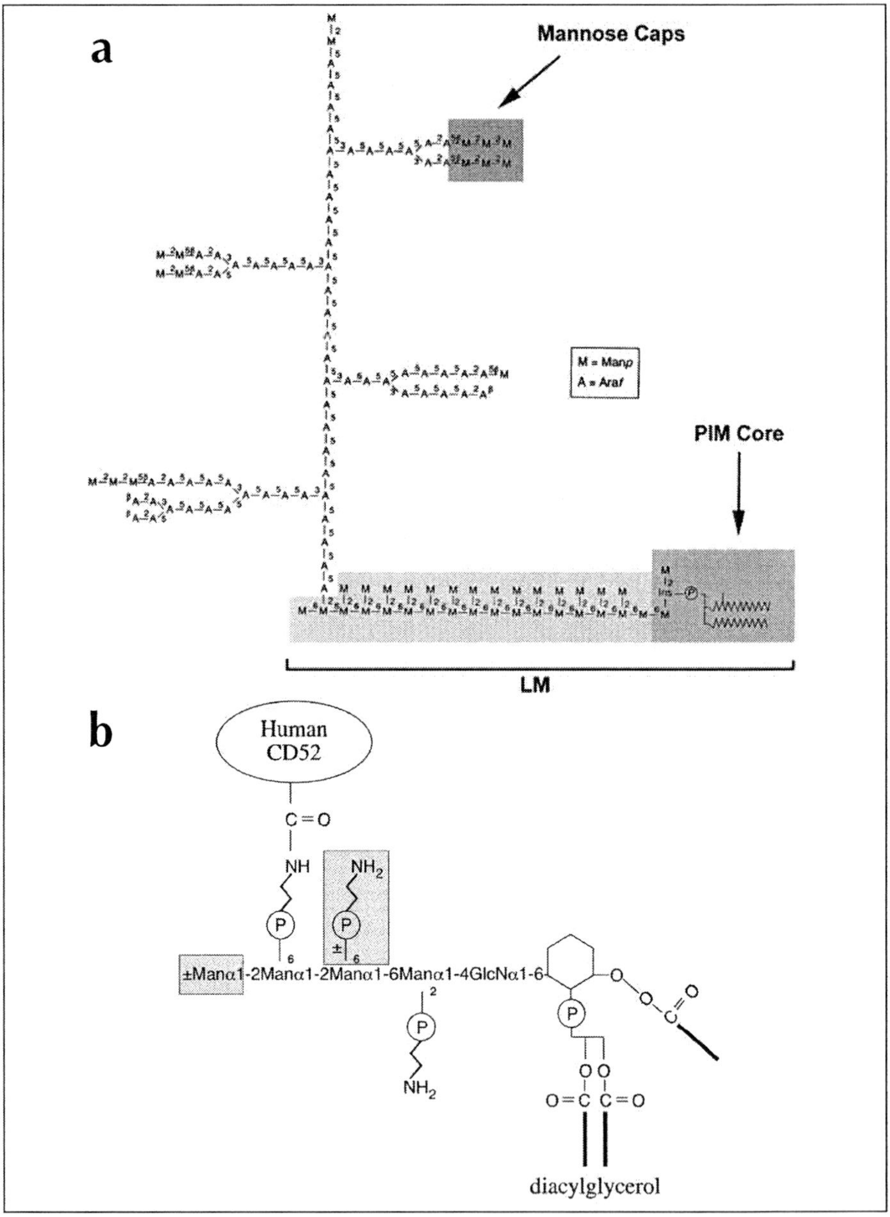

Figure 5. Schematic visualization of the structural relationship of mycobacterial LAM, LM and PIM_2, and mammalian GPI. (a) LAM is heterogeneous, but shares a tripartite amphipathic structure consisting of a phosphatidyl-myo-inositol anchor, which is indistinguishable from dimannosylated phosphatidylinositide (PIM_2), a mannan core, a branching arabinan polymer and a number of capping motifs that decorate the termini of the branched arabinan. The structural definition of LAM has resulted in two distinct arrangements occupying the terminal end: branched hexaarabinofuranosides (Ara_6) and linear tetraarbinofuranosides (Ara_4). However, in the case of LAM isolated from *M. tuberculosis*, these two types of arabinose termini are extensively capped with α-D-Manp residues, a product now called ManLAM. The mannose caps consist of monomers, dimers and trimers of α-D-Manp ($[\text{Man}\alpha1\_2\text{Man}\alpha1]_{0,1,2}$-) linked to the C-5 of the terminal β-D-Araf. (b) Key structural features of mammalian GPI. The structure of the GPI anchor of the human CD52 antigen is shown.

Phosphatidylinositol Mannoside Anchor

The current structural models for LAM stem largely from seminal studies performed by Brennan and colleagues during the late 1980s and early 1990s.[118,128] Detailed structural analysis of LAM showed that in addition to containing arabinose and mannose, it also contained glycerol, inositol, phosphate, lactate, succinate, palmitate and tuberculostearate, confirming the studies performed by Weber and Gray and others.[127] In addition, the phosphate group was shown to be in the form of an alkali-labile phosphatidyl-myo-inositol (PI) unit, and the two long-chain fatty acids were in the form of a diacylglycerol.

Recent studies have examined the PI portion of both LAM and LM and found them to be indistinguishable from the dimannosylated phosphatidylinositides (PIM_2), originally described in detail by Lee and Ballou.[129] The predominant acyl chains consist of palmitate (C_{16}) and 10-methyloctadecanoate (tuberculostearate, 10-methyl-C_{18}) with small amounts of C_{14}, 10-methyl-C_{16} and C_{18}. The higher PIMs are known to carry additional acyl chains on mannose residues.[130-132] The PI cores of LAM and LM from *M. tuberculosis* and *M. leprae* have also been shown to possess additional acylation of their PIM_2 core.[132]

Although LAM and LM are most frequently found to carry a PIM_2 anchor, it is clear that heterogeneity and species/strain variations in the positions and degree of acylation of these molecules also occur. Naturally occurring nonacylated forms have been reported, such as the mannan core typical of LAM or LM but lacking a PI anchor that was reported by Venisse et al.[133] Whether the absence of the anchor in such cases reflects experimental artefacts or partial degradation products of LAM and LM remains to be determined. More intriguing was the identification by the same group of a phosphoinositol-glyceroarabinomannan (PI-GAM) from *M. smegmatis*, which carries a nonacylated phosphoinositol-glycerol at the reducing end. More recently, Nigou et al demonstrated that there are four types of LAMs present in *M. bovis* BCG, which differ in the number and localization of fatty acids (from 1 to 4) esterifying the alkali labile phosphatidyl-myo-inositol anchor.[134] More work is needed to clarify the structural details and relative occurrence of the different acyl forms of LAM and LM.

Mannan Core

As mentioned earlier, Misaki et al and Azuma et al established the mannan core of LAM and LM as an α(1→6)-linked D-Manp backbone with short side chains of α(1→2)-linked D-Manp, followed by extensive α(1→5)-linked D-Araf residues.[124,135] More recent studies of LAM from both *M. tuberculosis* and *M. bovis* BCG estimated the mannan core to be around 20 mannose residues in total, with considerable heterogeneity with respect to the exact length and degree of branching.[133,136] The mannan core of LM appeared to be much longer than the corresponding LAM mannan core. In contrast, the LAM and LM from *M. smegmatis* have about 26 mannose residues with only about half of the α1→6 mannan backbone being further substituted at the 2-position.[137] Despite a number of elegant studies the exact position and the number of attachment sites for the arabinan chains remains to be determined along with the presence of additional phosphorylation on the mannan core.[133]

Arabinan Segments and Capping Motifs

The structural definition of LAM has resulted in two distinct arrangements occupying the terminal end: branched hexaarabinofuranosides (Ara_6) with the structure [β-D-Araf-(1→2)-α-D-Araf]$_2$-3,5-α-D-Araf-(1→5)-α-D-Araf and linear tetraarabinofuranosides (Ara_4) of the structure β-D-Araf-(1→2)-α-D-Araf-(1→5)-α-D-Araf-(1→5)-α-D-Araf.[138] However, in the case of LAM isolated from *M. tuberculosis*, these two types of arabinose termini are extensively capped with α-D-Manp residues, a product now called ManLAM.[139] The mannose caps consist of monomers, dimers and trimers of α-D-Manp linked to the C-5 of the terminal β-D-Araf. LAMs from *M. tuberculosis* strains (Erdman, H37Rv and H37Ra), *M. bovis* BCG and *M. leprae* were found to terminate with mannose caps. Intriguingly, a novel inositol phosphate-capping motif was found on the arabinan termini of AraLAM, which was previously thought to be uncapped.[132] Four such phosphoinositide motifs were found for each molecule of AraLAM, of which three were found to be mild alkali labile supporting the earlier observations of Weber and Gray.[127] Studies by Nigou and colleagues have addressed the location of succinyl and lactyl substituents.[134] As discussed above LAM from any single source is heterogeneous in size, branching, acylation, phosphorylation and capping. This most likely accounts for the broad diffuse banding of LAM that is observed when it is analyzed by polyacrylamide gel electrophoresis in the presence of sodium dodecyl sulfate (SDS-PAGE).[139]

The CD1b-dependent recognition of LAM involves uptake of LAM by the mannose receptor on monocyte-derived dendritic cells, which leads to the delivery of LAM to late endocytic compartments where presumably it is processed and becomes complexed to CD1b.[140] In some cases, T cells reactive to LAM also respond to the smaller acylated subunits of this antigen, including lipomannan (LM) and phosphatidyl-myo-inositol mannosides (PIMs) containing between two and six mannose residues.[46,51] These results suggest that uptake and enzymatic processing of LAM may occur in dendritic cells during the course of mycobacterial infection, thus leading to presentation of the lipidated core of LAM by CD1b and the generation of a specific T cell response. Whether such T cell responses have significant effects on the subsequent control of mycobacterial growth in vivo remains to be determined.

Several other studies have suggested that other glycolipids with the same basic glycosylphosphatidyl inositol structure that forms the lipid anchor of mycobacterial LAMs and PIMs may also bind to CD1d molecules and possibly serve as targets for T cell responses. For example, one study has shown that purified mouse CD1d1 contains associated lipids but no detectable peptides, and lipids were provisionally identified as glycosylphosphatidyl inositols (GPIs) on the basis of mass spectrometry data.[141] This result suggests that GPIs may be the major endogenous ligands of CD1d1, although it is not yet known if such endogenous ligands are capable of triggering T cell responses. Another isolated report has provided evidence that GPI-anchored antigens derived from Trypanosoma and Plasmodia were recognized by mouse NKT cells in a CD1d-restricted fashion. This provides further evidence that the GPI structure may be a common anchoring motif for glycolipids that associate with either group 1 or group 2 CD1 molecules.[142]

Cellular and Molecular Mechanisms of Antigen Presentation by CD1

There is evidence that CD1-restricted T cell recognition of nonpeptide antigens (i.e., lipids and glycolipids) is dependent in many cases on uptake and transport of the antigens to an intracellular compartment in APCs.[45,46,143] Additionally, studies on mycobacterial antigen presentation by human CD1b and presentation of αGalCer by murine CD1d1 have shown that the presentation of lipid antigens is prevented by agents that inhibit endosomal acidification (e.g., concanamycin A, chloroquine).[45,46,70] This indicates that, similar to peptide presentation by MHC class II, processing or loading in acidic compartments is in many cases a crucial step between lipid antigen uptake and its presentation on the cell surface. In contrast, the peptide transporter system TAP-1/2 is not required, either for the overall expression of CD1 proteins or their antigen-presenting functions.[45,48,144,145] Furthermore, the HLA-DM complexes that are necessary for normal antigen-presentation by MHC class II molecules seem not to be required by CD1 proteins. Taken together, these studies suggest substantial differences in the intracellular pathways leading to antigen association with CD1 on the one hand, and with MHC class I or class II on the other.

It is likely that both exogenous (taken up by endocytosis or phagocytosis) and endogenous (produced by intracellular pathogens within the antigen-presenting cell or APC) lipid antigens can enter the CD1 processing pathway. The mechanism for the uptake of exogenous lipid antigens has been studied in detail for CD1b-mediated presentation of mycobacterial LAM.[140] LAM is bound by the macrophage mannose receptor (MMR) and taken up by endocytosis, a process that can be blocked by soluble mannan and by antibodies directed against the MMR. CD1-transfectant cell lines that lack the MMR are not able to present LAM, but still can present smaller lipid or glycolipid antigens. LAM may serve as a prototype for CD1-presented glycolipid antigens that have very large hydrophilic caps, whereas other smaller and more hydrophobic glycolipid or lipid antigens, such as mycolic acids or GMM, may not require a specific receptor-mediated uptake for efficient presentation by CD1.

Several studies have also shown that CD1 molecules are delivered to endocytic compartments, and in some cases this has been found to be required for efficient antigen presentation.[71,146,147] In a variety of cell types including monocyte-derived dendritic cells, CD1b was also found predominantly in acidic endosomal compartments, including those where MHC class II molecules are usually loaded with peptides.[140,148] These MHC class II containing compartments (MIIC) are lipid-rich late endosomes or lysosomes having a multilamellar or multivesicular structure. CD1b is directed to MIICs and to other endocytic compartments by a targeting motif (YXXZ, where Y = tyrosine, X = any amino acid and Z = an amino acid with a bulky hydrophobic side chain) in the short cytoplasmic tail of the molecule.[148] This sequence interacts with one or several members of the family of clathrin adaptor protein complexes, thus directing CD1b into clathrin-coated pits and vesicles.[149] A YXXZ-targeting motif is also present in most other human and nonhuman CD1 molecules, but not in human CD1a, which does not localize to MIICs. Instead, CD1a seems to traffic independently within the recycling pathway of the early endocytotic system and does not require acidic endosomal compartments for presentation of mycobacterial lipid antigen.[143] Recent investigations on CD1c revealed that it is localized only to early and late endosomes, whereas CD1b is also directed to lysosomal compartments and MIIC.[150,151] The distinct localization of each of the various CD1 isoforms may indicate that these are specialized for surveying different parts of the cell for antigenic lipids. It is interesting to speculate that this may be one of the factors driving the diversification of CD1 proteins that has led to the multiple isoforms observed in humans and most other mammalian species.

In addition to the cytoplasmic tail targeting motif of most CD1 molecules, a second mechanism has also been recently suggested to contribute to the intracellular targeting and endosomal localization of CD1 proteins. Thus, it has been found that a fraction of mouse CD1d molecules becomes physically associated shortly after synthesis in the ER with the MHC class II invariant chain (Ii).[152] The Ii is well known to be a critical determinant of the targeting of newly synthesized MHC class II complexes to the MIIC, and it is therefore presumed that Ii may perform a similar targeting role for the cohort of CD1 molecules that it binds. Exactly how this alternative targeting strategy modifies the ultimate localization and function of CD1 molecules remains to be explored in detail. However, the data reported thus far already suggest a significant effect of Ii targeting on the recognition of mouse CD1d by certain NKT cells. The identification of this alternate mechanism for endosomal targeting of CD1 molecules indicates a further interesting parallel between the CD1 and MHC class II systems. It now appears that both types of antigen presenting molecules can use either cytoplasmic tail dependent or Ii dependent mechanisms to control their entry into the endocytic pathway (Fig. 6).[153]

The multilamellar or multivesicular MIICs may offer the optimal environment for loading of lipid antigens onto certain CD1 molecules, especially CD1b and CD1d, which usually seem to require a strongly acidified environment for this process to proceed efficiently. LAM colocalizes with CD1b in MIICs, and low pH enhances conformational changes of CD1b to facilitate the binding of this and other hydrophobic ligands, as shown by in vitro studies.[140,154] In addition the many degradative enzymes present in MIICs may process the glycan portion of LAM to make it suitable for presentation. However, the biochemical processing pathways that may modify lipid or glycolipid antigens within the MIICs of APCs are still unknown. One recent report using a modified synthetic form of α-GalCer has elegantly demonstrated the principle of enzymatic processing for glycolipid antigens prior to presentation by CD1 molecules.[155] However, in some of the relatively few examples of natural CD1-presented antigens that have been studied thus far, APCs appear not to be capable of enzymatic processing of common precursor forms of the antigen. For example, the mycobacterial glycolipid trehalose-6,6'-dimycolate, a cell wall component that contains the structures of both GMM and mycolic acids, is not

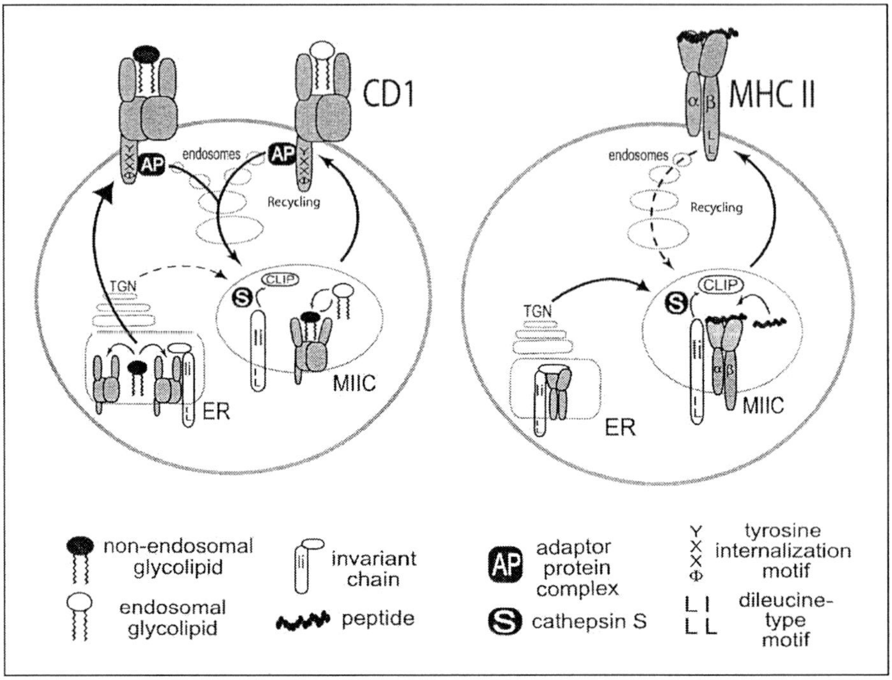

Figure 6. Endosomal pathways for presentation of antigens by CD1 and MHC class II. Key features of the trafficking of CD1 (left) and MHC class II (right) molecules to the MIIC are compared. In the case of MHC class II, associated invariant chain (Ii) blocks the antigen binding groove and directs MHC II from the trans-golgi network (TGN) to MHC class II compartments (MIIC). In the MIIC, the protease cathepsin S (S) cleaves Ii, releasing the MHC class II associated invariant chain peptide (CLIP) and exposing the groove for peptide exchange prior to release of MHC II to the cell surface (solid arrows). A second, less dominant pathway (dashed arrow) into the endosomes for MHC class II involves recycling from the cell surface as a result of the dileucine motif in the cytoplasmic tail of the β-chain. In contrast, most CD1 proteins move rapidly to the cell surface and are subsequently delivered to MIICs by the internalization and recycling pathway that is controlled by the cytoplasmic tail tyrosine-based motif and its interaction with clathrin adaptor protein complexes (AP) (solid arrows). Recent findings also indicate a second pathway to the MIIC for CD1 which is dependent on Ii and cathepsin S (dashed arrow). Reproduced with permission from Moody & Porcelli (2001).[153]

modified or processed into smaller components that can be recognized by T cells specific for either GMM or mycolic acids.[86] Thus, even activated macrophages do not appear to contain the enzymes required for efficient processing of this molecule by cleavage at the mycolate esters or the glycosidic linkage of trehalose.

There is evidence that antigen loading onto CD1b molecules can also occur via an endogenous pathway (i.e., antigens derived from intracellular pathogens). This has been shown by infecting CD1-expressing APCs in vitro with virulent *M. tuberculosis* and demonstrating that CD1b-restricted T cell lines recognize and lyse such infected target cells.[146,156] As in the exogenous pathway, the trafficking signal YXXZ is required for efficient antigen presentation, suggesting an endosomal site for loading of antigens produced within APCs. It is currently not clear which endocytic compartment is required. Since phagosomes containing live *M. tuberculosis* do not acidify normally, it may be more likely that CD1b is loaded with secreted or shed mycobacterial lipid antigens in a mycobacteria-free endosomal compartment to which the antigens have been translocated.[157-159]

Molecular Basis of Lipid Antigen Interactions with CD1 and TCRs

Based largely on X-ray crystallography data obtained on mouse CD1d1, the molecular mechanism by which amphipathic glycolipids interact with CD1 molecules is believed to require the binding of the lipid moiety to the hydrophobic antigen binding groove in the A' and F' pockets of CD1.[27] This interaction is believed to provide the binding energy that stabilizes the complex of ligand and CD1 in an aqueous environment. This binding is relatively nonspecific since it seems not to be dependent on the fine structure of the lipid tails (e.g., CD1b presents a variety of ligands that differ in the precise structure of the lipid tails). This model predicts that the orientation of the ligand is such that its polar head group is outside the binding groove in a relatively hydrophilic environment and available for direct contact with the TCR, thus enabling the specific recognition of lipid and glycolipid antigens by different T cells. Support for this comes from studies on the structural requirements of the CD1-restricted T cell recognition of GMM and of α-GalCer analogs.[70,85] This model accounts for the requirement of a hydrophobic portion detected in all antigens presented by CD1 molecules. Moreover, it explains the high specificity of CD1-restricted T cells for the fine structure of the polar head groups of lipid and glycolipid antigens, as well as the observed general lack of specificity for the hydrophobic alkyl tails of the antigens.

Potential Role of CD1 in Microbial Immunity

Substantial evidence that human CD1-restricted T cells are involved in the immune response to infectious diseases exists to date only in mycobacterial infections. However, it seems likely that other pathogenic microbes are also recognized by this system as well. In fact, initial reports have appeared which demonstrate T cell recognition of CD1-presented antigens from a Gram-negative bacterium, and natural and synthetic antigens typical for protozoan parasites.[69,89,142] In order to further evaluate this question, it is necessary to establish suitable animal models that will enable the study of presentation of lipid antigens from other pathogens in vivo. Since studies performed in the mouse model are difficult to interpret because of the absence of group 1 CD1 molecules in this species, the use of other animal species such as the guinea pig or rabbit may be required.[12,16] Nevertheless, there are already data that at least strongly suggest a role for CD1-restricted T cells in two mycobacterial infections in humans, namely leprosy and tuberculosis.

Leprosy is a disease with clinical manifestations that correlate with different types of cell mediated immune responses by the host, and is therefore valuable as a model for studying the role of different T cell subsets in the course of the disease. Clinical studies of human leprosy define a spectrum of disease, the extremes of which are defined by two markedly different forms of disease expression. One is tuberculoid leprosy, in which the growth of *M. leprae* can be controlled by a strong cell-mediated immune response to the pathogen resulting in a relatively benign clinical state. The other is lepromatous leprosy in which the infected host does not have an efficient T_H1-type cell-mediated immune response and suffers from uncontrolled growth of the bacilli and progressive disease. Between these two extremes, there is a spectrum of immunological reactions to *M. leprae* which present a mixture of these features. In a study examining the role of the CD1 system in leprosy, skin biopsy specimens of infiltrating granulomas from leprosy patients were found to contain mature dendritic cells expressing group I CD1 molecules.[160] The frequency of $CD1^+$ cells correlated with the level of cell-mediated immunity to *M. leprae*, being ten times higher in leprosy granulomas of tuberculoid leprosy patients than in those of patients with the immunologically unresponsive lepromatous leprosy. The frequency of $CD1^+$ cells may be influenced by the higher levels of GM-CSF in tuberculoid lesions, a cytokine that is responsible for the differentiation of dendritic cells and induction of group I CD1 on monocytes.[161,162] In granulomas from patients with lepromatous leprosy, GM-CSF secretion is inhibited by high levels of IL-10, which may be a key inhibitor of the CD1 system.[163] Administration of GM-CSF to leprosy patients results in infiltration of $CD1a^+$ cells into the skin lesions.[164] Another mechanism that may influence CD1 expression at the site of infection may be the down-regulation of CD1 on mycobacteria-infected APCs.[165]

Studies of CD1-restricted T cell responses to *M. tuberculosis* also support a role for CD1 in host defense against this pathogen. T cell lines specific for *M. tuberculosis* antigens and restricted by CD1 molecules have been derived from healthy individuals, patients with tuberculosis and from patients coinfected with *M. tuberculosis* and HIV.[45,47,48,156] There is also recent evidence that mycobacterial isoprenoid glycolipid antigens can be recognized by T cells from individuals who were recently infected by *M. tuberculosis*, but not by T cells from healthy uninfected individuals.[89] This suggests that these lipid and glycolipid antigens serve as recall antigens in the immune response against *M. tuberculosis*.

In vitro, APCs infected with virulent *M. tuberculosis* down-regulate their CD1 expression, but not the expression of MHC class I or II molecules.[165] Thus, *M. tuberculosis* may have evolved mechanisms to specifically inhibit CD1 expression. This may be an important immune evasion mechanism used by this pathogen and stresses the potential importance of the CD1 system as a component of host resistance against *M. tuberculosis*. In studies carried out in the mouse model, administration of CD1d1-blocking antibodies seemed to worsen the clinical outcome of mice infected with *M. tuberculosis*, although studies using CD1d-deficient mice have so far not shown a difference compared with control mice in the susceptibility to infection with *M. tuberculosis* [166-168]

CD1-restricted T cells may contribute directly to protective immunity to mycobacterial infection by at least two different mechanisms. First, human T cells recognizing CD1-presented mycobacterial antigens release high levels of IFN-γ but little or no IL-4, creating a T_H1-type response that is required by macrophages in order to kill intracellular bacteria and for the development of a robust cell-mediated immune defense against *M. tuberculosis*. Second, CD1-restricted T cells specific for mycobacterial antigens show a high level of cytolytic activity in vitro against APCs that were antigen-pulsed or previously infected with live mycobacteria. Lysis of chronically or productively infected macrophages contributes to host defense by direct killing of the bacteria or by making them available for uptake and destruction by freshly recruited activated macrophages. The destruction or the apoptotic death of mycobacteria-containing macrophages limits the reservoir of host cells for the pathogen. Apoptotic macrophages harboring mycobacteria can be phagocytosed by dendritic cells and processed in order to present additional antigens and generate new cytotoxic T cells.

Taken together, these findings strengthen the hypothesis that the recognition of lipid and glycolipid antigens may be an important part of the immune response of the host against mycobacterial infection. The discovery of the CD1 system as a novel mechanism for host response to infection provides new insight into the types of antigens that can activate the immune system. Importantly, since CD1-presented lipid and glycolipid antigens elicit specific responses of the adaptive immune system and are likely to serve as recall antigens, this pathway for antigen presentation may provide valuable new opportunities for vaccine development. It thus appears likely that future studies will reveal ways in which this potentially versatile system for immune recognition can be used to augment immunity against pathogens responsible for human diseases.

Acknowledgments

SAP is supported by grants from the NIH/NIAID, and by a grant from the Irene Diamond Foundation. GSB, who is currently a Lister Institute Jenner Research Fellow, acknowledges support from the Wellcome Trust, The Medical Research Council and the National Institutes of Health. The authors thank Dr. D. Branch Moody for many helpful discussions and for assistance with the figures. SAP is indebted to Ms. Giudita Pasta for her invaluable help in preparation of the manuscript.

References

1. Yu CY, Milstein C. A physical map linking the five CD1 human thymocyte differentiation antigen genes. EMBO J 1989; 8:3727-3732.
2. Shiina T, Ando A, Suto Y et al. Genomic anatomy of a premier major histocompatibility complex paralogous region on chromosome 1q21-q22. Genome Res 2001; 11:789-802.
3. Martin LH, Calabi F, Milstein C. Isolation of CD1 genes: A family of major histocompatibility complex- related differentiation antigens. Proc Natl Acad Sci USA 1986; 83:9154-9158.
4. Balk SP, Bleicher PA, Terhorst C. Isolation and characterization of a cDNA and gene coding for a fourth CD1 molecule. Proc Natl Acad Sci USA 1989; 86:252-256.
5. Porcelli SA. The CD1 family: A third lineage of antigen-presenting molecules. Adv Immunol 1995; 59:1-98.
6. Angenieux C, Salamero J, Fricker D et al. Characterization of CD1e, a third type of CD1 molecule expressed in dendritic cells. J Biol Chem 2000; 275:37757-37764.
7. van Agthoven A, Terhorst C. Further biochemical characterization of the human thymocyte differentiation antigen T6. J Immunol 1982; 128:426-432.
8. Amiot M, Dastot H, Fabbi M et al. Intermolecular complexes between three human CD1 molecules on normal thymus cells. Immunogenetics 1988; 27:187-195.
9. Oteo M, Parra JF, Mirones I et al. Single strand conformational polymorphism analysis of human CD1 genes in different ethnic groups. Tissue Antigens 1999; 53:545-550.
10. Mirones I, Oteo M, Parra-Cuadrado JF et al. Identification of two novel human CD1E alleles. Tissue Antigens 2000; 56:159-161.
11. Calabi F, Jarvis JM, Martin L et al. Two classes of CD1 genes. Eur J Immunol 1989; 19:285-292.
12. Dascher CC, Hiromatsu K, Naylor JW et al. Conservation of a CD1 multigene family in the guinea pig. J Immunol 1999; 163:5478-5488.
13. Matsuura A, Hashimoto Y, Kinebuchi M et al. Rat CD1 antigen: Structure, expression and function. Transplant Proc 1997; 29:1705-1706.
14. Hopkins J, Dutia BM, Rhind SM. Sheep CD1 genes and proteins. Vet Immunol Immunopathol 2000; 73:3-14.
15. Chun T, Wang K, Zuckermann FA et al. Molecular cloning and characterization of a novel CD1 gene from the pig. J Immunol 1999; 162:6562-6571.
16. Hayes SM, Knight KL. Group 1 CD1 genes in rabbit. J Immunol 2001; 166:403-410.
17. Calabi F, Belt KT, Yu CY et al. The rabbit CD1 and the evolutionary conservation of the CD1 gene family. Immunogenetics 1989; 30:370-377.
18. Bradbury A, Calabi F, Milstein C. Expression of CD1 in the mouse thymus. Eur J Immunol 1990; 20:1831-1836.
19. Balk SP, Bleicher PA, Terhorst C. Isolation and expression of cDNA encoding the murine homologues of CD1. J Immunol 1991; 146:768-774.
20. Bendelac A, Lantz O, Quimby ME et al. CD1 recognition by mouse NK1+ T lymphocytes. Science 1995; 268:863-865.
21. Mendiratta SK, Martin WD, Hong S et al. CD1d1 mutant mice are deficient in natural T cells that promptly produce IL-4. Immunity 1997; 6:469-477.
22. Chen YH, Wang B, Chun T et al. Expression of CD1d2 on thymocytes is not sufficient for the development of NK T cells in CD1d1-deficient mice. J Immunol 1999; 162:4560-4566.
23. Moseley WS, Watson ML, Kingsmore SF et al. CD1 defines conserved linkage group border between human chromosome 1 and mouse chromosomes 1 and 3. Immunogenetics 1989; 30:378-382.
24. Balk SP. MHC evolution. Nature 1995; 374:505-506.
25. Lerch PG, van de Rijn M, Smart JE et al. Isolation and purification of the human thymocyte antigens T6 and M241. Mol Immunol 1986; 23:131-139.
26. van de Rijn M, Lerch PG, Knowles RW et al. The thymic differentiation markers T6 and M241 are two unusual MHC class I antigens. J Immunol 1983; 131:851-855.
27. Zeng Z, Castano AR, Segelke BW et al. Crystal structure of mouse CD1: An MHC-like fold with a large hydrophobic binding groove. Science 1997; 277:339-345.
28. McMichael A, Pilch JR, Galfre G et al. A human thymocyte antigen defined by a hybrid myeloma monoclonal antibody. Eur J Immunol 1979; 9:205-210.
29. Porcelli SA, Modlin RL. The CD1 system: Antigen-presenting molecules for T cell recognition of lipids and glycolipids. Ann Rev Immunol 1999; 17:297-329.
30. Kasinrerk W, Baumruker T, Majdic O et al. CD1 molecule expression on human monocytes induced by granulocyte- macrophage colony-stimulating factor. J Immunol 1993; 150:579-584.
31. Small TN, Knowles RW, Keever C et al. M241 (CD1) expression on B lymphocytes. J Immunol 1987; 138:2864-2868.

32. Plebani A, Proserpio AR, Guarneri D et al. B and T lymphocyte subsets in fetal and cord blood: Age-related modulation of CD1c expression. Biol Neonate 1993; 63:1-7.
33. Blumberg RS, Colgan SP, Balk SP. CD1d: Outside-in antigen presentation in the intestinal epithelium? Clin Exp Immunol 1997; 109:223-225.
34. Mandal M, Chen XR, Alegre ML et al. Tissue distribution, regulation and intracellular localization of murine CD1 molecules. Mol Immunol 1998; 35:525-536.
35. Brossay L, Jullien D, Cardell S et al. Mouse CD1 is mainly expressed on hemopoietic-derived cells. J Immunol 1997; 159:1216-1224.
36. Spada FM, Borriello F, Sugita M et al. Low expression level but potent antigen presenting function of CD1d on monocyte lineage cells. Eur J Immunol 2000; 30:3468-3477.
37. Shimamura M, Ohteki T, Launois P et al. Thymus-independent generation of NK1+ T cells in vitro from fetal liver precursors. J Immunol 1997; 158:3682-3689.
38. Park SH, Roark JH, Bendelac A. Tissue-specific recognition of mouse CD1 molecules. J Immunol 1998; 160:3128-3134
39. Roark JH, Park SH, Jayawardena J et al. CD1.1 expression by mouse antigen-presenting cells and marginal zone B cells. J Immunol 1998; 160:3121-3127.
40. Porcelli S, Brenner MB, Greenstein JL et al. Recognition of cluster of differentiation 1 antigens by human CD4-CD8-cytolytic T lymphocytes. Nature 1989; 341:447-450.
41. Faure F, Jitsukawa S, Miossec C et al. CD1c as a target recognition structure for human T lymphocytes: Analysis with peripheral blood gamma/delta cells. Eur J Immunol 1990; 20:703-706.
42. Balk SP, Ebert EC, Blumenthal RL et al. Oligoclonal expansion and CD1 recognition by human intestinal intraepithelial lymphocytes. Science 1991; 253:1411-1415.
43. Spada FM, Grant EP, Peters PJ et al. Self-recognition of CD1 by gamma/delta T cells: Implications for innate immunity. J Exp Med 2000; 191:937-948.
44. Cardell S, Tangri S, Chan S et al. CD1-restricted CD4+ T cells in major histocompatibility complex class II-deficient mice. J Exp Med 1995; 182:993-1004.
45. Porcelli S, Morita CT, Brenner MB. CD1b restricts the response of human CD4-8- T lymphocytes to a microbial antigen. Nature 1992; 360:593-597.
46. Sieling PA, Chatterjee D, Porcelli SA et al. CD1-restricted T cell recognition of microbial lipoglycan antigens. Science 1995; 269:227-230.
47. Gong J, Stenger S, Zack JA et al. Isolation of mycobacterium-reactive CD1-restricted T cells from patients with human immunodeficiency virus infection. J Clin Invest 1998; 101:383-389.
48. Beckman EM, Melian A, Behar SM et al. CD1c restricts responses of mycobacteria-specific T cells. Evidence for antigen presentation by a second member of the human CD1 family. J Immunol 1996; 157:2795-2803.
49. Rosat JP, Grant EP, Beckman EM et al. CD1-Restricted microbial lipid antigen-specific recognition found in the CD8+ $\alpha\beta$ T cell pool. J Immunol 1999; 162:366-371.
50. Grant EP, Degano M, Rosat JP et al. Molecular recognition of lipid antigens by T cell receptors. J Exp Med 1999; 189:195-205.
51. Sieling PA, Ochoa MT, Jullien D et al. Evidence for human CD4+ T cells in the CD1-restricted repertoire: Derivation of mycobacteria-reactive T cells from leprosy lesions. J Immunol 2000; 164:4790-4796.
52. Teitell M, Holcombe HR, Brossay L et al. Nonclassical behavior of the mouse CD1 class I-like molecule. J Immunol 1997; 158:2143-2149.
53. Bendelac A, Rivera MN, Park SH et al. Mouse CD1-specific NK1 T cells: Development, specificity, and function. Annu Rev Immunol 1997; 15:535-562.
54. Lantz O, Bendelac A. An invariant T cell receptor α chain is used by a unique subset of major histocompatibility complex class I-specific CD4+ and CD4-8- T cells in mice and humans. J Exp Med 1994; 180:1097-1106.
55. Godfrey DI, Hammond KJ, Poulton LD et al. NKT cells: Facts, functions and fallacies. Immunol Today 2000; 21:573-583.
56. Cui J, Shin T, Kawano T et al. Requirement for Vα14 NKT cells in IL-12-mediated rejection of tumors. Science 1997; 278:1623-1626.
57. Smyth MJ, Crowe NY, Godfrey DI. NK cells and NKT cells collaborate in host protection from methylcholanthrene-induced fibrosarcoma. Int Immunol 2001; 13:459-463.
58. Smyth MJ, Godfrey DI. NKT cells and tumor immunity: A double-edged sword. Nat Immunol 2000; 1:459-460.
59. Behar SM, Podrebarac TA, Roy CJ et al. Diverse TCRs recognize murine CD1. J Immunol 1999; 162:161-167.
60. Exley M, Garcia J, Balk SP et al. Requirements for CD1d Recognition by Human Invariant Vα24+ CD4-CD8- T Cells. J Exp Med 1997; 186:109-120.

61. Davodeau F, Peyrat M-A, Necker A et al. Close phenotypic and functional similarities between human and murine αβ T cells expressing invariant TCR α-chains. J Immunol 1997; 158:5603-5611.
62. Prussin C, Foster B. TCR Vα24 and Vβ11 coexpression defines a human NK1 T cell analog containing a unique Th0 subpopulation. J Immunol 1997; 159:5862-5870.
63. Wilson SB, Kent SC, Patton KT et al. Extreme Th1 bias of invariant Valpha24JalphaQ T cells in type 1 diabetes. Nature 1998; 391:177-181.
64. Oishi Y, Sumida T, Sakamoto A et al. Selective reduction and recovery of invariant Valpha24JalphaQ T cell receptor T cells in correlation with disease activity in patients with systemic lupus erythematosus. J Rheumatol 2001; 28:275-283.
65. Kojo S, Adachi Y, Keino H et al. Dysfunction of T cell receptor AV24AJ18+, BV11+ double-negative regulatory natural killer T cells in autoimmune diseases. Arthritis Rheum 2001; 44:1127-1138.
66. Beckman EM, Porcelli SA, Morita CT et al. Recognition of a lipid antigen by CD1-restricted αβ+ T cells. Nature 1994; 372:691-694.
67. Castano AR, Tangri S, Miller JE et al. Peptide binding and presentation by mouse CD1. Science 1995; 269:223-226.
68. Lee DJ, Abeyratne A, Carson DA et al. Induction of an antigen-specific, CD1-restricted cytotoxic T lymphocyte response In vivo. J Exp Med 1998; 187:433-438.
69. Fairhurst RM, Wang CX, Sieling PA et al. CD1 presents antigens from a gram-negative bacterium, Haemophilus influenzae type B. Infect Immun 1998; 66:3523-3526.
70. Kawano T, Cui J, Koezuka Y et al. CD1d-restricted and TCR-mediated activation of valpha14 NKT cells by glycosylceramides. Science 1997; 278:1626-1629.
71. Spada F, Koezuka Y, Porcelli SA. CD1d-restricted recognition of synthetic glycolipid antigens by human natural killer T cells. J Exp Med 1998; 188:1-6.
72. Burdin N, Brossay L, Degano M et al. Structural requirements for antigen presentation by mouse CD1. Proc Natl Acad Sci USA 2000; 97:10156-10161.
73. Naidenko OV, Maher JK, Ernst WA et al. Binding and antigen presentation of ceramide-containing glycolipids by soluble mouse and human CD1d molecules. J Exp Med 1999; 190:1069-1080.
74. Shamshiev A, Donda A, Carena I et al. Self glycolipids as T-cell autoantigens. Eur J Immunol 1999; 29:1667-1675.
75. Shamshiev A, Donda A, Prigozy TI et al. The alpha/beta T cell response to self-glycolipids shows a novel mechanism of CD1b loading and a requirement for complex oligosaccharides. Immunity 2000; 13:255-264.
76. Altamirano MM, Woolfson A, Donda A et al. Ligand-independent assembly of recombinant human CD1 by using oxidative refolding chromatography. Proc Natl Acad Sci USA 2001; 98:3288-3293.
77. Minnikin D.E. Lipids: Complex lipids, their chemistry, biosynthesis and roles. In: Ratledge C, Stanford J, eds. The Biology of Mycobacteria. London: Academic Press, 1982:95-184.
78. Brennan PJ, Nikaido H. The envelope of mycobacteria. Annu Rev Biochem 1995; 64:29-63.
79. Krembel J, JEtémadi A.H. Sur la structure d'un nouveau type d'acides mycoliques de Mycobacterium smegmatis. Tetrahedron 1966; 22:1113-1119.
80. Kusaka T, Mori T. Pyrolysis gas chromatography-mass spectrometry of mycobacterial mycolic acid methyl esters and its application to the identification of Mycobacterium leprae. J Gen Microbiol 1986; 132:3404-3406.
81. Kaneda K, Imaizumi S, Yano I. Distribution of C_{22}-, C_{24}- and C_{26}-alpha-unit-containing mycolic acid homologues in mycobacteria. Microbiol Immunol 1995; 39:563-570.
82. Kaneda K, Naito S, Imaizumi S. et al. Determination of molecular species composition of C80 or longer-chain alpha-mycolic acids in Mycobacterium spp. by gas chromatography-mass spectrometry and mass chromatography. J Clin Microbiol 1986; 24:1060 1070.
83. Yuan Y, Crane DC, Musser JM et al. MMAS-1, the branch point between cis- and trans-cyclopropane-containing oxygenated mycolates in Mycobacterium tuberculosis. J Biol Chem 1997; 272:10041-10049.
84. Danielson SJ, Gray GR. Structures of the two homologous series of dialkene mycolic acids from Mycobacterium smegmatis. J Biol Chem 1982; 257:12196-12203.
85. Moody DB, Reinhold BB, Guy MR et al. Structural requirements for glycolipid antigen recognition by CD1b-restricted T cells. Science 1997; 278:283-286.
86. Moody DB, Reinhold BB, Reinhold VN et al. Uptake and processing of glycosylated mycolates for presentation to CD1b-restricted T cells. Immunol Lett 1999; 65:85-91.
87. Moody DB, Guy MR, Grant E et al. CD1b-mediated T cell recognition of a glycolipid antigen generated from mycobacterial lipid and host carbohydrate during infection. J Exp Med 2000; 192:965-976.

88. Moody DB, Besra GS, Wilson IA et al. The molecular basis of CD1-mediated presentation of lipid antigens. Immunol Rev 1999; 172:285-296.
89. Moody DB, Ulrichs T, Muhlecker W et al. CD1c-mediated T-cell recognition of isoprenoid glycolipids in Mycobacterium tuberculosis infection. Nature 2000; 404:884-888.
90. Besra GS, Sievert T, Lee RE et al. Identification of the apparent carrier in mycolic acid synthesis. Proc Natl Acad Sci USA 1994; 91:12735-12739.
91. Besra GS, Chatterjee D. Lipids and Carbohydrates of Mycobacterium tuberculosis. In: Bloom BR, ed. Tuberculosis: Pathogenesis, Protection and Control. Washington, DC: ASM Press, 1994:285-306.
92. Takayama K, Schnoes HK, Semmler EJ. Characterization of the alkali-stable mannophospholipids of Mycobacterium smegmatis. Biochim Biophys Acta 1973; 316:212-221.
93. Bugg TD, Brandish PE. From peptidoglycan to glycoproteins: Common features of lipid-linked oligosaccharide biosynthesis. FEMS Microbiol Lett 1994; 119:255-262.
94. Colussi PA, Taron CH, Mack JC et al. Human and Saccharomyces cerevisiae dolichol phosphate mannose synthases represent two classes of the enzyme, but both function in Schizosaccharomyces pombe. Proc Natl Acad Sci USA 1997; 94:7873-7878.
95. Strahl-Bolsinger S, Gentzsch M, Tanner W. Protein O-mannosylation. Biochim Biophys Acta 1999; 1426:297-307.
96. Maeda Y, Tanaka S, Hino J et al. Human dolichol-phosphate-mannose synthase consists of three subunits, DPM1, DPM2 and DPM3. EMBO J 2000; 19:2475-2482.
97. Besra GS, Morehouse CB, Rittner CM et al. Biosynthesis of mycobacterial lipoarabinomannan. J Biol Chem 1997; 272:18460-18466.
98. Mikusova K, Mikus M, Besra GS et al. Biosynthesis of the linkage region of the mycobacterial cell wall. J Biol Chem 1996; 271:7820-7828.
99. Rip JW, Rupar CA, Ravi K et al. Distribution, metabolism and function of dolichol and polyprenols. Progress in Lipid Research 1985; 24:269-309.
100. Rilling HC. The mechanism of the condensation reactions of cholesterol biosynthesis. Fourth Morton lecture. Biochem Soc Trans 1985; 13:997-1003.
101. Quellhorst Jr GJ, Hall CW, Robbins AR et al. Synthesis of dolichol in a polyprenol reductase mutant is restored by elevation of cis-prenyl transferase activity. Arch Biochem Biophys 1997; 343:19-26.
102. Pennock JF, Hemming FW, Morton RA. Dolichol: A naturally occurring isoprenoid alcohol. Nature 1960; 186:470-472.
103. Quellhorst Jr GJ, Piotrowski JS, Steffen SE et al. Identification of Schizosaccharomyces pombe prenol as dolichol-16,17. Biochem Biophys Res Commun 1998; 244:546-550.
104. Couto AS, Kimura EA, Peres VJ et al. Active isoprenoid pathway in the intra-erythrocytic stages of Plasmodium falciparum: Presence of dolichols of 11 and 12 isoprene units. Biochem J 1999; 341(Pt 3):629-637.
105. Zhou D, White RH. Early steps of isoprenoid biosynthesis in Escherichia coli. Biochem J 1991; 273(Pt 3):627-634.
106. Horbach S, Sahm H, Welle R. Isoprenoid biosynthesis in bacteria: Two different pathways? FEMS Microbiol Lett 1993; 111:135-140.
107. Rohmer M. The discovery of a mevalonate-independent pathway for isoprenoid biosynthesis in bacteria, algae and higher plants. Nat Prod Rep 1999; 16:565-574.
108. Rohmer M, Knani M, Simonin P et al. Isoprenoid biosynthesis in bacteria: A novel pathway for the early steps leading to isopentenyl diphosphate. Biochem J 1993; 295(Pt 2):517-524.
109. Lois LM, Campos N, Putra SR et al. Cloning and characterization of a gene from Escherichia coli encoding a transketolase-like enzyme that catalyzes the synthesis of D-1- deoxyxylulose 5-phosphate, a common precursor for isoprenoid, thiamin, and pyridoxol biosynthesis. Proc Natl Acad Sci USA 1998; 95:2105-2110.
110. Putra SR, Disch A, Bravo JM et al. Distribution of mevalonate and glyceraldehyde 3-phosphate/pyruvate routes for isoprenoid biosynthesis in some gram-negative bacteria and mycobacteria. FEMS Microbiol Lett 1998; 164:169-175.
111. Crick DC, Mahapatra S, Brennan PJ. Biosynthesis of the arabinogalactan-peptidoglycan complex of Mycobacterium tuberculosis. Glycobiology 2001; 11:107R-118R.
112. De Rosa M, Gambacorta A, Gliozzi A. Structure, biosynthesis, and physicochemical properties of archaebacterial lipids. Microbiol Rev 1986; 50:70-80.
113. Wolucka BA, McNeil MR, de Hoffmann E et al. Recognition of the lipid intermediate for arabinogalactan/arabinomannan biosynthesis and its relation to the mode of action of ethambutol on mycobacteria. J Biol Chem 1994; 269:23328-23335.
114. Wolucka BA, de Hoffmann E. Isolation and characterization of the major form of polyprenyl-phospho-mannose from Mycobacterium smegmatis. Glycobiology 1998; 8:955-962.

115. Takayama K, Goldman DS. Pathway for the synthesis of mannophospholipids in Mycobacterium tuberculosis. Biochim Biophys Acta 1969; 176:196-198.
116. Scherman MS, Kalbe-Bournonville L, Bush D et al. Polyprenylphosphate-pentoses in mycobacteria are synthesized from 5- phosphoribose pyrophosphate. J Biol Chem 1996; 271:29652-29658.
117. Cole ST, Brosch R, Parkhill J et al. Deciphering the biology of Mycobacterium tuberculosis from the complete genome sequence. Nature 1998; 393:537-544.
118. Chatterjee D, Khoo KH. Mycobacterial lipoarabinomannan: An extraordinary lipoheteroglycan with profound physiological effects. Glycobiology 1998; 8:113-120.
119. Vercellone A, Nigou J, Puzo G. Relationships between the structure and the roles of lipoarabinomannans and related glycoconjugates in tuberculosis pathogenesis. Front Biosci 1998; 3:e149-e163.
120. Chargaff E, Schaefer W. A specific polysaccharide from the bacillus Calmette Guérin. J Biol Chem 1935; 112:393-405.
121. Menzel AEO, Heidelberger M. Specific and nonspecific cell polysaccharides of bovine strain tubercle bacillus. J Biol Chem 1939; 127:221-236.
122. Siebert FB, Watson DW. Isolation of the polysaccharides and nucleic acid of tuberculin by electrophoresis. J Biol Chem 1941; 140:55-69.
123. Misaki A, Seto N, Azuma I. Structure and immunological properties of D-arabino-D-galactans isolated from cell walls of Mycobacterium species. J Biochem (Tokyo) 1974; 76:15-27.
124. Azuma I, Kimura H, Niinaka T et al. Chemical and immunological studies on mycobacterial polysaccharides. 1. Purification and properties of polysaccharides from human tubercle bacilli. J Bacteriol 1968; 95:263-271.
125. Tsumita T, Matsumoto R, Mizuno D. Chemical and biological properties of the haemaglutination antigen, a lipopolysaccharide of Mycobacterium tuberculosis. Japanese J Med Sci 1960; 13:131-138.
126. Ohashi M. Studies on the chemical structure of serologically active arabinomannan from mycobacteria. Japanese J Exp Med 1970; 40:1-14.
127. Weber PL, Gray GR. Structural and immunochemical characterization of the acidic arabinomannan of Mycobacterium smegmatis. Carbohydrate Res 1979; 74:259-278.
128. Belanger AE, Inamine JM. Genetics of cell wall biosynthesis. In: Hatfull GF , Jacobs Jr WR, eds. Molecular Genetics of Mycobacteria. Washington, DC: America Society of Microbiology Press, 2000:191-202.
129. Lee YC, Ballou CE. Structural studies on the myoinositol mannosides from the glycolipids of Mycobacterium tuberculosis and Mycobacterium phlei. J Biol Chem 1964; 239:1316-1327.
130. Pangborn MC, McKinney JA. Purification of serologically active phosphoinositides of Mycobacterium tuberculosis. J Lipid Res 1966; 7:627-633.
131. Brennan PJ, Ballou CE. Biosynthesis of mannophosphoinositides by Mycobacterium phlei. J Biol Chem 1967; 242:3046-3056.
132. Khoo K-H, Dell A, Morris HR et al. Structural definition of acylated phosphatidylinositol mannosides from Mycobacterium tuberculosis: Definition of a common anchor for lipomannan and lipoarabinomannan. Glycobiology 1995; 5:117-127.
133. Venisse A, Riviere M, Vercauteren J et al. Structural analysis of the mannan region of lipoarabinomannan from Mycobacterium bovis BCG. Heterogeneity in phosphorylation state. J Biol Chem 1995; 270:15012-15021.
134. Nigou J, Gilleron M, Cahuzac B et al. The phosphatidyl-myo-inositol anchor of the lipoarabinomannans from Mycobacterium bovis bacillus Calmette Guerin. Heterogeneity, structure, and role in the regulation of cytokine secretion. J Biol Chem 1997; 272:23094-23103.
135. Misaki A, Azuma I, Yamamura Y. Structural and immunochemical studies on D-arabino-D-mannans and D mannans of Mycobacterium tuberculosis and other Mycobacterium species. J Biochem (Tokyo) 1977; 82:1759-1770.
136. Chatterjee D, Khoo KH, McNeil MR et al. Structural definition of the nonreducing termini of mannose- capped LAM from Mycobacterium tuberculosis through selective enzymatic degradation and fast atom bombardment-mass spectrometry. Glycobiology 1993; 3:497-506.
137. Khoo KH, Douglas E, Azadi P et al. Truncated structural variants of lipoarabinomannan in ethambutol drug- resistant strains of Mycobacterium smegmatis. Inhibition of arabinan biosynthesis by ethambutol. J Biol Chem 1996; 271:28682-28690.
138. Chatterjee D, Bozic CM, McNeil M et al. Structural features of the arabinan component of the lipoarabinomannan of Mycobacterium tuberculosis. J Biol Chem 1991; 266:9652-9660.
139. Chatterjee D, Hunter SW, McNeil M et al. Lipoarabinomannan. Multiglycosylated form of the mycobacterial mannosylphosphatidylinositols. J Biol Chem 1992; 267:6228-6233.
140. Prigozy TI, Sieling PA, Clemens D et al. The mannose receptor delivers lipoglycan antigens to endosomes for presentation to T cells by CD1b molecules. Immunity 1997; 6:187-197.

141. Joyce S, Woods AS, Yewdell JW et al. Natural ligand of mouse CD1d1: Cellular glycosylphosphatidy linositol. Science 1998; 279:1541-1544.
142. Schofield L, McConville MJ, Hansen D et al. CD1d-restricted immunoglobulin G formation to GPI-anchored antigens mediated by NKT cells. Science 1999; 283:225-229.
143. Sugita M, Grant EP, van Donselaar E et al. Separate pathways for antigen presentation by CD1 molecules. Immunity 1999; 11:743-752.
144. Hanau D, Fricker D, Bieber T et al. CD1 expression is not affected by human peptide transporter deficiency. Hum Immunol 1994; 41:61-68.
145. Brutkiewicz RR, Bennink JR, Yewdell JW et al. TAP-independent, beta 2-microglobulin-dependent surface expression of functional mouse CD1.1. J Exp Med 1995; 182:1913-1919.
146. Jackman RM, Stenger S, Lee A et al. The tyrosine-containing cytoplasmic tail of CD1b is essential for its efficient presentation of bacterial lipid antigens. Immunity 1998; 8:341-351.
147. Chiu YH, Park SH, Benlagha K et al. Multiple defects in antigen presentation and T cell development by mice expressing cytoplasmic tail-truncated CD1d. Nat Immunol 2002; 3:55-60.
148. Sugita M, Jackman RM, van Donselaar E et al. Cytoplasmic tail-dependent localization of CD1b antigen-presenting molecules to MIICs. Science 1996; 273:349-352.
149. Robinson MS, Bonifacino JS. Adaptor-related proteins. Curr Opin Cell Biol 2001; 13:444-453.
150. Briken V, Jackman RM, Watts GF et al. Human CD1b and CD1c isoforms survey different intracellular compartments for the presentation of microbial lipid antigens. J Exp Med 2000; 192:281-288.
151. Briken V, Moody DB, Porcelli SA. Diversification of CD1 proteins: Sampling the lipid content of different cellular compartments. Semin Immunol 2000; 12:517-525.
152. Jayawardena-Wolf J, Benlagha K, Chiu Y-H et al. CD1d Endosomal Trafficking Is Independently Regulated by an Intrinsic CD1d-Encoded Tyrosine Motif and by the Invariant Chain. Immunity 2001; 15:897-908.
153. Moody DB, Porcelli SA. CD1 Trafficking: Invariant Chain Gives a New Twist to the Tale. Immunity 2001; 15:861-865.
154. Ernst WA, Maher J, Cho S et al. Molecular interaction of CD1b with lipoglycan antigens. Immunity 1998; 8:331-340.
155. Prigozy TI, Naidenko OV, Qasba P et al. Glycolipid antigen processing for presentation by CD1d molecules. Science 2001; 291:664-667.
156. Stenger S, Mazzaccaro RJ, Uyemura K et al. Differential effects of cytolytic T cell subsets on intracellular infection. Science 1997; 276:1684-1687.
157. Xu S, Cooper A, Sturgill-Koszycki S et al. Intracellular trafficking in Mycobacterium tuberculosis and Mycobacterium avium-infected macrophages. J Immunol 1994; 153:2568-2578.
158. Schaible UE, Hagens K, Fischer K et al. Intersection of group I CD1 molecules and mycobacteria in different intracellular compartments of dendritic cells. J Immunol 2000; 164:4843-4852.
159. Beatty WL, Rhoades ER, Ullrich HJ et al. Trafficking and release of mycobacterial lipids from infected macrophages. Traffic 2000; 1:235-247.
160. Sieling PA, Jullien D, Dahlem M et al. CD1 expression by dendritic cells in human leprosy lesions: Correlation with effective host immunity. J Immunol 1999; 162:1851-1858.
161. Yamamura M, Uyemura K, Deans RJ et al. Defining protective responses to pathogens: Cytokine profiles in leprosy lesions. Science 1991; 254:277-279.
162. Yamamura M, Wang X-H, Ohmen JD et al. Cytokine patterns of immunologically mediated tissue damage. J Immunol 1992; 149:1470-1475.
163. Thomssen H, Kahan M, Londei M. Differential effects of interleukin-10 on the expression of HLA class II and CD1 molecules induced by granulocyte/macrophage colony- stimulating factor/ interleukin-4. Eur J Immunol 1995; 25:2465-2470.
164. Kaplan G, Walsh G, Guido LS et al. Novel responses of human skin to intradermal recombinant granulocyte/macrophage-colony-stimulating factor: Langerhans cell recruitment, keratinocyte growth, and enhanced wound healing. J Exp Med 1992; 175:1717-1728.
165. Stenger S, Niazi KR, Modlin RL. Downregulation of CD1 expression on human monocyte-derived cells by mycobacterial infection. J Immunol 1998; 161:3582-3258.
166. Szalay G, Zugel U, Ladel CH et al. Participation of group 2 CD1 molecules in the control of murine tuberculosis. Microbes Infect 1999; 1:1153-1157.
167. Behar SM, Dascher CC, Grusby MJ et al. Susceptibility of mice deficient in CD1D or TAP1 to infection with Mycobacterium tuberculosis. J Exp Med 1999; 189:1973-1980.
168. Sousa AO, Mazzaccaro RJ, Russell RG et al. Relative contributions of distinct MHC class I-dependent cell populations in protection to tuberculosis infection in mice. Proc Natl Acad Sci USA 2000; 97:4204-4208.
169. Porcelli SA, Segelke BW, Sugita M et al. The CD1 family of lipid antigen presenting molecules. Immunol Today 1998; 19:362-368.

Chapter 11

Processing and Presentation of Glycoproteins in the MHC Class I and II Antigen Presentation Pathways

Denise Golgher, Tim Elliott and Mark Howarth

Abstract

$CD8^+$ and $CD4^+$ T cells are stimulated by peptides bound to MHC class I and II respectively. The processing pathways for the generation of class I or class II binding peptides are specialized in surveying different intracellular compartments. While class I peptide loading occurs in the ER, class II loading occurs in the endocytic compartments. Many of the peptides generated in either compartment are derived from glycoproteins from normal or malignant cells or from intracellular or extracellular pathogens. In a given polypeptide there are potentially many T cell epitopes but only a few actually bind to the MHC molecules and generate a T cell response, while many others are cryptic. The carbohydrate moiety present in glycoproteins has been shown to influence the antigen processing pathway and generation of T cell epitopes in different ways. It can stabilize a certain three-dimensional conformation of the glycoprotein determining which sites are more accessible to proteases, it can hinder the access of proteases or block certain sites, it can target exogenous antigen to different antigen presenting cells and it can also be part of the epitope that will stimulate T cells. Changes in glycosylation patterns are common in malignancies, age and pathological mechanisms. These changes have the potential to alter the hierarchy of peptides that will bind to MHC molecules and induce a T cell response that would otherwise be cryptic, causing the onset of undesirable immune responses as is the case for autoimmune processes.

Introduction

T cell mediated responses are pivotal in the defense against infectious agents,[1,2] against tumors[3] and in the onset of autoimmune processes.[4,5] Many of the peptides that stimulate $CD8^+$ or $CD4^+$ T cells are derived from glycoproteins present in normal, virally infected or malignant cells as well as glycoproteins derived from extracellular pathogens. The presence of glycans in proteins may influence the recognition of epitopes by T cells directly or indirectly. Carbohydrates have a direct effect when the carbohydrate moiety is actually part of the epitope recognized by T cells.[6–12] Alternatively, carbohydrates have an indirect effect when the carbohydrate moiety influences the processing and generation of the peptides that will bind to the MHC.[13–17] N-linked glycans are large (approximately 30Å, similar to an immunoglobulin domain) and mobile,[18] so they can shield large areas of a protein surface (see, for example, the movie at http://www.biochem.arizona.edu/classes/bioc462/462a/NOTES/CARBO/mobility.html) and thus have a large effect on protein-protein interactions, including degradation by proteases. In this chapter we discuss what is currently known about the processing and

Immunobiology of Carbohydrates, edited by Simon Y.C. Wong and Gemma Arsequell.
©2003 Eurekah.com and Kluwer Academic / Plenum Publishers.

presentation of glycoproteins in the MHC class I and class II antigen presentation pathways and how their carbohydrate moieties can contribute to the diversity of T cell responses. We also point out what is not known in this field and suggest directions for future work.

Where Does Glycosylation Occur?

Glycosylation is divided into N-linked and O-linked modification. N-linked glycosylation is the dominant form of glycosylation for secretory and membrane-bound proteins and occurs on certain Asn residues present in the motif Asn-X-Ser/Thr. A core $GlcNAc_2Man_9Glc_3$ structure is transferred en bloc to the Asn in the ER. This core glycan can then be modified extensively in the ER and as the protein moves through the Golgi apparatus. O-linked glycosylation occurs principally on Ser and Thr but also on Hydroxyproline and Hydroxylysine, present in collagen. O glycosylation mainly occurs in the ER and Golgi but more recently it has been found that a number of cytosolic and nuclear proteins possess an O-GlcNAc modification.[19] Key things to bear in mind are that it is very difficult to predict from protein sequences alone if a protein will be glycosylated and how it will be glycosylated. Also, a range of different glycans may be present at a single glycosylation site in a protein.[20]

The MHC Class I Processing Pathway

Major Histocompatibility Complex (MHC) class I processing is one of the most intensively studied systems in cell biology and is described in detail in several excellent reviews.[21,22] We give only an outline of the class I processing pathway here, before focussing on how this pathway can be affected by glycosylation of the antigenic protein.

Outline of the Class I Processing Pathway (Fig. 1)

Assembled MHC class I molecules consist of three components: heavy chain which is a glycoprotein with a single transmembrane helix, the non-glycosylated soluble protein β2–microglobulin (β2m) and peptide. MHC class I is expressed on the surface of nearly every nucleated cell in the body. MHC class I binds peptides which are predominantly 8–10 residues in length with both their N- and C- termini buried within the peptide binding groove. These peptides are principally generated by degradation of proteins in the cytosol by the proteasome. The 26S proteasome in the cytosol adds ubiquitin to proteins, unfolds them and degrades them into peptides of 4–25 residues.[23] These peptides can be subsequently trimmed by other more recently discovered cytosolic proteases.[22] The Transporter associated with Antigen Processing (TAP), a member of the ATP Binding Cassette superfamily, transports peptides of 8–16 residues from the cytosol into the ER where the peptides are loaded onto the heavy chain:β2m complex. Loading of peptides onto class I occurs when the heavy chain: β2m complex is associated with TAP, calreticulin, ERp57 and the specialized TAP associated glycoprotein called tapasin. The formation of this peptide loading complex seems to be crucial for the assembly of class I with high affinity stable-binding peptide.[24,25] This may be because of some catalysis of dissociation of fast off-rate peptides, or because of trimming of extended peptides to an optimal length for occupying the class I binding groove.[26] After loading with peptide, class I can traffic through the Golgi apparatus to the cell surface, where it can be recognized by $CD8^+$ T cells.

How Does Glycosylation of the Protein Antigen Affect MHC Class I Processing?

Proteins for degradation by the proteasome may come from a number of sources. Conventionally they are fully folded cytosolic or nuclear proteins which may well have been O-glycosylated. Alternatively, in 2001 it was demonstrated by two groups that a major substrate for the proteasome was defective ribosomal products (DRiPs)—polypeptides which had just been produced by ribosomes and had been mistranslated or failed to fold.[27,28] DRiPs produced from genes encoding glycoproteins of the secretory pathway will not have been

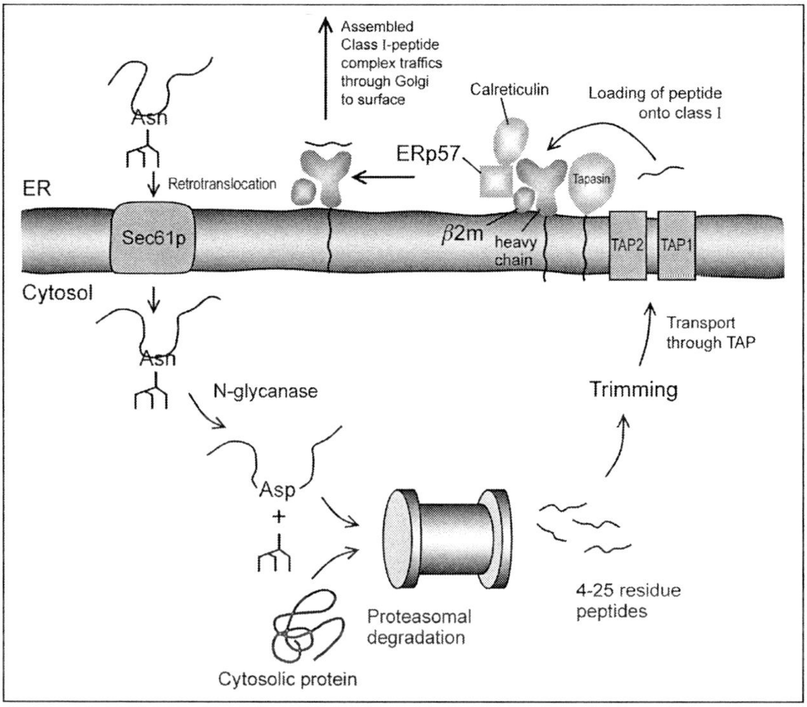

Figure 1. The MHC class I processing pathway.

transported to the ER and so will not be glycosylated. A third source of proteasomal substrates comes from proteins in the secretory pathway, which have been retrotranslocated from the ER to the cytosol via Sec61p,[16,29,30] as part of the ER-associated degradation pathway. A large proportion of proteins in the secretory pathway are glycosylated and when they are translocated back into the cytosol their N-glycans can be removed by N-glycanase.[31] This enzyme hydrolyses the bond between Asn and the first GlcNAc of the glycan, to leave a tell-tale Asp in place of the Asn that was originally translated.

How Significant Are Epitopes Deamidated by N-Glycanase?

The first CTL epitope found to be modified by deamidation came from the search for melanoma tumor antigens important in generating $CD8^+$ responses in cancer patients. There are two different approaches to identify tumor antigens. The genetic approach involves the transfection of pools of cDNA made from the tumor into COS cells that express an appropriate MHC molecule. A positive pool will be identified using $CD8^+$ T cells specific to the tumor.[32] The biochemical approach involves eluting peptides from the surface of the tumor.[33] The peptides are fractionated by HPLC and the fractions are tested for the ability to sensitize targets for lysis using tumor-specific $CD8^+$ T cells. The positive fraction is then sequenced by mass spectrometry. The epitope derived from a melanocyte differentiation antigen, tyrosinase, was identified using a genetic approach[34] as residues 368–376 of tyrosinase. On the other hand, the biochemical approach indicated that a mutated form of 368–376 was the epitope: Asp at position three was found instead of Asn. It was subsequently demonstrated that the unmutated tyrosinase had undergone a posttranslational modification that changed the genetically encoded Asn into Asp.[35] Although not formally proved, the likelihood was that this epitope was derived from an enzymatic reaction, given that this Asn was part of a glycosylation motif.

Later work has demonstrated that presentation of posttranslationally deamidated epitopes is not restricted to self/tumor antigens but is also important for viral glycoproteins. In an attempt to map epitopes from the Hepatitis C virus envelope glycoprotein E1 important in stimulating $CD8^+$ T cells in infected chimpanzees, Selby et al[36] generated an E1-specific CTL line and tested synthetic epitopes spanning the whole glycoprotein for CTL activity. One such peptide did stimulate CTL, 233-GNASRCWVA-241, but the peptide concentration required to sensitize target cells for lysis was surprisingly high (>200nM). Given that the Asn residue from this peptide was part of a known glycosylated site, they tested a deamidated version of the peptide, with the Asn at Asn-234 changed to Asp. With this deamidated peptide they found that as little as 0.001 nM of peptide could now sensitize target cells for lysis using the same CTL. CTL against this epitope were still present in chimpanzees 5 years after infection. It was demonstrated that this viral glycoprotein had to be glycosylated, retrotranslocated into the cytosol and the deamidated processed peptide transported back via TAP into the ER.[36] The same posttranslational modification has been described for another TAP-dependent epitope from the HIV-1 envelope glycoprotein.[37]

Deamidation and the dominant source of proteasome substrates from glycoproteins were explored further in a study on Lymphocytic Choriomeningitis Virus (LCMV).[38] LCMV Armstrong glycoprotein encodes a peptide, GP92, which has the sequence 92-CSANNSHHYI-101. This peptide bound well to $H\text{-}2D^b$ but produced a weak immune response. It was suggested that this weak response was because Asn-95 was glycosylated in the mature viral protein. At the cell surface they found both the unglycosylated peptide (Asn-95), likely produced from DRiPs, and the deglycosylated peptide (Asp-95), produced after retrotranslocation and cleavage by N-glycanase, but not the N-glycosylated peptide (they analyzed the peptide with a single N-linked GlcNAc residue attached). All three forms of the peptide bound equally to $H\text{-}2D^b$ and could elicit CTL when mice were immunized with the respective peptides but the study was complicated by the cross-reactivity exhibited by these CTL. Presentation of the deglycosylated peptide provided further evidence for the significance of retrotranslocation in the generation of epitopes from glycoproteins. Presentation of the unglycosylated peptide is consistent with the DRiP hypothesis, although other possibilities, such as incomplete occupancy of the glycosylation site in the native viral protein, are also possible. The lack of presentation of the unglycosylated peptide derived from tyrosinase, as shown by mass spectrometry, with only the de-N-glycosylated peptide reaching the cell surface, even though both bind equally well to HLA-A2, suggests that, at least for this protein, epitope production from DRiPs is insignificant.[39]

All these examples mean that when searching for class I binding motifs or molecular mimics, it would be wise to consider the possibility that if the gene sequence of a secretory protein encodes Asn-X-Ser/Thr, Asp-X-Ser/Thr may actually be generated in the cytosol and be presented at the cell surface.

How Do Cytosolic Proteases Deal with the Glycosylation of Their Substrates?

There is very little work addressing the proteolysis of glycosylated substrates in the cytosol. These glycosylated substrates are likely to be principally O-glycosylated since N-glycanase would remove most N-glycans. It is unlikely that a large glycan would be able to enter the proteasome since the size of the pore of the proteasome is 13Å[40] and so this may limit access to the proteasome of peptides even with glycans as small as monosaccharides attached. Proteasome degradation of the transcription factor Sp1 has been shown to be inhibited by the presence of O-GlcNAc at multiple sites.[41] This glycosylation was reduced when there was glucose starvation and Sp1 became susceptible to rapid proteasomal degradation. This result needs to be confirmed with proteasomes in vitro to establish whether O-glycosylation causes the block in degradation and does not merely correlate with it.

How Does Glycosylation Affect the Generation of Class I Epitopes by Proteolysis in the ER?

Degradation of longer polypeptides in the ER, without retrotranslocation to the cytosol, can also generate epitopes that bind to class I.[42,43] We have explored the influence of glycans on this degradation using Influenza nucleoprotein (NP) targeted to the ER with a signal sequence.[17] This causes NP to be glycosylated at two sites and abolishes presentation of the immunodominant D^b- or K^k-restricted epitopes in TAP-deficient cells. However, blocking glycosylation with tunicamycin or by mutating out the glycosylation sites from NP restored presentation of these epitopes. Further analysis of the rate of degradation of the different forms of NP led to the conclusion that N-glycosylation could abolish the proteolytic generation of antigenic peptide in the ER independent of the effect of glycosylation on the stability of the protein.

How Well Does TAP Transport Glycosylated Peptides?

Our group has shown that addition of an O-GlcNAc residue had almost no effect on the strength of binding to TAP, measured by the extent to which the glycopeptide competed with the transport of another peptide.[44] However, when the efficiency of transport of the glycopeptide into the ER was measured, transport of the glycosylated peptide was still possible but was reduced by approximately 40% compared to the unglycosylated peptide. TAP-transport of glycopeptides is consistent with the work of Gromme et al who have shown that peptides containing side-chains with an extended size of 70Å, equivalent to the size of a 21-mer peptide, can still be efficiently transported by TAP.[45] Despite this result it would still be worth testing whether peptides with glycans larger than monosaccharides can be efficiently transported by TAP.

Does Glycosylation Affect Other Steps in the MHC Class I Pathway?

After assembly of MHC class I with peptide, it exits the ER and traffics through the Golgi apparatus to the cell surface. The influence of peptide glycosylation on the peptide loading complex or trafficking through the secretory pathway has not yet been investigated.

The MHC Class II Processing Pathway

MHC Class II molecules are formed by two non-covalently associated transmembrane proteins, α and β. In the mouse the class II molecules are I-E and I-A and in humans HLA-DR, DQ and DP. Class II is expressed mostly by professional antigen presenting cells (APCs): B cells, macrophages and dendritic cells. Class II molecules evolved to present peptides derived from proteins located in the lysosomal/endosomal compartments. These could be resident membrane and soluble proteins as well as derived from exogenous antigen that can be concentrated in such compartments by different internalization mechanisms.[46,47] While class I, whose peptide binding groove is closed at each end, binds the N- and C-termini of the peptide, the peptide binding groove of class II is open at each end. As a result, class II molecules have the ability to bind to much larger peptides than MHC class I, but most of the peptides eluted from class II molecules range in size from 12 to 19 residues.[48] After synthesis and translocation into the ER, the α and β chains associate with a third transmembrane glycoprotein, the invariant chain Ii (Figure 2). Ii has three major functions: (i) it acts as a chaperone aiding in the correct folding of MHC class II molecules, (ii) one part of Ii, CLIP (the class II-associated Ii peptide), binds to the peptide binding groove and impedes the binding of peptides and polypeptides to the class II in the ER, and (iii) it has a targeting signal which will direct the class II molecule to endocytic compartments. A nonameric complex of $(\alpha\beta Ii)_3$ is formed and is exported from the ER but cannot yet bind to peptide ligands since the CLIP region occupies the binding groove. Two to three hours are necessary between transit through the Golgi and expression of class II as mature molecules at the cell surface.[49] During this period intersection of this nonameric complex with the endocytic pathway enables mixing of endocytosed antigen and the loading of class II with stable peptides. The invariant chain is removed by the combined action of

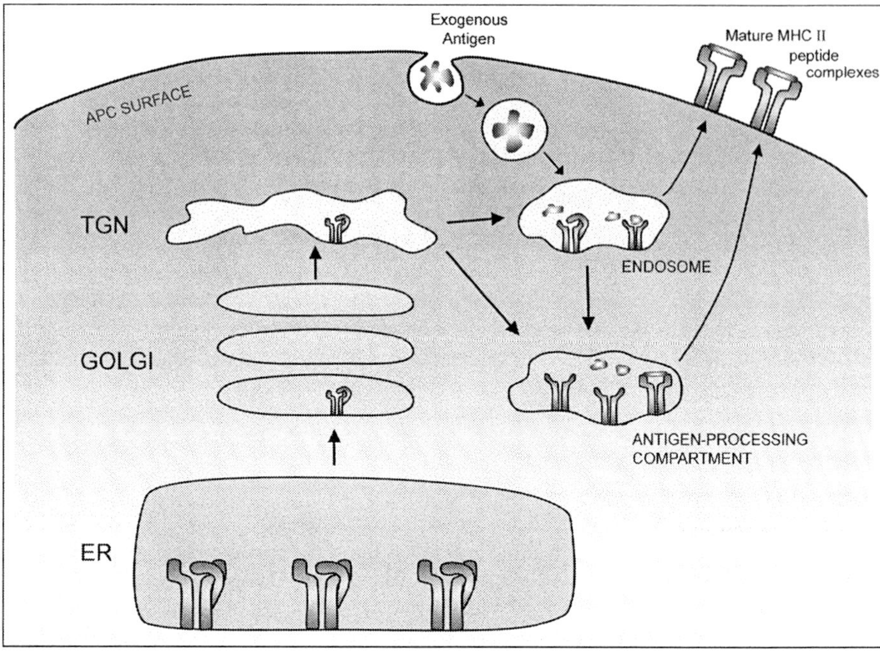

Figure 2. The MHC class II processing pathway.

proteases and acidic pH, with the help of another protein, HLA-DM (human) or H-2M (mouse). DM not only exchanges CLIP for cognate peptide but also exchanges low-stability peptides (with a high off-rate) for peptides that bind class II molecules more stably.[50] In B cells another molecule, HLA-DO/H-2O, participates in this editing process affecting the repertoire of peptides presented by class II molecules by a mechanism which is as yet unknown.[50] The precise trafficking pathway used by class II αβ–Ii complexes to access peptide-rich antigen processing compartments is a matter of debate, but class II can be found throughout the endocytic system.[51] A special compartment referred as the MHC class II compartment (MIIC) has been isolated and identified as the class II loading compartment.[52] It is likely that this pre-lysosomal/ lysosomal antigen-processing compartment is the major compartment for peptide loading, since it is the most denaturing to foreign antigens, is the most proteolytic, and contains the highest concentration of the CLIP-removing accessory molecule HLA-DM. Given that class II molecules are detected throughout the endocytic system, peptide loading can occur in different compartments.[53] The balance between the availability of the peptide as unfolding and proteolysis of the antigen is initiated and the availability of a class II molecule to capture this peptide will determine the compartment in which class II is loaded. The open ends of the class II MHC peptide-binding groove are well suited to the capture of unfolded and extended antigen domains and this may be essential to avoid overdigestion and destruction of T cell epitopes.[54] A stable complex of peptide:class II will be exported to the cell surface and stimulate the cognate $CD4^+$ T cells.

How Does Glycosylation Affect Class II Processing?

Glycans could protect proteins from degradation by proteases, either by hindering access of proteases to a given T cell determinant or perhaps by inducing conformational changes in the protein's three-dimensional structure.[55–57] In some cases the formation of epitopes is constrained by heavy glycosylation of the protein. Protein deglycosylation may increase susceptibility to proteases. It is interesting to note that the priming of the T cell response for the tumor antigen mucin was much more efficient when its glycosylation was reduced or removed.[58] For influenza virus hemagglutinin (HA), glycosylation has been shown to hinder the generation of $CD4^+$ T cell epitopes either from HA that had been taken up exogenously by APC[14] or from HA endogenously expressed by the APC.[13] Gelder et al[59] conducted an extensive analysis of HA responses in immunized volunteers. Using synthetic peptides he tried to establish a correlation between the strength with which certain peptides would bind to class II and the ability to generate a $CD4^+$ response. Many of the peptides that did bind well to class II but did not generate $CD4^+$ responses corresponded to glycosylated regions of HA.[59] In the case of the tumor antigen MUC 1, glycosylation seems to target and localize this heavily glycosylated protein in early endosomes and no antigen processing occurs, possibly because the antigen recycles back to the cell surface and is released before being processed for presentation. On the other hand, deglycosylated mucin traffics to late endosomes or the class II loading compartment and gets processed rapidly.[60] However, for other T cell determinants the glycosylation of the protein was essential for the generation of the epitope, even though the epitope itself was unglycosylated.[14,15] Here the presence of the N-glycans was more likely to alter the pathway of glycoprotein processing. The discovery that the major protease involved in the processing of a microbial antigen presented by class II is an asparaginyl endopeptidase provided direct evidence that glycosylation can have a major role in antigen processing.[61] Asparaginyl endopeptidase was isolated from lysosomes of a human B cell line but is also present in other APCs. The glycosylation of an Asn from a domain of tetanus toxin protein blocked cleavage by asparaginyl endopeptidase and thus prevented the generation of a class II epitope from this protein.

There are some examples in the literature that show N-glycans to be essential for T cell recognition, but since the T cell determinants have not been identified, the mechanism by which glycosylation is important is not clear. Dudler et al[62] demonstrated that clones derived from allergic patients were specific for a glycosylated form of the bee venom allergen phospholipase A_2 (PLA_2). These clones were not stimulated by full-length or truncated recombinant PLA_2 produced in *E.coli*, nor by a set of overlapping peptides. The authors argue that a glycosylated epitope was being recognized by the T cell clones but no formal evidence was provided for this, so a role for the carbohydrate moiety in influencing antigen processing cannot be excluded. Also, a deamidated form of the peptide was not tested. Interestingly, treatment of PLA_2 with mannosidase also abolished its recognition. This indicates that if the carbohydrate moiety were part of the epitope at least 5 sugar residues were being recognized by the TCR. This seems unlikely according to the crystal structure of glycopeptide:MHC complexes (see below). Mannosidase treatment also reduced $CD4^+$ T cell stimulation by a *Mycobacterium tuberculosis* antigen complex.[63] It is important to note here that this also could be caused by reduced uptake of deglycosylated antigen by APCs by, for example, the mannose receptor[65] or differential targeting of the antigen into a different endocytic loading compartment through a lectin interaction (not yet identified) where the antigen processing environment is different.

The importance of glycans in recognition by $CD4^+$ clones specific for two different tumor antigens has been reported. We isolated three different anti-tumor clones from mice that had been previously vaccinated against the colon carcinoma, CT26. These clones were stimulated by the ER-resident precursor of gp70, the glycoprotein of the endogenous murine leukemia virus, which is gp90. The recognition of gp90 was both glycosylation- and conformation-dependent. As with PLA_2, synthetic peptides spanning the whole of gp90 failed to stimulate the clones, raising the question as to whether the class II epitope itself was glycosylated. In this case we also tested the deamidated forms of peptides, which failed to stimulate the clones.[64] For the melanoma antigen tyrosinase, the stimulation of $CD4^+$ T cells isolated from melanoma patients was also dependent on glycosylation of tyrosinase.[66] Again, the epitope could not be identified from a set of overlapping synthetic peptides but point mutations in the glycoprotein indicated that four out of seven potential glycosylation sites were important for the recognition of tyrosinase by the clone, suggesting an important role for the

N-glycans in affecting presentation of tyrosinase. For both gp90 and tyrosinase it seems that glycosylation might have both a direct and an indirect effect on processing and presentation, where the epitope itself is probably glycosylated (given that overlapping synthetic peptides were not stimulatory) and the generation of this glycopeptide epitope is dependent on the three-dimensional conformation of the protein. It is also possible that none of the synthetic peptides was optimal and that glycans have only an indirect effect. Given the difficulty in synthesizing a whole range of different peptides when the stimulatory epitope cannot at least be mapped to a certain region of the protein, it seems that the only solution would be to directly identify the peptides biochemically. This would involve pulsing APCs with the glycosylated protein, eluting peptides from the restricting class II molecule, and sequencing the immunogenic peptide.

Glycopeptides As T Cell Epitopes

Although the experiments described above showed that glycosylation affects the generation of class I and class II epitopes, they did not clearly demonstrate that the epitope recognized by T cells was itself glycosylated. For a long time it was generally believed that only unmodified peptides could bind to MHC class I and II molecules. The immune response to polysaccharides was T cell-independent and monosaccharides or polysaccharides could not compete or block binding of peptides to class II molecules.[67] Despite this, studies with conjugation of haptens (e.g., trinitrophenyl or fluorescein derivatives) to polypeptides had clearly demonstrated that T cell responses could be primed against the haptenated peptide and subsequently generate hapten-specific responses and responses to the unmodified peptide.[68] This provided a foundation for studies to test if glycopeptides would also affect T cell responses, given that glycosylation is a physiologically relevant modification (for a review of T cell responses to other posttranslational modifications see Kastrup et al[69]).

Initial studies demonstrating that glycopeptides could bind to MHC molecules were done using synthetic peptides. Ishioka et al[6] synthesized a series of glycopeptides, in which a GlcNAc N-linked to an Asn residue was placed at various positions throughout the sequence of a model T cell epitope, ovalbumin. The glycopeptides that bound stably to the MHC class II I-A^d were used to immunize mice. The results showed that in most instances the glycopeptide was less immunogenic than its non-glycosylated analog, but that a detectable glycopeptide-specific response could be obtained. The substitution of GlcNAc by a structurally related analog ablated T cell recognition, indicating that the carbohydrate moiety was important.[6] Harding et al[7] did the same set of experiments using different glycopeptides with affinity for a different murine class II molecule, I-A^k. Mice were immunized with a peptide from hen egg lysozyme (HEL) that had the disaccharide galabiose linked to the amino terminus. In vitro assays to analyze the fine specificity of glycopeptide-specific hybridomas from the immunized mice indicated that the T cells were not sensitive to small changes in the carbohydrate moiety, but they were sensitive to its location in the peptide sequence and changes to the peptide sequence itself. This indicated that the carbohydrate moiety was influencing T cell recognition by changing the conformation of the peptide backbone.[7] When the disaccharide was attached to the middle of the same HEL-derived peptide and used to immunize mice a very different result was obtained. In this case the fine structure of the sugar moiety was important in creating the T cell determinant and sugar substitutions stopped recognition by the TCR.[8] Two of these galabiose-specific T cells were analyzed further regarding their TCR fine specificity for the glycopeptide. Two different glycopeptides were engineered: the hydroxyl group at the C-6 position was removed, either from the Gal distal to the anchoring amino acid, or from the proximal Gal. Removal of the hydroxyl group from the proximal Gal was tolerated but not from the distal Gal. Although the proximal hydroxyl group could be removed the substitution of the proximal Gal by Glc was not tolerated. This indicated that the outer Gal was in direct contact with the TCR but the proximal Gal was also important in maintaining T cell recognition. The identity of the peptide side-chains was also essential for T cell recognition.[70]

Otvos et al[71] had isolated T cell clones specific for an epitope on the rabies virus glycoprotein 29-VVEDEGCTNLSGF-41, and tested the glycosylated version of this peptide for T–cell stimulatory activity. The rationale for these studies came from the fact that different strains of rabies-related viruses can have mutations that favor differential glycosylation patterns and this specific peptide contained a glycosylation motif. While β-N-glycosylation (GlcNAc) of the sequence fully abolished the peptide's ability to bind to class II and to stimulate a $CD4^+$ specific hybridoma, incorporation of an α-linked GalNAc did not.

While the previous studies used glycopeptides in which the peptide in itself was immunogenic in mice, another group tested the same approach but using a non-immunogenic self peptide known to bind well to the murine class II $I\text{-}E^k$.[72] The aim of this study was to test whether the glycosylation of a non-immunogenic peptide could make it immunogenic. The carbohydrate chosen was the Tn antigen, a small mucin type O-linked α–D-GalNAc that is expressed in 90% of carcinomas.[73] Unfortunately analysis of bulk T cell responses from mice immunized with the glycopeptide showed that changing residue 72 to Thr or Ser, in order to to allow O-linkage of the Tn antigen, made the peptide very immunogenic by itself. The same group synthesized glycopeptides of higher complexity and demonstrated that while glycopeptides containing a trisaccharide unit could still bind well to MHC class II they were not as immunogenic as peptides containing one or two sugar units.[12] Using bulk T cell cultures from immunized mice they could demonstrate that T cells primed with glycopeptides having one or two sugar units were highly specific for the identity of the glycan group. The important conclusion as to whether a self non-immunogenic peptide could be converted into an immunogenic one could not be made, given that all immunogenic glycopeptides had one amino acid change from the wild-type sequence. The analysis of the fine specificity of these Tn-peptides was done after generating clones of T cell-hybridomas from these bulk cultures derived from immunized mice.[74] Although partial cross-reactivity was observed in some hybridomas, several of the hybridomas were highly specific. For example, 17 out of 19 hybridomas that were specific for the glycopeptide 67-VITAFTEGLK-76 with the α–D-GalNAc group attached to Thr-72 did not respond to the same peptide with α–D-GlcNAc attached to the same position. The only difference between these sugar residues is the orientation of the hydroxyl group located at the fourth carbon atom in the carbohydrate ring structure, indicating that the glycan is recognized by the TCR.[74] Interestingly, although these two glycopeptides generate completely different responses from the TCR, molecular modelling indicates that the conformation of the glycopeptide relative to the TCR is identical.[75]

All the studies described above involved testing glycopeptides as epitopes for $CD4^+$ T cells. For epitopes recognized by $CD8^+$ T cells the candidate glycosylation was the cytosolic addition of O-GlcNAc sugar residues. O-GlcNAc addition to cytosolic proteins has been identified by Wells et al.[19] This modification is thought to be important in signal transduction pathways since phosphorylation/O-GlcNAc additions occur in the same residue in a given protein in an alternate manner. O-GlcNAc addition is thought to "protect" sites from being phosphorylated.[19] It was therefore conceivable that some peptides that bind to class I would have this sugar residue. To test if an O-GlcNAc addition could generate a new T cell epitope, Haurum et al[9] added this sugar residue to peptides related to the Sendai virus nucleoprotein 324-FAPGNYPAL-332 that can bind to both MHC class I D^b and K^b. Those glycopeptides that bound well to both class I molecules were used to immunize mice. Glycopeptide-specific $CD8^+$ T cell clones could be isolated that were highly specific for the O-GlcNAc since they could not be stimulated by the same peptide without an O-GalNAc attached to it.[9]

From these studies with synthetic glycopeptides some important conclusions could be made:

i) glycosylation of epitopes can inhibit, have no effect on or in rare cases enhance binding to MHC molecules,
ii) glycopeptide-specific T cell responses can be elicited,
iii) peptides linked to monosaccharides or dissacharides are most likely to generate a glycopeptide-specific response.

iv) Depending on where the sugar is located, the TCR can directly recognize the sugar residue and some amino acids in the peptide backbone, or it recognizes the peptide backbone under a different conformation induced by the carbohydrate moiety. Therefore the potential for glycosylation to contribute directly to epitope diversity is enormous, but do naturally processed glycopeptide epitopes exist?

Naturally Processed Glycopeptide Epitopes

Naturally processed glycopeptides have been identified by mass spectrometry in a pool of peptides bound to human MHC class I and class II alleles.[44,76–78] Using mass spectrometry, a peptide with an N-GlcNAc modification was identified from HLA-DR1 and a peptide with an N-GlcNAc-GlcNAc modification was identified from DQ8.[76,77] Dustin et al[79] showed that a significant percentage of peptides (0.75%) eluted from HLA-DR from human cell lines contained mannose 6-phosphate. Phosphorylation of mannose residues to generate Man-6-P is carried on the outer mannose residues of the N-glycan and targets proteins to lysosomes, so this MHC-bound glycan should have at least 9 sugar residues. Several attempts to sequence peptides containing Man-6-P were not successful, but one of the peptides could be identified, after 15 cycles of microsequencing, as deriving from lysosomal acid lipase. There was no direct identification of the N-glycan (probably lost during the cycles of microsequencing) but the Asn was thought to be located outside the class II binding groove.[79]

To identify naturally processed glycopeptides bound to class I, peptides were extracted from normal human spleen, fractionated by reverse phase HPLC and eluted fractions were subjected to galactosyltransferase-mediated labelling with [^{3}H]-Gal, which will specifically label terminal O–β–GlcNAc residues.[80] Approximately 0.1% of peptides were positively labelled. As an alternative approach, peptides were eluted from human class I molecules and subjected to a series of lectin affinity columns. As opposed to Concanavalin A or *Arachis hypogea* lectin that did not retain any significant fraction of the peptides, the wheat germ agglutinin (WGA) column (specific for terminal GlcNAc residues) retained approximately 1% of the peptides. Sequencing by Edman degradation of the pool of peptides bound to the WGA column showed an increased amount of the canonical anchor motif of the appropriate MHC I as well as Ser and Thr which might be expected for peptides with O-GlcNAc addition.[44,78] Although these reports show that naturally processed glycopeptides can be eluted from surface class I and II molecules, there have been no reports of naturally occurring TCRs specific for these glycopeptides.

The first example of a naturally glycosylated T cell determinant came from studies of collagen-induced arthritis in mice.[10] The immunodominant epitope was located within a cyanogen bromide-cleaved fragment of rat type II collagen (CII), containing residues 256–270. Further analysis of this fragment through tryptic digestion and RPHPLC separation showed that naturally glycosylated peptides could bind to class II and stimulate the CII-specific T cells. Mass spectrometry of this CII fragment revealed that it was variably glycosylated, containing up to five sugar units. It was later demonstrated that most of the CII-specific hybridomas recognized the form of peptide with Gal on residue 264.[81] Although T cells with specificity for the deglycosylated form were also detected, the presence of the carbohydrate moiety played a clear role in the development of arthritis, since deglycosylated CII peptide induced arthritis with a later onset, lower incidence and milder symptoms.[10]

There is only one example of a naturally occurring glycopeptide-specific class I restricted T cell response but it provides evidence that inducing responses to glycopeptides could have an important role in future tumor immunotherapy protocols. This T cell response was found by Zhao and Cheung[82] after immunizing mice with irradiated EL4 lymphoma cells. The paper is complicated by the fact that they were not able to determine the molecular identity of the immunogenic glycopeptide, associating its presentation on various tumor cell-lines with the presence of the glycolipid disialoganglioside GD_2. They showed that the response is $CD8^+$ T cell-dependent, αβ TCR-dependent, TAP-dependent, H-2^b-restricted and required GD_2 expression. Antibodies to GD_2 or H-$2K^b$/H-$2D^b$ could prevent lysis but soluble GD_2 could not.

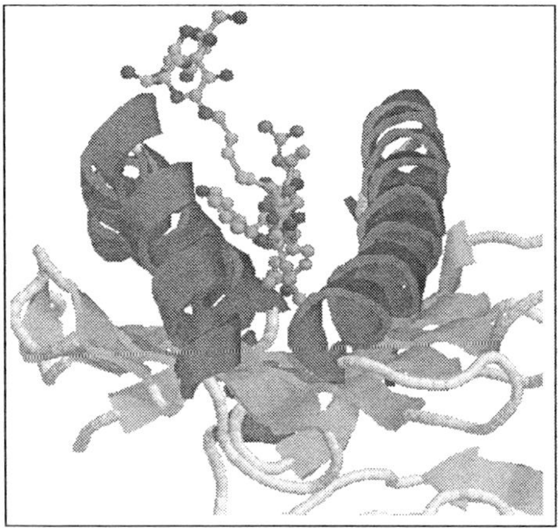

Figure 3. The crystal structure of RGY8-6H-Gal_2 bound to H-$2K^b$ (taken from Entrez Structure file 1KBG).

It is unlikely that class I could directly present a glycolipid to T cells (although CD1 molecules can do this) and so we interpret their results to mean that the altered glycolipid glycosylation also led to altered glycosylation of certain proteins and thus generated one or several new immunogenic glycopeptide epitopes. Unfortunately the authors did not try to inhibit lysis separately with antibodies to H-$2K^b$ or H-$2D^b$ to see if the immune response were directed to more than one glycopeptide epitope. The presence of GD_2 on a wide range of human tumors suggests that this GD_2-associated immune response would be a productive area for further investigation.

Crystal Structures of MHC-Glycopeptide Complexes

Three crystal structures of glycopeptides bound to MHC molecules have been determined, all published in the same issue of Immunity in 1999. Speir et al[83] solved the structure of H-$2K^b$ bound to a modified version of the Vesicular Stomatitis Virus nucleoprotein peptide RGYVYQGL (RGY8), where the disaccharide Gal–α(1,4)Gal–β is connected with a homocysteine linker to position 6 of the peptide (RGY8-6H-Gal_2). The structure of the peptide, heavy chain and β2m is nearly identical for RGY8-6H-Gal_2-K^b as for the previously determined structure of RGY8-K^b.[84] This is consistent with the peptide binding affinity to K^b being unaltered by the glycan modification.[85] It indicates that the carbohydrate specificity of the T cells depends on direct recognition of the glycan and not on a conformational change in the bound peptide induced by the glycan, as had been suggested by Harding et al for class II.[7] The glycan extended 12Å above the peptide backbone (Figure 3), presenting a substantial solvent-exposed structure and resting next to the α2-helix of K^b. The conformation of the glycan was very similar to that of the disaccharide when free in solution.[86] The exposure of the glycan above the peptide backbone in this structure may be the maximum that is consistent with an αβ TCR-mediated glycan-specific and peptide-specific response. Peptides linked to trisaccharides produced a γδ rather than an αβ T cell response[85] and even disaccharides on peptides bound to D^b are poorly immunogenic, possibly because they project further up towards the TCR as the peptide mainchain of D^b peptides arches up towards the TCR too. The γδ TCR response to a glycan occurred independently of whether the glycan is presented by an MHC-bound peptide or on a glycolipid.[85]

We have solved the structures of two O-GlcNAc substituted derivatives of the Sendai virus nucleoprotein peptide FAPGNYPAL bound to H-$2D^b$.[87] The results with FAPS(O-GlcNAc)NYPAL produced similar results to those of Speir et al:[83] the structure of heavy chain, β2m and the peptide backbone was essentially unchanged by the glycan modification.

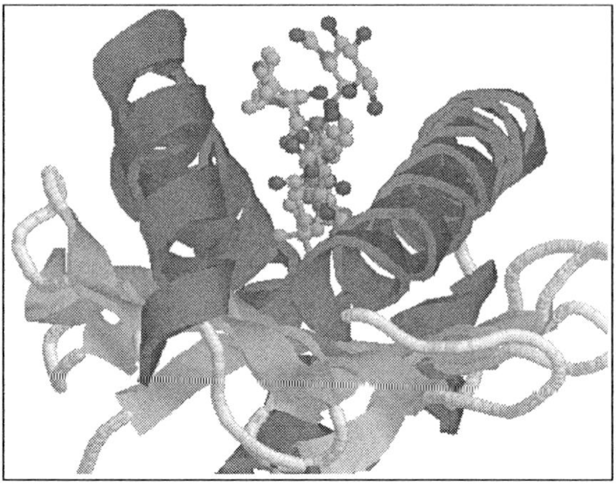

Figure 4. The crystal structure of FAPS(O-GlcNAc)NYPAL bound to H-2D^b (taken from Entrez Structure file 1QLF).

Again the glycan was highly exposed (Figure 4) with about half of the exposed surface area accounted for by the GlcNAc. The sidechain orientations in the glycopeptides were unaltered except for the Tyr at P6 which bends to stack with the sugar.

The structure of D^b in complex with FAPGS(O-GlcNAc)YPAL was, on the other hand, a whole new kettle of fish. In this case the glycan occurred where there would normally be an anchor residue Asn. The glycan cannot fit within the pocket usually filled by Asn and the peptide backbone was forced to rotate at P4-P7 by 180°, leaving the glycan exposed to solvent. The peptide backbone also has increased flexibility. This was the first structure of any MHC class I-peptide complex where a main anchor pocket was not filled but nevertheless the structure of the peptide binding groove was unaltered from a structure where all the anchor pockets are filled.[88] This has implications for any theory of peptide-induced conformational changes in class I which could regulate release of class I from the peptide loading complex, allowing export of class I to the cell surface.[89,90] P5 Asn normally forms two hydrogen bonds to Gln-97 in the pocket of D^b but in this structure two ordered water molecules formed these hydrogen bonds. Without a P5 anchor, the glycopeptide had enhanced mobility.

It should also be mentioned that Jensen et al modelled the structure of a hemoglobin peptide glycosylated with either Tn (α-D-GalNAc) or α-D-GlcNAc. The resulting structures were essentially identical, with only the difference in the position of the C4 hydroxy group of the sugar explaining the differential T cell reactivity against the two glycopeptides.[75]

How Does the T Cell Receptor Interact with MHC-Glycopeptide Complexes?

A crystal structure of an MHC-glycopeptide-TCR complex is to be eagerly awaited but at present one can only model how glycosylation would affect the interaction with the TCR. Our group has modelled the interaction of the TCR with the two D^b-glycopeptide complexes whose structures are described above.[87] This suggested that each glycan fits into a cavity between CDRα and CDRβ of the TCR and showed how conserved CDR3β residues could make hydrogen bonds with sugar hydroxyls to ensure the specificity for GlcNAc over GalNAc substitution of the peptide. With the glycan between CDR3α and CDR3β, the tips of the CDR3 loops can still scrutinize the exposed side-chains of the peptide. In each Db-glycopeptide complex the electron density indicated that the glycan had a set of distinct conformations which was more variable than for previously reported amino acid side-chains. However, oscillation between these glycan conformations was not detected by nuclear magnetic resonance relaxation rate studies. Mobility of the glycan may have an impact on the thermodynamics and kinetics or TCR binding, because of the entropic cost of restricting the glycan to a single conformation and it would be interesting to test this using surface plasmon resonance.

Sequencing of many CDR3 sequences from TCRs that recognize glycopeptides has found an abundance of small polar amino acids (such as Asn, Ser, Glu) which may be ideal for hydrogen bonding with sugar hydroxyls. Aromatic residues were also common in these CDR3 regions and may be involved in hydrophobic and stacking interactions with the sugar.[91,87]

Glycosylation in Autoimmune and Anti-Tumor T Cell Responses

In a given polypeptide there are potentially many T cell epitopes but only a few actually bind to the MHC molecules and generate a T cell response: a phenomenon known as immunodominance. Whether the influence of carbohydrates on T cell recognition is direct or indirect, changes in glycosylation patterns of certain antigens have the potential to change the hierarchy of dominant/cryptic determinants from a given antigen. During some autoimmune processes and viral infections, T cells specific for epitopes that are no longer cryptic have an important role.[4,92,93] This comes from the fact that if epitopes are cryptic during T cell development of their cognate TCRs, these T cells may escape thymic deletion and be exported to the periphery. With time or some pathological process this epitope may be exposed and elicit an undesired T cell response that can trigger an autoimmune process. As discussed before, deglycosylation can certainly uncover cryptic epitopes generated in the ER[17] but whether natural changes in the glycosylation pattern play a role in regulating hierarchy of dominant/cryptic has not been investigated. Nevertheless the best known example of a naturally occurring glycopeptide-specific T cell response comes from the epitope from collagen that is involved in experimental arthritis. Posttranslational modifications of CII can be influenced by factors such as age and hormones[11] so it is possible that collagen specific epitopes presented to T cells could change with these events too. Indeed it has been demonstrated that in patients with rheumatoid arthritis the glycosylation patterns in IgG changes.[94] The analysis of self-peptides bound to the type I diabetes associated class II molecule HLA-DQ8 identified a naturally processed glycopeptide derived from HLA-DQ1 that had two N-GlcNAc units. This peptide was 13 amino acids long and the disaccharide was attached to Asn81 75-IVIKRSNSTAATN-87. Interestingly, the unglycosylated version of this peptide could not compete for binding to the class II molecule indicating that the glycan moiety contributed to the binding energy of the glycopeptide to class II.[77] Also of interest is the fact that the $CD4^+$ T cell epitope in patients with celiac disease is a deamidated epitope.[95] Although this epitope was not generated through N-glycanase activity because it was not part of a glycosylation motif, this example certainly shows that deamidation can generate epitopes important in autoimmune processes. This has been discussed in a recent paper by Manoury et al[96] This paper shows that myelin basic protein contains a major processing site for asparaginyl endopeptidase so the generation of a dominant $CD4^+$ epitope only occurs if this site is not cleaved. As discussed by the author, with time Asn residues in myelin basic protein may suffer spontaneous deamidation, destroying the cleavage site for asparaginyl endopeptidase and as a result the dominant epitope can be generated and initiate an undesirable T cell response. Many deamidated $CD8^+$ epitopes generated through N-glycanase activity in the cytosol have been identified. Although deamidated $CD4^+$ epitopes certainly exist[95,97] this deamidation is not caused by deglycosylation but by a natural deamidation reaction. It would be of great interest to determine if N-glycanase activity is also present in MHC class II loading compartments and if this has any importance in the generation of $CD4^+$ deamidated epitopes.

In many instances undesirable autoimmune responses can be desirable anti-tumor responses. For example many of the $CD8^+$ T cell responses in patients with melanoma are directed to melanocyte associated antigens such as tyrosinase and gp100.[98] Given the history of differential glycosylation patterns in different tumors and with the progression of a tumor[99,100] it is likely that generation of T cell epitopes will change as the malignant process evolves. As previously mentioned, the generation of tumor-specific $CD4^+$ T cell responses to two different tumor antigens, gp90 and tyrosinase is dependent on the presence of N- glycans.[64,66] Another tumor antigen that has been extensively studied for its therapeutic potential is mucin. In particular, a member of the mucin family, MUC 1, which is a transmembrane glycoprotein expressed at the apical surface of normal glandular epithelia of many organs and is expressed in some hematopoietic cells (activated T cells, plasma cells, B cells). During malignant processes different things happen to mucin: the distribution of mucin is not polarized as it is on normal epithelial cells, its expression is greatly increased and the glycosylation pattern changes so that there are fewer sites glycosylated and the glycans are much shorter. As a result of this, new epitopes are exposed both on the polypeptide backbone and on the carbohydrate chains. The Tn antigen, for example, is exposed as a result of a blockage in mucin O-glycan elongation, a characteristic feature of human carcinomas. Immune responses against MUC 1 have been detected in cancer patients and this stimulated a great deal of research to explore the potential of this antigen in vaccine models.[101,102] Several pre-clinical trial experiments demonstrated that an effective cellular anti-tumor response could be obtained by vaccination against deglycosylated MUC1. The results from phase I trials showed a small increase in MUC 1-specific CTL in the peripheral blood in response to vaccination with deglycosylated MUC 1 peptide[103] but in most cases vaccination was associated with a predominant antibody response.[104]

Conclusions and Perspectives

As well pointed out by McAdam et al,[97] "The frequency and biological importance of posttranslationally modified T cell epitopes in immunology may well be underappreciated." There is little doubt that the importance of glycopeptide-specific responses found when studying collagen-induced arthritis and bee venom models can be extended to a wide number of conditions. Any hunt for an antigenic epitope, having identified a T cell response, is difficult, but it is many times harder if this epitope is glycosylated because of difficulties in analysis and synthesis of candidate epitopes. It is dispiriting to grow up $5x10^{10}$ cells (50 litres of cells if they are at 10^6 per ml) and to isolate only 15ng glycopeptide for analysis (Haurum and Elliott, unpublished results). This makes the study of glycopeptide-specific immune responses an area where the developments in sensitivity from nanoflow HPLC and electrospray ionization mass spectrometry will be especially crucial. These methods have already proved themselves in the identification of phosphorylated peptides presented by class I,[105] with detection limits down to 2 attomole, or in one case peptides present on the cell surface at an average of less than one copy per cell.[106] Fragmentation of the peptide with a secondary mass spectrometer and database searching can allow identification of the modified peptide sequence. The initiation of Glycomics projects, including a Consortium for Functional Glycomics, (http://glycomics.scripps.edu), with the long-term goal of understanding the glycosylation of every glycoprotein in the proteome, will also provide an enormous stimulus to this field.[107] Microarrays observing the change in expression of glycosyltransferases and carbohydrate binding proteins in malignant cells, for example, are promised. 2D-gel profiles of proteins that have a given glycosylation pattern, e.g., O-GlcNAc residues, will also be possible.[19,20] Also, neural networks which can scan sequences for likely O-GlcNAc modification are being developed.[108] This will give an enormous boost to looking for candidate glycopeptides as antigens in various conditions, notably cancer. Also, the ability to obtain various glycosylated peptides, a challenge to synthesize even for experienced organic chemists, will increase, which has been in many cases a limiting factor. The consortium also aims to generate mice with genes for glycosyltransferases and carbohydrate binding proteins knocked out. Transplantation of cells from wild-type mice to these knockout mice could generate a T cell-mediated immune response that could lead to the identification of glycopeptides presented by MHC molecules on wild-type cells but which these mutant mice have not become tolerant to. A knockout mouse for α-mannosidase II has already been generated by Chui et al[109] and interestingly enough it develops, by a yet unknown mechanism, a systemic autoimmune disease similar to human systemic lupus erythematosus.

From the examples in the literature it seems that bulky N-glycans can have a major influence on both class I and class II antigen processing but are less likely to be directly recognized by T cells. There are many examples of N-glycosylation leading to generation of deamidated epitopes which are seen by $CD8^+$ T cells but no examples of recognition of $CD4^+$ epitopes deamidated through N-glycanase activity. This could be simply a result of more research being done on $CD8^+$ epitopes or alternatively, we would like to speculate, a reflection of more efficient N-glycanase cleavage in the cytosol (where most class I epitopes are generated) than in the MHC class II loading compartment. Although glycosidases present in the endosomal/lysosomal compartment have been identified,[110] efficient glycan trimming and removal may not happen where and when the loading onto class II occurs. It would be of major interest to determine in which endocytic compartments the generation of N-glycosylation dependent epitopes occurs and analyze this loading compartment for the presence of different glycosidases.

References

1. O'Garra A, Macatonia SE, Hsieh CS et al. Regulatory role of IL4 and other cytokines in T helper cell development in an alpha beta TCR transgenic mouse system. Res Immunol 1993; 144(8):620-5.
2. Constant SL, Bottomly K. Induction of Th1 and Th2 CD4+ T cell responses: the alternative approaches. Annu Rev Immunol 1997; 15:297-322.
3. Toes RE, Ossendorp F, Offringa R et al. CD4 T cells and their role in antitumor immune responses. J Exp Med 1999; 189(5):753-756.
4. Sercarz EE, Lehmann PV, Ametani A et al. Dominance and crypticity of T cell antigenic determinants. Annu Rev Immunol 1993; 11:729-766.
5. von Herrath MG, Dockter J, Oldstone MB. How virus induces a rapid or slow onset insulin-dependent diabetes mellitus in a transgenic model. Immunity 1994; 1(3):231-242.
6. Ishioka GY, Lamont AG, Thomson D et al. MHC interaction and T cell recognition of carbohydrates and glycopeptides. J Immunol 1992; 148(8):2446-2451.
7. Harding CV, Kihlberg J, Elofsson M et al. Glycopeptides bind MHC molecules and elicit specific T cell responses. J Immunol 1993; 151(5):2419-2425.
8. Deck B, Elofsson M, Kihlberg J et al. Specificity of glycopeptide-specific T cells. J Immunol 1995; 155(3):1074-8.
9. Haurum JS, Arsequell G, Lellouch AC et al. Recognition of carbohydrate by major histocompatibility complex class I-restricted, glycopeptide-specific cytotoxic T lymphocytes. J Exp Med 1994; 180(2):739-44
10. Michaelsson E, Malmstrom V, Reis S et al. T cell recognition of carbohydrates on type II collagen. J Exp Med 1994; 180(2):745-9.
11. Michaelsson E, Broddefalk J, Engstrom A et al. Antigen processing and presentation of a naturally glycosylated protein elicits major histocompatibility complex class II-restricted, carbohydrate-specific T cells. Eur J Immunol 1996; 26(8):1906-10.
12. Galli-Stampino L, Meinjohanns E, Frische K et al. T-cell recognition of tumor-associated carbohydrates: the nature of the glycan moiety plays a decisive role in determining glycopeptide immunogenicity. Cancer Res 1997; 57(15):3214-22.
13. Thomas DB, Hodgson J, Riska PF et al. The role of the endoplasmic reticulum in antigen processing. N-glycosylation of influenza hemagglutinin abrogates CD4+ cytotoxic T cell recognition of endogenously processed antigen. J Immunol 1990; 144(7):2789-94.
14. Drummer HE, Jackson DC, Brown LE. Modulation of CD4+ T-cell recognition of influenza hemagglutinin by carbohydrate side chains located outside a T-cell determinant. Virology 1993; 192(1):282-9.
15. Sjolander S, Bolmstedt A, Akerblom L et al. N-linked glycans in the CD4-binding domain of human immunodeficiency virus type 1 envelope glycoprotein gp160 are essential for the in vivo priming of T cells recognizing an epitope located in their vicinity. Virology 1996; 215(2):124-33.
16. Bacik I, Snyder HL, Anton LC et al. Introduction of a glycosylation site into a secreted protein provides evidence for an alternative antigen processing pathway: transport of precursors of major histocompatibility complex class I-restricted peptides from the endoplasmic reticulum to the cytosol. J Exp Med 1997; 186(4):479-87.
17. Wood P, Elliott T. Glycan-regulated antigen processing of a protein in the endoplasmic reticulum can uncover cryptic cytotoxic T cell epitopes. J Exp Med 1998; 188(4):773-8.
18. Rudd PM, Wormald MR, Stanfield RL et al. Roles for glycosylation of cell surface receptors involved in cellular immune recognition. J Mol Biol 1999; 293(2):351-66.

19. Wells L, Vosseller K, Hart GW. Glycosylation of nucleocytoplasmic proteins: signal transduction and O-GlcNAc. Science 2001; 291(5512):2376-8.
20. Dwek MV, Ross HA, Leathem AJ. Proteome and glycosylation mapping identifies post-translational modifications associated with aggressive breast cancer. Proteomics 2001; 1(6):756-62.
21. Pamer E, Cresswell P. Mechanisms of MHC class I—restricted antigen processing. Annu Rev Immunol 1998; 16:323-58.
22. Yewdell JW, Bennink JR. Cut and trim: generating MHC class I peptide ligands. Curr Opin Immunol 2001; 13(1):13-8.
23. Kisselev AF, Akopian TN, Woo KM et al. The sizes of peptides generated from protein by mammalian 26 and 20 S proteasomes. Implications for understanding the degradative mechanism and antigen presentation. J Biol Chem 1999; 274(6):3363-71.
24. Lewis JW, Elliott T. Evidence for successive peptide binding and quality control stages during MHC class I assembly. Curr Biol 1998; 8(12):717-20.
25. Peh CA, Burrows SR, Barnden M et al. HLA-B27-restricted antigen presentation in the absence of tapasin reveals polymorphism in mechanisms of HLA class I peptide loading. Immunity 1998; 8(5):531-42.
26. Brouwenstijn N, Serwold T, Shastri N. MHC class I molecules can direct proteolytic cleavage of antigenic precursors in the endoplasmic reticulum. Immunity 2001; 15(1):95-104.
27. Schubert U, Anton LC, Gibbs J et al. Rapid degradation of a large fraction of newly synthesized proteins by proteasomes. Nature 2000; 404(6779):770-774.
28. Reits EA, Vos JC, Gromme M et al. The major substrates for TAP in vivo are derived from newly synthesized proteins. Nature 2000; 404(6779):774-778.
29. Wiertz EJ, Tortorella D, Bogyo M et al. Sec61-mediated transfer of a membrane protein from the endoplasmic reticulum to the proteasome for destruction. Nature 1996; 384(6608):432-438.
30. Plemper RK, Bohmler S, Bordallo J et al. Mutant analysis links the translocon and BiP to retrograde protein transport for ER degradation. Nature 1997; 388(6645):891-895.
31. Suzuki T, Seko A, Kitajima K et al. Identification of peptide:N-glycanase activity in mammalian-derived cultured cells. Biochem Biophys Res Commun 1993; 194(3):1124-1130.
32. Boel P, Wildmann C, Sensi ML et al. BAGE: a new gene encoding an antigen recognized on human melanomas by cytolytic T lymphocytes. Immunity 1995; 2(2):167-175.
33. Cox AL, Skipper J, Chen Y et al. Identification of a peptide recognized by five melanoma-specific human cytotoxic T cell lines. Science 1994; 264(5159):716-719.
34. Wolfel T, Van Pel A, Brichard V et al. Two tyrosinase nonapeptides recognized on HLA-A2 melanomas by autologous cytolytic T lymphocytes. Eur J Immunol 1994; 24(3):759-764.
35. Skipper JC, Hendrickson RC, Gulden PH et al. An HLA-A2-restricted tyrosinase antigen on melanoma cells results from posttranslational modification and suggests a novel pathway for processing of membrane proteins. J Exp Med 1996; 183(2):527-534.
36. Selby M, Erickson A, Dong C et al. Hepatitis C virus envelope glycoprotein E1 originates in the endoplasmic reticulum and requires cytoplasmic processing for presentation by class I MHC molecules. J Immunol 1999; 162(2):669-676.
37. Ferris RL, Hall C, Sipsas NV et al. Processing of HIV-1 envelope glycoprotein for class I-restricted recognition: dependence on TAP1/2 and mechanisms for cytosolic localization. J Immunol 1999; 162(3):1324-1332.
38. Hudrisier D, Riond J, Mazarguil H et al. Genetically encoded and post-translationally modified forms of a major histocompatibility complex class I-restricted antigen bearing a glycosylation motif are independently processed and co-presented to cytotoxic T lymphocytes. J Biol Chem 1999; 274(51):36274-36280.
39. Mosse CA, Meadows L, Luckey CJ et al. The class I antigen-processing pathway for the membrane protein tyrosinase involves translation in the endoplasmic reticulum and processing in the cytosol. J Exp Med 1998; 187(1):37-48.
40. Rudd PM, Elliott T, Cresswell P et al. Glycosylation and the immune system. Science 2001; 291(5512):2370-6.
41. Han I, Kudlow JE. Reduced O glycosylation of Sp1 is associated with increased proteasome susceptibility. Mol Cell Biol 1997; 17(5):2550-8.
42. Elliott T, Willis A, Cerundolo V et al. Processing of major histocompatibility class I-restricted antigens in the endoplasmic reticulum. J Exp Med 1995; 181(4):1481-91.
43. Snyder HL, Yewdell JW, Bennink JR. Trimming of antigenic peptides in an early secretory compartment. J Exp Med 1994; 180(6):2389-94.
44. Haurum JS, Hoier IB, Arsequell G et al. Presentation of cytosolic glycosylated peptides by human class I major histocompatibility complex molecules in vivo. J Exp Med 1999; 190(1):145-50.

45. Gromme M, van der Valk R, Sliedregt K et al. The rational design of TAP inhibitors using peptide substrate modifications and peptidomimetics. Eur J Immunol 1997; 27(4):898-904.
46. Castellino F, Zhong G, Germain RN. Antigen presentation by MHC class II molecules: invariant chain function, protein trafficking, and the molecular basis of diverse determinant capture. Hum Immunol 1997; 54(2):159-69.
47. Engelhard VH. Structure of peptides associated with class I and class II MHC molecules. Annu Rev Immunol 1994; 12:181-207.
48. Rudensky A, Preston-Hurlburt P, Hong SC, Barlow A, Janeway CA, Jr. Sequence analysis of peptides bound to MHC class II molecules. Nature 1991; 353(6345):622-7.
49. Cresswell P. Assembly, transport, and function of MHC class II molecules. Annu Rev Immunol 1994; 12:259-93.
50. Brocke P, Garbi N, Momburg F et al. HLA-DM, HLA-DO and tapasin: functional similarities and differences. Curr Opin Immunol 2002; 14(1):22-9.
51. Hiltbold EM, Roche PA. Trafficking of MHC class II molecules in the late secretory pathway. Curr Opin Immunol 2002; 14(1):30-5.
52. Amigorena S, Drake JR, Webster P et al. Transient accumulation of new class II MHC molecules in a novel endocytic compartment in B lymphocytes. Nature 1994; 369(6476):113-20.
53. Pinet V, Malnati MS, Long EO. Two processing pathways for the MHC class II-restricted presentation of exogenous influenza virus antigen. J Immunol 1994; 152(10):4852-4860.
54. Deng H, Apple R, Clare-Salzler M et al. Determinant capture as a possible mechanism of protection afforded by major histocompatibility complex class II molecules in autoimmune disease. J Exp Med 1993; 178(5):1675-1680.
55. Zilberstein A, Snider MD, Lodish HF. Synthesis and assembly of the vesicular stomatitis virus glycoprotein. Cold Spring Harb Symp Quant Biol 1982; 46(Pt 2):785-795.
56. Gallagher P, Henneberry J, Wilson I et al. Addition of carbohydrate side chains at novel sites on influenza virus hemagglutinin can modulate the folding, transport, and activity of the molecule. J Cell Biol 1988; 107(6 Pt 1):2059-2073.
57. Li Z, Pinter A, Kayman SC. The critical N-linked glycan of murine leukemia virus envelope protein promotes both folding of the C-terminal domains of the precursor polyprotein and stability of the postcleavage envelope complex. J Virol 1997; 71(9):7012-7019.
58. Hiltbold EM, Alter MD, Ciborowski P et al. Presentation of MUC1 tumor antigen by class I MHC and CTL function correlate with the glycosylation state of the protein taken up by dendritic cells. Cell Immunol 1999; 194(2):143-149.
59. Gelder C, Davenport M, Barnardo M et al. Six unrelated HLA-DR-matched adults recognize identical CD4+ T cell epitopes from influenza A haemagglutinin that are not simply peptides with high HLA-DR binding affinities. Int Immunol 1998; 10(2):211-222.
60. Hiltbold EM, Vlad AM, Ciborowski P et al. The mechanism of unresponsiveness to circulating tumor antigen MUC1 is a block in intracellular sorting and processing by dendritic cells. J Immunol 2000; 165(7):3730-3741.
61. Manoury B, Hewitt EW, Morrice N et al. An asparaginyl endopeptidase processes a microbial antigen for class II MHC presentation. Nature 1998; 396(6712):695-699.
62. Dudler T, Altmann F, Carballido JM et al. Carbohydrate-dependent, HLA class II-restricted, human T cell response to the bee venom allergen phospholipase A2 in allergic patients. Eur J Immunol 1995; 25(2):538-542.
63. Romain F, Horn C, Pescher P et al. Deglycosylation of the 45/47-kilodalton antigen complex of Mycobacterium tuberculosis decreases its capacity to elicit in vivo or in vitro cellular immune responses. Infect Immun 1999; 67(11):5567-5572.
64. Golgher D, Korangy F, Gao B et al. An immunodominant MHC class II-restricted tumor antigen is conformation dependent and binds to the endoplasmic reticulum chaperone, calreticulin. J Immunol 2001; 167(1):147-155.
65. Sallusto F, Cella M, Danieli C et al. Dendritic cells use macropinocytosis and the mannose receptor to concentrate macromolecules in the major histocompatibility complex class II compartment: downregulation by cytokines and bacterial products. J Exp Med 1995; 182(2):389-400.
66. Housseau F, Moorthy A, Langer DA, Robbins PF, Gonzales MI, Topalian SL. N-linked carbohydrates in tyrosinase are required for its recognition by human MHC class II-restricted CD4(+) T cells. Eur J Immunol 2001; 31(9):2690-701.
67. Harding CV, Roof RW, Allen PM et al. Effects of pH and polysaccharides on peptide binding to class II major histocompatibility complex molecules. Proc Natl Acad Sci U S A 1991; 88(7):2740-4.
68. Berd D. Autologous, hapten-modified vaccine as a treatment for human cancers. Vaccine 2001; 19(17-19):2565-70.

69. Kastrup IB, Andersen MH, Elliott T et al. MHC-restricted T cell responses against posttranslationally modified peptide antigens. Adv Immunol 2001; 78:267-89.
70. Deck MB, Sjolin P, Unanue ER, Kihlberg J. MHC-restricted, glycopeptide-specific T cells show specificity for both carbohydrate and peptide residues. J Immunol 1999; 162(8):4740-4.
71. Otvos L, Jr., Krivulka GR, Urge L et al. Comparison of the effects of amino acid substitutions and beta-N- vs. alpha-O-glycosylation on the T-cell stimulatory activity and conformation of an epitope on the rabies virus glycoprotein. Biochim Biophys Acta 1995; 1267(1):55-64.
72. Jensen T, Galli-Stampino L, Mouritsen S et al. T cell recognition of Tn-glycosylated peptide antigens. Eur J Immunol 1996; 26(6):1342-9.
73. Springer GF. Immunoreactive T and Tn epitopes in cancer diagnosis, prognosis, and immunotherapy. J Mol Med 1997; 75(8):594-602.
74. Jensen T, Hansen P, Galli-Stampino L et al. Carbohydrate and peptide specificity of MHC class II-restricted T cell hybridomas raised against an O-glycosylated self peptide. J Immunol 1997; 158(8):3769-78.
75. Jensen T, Nielsen M, Gad M et al. Radically altered T cell receptor signaling in glycopeptide-specific T cell hybridoma induced by antigen with minimal differences in the glycan group. Eur J Immunol 2001; 31(11):3197-206.
76. Chicz RM, Urban RG, Gorga JC et al. Specificity and promiscuity among naturally processed peptides bound to HLA-DR alleles. J Exp Med 1993; 178(1):27-47.
77. Chicz RM, Lane WS, Robinson RA et al. Self-peptides bound to the type I diabetes associated class II MHC molecules HLA-DQ1 and HLA-DQ8. Int Immunol 1994; 6(11):1639-49.
78. Kastrup IB, Stevanovic S, Arsequell G et al. Lectin purified human class I MHC-derived peptides: evidence for presentation of glycopeptides in vivo. Tissue Antigens 2000; 56(2):129-35.
79. Dustin ML, McCourt DW, Kornfeld S. A mannose 6-phosphate-containing N-linked glycopeptide derived from lysosomal acid lipase is bound to MHC class II in B lymphoblastoid cell lines. J Immunol 1996; 156(5):1841-7.
80. Roquemore EP, Chou TY, Hart GW. Detection of O-linked N-acetylglucosamine (O-GlcNAc) on cytoplasmic and nuclear proteins. Methods Enzymol 1994; 230:443-60.
81. Corthay A, Backlund J, Broddefalk J et al. Epitope glycosylation plays a critical role for T cell recognition of type II collagen in collagen-induced arthritis. Eur J Immunol 1998; 28(8):2580-90.
82. Zhao XJ, Cheung NK. GD2 oligosaccharide: target for cytotoxic T lymphocytes. J Exp Med 1995; 182(1):67-74.
83. Speir JA, Abdel-Motal UM, Jondal M, Wilson IA. Crystal structure of an MHC class I presented glycopeptide that generates carbohydrate-specific CTL. Immunity 1999; 10(1):51-61.
84. Fremont DH, Matsumura M, Stura EA et al. Crystal structures of two viral peptides in complex with murine MHC class I H-2Kb. Science 1992; 257(5072):919-27.
85. Abdel-Motal UM, Berg L, Rosen A et al. Immunization with glycosylated Kb-binding peptides generates carbohydrate-specific, unrestricted cytotoxic T cells. Eur J Immunol 1996; 26(3):544-51.
86. Svensson G, Albertsson J, Svensson C et al. X-ray crystal structure of galabiose, O-alpha-D-galactopyranosyl-(1—4)-D-galactopyranose. Carbohydr Res 1986; 146(1):29-38.
87. Glithero A, Tormo J, Haurum JS et al. Crystal structures of two H-2Db/glycopeptide complexes suggest a molecular basis for CTL cross-reactivity. Immunity 1999; 10(1):63-74.
88. Apostolopoulos V, McKenzie IF, Wilson IA. Getting into the groove: unusual features of peptide binding to MHC class I molecules and implications in vaccine design. Front Biosci 2001; 6:D1311-20.
89. Catipovic B, Talluri G, Oh J et al. Analysis of the structure of empty and peptide-loaded major histocompatibility complex molecules at the cell surface. J Exp Med 1994; 180(5):1753-61.
90. Elliott T. How does TAP associate with MHC class I molecules? Immunol Today 1997; 18(8):375-9.
91. Jensen T, Hansen P, Faurskov Nielsen A, Meldal M, Komba S, Werdelin O. Shared structural motifs in TCR of glycopeptide-recognizing T cell hybridomas. Eur J Immunol 1999; 29(9):2759-68.
92. Lanzavecchia A. How can cryptic epitopes trigger autoimmunity? J Exp Med 1995; 181(6):1945-8.
93. Gallimore A, Hombach J, Dumrese T et al. A protective cytotoxic T cell response to a subdominant epitope is influenced by the stability of the MHC class I/peptide complex and the overall spectrum of viral peptides generated within infected cells. Eur J Immunol 1998; 28(10):3301-11.
94. Malhotra R, Wormald MR, Rudd PM et al. Glycosylation changes of IgG associated with rheumatoid arthritis can activate complement via the mannose-binding protein. Nat Med 1995; 1(3):237-243.
95. Anderson RP, Degano P, Godkin AJ, Jewell DP, Hill AV. In vivo antigen challenge in celiac disease identifies a single transglutaminase-modified peptide as the dominant A-gliadin T-cell epitope. Nat Med 2000; 6(3):337-342.

96. Manoury B, Mazzeo D, Fugger L, Viner N, Ponsford M, Streeter H et al. Destructive processing by asparagine endopeptidase limits presentation of a dominant T cell epitope in MBP. Nat Immunol 2002; 3(2):169-74.
97. McAdam SN, Fleckenstein B, Rasmussen IB et al. T cell recognition of the dominant I-A(k)-restricted hen egg lysozyme epitope: critical role for asparagine deamidation. J Exp Med 2001; 193(11):1239-46.
98. Boon T, Coulie PG, Van den Eynde B. Tumor antigens recognized by T cells. Immunol Today 1997; 18(6):267-8.
99. Feizi T. Carbohydrate antigens in human cancer. Cancer Surv 1985; 4(1):245-69.
100. Kim YJ, Varki A. Perspectives on the significance of altered glycosylation of glycoproteins in cancer. Glycoconj J 1997; 14(5):569-76.
101. Takahashi T, Makiguchi Y, Hinoda Y, Kakiuchi H, Nakagawa N, Imai K et al. Expression of MUC1 on myeloma cells and induction of HLA-unrestricted CTL against MUC1 from a multiple myeloma patient. J Immunol 1994; 153(5):2102-9.
102. Jerome KR, Barnd DL, Bendt KM, Boyer CM, Taylor-Papadimitriou J, McKenzie IF et al. Cytotoxic T-lymphocytes derived from patients with breast adenocarcinoma recognize an epitope present on the protein core of a mucin molecule preferentially expressed by malignant cells. Cancer Res 1991; 51(11):2908-2916.
103. Barratt-Boyes SM. Making the most of mucin: a novel target for tumor immunotherapy. Cancer Immunol Immunother 1996; 43(3):142-151.
104. Apostolopoulos V, Osinski C, McKenzie IF. MUC1 cross-reactive Gal alpha(1,3)Gal antibodies in humans switch immune responses from cellular to humoral. Nat Med 1998; 4(3):315-320.
105. Zarling AL, Ficarro SB, White FM et al. Phosphorylated peptides are naturally processed and presented by major histocompatibility complex class I molecules in vivo. J Exp Med 2000; 192(12):1755-1762.
106. Crotzer VL, Christian RE, Brooks JM et al. Immunodominance among EBV-derived epitopes restricted by HLA-B27 does not correlate with epitope abundance in EBV-transformed B-lymphoblastoid cell lines. J Immunol 2000; 164(12):6120-6129.
107. Hirabayashi J, Arata Y, Kasai K. Glycome project: concept, strategy and preliminary application to Caenorhabditis elegans. Proteomics 2001; 1(2):295-303.
108. Gupta R, Jung E, Gooley AA et al. Scanning the available Dictyostelium discoideum proteome for O-linked GlcNAc glycosylation sites using neural networks. Glycobiology 1999; 9(10):1009-22.
109. Chui D, Sellakumar G, Green R et al. Genetic remodeling of protein glycosylation in vivo induces autoimmune disease. Proc Natl Acad Sci USA 2001; 98(3):1142-47.
110. Aronson NN, Jr., Kuranda MJ. Lysosomal degradation of Asn-linked glycoproteins.Faseb 1989; 3(14):2615-2622.

Chapter 12

Chemical Synthesis of Bacterial Carbohydrates

Vince Pozsgay

Introduction

Carbohydrates, in the form of complex oligo- and polysaccharides (PSs), are ubiquitous components of the cell surface of most human pathogenic bacteria. Major groups are: capsular polysaccharide (CPS), teichoic acid (C-polysaccharide or C-antigen), lipoteichoic acid (Forssman or F-antigen), lipopolysaccharide (LPS), and lipooligosaccharide (LOS). Medical interest in bacterial polysaccharides dates back to the early work of Heidelberger and Avery who showed that the "type-specific substance" in culture filtrates of pneumococci was a polysaccharide that was immunogenic in animals.[1] This was the first instance to demonstrate that a nonproteinaceous material can be immunogenic. Subsequent work established that each pneumococcal type carries its own type-specific capsular material that alone defines serological specificity. This is possible not only because of the potentially large number of structures that can be constructed even from a few monosaccharides but also because of the enormous structural variability of monosaccharides found in bacteria compared to those in mammals whose oligo- and polysaccharides are constructed of only nine monosaccharides. The diversity is increased by the numerous noncarbohydrate substituents (eg. acetyl, pyruvate acetal, methyl) attached to the saccharides' functional groups. In many bacteria, for example in Gram-positive pneumococci[2] and in Gram-negative bacteria such as *Haemophilus influenzae* type b and *Salmonella typhi*, the expression of capsules is essential for virulence. In addition to being a virulence factor, CPS provides mechanical protection and guards the cell against the intracellular osmotic pressure. Moreover, it impairs phagocytosis and activates the alternate pathway of complement activation.[3] CPSs have a high molecular mass (up to 10^6 daltons) and consist of regular structures in that they are polymers of linear or branched oligosaccharides (repeating units) consisting of up to seven monosaccharide residues. The repeating units may be interconnected through the usual, O-glycosidic bonds or often through anomerically located phosphodiester linkages. Gram-negative bacteria may or may not have capsules but they invariably express highly complex LPS or LOS that are embedded in the cell membrane through their innermost domains referred to as Lipid A.[4] Structurally, this portion is a disaccharide consisting of two β-(1→6)-linked 2-amino-2-deoxy-D-glucose residues that carry lipid acyl groups. To this unit is attached a conserved region of some ten monosaccharide residues termed the core domain featuring KDO (3-deoxy-D-*manno*-2-ulosonic acid) and heptose moieties in addition to more common sugars such as galactose, N-acetyl-glucosamine and glucose. In fully developed LPS the outer part of the core region carries a structurally diverse polysaccharide termed O-specific polysaccharide (O-SP) that, similarly to CPS, is composed of oligosaccharide repeating units. The physical appearance of a LPS- producing organism is indicative of the developmental status of the O-SP: those that have a smooth (S) appearance have their O-SPs fully developed whereas those that lack the O-specific side chain appear irregular (R for rough). Increasing evidence tends to indicate that LOSs are not R-type LPSs but separate entities.[5] In striking similarity to the CPSs, only few bacteria are known that express the same O-SPs, e.g.,

Immunobiology of Carbohydrates, edited by Simon Y.C. Wong and Gemma Arsequell.
©2003 Eurekah.com and Kluwer Academic / Plenum Publishers.

Shigella sonnei and *Plesiomonas aeruginosa* O7. Therefore, just as with CPSs, the O-SPs alone define serotype specificity and are therefore diagnostic markers. Whilst LPSs (endotoxins) are immunogenic and with their Lipid A components being amongst the most toxic substances to humans, the O-SPs have no known pharmacological effect and are devoid of immunogenicity, presumably because of their low molecular weight (up to 30 kDa).

Purified CPSs elicit anti-polysaccharide antibodies in healthy adults that provide specific immunity against the homologous organism that may last for up to eight years.[6,7] Although a four-valent CPS vaccine against pneumococcal infections was licensed in the USA for medical use already in 1945 the overwhelming success of the newly discovered antibiotics in the fifties suppressed enthusiasm in preventive medicine. The unanticipated emergence of drug-resistant bacterial strains shortly thereafter revived interest in polysaccharide vaccines that led to the licensure of a 14-valent pneumococcal vaccine in 1978. The currently available vaccine, marketed under the trade-name Pnu Immune 23 (Wyeth Lederle), contains 25 μg of the CPS of each of the 23 most invasive pneumococci and offers approximately 90 % protection against all pneumococcal infections in immunocompetent adults in the United States. Reinjection is recommended after five years. Other CPS vaccines are licensed against groups A, C, W-135, and Y *Neisseria meningitidis* (Menomune, Aventis-Pasteur) and against *Salmonella typhi* causing typhoid fever (Typhim Vi, Aventis-Pasteur). The limitations of CPSs as vaccines are that, with few exceptions such as *Neisseria meningitidis* group A, they are poorly immunogenic in infants and young children under two years of age and produce no booster response (lack of immunological memory) in any age. Polysaccharide vaccines produce mainly IgM antibodies in addition to a smaller amount of IgG. Once induced, the IgM level usually drops rapidly and IgM antibodies do not affect bacterial carriage in the nasopharynx. In comparison, IgG antibodies persist for many years.

Contrary to proteins that are thymus-dependent (TD) immunogens and produce immune response at any age, the human immune response to polysaccharides is thymus-independent (TI). The TI antibody response is developmentally regulated and is poor until the age of two years. Avery and Goebel demonstrated in the thirties that the type 3 pneumococcal CPS could be converted to an efficient immunogen by coupling covalently with horse serum albumin.[8] This approach was based on the pioneering studies of Landsteiner who showed that nonimmunogenic small organic compounds can be rendered immunogenic by their covalent attachment to proteins. Avery and Goebel were the first to demonstrate that protein conjugates of oligosaccharides instead of polysaccharides could also elicit protective antibodies: rabbits immunized with a disaccharide fragment of type 3 pneumococcal CPS covalently linked to a carrier protein were protected against challenge by the homologous organism. These observations constitute the historical foundation of the concept of conjugate vaccine technology. Currently, commercial glycoconjugate vaccines are in medical use against infections caused by *Haemophilus* influenzae type b (Hib), *Neisseria meningitidis* type C, and *Streptococcus pneumoniae*.

The O-SPs of Gram-negative bacteria fulfill a role analogous to that of the CPSs in that they are essential for bacterial virulence in LPS-producing bacteria and express serotype specificity. Antibodies to the O-SPs of *Shigellae*,[9] and *E. coli*[10] were shown to confer serotype-specific immunity. Historically, this was first documented with protein conjugates of oligosaccharides obtained by specific degradation of the native O-SP of *Salmonella*.[11] Protein conjugates of detoxified (lipid free) LPSs or their O-SP domains of *Salmonella enteritica* have also been shown to elicit anti-O-SP antibodies that conferred protection against homologous organisms.[12]

Despite the overwhelming medical success of polysaccharide-protein conjugate vaccines, there are a number of impediments that need to be overcome for developing a successful glycoconjugate vaccine. These include (a) possibility of biological contamination, (b) difficulties in polysaccharide purification, (c) analytical (quality assurance) problems with both the polysaccharide and the conjugate, (d) possibility of residual toxicity with LPS-derived polysaccharides, (e) our ignorance of a correlate between structure and immunogenicity. The recognition that fragments of native polysaccharides as haptens in the form of protein conjugates may elicit polysaccharide-specific antibodies raised the hope that synthetic oligosaccharides of well-defined structure may replace polysaccharides as conjugate vaccine components. Potential advantages include: (a) characterization of OS haptens is possible by rigorous spectroscopic methods, (b) controlled structural modification is feasible with synthetic oligosaccharides, (c) defined conjugation schemes may be used, (d) study of the effect of structural variables (oligosaccharide size, identity of terminal saccharide, length and type of the linker, saccharide/protein molar ratio, number of attachment points on the protein) is possible.

Early studies on the antigenicity (recognition by antibodies) of oligosaccharides and on the immunogenicity (ability to induce an immune response) of their protein conjugates employed oligosaccharides obtained from the native polysaccharides by controlled acid hydrolysis or by phage-associated endoglycanase degradation.[13] Following the preparation in the sixties of unique dideoxy glycosides that are the immunodominant monosaccharides of *Salmonella* serogroups, chemical synthesis of OS fragments of bacterial PSs were initiated in the early seventies by the groups of Lindberg and Garegg. They prepared a range of disaccharides corresponding to *Salmonella* O-SPs having either simple alkyl aglycons or more complex linkers that made them suitable for covalent attachment to a protein carrier.[13] The last two decades have witnessed the development of a number of highly efficient glycosylation methods together with new protecting groups that allowed the chemical synthesis of medium-sized oligosaccharides on a routine basis. The synthesis of bacterial oligosaccharides above 20-mers also became possible.[14]

As pointed out recently by Kamerling in his comprehensive review of the chemical aspects of *Streptococcal* PSs, most of the synthetic pneumococcal PS fragments "were used in inhibition studies".[2] This is certainly true also for many of the chemically synthesized fragments of other CPSs and LPSs where the major goal was the mapping of the carbohydrate binding specificities of anti-PS (monoclonal) antibodies. The driving force of most of the current synthetic efforts towards oligosaccharides related to bacterial polysaccharides is at least twofold. First, it was recognized that they are essential reagents in the elucidation of the biological and physicochemical roles of cell-surface oligo- and polysaccharides. Second, increasing evidence indicates that protein conjugates of oligosaccharide fragments of the CPS's and O-SP's instead of the native polymers may be used as vaccines for the prevention of infections.

We have recently reviewed the synthesis and evaluation of the immunogenicity of homogeneous oligosaccharides of well-defined structure obtained either by chemical synthesis or by specific degradation.[13] Here we review chemical synthesis of bacterial oligosaccharides published from 1997 until the early part of 2002. When more than one oligosaccharide is described in a paper, a typical sequence is selected for inclusion here and reference is made to others. Some of the reaction schemes are presented in an abbreviated form and require comparisons with the reported transformations. The chapter is organized according to the alphabetical order of the bacteria from which the oligosaccharides synthesized are derived. Also included are references to their immunological evaluation. For earlier reviews of synthetic oligosaccharides related to selected bacterial polysaccharides the reader is referred to the papers of Bundle (O-antigens),[15] Kosma (core structures),[16] Kusomoto (Lipid A),[17] Lipták (mycobacterial oligosaccharides),[18] and Oscarson (core structures, phosphodiester-linked OSs, uronic acids).[19,20]

Synthesis of Bacterial Oligosaccharides

Acinetobacter

Acinetobacter are Gram-negative bacteria occurring in soil and water. Some are human pathogens showing increasing resistance to available antibiotics and are responsible for many hospital acquired infections, including skin infections, meningitis and pneumonia. A unique feature of the inner core region of LPS of the species *A. calcoaceticus* is the rare octulosonic acid **1** (KO) that differs from the usual KDO residue by the presence of a hydroxyl group at C-3. The inner core region of this bacterium contains α-(2→4)-linked KDO-KO disaccharide **8** synthesis of which was reported by Kosma and coworkers.[21,22] First, the KDO glycal **2** was prepared from

KDO and was oxidized with m-chloroperbenzoic acid to afford the epoxide **3**. Trans-diaxial ring opening with moist silica gel furnished the KO diol **4**. This was converted into the carbonate-protected allyl glycoside triol **5** in several steps involving protection of the anomeric hydroxyl group with an allyl group, deacetylation followed by reaction with the diphosgene-collidine reagent that introduced the carbonate moiety. Condensation of **5** with the KDO bromide donor **6** occurred exclusively at the HO-4 hydroxyl group to yield the α-(2→4)-linked disaccharide 7 in 18 % yield from which the allyl glycoside **8** was obtained after deacetylation (Zemplén) and saponification by sodium hydroxide. Conversion into the spacered ligand **9** was performed with cysteamine under UV irradiation.[22]

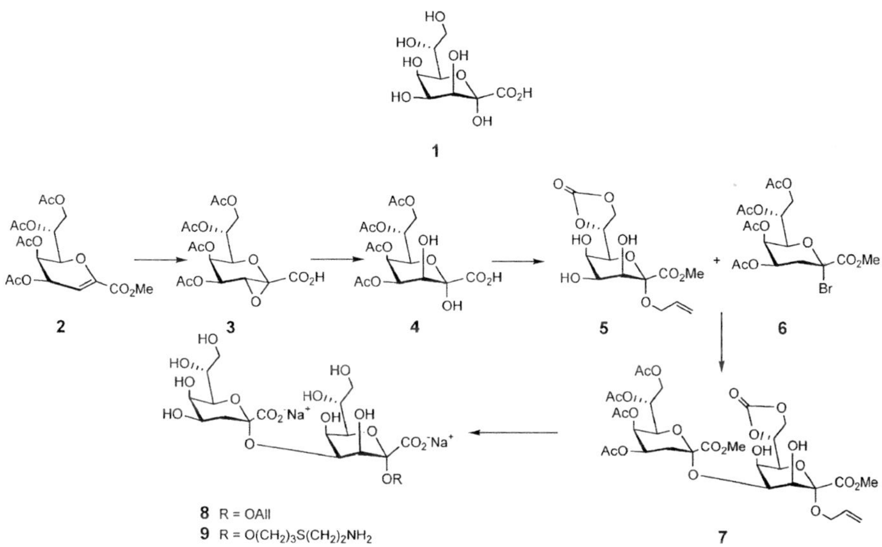

Arthrobacter sp CE-17

Arthrobacter is a soil bacterium whose exopolysaccharide consists of the double branched acidic octasaccharide repeating unit **10** bearing the rare pyruvate acetal moiety. The synthesis of the linear tetrasaccharide segment of this polysaccharide as its 5-aminopentyl glycoside **24** using two disaccharides **16** and **22** as key intermediates was reported by Lemanski and Ziegler.[23] The synthetic sequence employed here and several others described later successfully exploited the aglycon-dependent differential reactivities of thioglycosides, first examples of which were reported in a conference lecture by this author.[24] Starting from ethylthio mannoside **11**, phenylthio glucoside **12**, and an interconnecting moiety **13**, the tethered disaccharide **14** was prepared. First the tether unit **13** was linked to the alcohol **11** followed by removal of the tert-butyl protecting group. The intermediate acid so obtained was condensed with compound **12** in the presence of DCC to afford the interconnected dimer **14** in 81 % yield. Next, the acetal moiety in **14** was reductively opened using sodium cyanoborohydride/HCl and the resulting alcohol **15** (71 %) was subjected to intramolecular glycosylation using methyl triflate as the thiophilic activator to afford the bridged Manp-Glcp disaccharide donor **16** in 69 % yield. A starting compound for the second half (**21**) of the target tetrasaccharide was the diol **17** that was converted to the partially protected glycoside **18** by phase transfer-catalyzed benzylation in 53 % yield. The isomeric monobenzyl ether was obtained in 21 % yield. The donor **20** was prepared from the isopropylidene derivative **19** in a sequence involving chloroacetylation (81 %), acidic hydrolysis of the isopropylidene group (88 %), followed by benzoylation (83 %). Condensation of the alcohol **18** with the thiorhamnoside **20** was carried

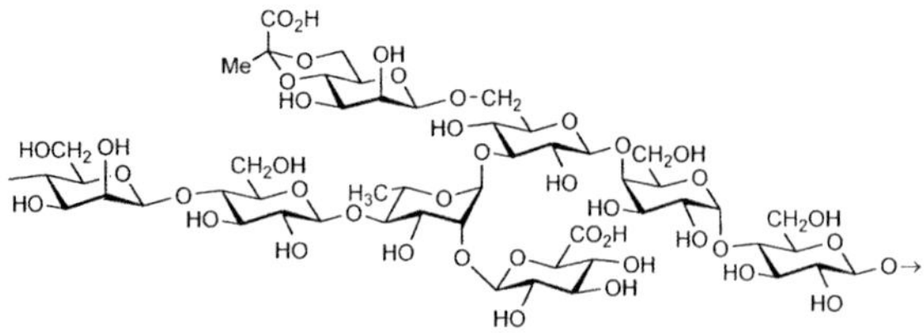

10

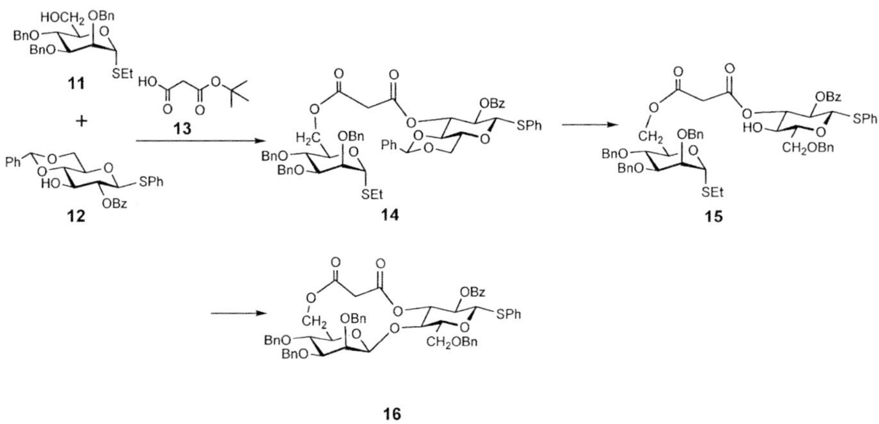

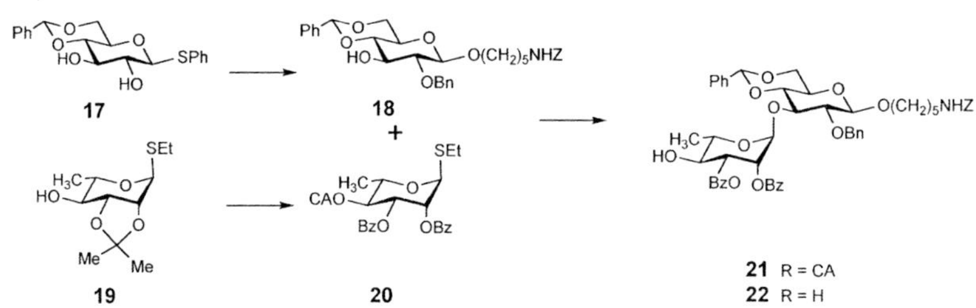

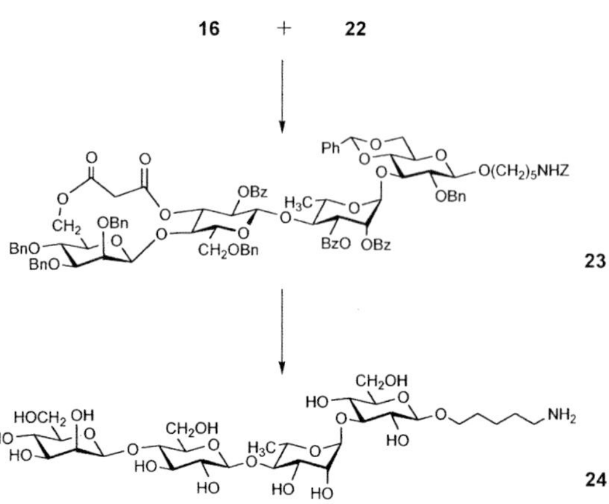

out with the NIS/TMSOTf reagent to afford the disaccharide **21** that was deprotected by treatment with thiourea to furnish the alcohol **22**.

Condensation of the tethered thiodisaccharide **16** and the acceptor **22** was successful with the NIS/TMSOTf reagent under rigorous temperature control to produce the fully protected tetrasaccharide **23** in 64 % yield. Several other methods of thioglycoside activation using NIS/TfOH, Br_2/AgOTf, and DMTST failed. Compound **23** was conventionally deprotected in a two-step sequence to afford **24** in which the amino group in the spacer portion may be used for covalent attachment to proteins.

Bordetella pertussis

Bordetella are Gram-negative, strictly aerobic coccobacilli that are among the most fastidious pathogens. *B. pertussis* is the causative agent of whopping cough and is pathogenic to humans only. While the entire structure of the LPS of this organism is unknown, the presence of the partial structure **25** was demonstrated, although without determining the absolute configurations of the monosaccharide residues. A synthesis of the spacer-linked acidic disaccharide fragment **35** was reported by Nilsson and Norberg.[25] Thioglucoside diol **17** was mono-tosylated and the product **26** (50 %) was subjected to base-treatment to afford the epoxide **27** from which the azido-altropyranoside **28** was prepared (82 % overall yield) using sodium azide for the trans-diaxial ring opening. Next, a second azido group was introduced at C-3 through triflate-mediated configurational inversion (→ **29**, 52 %) then the spacer moiety was attached using 2-(trifluoroacetamidophenyl)ethanol as the aglycon under DMTST activation followed by reductive opening ($NaCNBH_3$/HBF_4) of the benzylidene acetal ring to furnish the diazido-alcohol **30**.

GlcpNAc-(1→4)-ManpA-2,3-(NAc)$_2$-(1→3)-FucpNAcMe-(1→

25

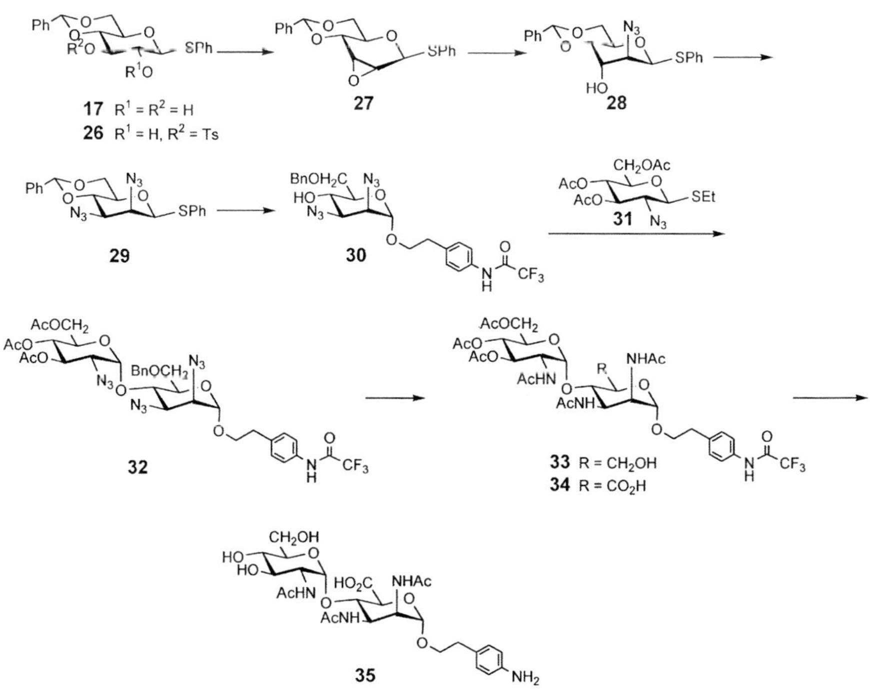

This was coupled with the azido-glucose derivative **31** in the presence of DMTST to give the disaccharide **32** together with the β-anomer in 86 % combined yield. In the final stages, the three azido groups were converted to acetamido functions using hydrogen sulfide/pyridine for reduction and acetic anhydride/pyridine for the N-acetylation followed by the removal of the benzyl protecting group by hydrogenolysis to afford the alcohol **33**. A two-step oxidation sequence involving treatment with DMSO/Ac_2O followed by iodine installed the carboxyl group in the reducing-end moiety (→**34**, 39 %). Inferior results were obtained with the use of the TEMPO and the pyridinium dichromate oxidation protocols. Treatment of the acid **34** with potassium hydroxide removed all the protecting groups to give the disaccharide as the 2-(p-aminophenyl)ethyl glycoside **35** suitable for covalent attachment to proteins.

Burholderia cepacia

First recognized as a phytopathogen, *B. cepacia* is now considered as an emerging opportunistic human pathogen, especially for hospitalized and immunocompromised patients suffering from cystic fibrosis. Interestingly, several strains of this bacterium are potential biological control agents. Using chemical approaches similar to that outlined above for *Acinetobacter calcoaceticus* Kosma's group reported the synthesis of disaccharides containing α-(2→ 4)-linked KDO and KO residues **8** and **36** as allyl glycosides corresponding to the inner core units of both *B. cepacia* and *Acinetobacter haemolyticus*. The allyl group was installed for covalent attachment of the saccharide constructs to proteins through their cysteine adducts **9** and **37** respectively.[22]

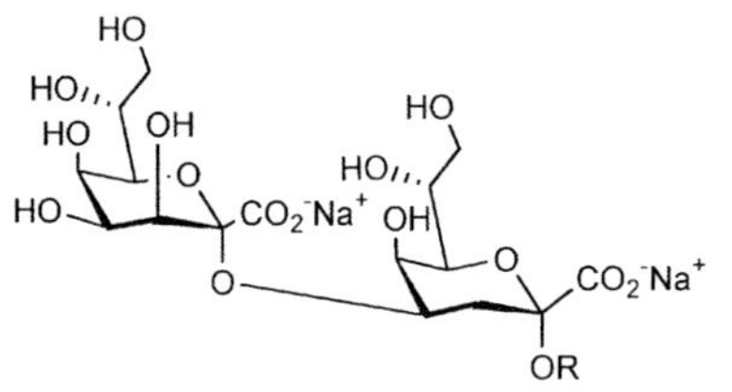

36 R = OAll
37 R = $O(CH_2)_3S(CH_2)_2NH_2$

Chlamydiae

The outer membranes of all chlamydial species contain a trimer of KDO that is linked to the Lipid A part. Following the synthesis of numerous derivatives of the KDO trimer, Kosma and coworkers described the preparation of deoxy analogues of (2→8)-linked KDO disaccharides.[26] For example, disaccharide **40** was obtained by condensation of the deoxy-donor **38** and the KDO allyl glycoside **39** in the presence of $Hg(CN)_2$ followed by removal of the protecting groups under Zemplén conditions then with sodium hydroxide. Under the glycosylation conditions elimination reactions also occurred to give, among other products, the unsaturated derivative **41**. Other disaccharides synthesized include propyl glycosides **42**, **43**, and **44**. For the synthesis of the *Chlamydia*-specific trisaccharide **45** see reference 21. Monoclonal antibodies were raised against a synthetic glycoconjugate prepared with the trisaccharide epitope **46** and their carbohydrate-binding specificities were evaluated.[27] Syntheses of tri- **47**, and tetrasaccharides **49** and **51** corresponding to *Chlamydophila psittaci* were also reported.[28] These oligosaccharides were converted to their cysteamine adducts **48**, **50**, and **52** that were coupled to bovine serum albumin. The neoglycoproteins so obtained were used to study the carbohydrate binding specificities of monoclonal antibodies raised against chlamydial LPS. The binding data indicate that the 8-substituted KDO residue is necessary for binding. The role of conformational differences between the oligosaccharides was emphasized.[28]

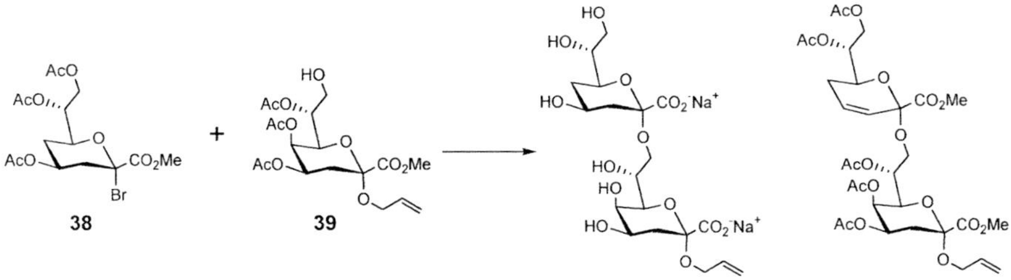

38 **39** **40** **41**

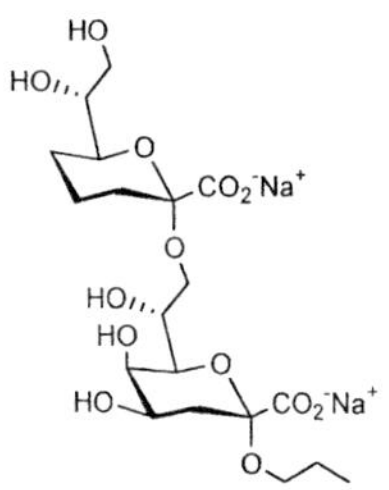

42

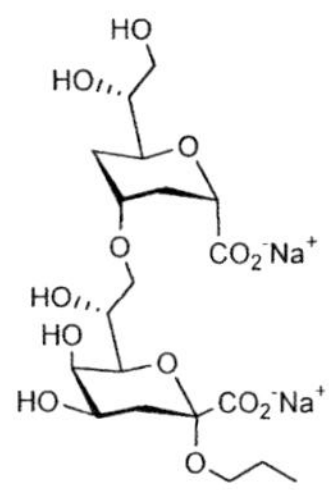

43

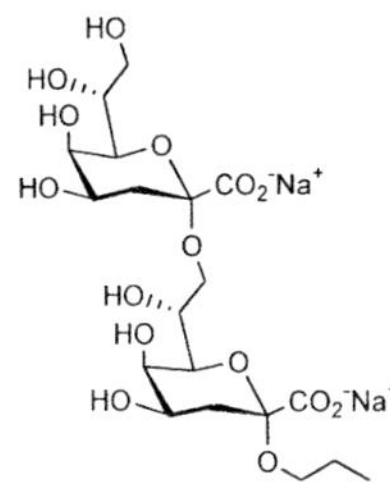

44

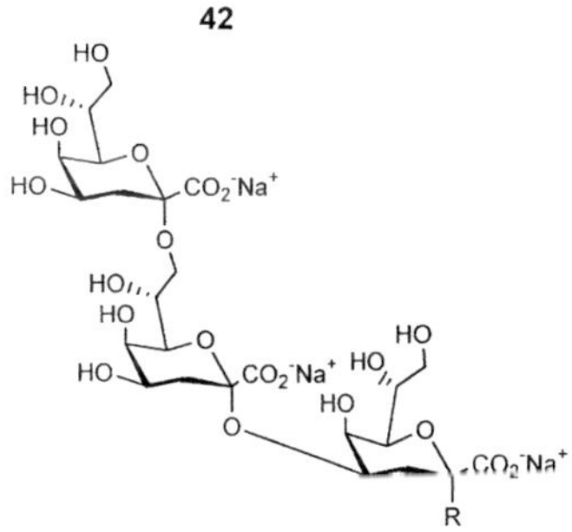

45 R = OAll
46 R = $O(CH_2)_3S(CH_2)_2NH_2$

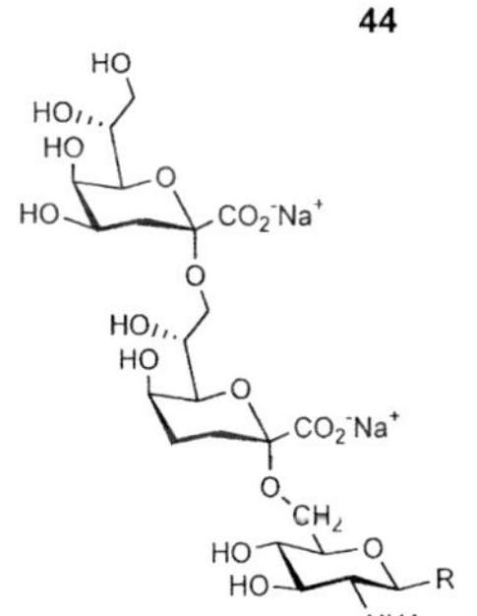

47 R = OAll
48 R = $O(CH_2)_3S(CH_2)_2NH_2$

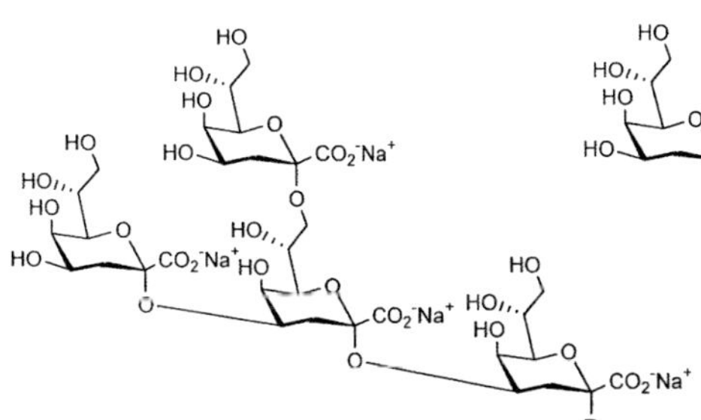

49 R = OAll
50 R = $O(CH_2)_3S(CH_2)_2NH_2$

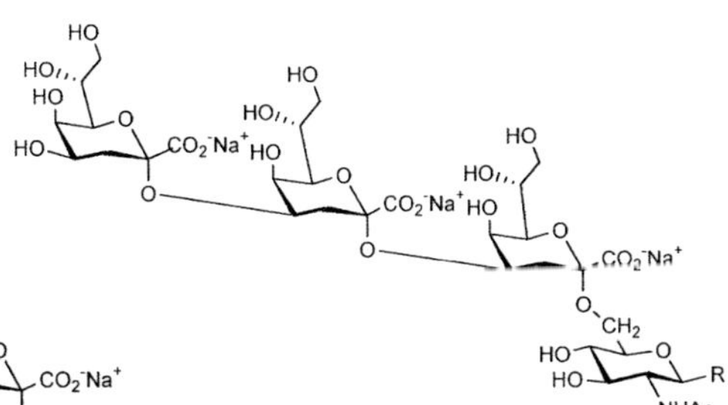

51 R = OAll
52 R = $O(CH_2)_3S(CH_2)_2NH_2$

Citrobacter freundii

Citrobacter are a genus of Gram-negative bacteria. Since *Citrobacter* colonies are part of the normal gut flora, they are not considered enteric pathogens, although *C. freundii* may cause diarrhea and other maladies. The O-SP of *C. freundii* O-28 (1c) consists of the trisaccharide repeating unit **53** containing one D-ribose and two L-rhamnose residues. Horooka and associates synthesized oligosaccharides related to **53** by employing a rarely used method of glycosylation, a key feature of which is the activation of glycose hemiacetals with TMSBr, $CoBr_2$, and Bu_4NBr.[29] For the synthesis of the complete repeating unit the key starting material **54** was converted to the allylated derivative **55** (83 %) that was hydrolyzed by the action of sulfuric acid in acetic acid at 85 °C to afford the hemiacetal **56** in 60 % yield. Acceptor **54** and the donor **56** were condensed in the presence of the activating reagents to furnish the disaccharide **57** in 50 % yield. This was subjected to the action of palladium chloride and the resulting propenyl derivative was hydrolyzed in acetate buffer. The alcohol **58** (81 %) obtained was condensed with the benzylated ribose hemiacetal **59** under conditions described above for the hemiacetal activation to yield the protected trisaccharide **60** in 45 % yield. Removal of the protecting groups by catalytic hydrogenolysis afforded the trisaccharide methyl glycoside **61**. The disaccharide methyl glycosides **62** and **63** were prepared in analogous reactions.

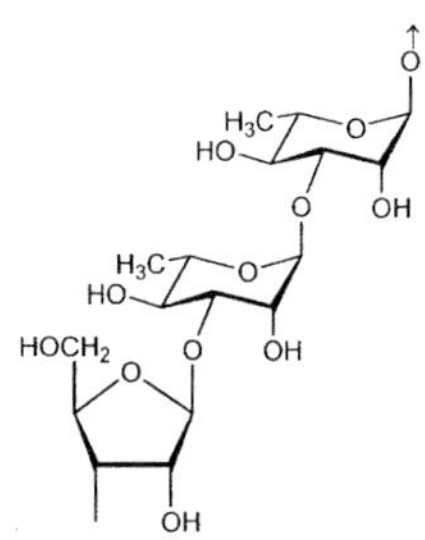

53

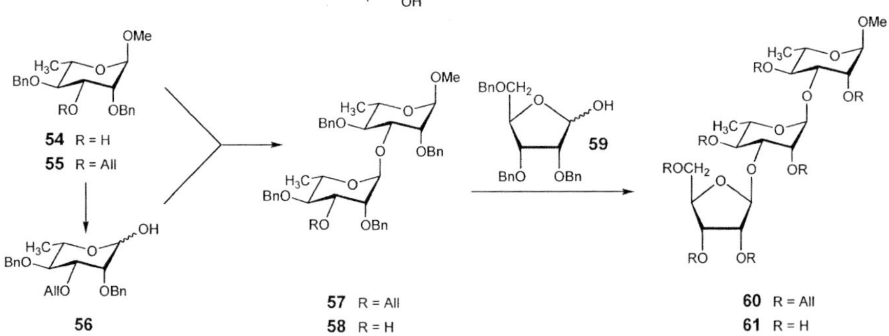

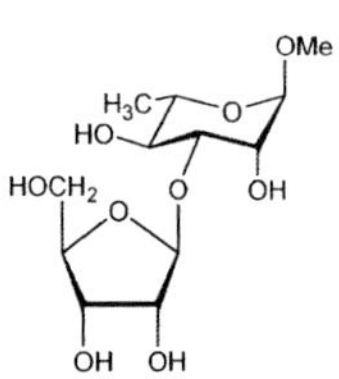

62

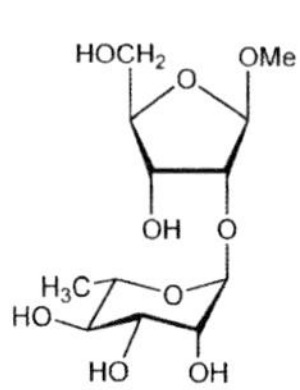

63

Cryptococcus neoformans

Cryptococci are opportunistic pathogens that are among the major causative organisms of death of AIDS patients. Although *Cryptococci* are fungi, and therefore would not qualify for inclusion in this review, they bear some similarities with encapsulated bacteria in that their polysaccharide capsules are important virulence factors and determine type specificity.

β-D-Xyl*p*
1
↓
2
α-D-Man*p*-(1→3)-α-D-Man*p*-(1→3)-α-D-Man*p*-(1→O)CH_3 **64**

β-D-Glc*p*A
1
↓
2
α-D-Man*p*-(1→3)-α-D-Man*p*-(1→3)-α-D-Man*p*-(1→O)CH_3 **65**

Garegg's group reported the synthesis of tetrasaccharide methyl glycosides **64** and **65** that represent partial structures of the capsular material of this bacterium. Key starting materials were the benzylidene acetal-protected thiomannoside **66** and the alcohol **67** that were condensed in the presence of DMTST to yield the fully protected disaccharide **68** in 87 % yield. Removal of the benzoyl groups under Zemplén conditions afforded the diol **69**. Next, it was converted to a stannylidene acetal intermediate by the action of dibutyltin oxide in methanol to enhance the nucleophilicity of HO-3'. The subsequent regioselective condensation with donor **66** occurred under activation by DMTST to afford trisaccharide in 40 % yield (from **68**). Also detected was the corresponding tetrasaccharide arising from bis-glycosylation at HO-2' and 3'. However, the 2'-linked trisaccharide was not observed. A further chain extension step, using the xylosyl bromide **71** as the donor and silver triflate as the activator gave the protected tetrasaccharide derivative **72** (79 %). Conversion into the target tetrasaccharide **64** was effected by the action of NaOMe in MeOH followed by hydrogenolysis (77 % in two steps).

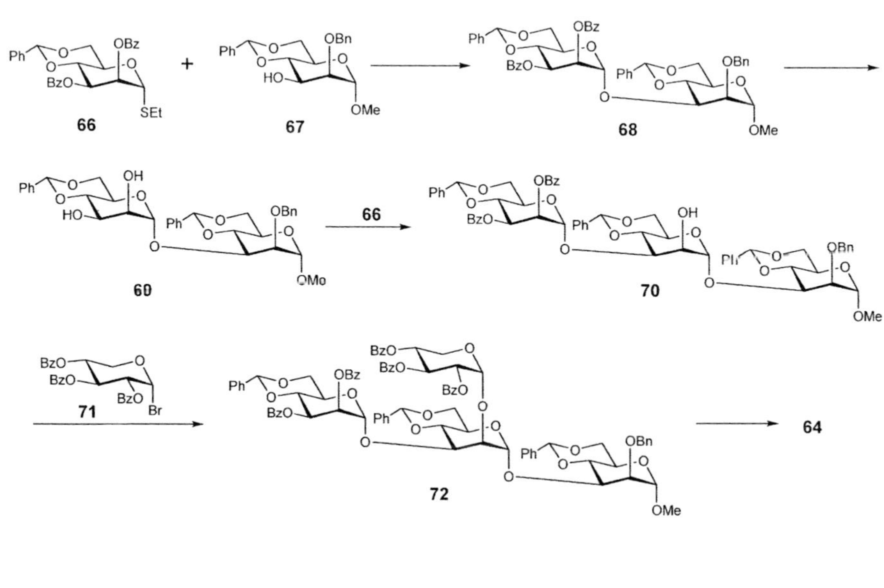

Using the trisaccharide alcohol **70** as the acceptor and the glucuronyl donor **73** as the donor under DMTST activation, tetrasaccharide **74** was obtained in 30 % yield.[30] Removal of the protecting groups to afford **75** required catalytic hydrogenolysis followed by Zemplén deacetylation and saponification with sodium hydroxide (68 % in three steps). In the native capsular polysaccharide of *Cryptococcus* the 6-position of the mannose residues is frequently acetylated. In order to make partially acetylated probes available, Garegg and associates synthesized the partially O-acetylated tetrasaccharides **76** and **77**. Thus, tetrasaccharide **72** was subjected to a series of reactions involving replacement of the benzoyl groups by benzyls followed by removal of the benzylidene acetal moieties (acetic acid), partial acetylation (acetyl chloride/collidine), and catalytic hydrogenolysis. Compounds **76** and **77** were isolated in 35 and 21 % yields, respectively, from **72**.

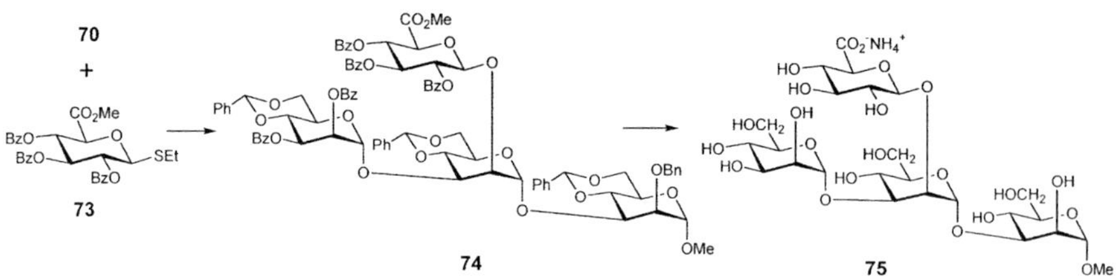

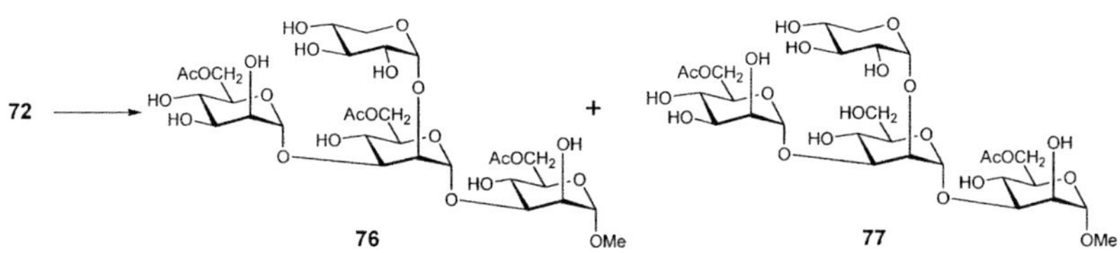

Escherichia coli

The first total synthesis of the Retype LPS (**90**) of *E. coli* was reported by Kusumoto's group in 2001. Lipotetrasaccharide **90** contains two glucosamine and two KDO residues in addition to six lipid chains and two phosphate residues one of which is anomerically located. The synthesis was conducted in a stepwise fashion and started by coupling the protected glucosamine trichloroacetimidate donor **78** with the acceptor **79** under promotion by the Lewis acid Nafion-TMS in a mixture of dichloromethane and perfluorohexane. Disaccharide **80** was obtained in 73 % yield. This was converted to the diol **81** that already has all four lipid chains in place in several steps including (i) acylation at OH-3 with the (trifluoromethyl)benzyl-protected fatty acid **93** in the presence of DCC, (ii) removal of the Troc group in the presence of zinc-copper alloy, (iii) N-acylation with (tetradecanoyloxy)tetradecanoic acid **91** (DCC), and (iv) removal of the benzylidene acetal group with aqueous trifluoroacetic acid. The combined yield for these transformations was 65 %. Next, the diol **81** was regioselectively silylated at HO-6 with triethylsilyl chloride to increase the nucleophilicity of the O-6 atom. Exposure of the silylated disaccharide **82** to KDO fluoride **83** in the presence of boron trifluoride etherate afforded the trisaccharide **84** in 89 % yield. Subsequently, the phosphate moiety was installed at O-4 of the middle residue using the reagent **85** followed by oxidation with MCPBA in 96 % (for two steps) and desilylation with hydrogen fluoride (89 %). The diol **86** so obtained was regioselectively triethylsilylated at O-4 of the KDO residue using triethylsilyl chloride, to enhance its nucleophilicity at this site in 85 % to afford the acceptor **87** for the next glycosylation step. Its reaction with the isopropylidene-protected KDO donor **88** in the presence of boron trifluoride etherate yielded the tetrasaccharide **89** in 75 % yield. The use of donor **83** instead of **88** gave a much lower yield, presumably as a consequence of unfavorable steric interactions.

Next, the isopropylidene group was removed by the action of trifluoroacetic acid and the allyl groups were cleaved by a cationic iridium complex then treatment with iodine. Subsequent phosphorylation at the anomeric position of the reducing end moiety was carried out by using tetrabenzyl diphosphate and lithium bistrimethylsilylamide. Finally, the benzyl protecting groups were removed by hydrogenolysis to afford glycolipid **90**. The synthesis of additional Lipid A derivatives may be found in reference 31.

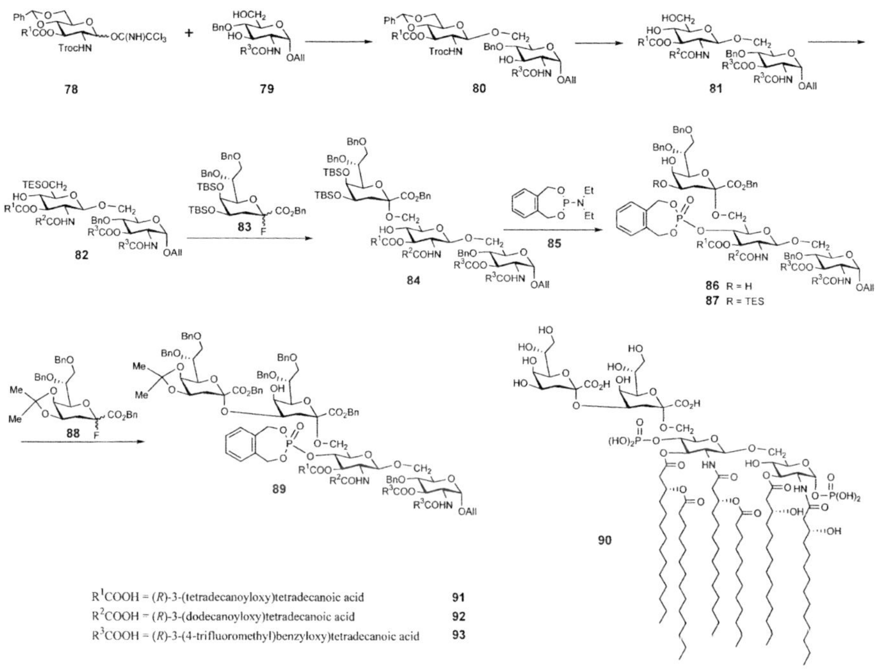

R^1COOH = (*R*)-3-(tetradecanoyloxy)tetradecanoic acid **91**
R^2COOH = (*R*)-3-(dodecanoyloxy)tetradecanoic acid **92**
R^3COOH = (*R*)-3-(4-trifluoromethyl)benzyloxy)tetradecanoic acid **93**

E. coli K12

The synthesis of di- and trisaccharides corresponding to the nonreducing terminus of the *E. coli* K-12 LPS core oligosaccharide was reported by Zamojski's group as allyl glycosides **101** and **107**.[32] In case of the disaccharide **101** the thioglucoside **94** and the L-*glycero*-D-*manno* heptoside **95** were condensed in the presence of methyl triflate to afford the benzyl bioside **96** in 74 % yield. This was debenzylated by catalytic hydrogenolysis over palladium-on-charcoal followed by O-acetylation with acetic anhydride to afford peracetate **97**. Exposure of **97** to the action of hydrazine in DMF selectively removed the anomeric acetyl group to yield the hemiacetal **98** in 52 % overall yield. Compound **98** was converted to the trichloroacetimidate **99** with trichloroacetonitrile in the presence of potassium carbonate. Subsequent reaction with allyl alcohol using tin tetrachloride as the promoter produced the allyl glycoside **100** in 67 % yield. The phthalimido group in this intermediate was converted to an acetamido group using ethylenediamine in butanol followed by acetylation with acetic anhydride. In the final step the O-acetyl groups were removed by exposure to sodium bicarbonate in methanol to yield the unprotected disaccharide **101**. Precursors for the trisaccharide **107** were the allyl glycoside **102** and the heptosyl trichloroacetimidate **103** that were combined in the presence of trimethylsilyl triflate to afford the disaccharide **104**. The partially protected derivative **105** was obtained in four steps involving (i) removal of all acetyl groups, (ii) silylation at O-6' with tert-butyldimethylsilyl chloride in pyridine (iii) acetylation with acetic anhydride and (iv) desilylation with hydrogen fluoride in a combined yield of 75 %.

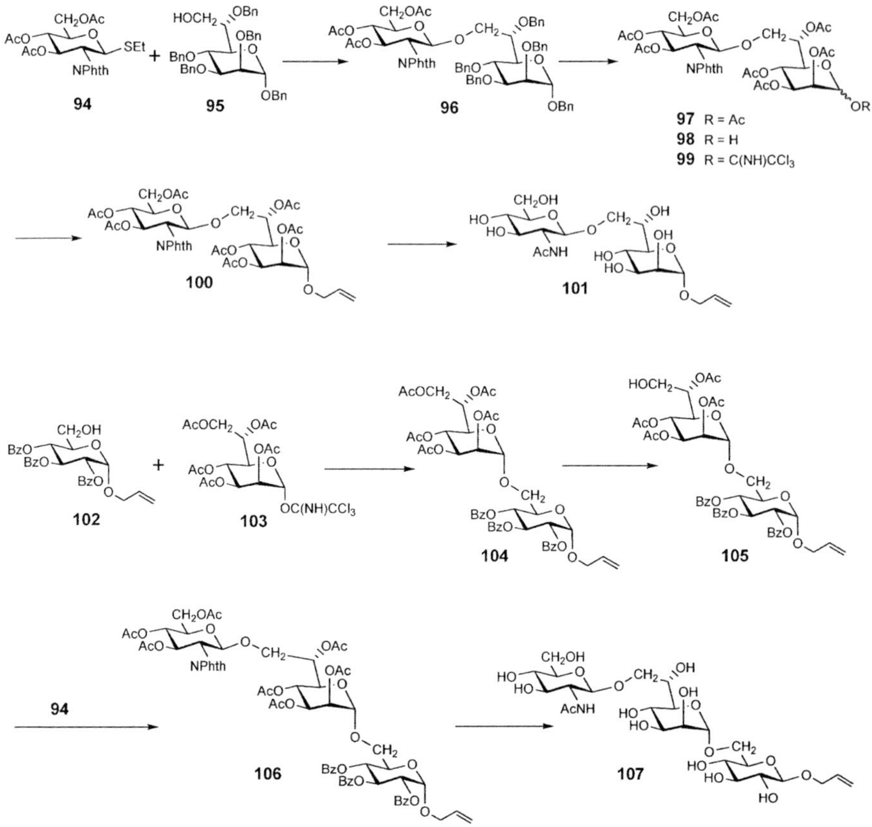

Conventional desilylation with tetrabutylammonium fluoride gave inferior results due to acyl migration. The alcohol **105** was glycosylated with the glucosamine donor **94** as above to yield the protected trisaccharide **106** (70 %) that was deprotected as outlined above (→**107**). The allyl glycosides **101** and **107** were derivatized with cysteamine under radical conditions and the constructs so obtained were coupled to bovine serum albumin through isothiocyanate intermediates. Whilst several monoclonal antibodies failed to recognize the glycoconjugates containing terminally located glucosamine residues, these residues were recognized by a polyclonal antibody pool raised against the complete core *E. coli* K-12 LPS indicating the presence of antibodies with specificities to terminal glucosamine-containing epitopes.

E. coli O35

The O-antigenic polysaccharide of this organism is composed of a hexasaccharide repeating unit containing the branching trisaccharide **110** that constitutes the nonreducing end of the polysaccharide.[33] The key monosaccharide precursors towards **110** were compounds **111**, **112**, and **114**. Thus, condensation of the thioglucoside **111** with the rhamnose diol **112** under NIS/AgOTf activation afforded, predominantly, the β-(1→3)-linked disaccharide **113** in 59 % yield. Subsequent glycosylation with the azido-galactose imidate **114** using TMSOTf as the activator yielded the trisaccharide **115** in high stereoselectivity, in 90 % yield. To replace the phthalimido for an acetamido group, compound **115** was first treated with ethylenediamine, then the product was acetylated with acetic anhydride. Next, the azido group was converted to an acetamido group by the action of thioacetic acid in pyridine (90 %). Subsequent de-O-acetylation afforded the trisaccharide **116** that, upon treatment with 2M HCl gave the

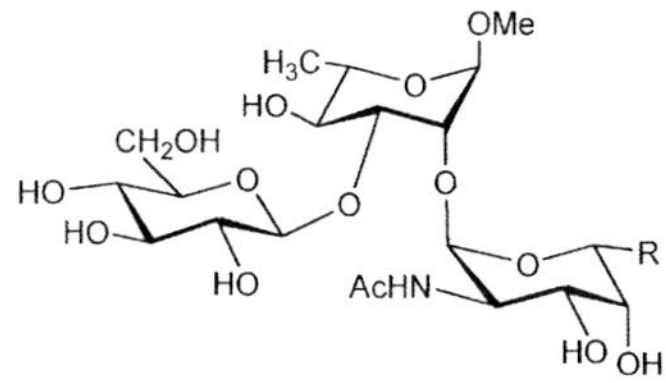

108 R = CH_2OH
109 R = COOH
110 R = $CONH_2$

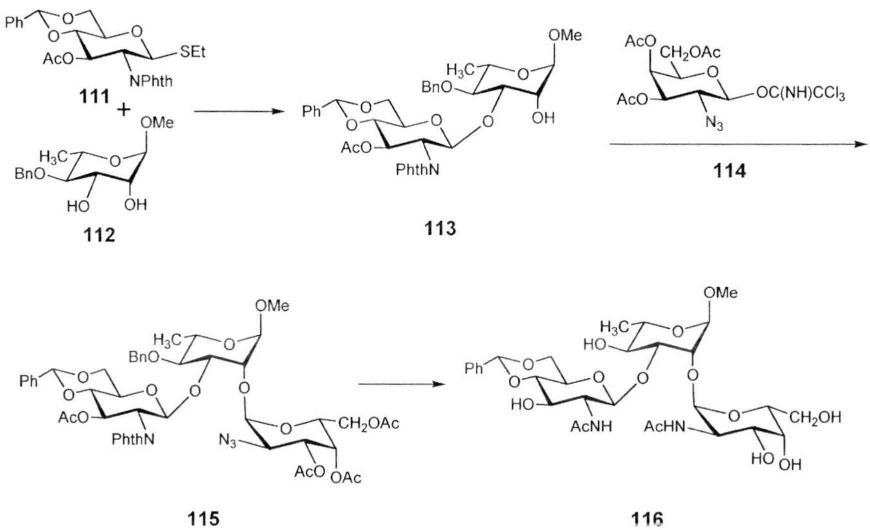

unprotected trisaccharide **108**. Alternatively, the acetal-protected trisaccharide **116** was treated with the TEMPO/NaOCl reagent followed by removal of the benzylidene moiety to afford the acidic trisaccharide **109** in 80 % combined yield. The trisaccharide carboxamide **110** was prepared from the acid **109** by reaction with ammonia in the presence of a peptide-coupling reagent in 90 % yield.

E. coli O55

Compound **660** in the section on *Vibrio cholera* O139 is also part of the O-antigen of *E. coli* O55.[34]

E. coli O101

The polysaccharide of *E. coli* O101:K103:H$^-$ is composed of the trisaccharide repeating unit **117**. A key feature of this polysaccharide is the highly labile pyruvic acid ketal formed on the cis-vicinal hydroxyl groups of a galactose residue. The synthesis of the trisaccharide analog **128** from monosaccharide precursors was reported by Roy's group.[35] Condensation of the thiogalactoside donor **118** and the rhamnoside acceptor **119** under activation by cupric bromide and tetrabutylammonium bromide afforded the disaccharide **120** in 80 % yield. Next, the methoxybenzyl groups of the galactose moiety were replaced by acetyls by oxidation with DDQ followed by acetylation to afford **121**. This was subjected to reductive acetal ring opening using sodium cyanoborohydride and hydrogen chloride to yield the alcohol **122**, having HO-4' unprotected, in 79 % combined yield. Galactosylation of the disaccharide **122** with the thiogalactoside donor **123** using methyl triflate as the activator produced the trisaccharide **124**

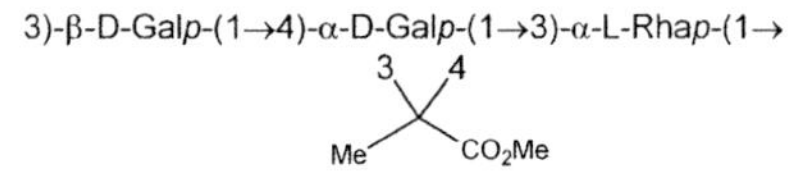

117

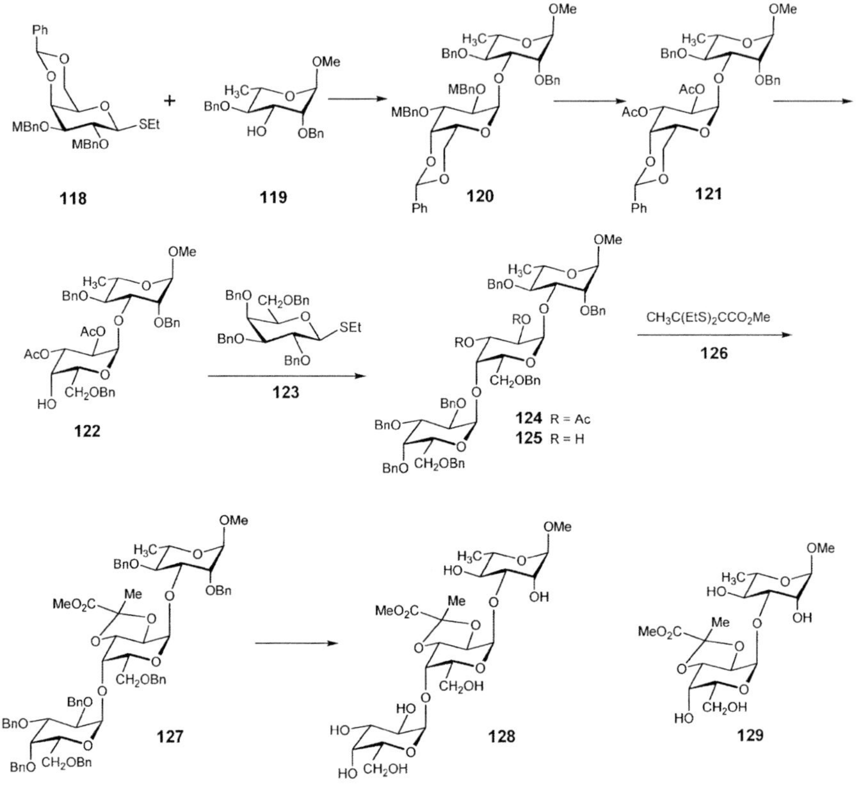

(79 %). Treatment of **124** under Zemplén conditions afforded the diol **125**. This was reacted with diethylthio-propionate **126** to produce the pyruvate intermediate **127** in 51 % yield, using triflic acid/sulfuryl chloride as the activator for such ketalation as suggested by Lipták.[36] Removal of the benzyl protecting groups by hydrogenolysis afforded the target trisaccharide ketal **128**. The synthesis of the disaccharide **129** has also been described.[35]

E. coli O126

The repeating unit of the O-SP of this bacterium consists of the branched tetrasaccharide **130**. Disaccharides related to this PS were prepared by Sengupta's group.[37] Starting from the acetal-protected D-galactoside **132** and the benzylated L-thiofucoside **131**, the disaccharide fragment **133** was obtained under promotion by the NIS/TfOH reagent followed by deprotection in several steps. Using similar chemistry, the stereoisomers **134**, **135**, and **136** have also been prepared. The synthesis of the tetrasaccharide **137** corresponding to a frame-shifted analog of **130** was reported by Roy's group.[38]

β-L-Fuc*p*
1
↓
2
2)-β-D-Man*p*-(1→3)-α-D-Gal*p*-(1→3)-β-D-Glc*p*NAc-(1→ **130**

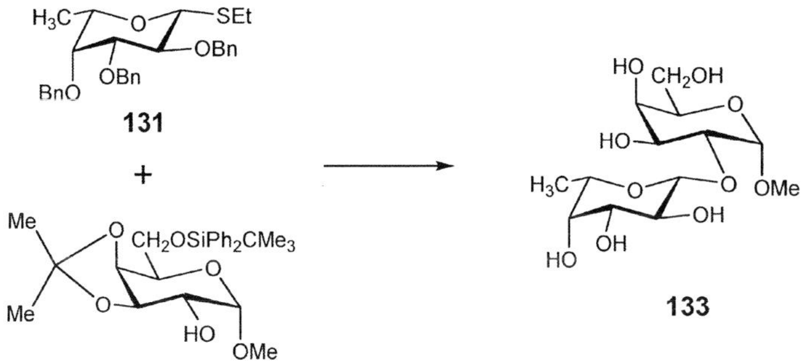

131 + **132** → **133**

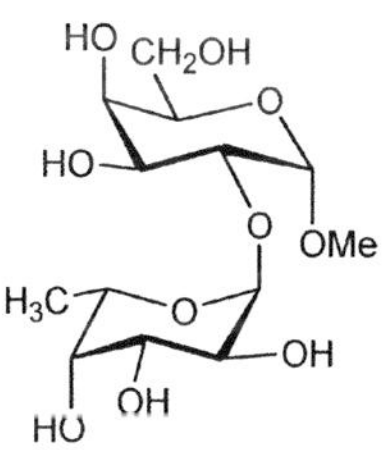

134

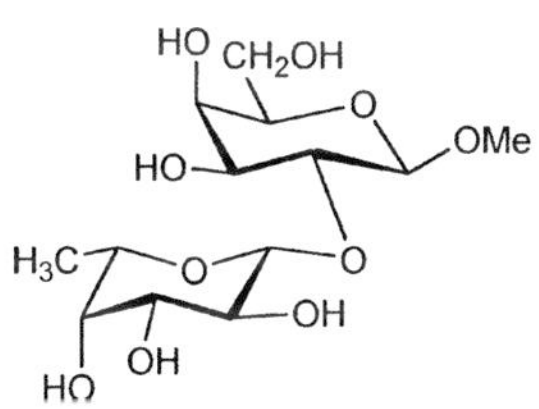

135

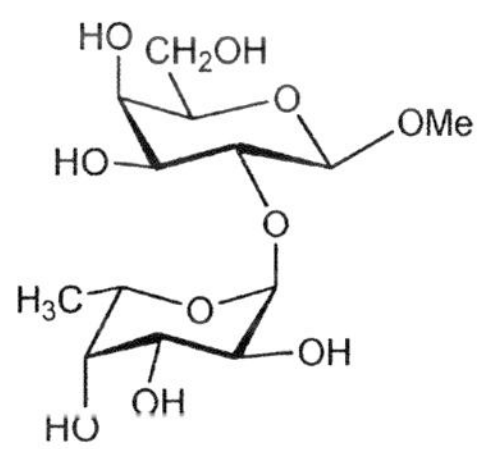

136

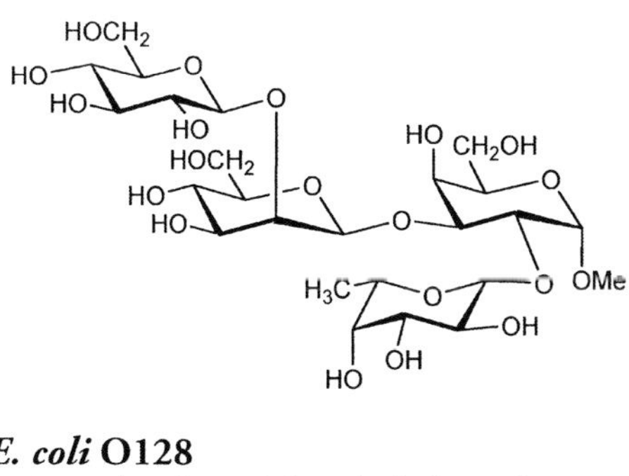

137

E. coli O128

Disaccharide **136** described above also represents the branchpoint of the O-SP of *E. coli* O128.[37]

Haemophilus ducreyi

The LPS of this bacterium contains a linear hexasaccharide corresponding to **146** that contains both D- and L-*glycero*-α-D-*manno*-heptopyranosyl residues. The synthesis of **146** was reported by Oscarson's group.[39] Using the phthalimido glucose **138** as the acceptor and the galactosyl bromide **139** as the donor, the lactosamine derivative **140** was assembled under promotion by silver triflate in 93 % yield. The thioglycoside donor **140** was then coupled with the galactose acceptor **141** in the presence of NIS/TfOH to afford the trisaccharide **142** as a trimethylsilylethyl glycoside in 70 % yield. This intermediate was converted to the anomeric acetate **143** in one step by the action of acetic anhydride and boron trifluoride etherate. Next, trisaccharide **143** was converted to the thioglycoside donor derivative **144** (73 %) using ethanethiol as the thiol source and BF_3-etherate as the promoter. The thiotrisaccharide donor **144** was coupled with the trisaccharide acceptor **145** containing two heptose residues separated by a glucose moiety under NIS/TfOH activation to produce the fully protected hexasaccharide **146** as a trifluoroacetamidophenyl glycoside in 45 % yield. The aglycon makes this assembly suitable for conjugation to immunogenic carriers. In agreement with earlier observations, higher yields were observed in glycosylation reactions involving acetamido sugars when the acetamido groups were replaced by diacetamido moieties.

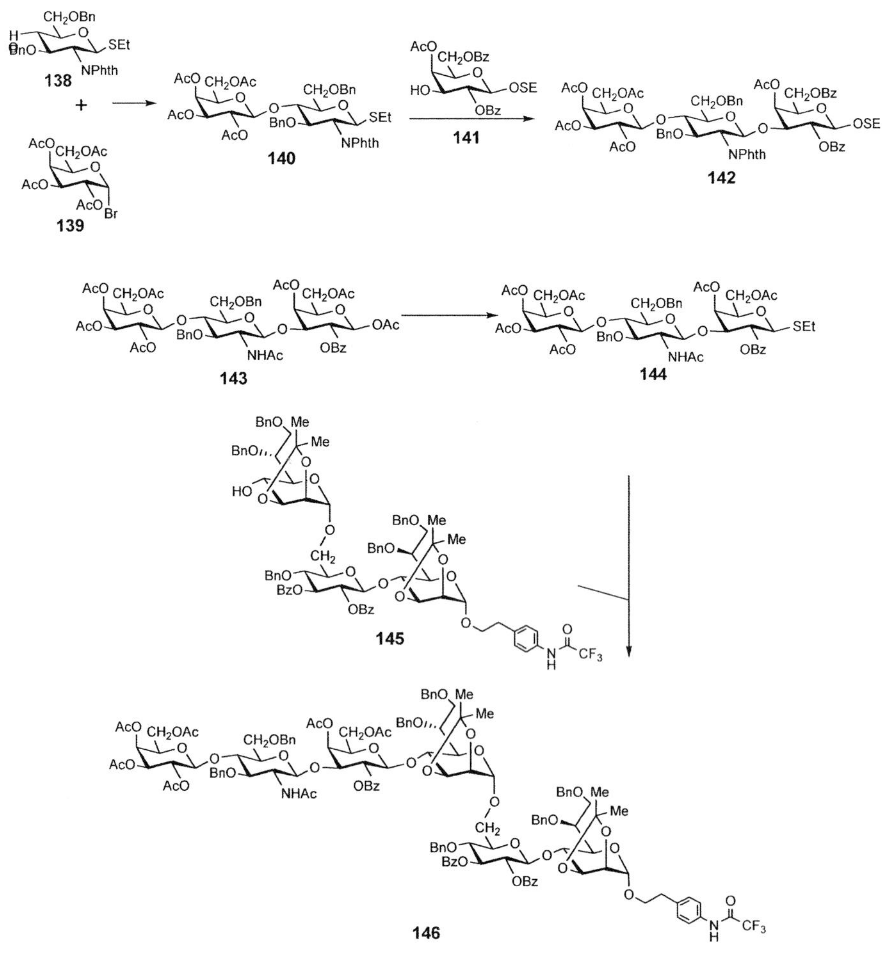

Haemophilus influenzae

H. influenzae are Gram-negative bacteria classified into six serotypes based on their type-specific CPSs. Serotype b (Hib) is the most virulent, causing meningitis, pneumonia, and otitis. As noted, a CPS-based vaccine was licensed against this bacterium in 1975 followed by conjugate vaccines containing the native CPS or fragments thereof covalently attached to medically acceptable proteins. In countries where this vaccine is routinely used, diseases caused by *H. influenzae* type b were almost completely eliminated.[40]

H. influenzae do not express fully developed LPSs characteristic of many enteropathogens. It lacks the O-SP side chains and the incomplete LPS is, in fact, better termed as LOS. Syntheses of L-*glycero*-D-*manno*-heptopyranose-containing tri- (**147**) and tetrasaccharide (**148**) components of the LOSs of *H. influenzae* were reported by Bernlind and Oscarson.[41] Key building block was the fully benzylated heptose orthoester **149** that was subjected to ethanethiol in the presence of TMSOTf to yield thioheptoside **150** (76 %). Treatment of **150** with sodium methoxide provided the key acceptor unit **151**. The assembly of **148** proceeded with the condensation of thioheptoside **151** with benzobromo-galactoside **152** in the presence of silver triflate to afford the disaccharide **153** in 67 % yield. This was coupled with the spacer-linked diheptoside alcohol **154** through thioglycoside activation by DMTST to produce the tetrasaccharide **155** containing three heptose moieties, in 73 % yield. The protecting groups were removed in two steps involving deacetylation by sodium methoxide followed by catalytic hydrogenolysis to afford the target tetrasaccharide **156** that may be converted into bioconjugatable form. Using similar approaches, the branched tri- (**157** and **158**) and tetrasaccharides (**159** and **160**) were also synthesized in a form that allows their covalent attachment to proteins.[42] It is noted that oligosaccharides **158** and **160** also occur in the LOS of *H. ducreyi*.

The inner core structure of *H. influenzae* contains the linear trisaccharide analogue[43] **170** consisting of heptose, KDO, and acetamido-glucopyranose moieties. An important feature of this structure is a phosphate residue in position 4 of the KDO moiety. A phosphate residue at position 5 of the same moiety is also quite common. Trisaccharide analogue **170** also occurs in the LPS of other bacteria including *Bordetella pertussis*, *H. ducreyi*, *Vibrio cholera*, and *Vibrio salmonicida*. Key precursors in its synthesis were the L-*glycero*-D-*manno*-heptopyranose donor **161** and the KDO acceptor **162** condensation of which under Schmidt-conditions afforded the disaccharide **163** in 64 % yield. Removal of the silyl protecting group by tetrabutylammonium fluoride followed by treatment with sodium methoxide in methanol, then acetylation with acetic anhydride provided the disaccharide as the methyl ester **164** in 73 % yield. Next, the benzyl groups were cleaved by hydrogenolysis quantitatively, and the diol so formed was treated with chloroacetic anhydride followed by exposure to titanium tetrabromide that produced the disaccharide bromide **165** (71 %). This was condensed with the glucosamine acceptor **166** under promotion by mercuric salts to afford the fully protected trisaccharide **167** (28 %).

The difficulties in this glycosylation reaction are indicated by the formation of the glycal derivative **168** in 50 % yield due to the competing elimination of hydrogen bromide from the disaccharide donor **165**. Next the phosphate moiety was introduced at O-4 of the middle residue by reaction with diisopropylphosphoramidite followed by *m*-chloroperbenzoic acid-oxidation (→**169**). In the final stage the protecting groups were removed in three steps by sequential treatment with tetrabutylammonium fluoride, sodium methoxide, and sodium hydroxide to afford the allyl glycoside **170** in 68 % overall yield. The aglycon was extended by reaction with cysteamine under UV light and the resulting ω-aminoalkyl glycoside **171** was reacted with thiophosgene to afford a highly reactive isothiocyanate that was coupled to bovine serum albumin (BSA). The neoglycoconjugates prepared contained 8-11 saccharide chains per mol of BSA. The disaccharide **172** was also synthesized. Attempted synthesis of its ω-amino counterpart **173** proved to be abortive because of the lability of its phosphate group during chromatography on Bio-Gel P2.

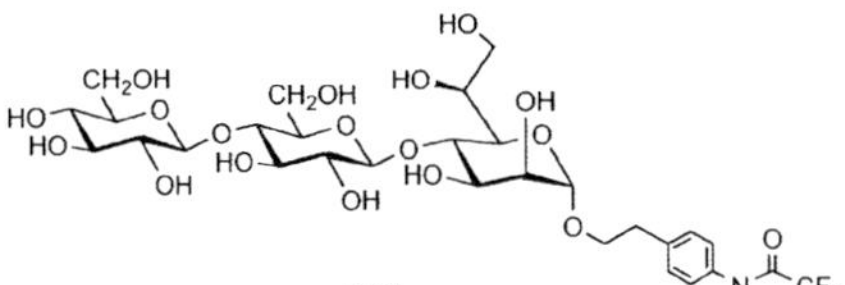

147

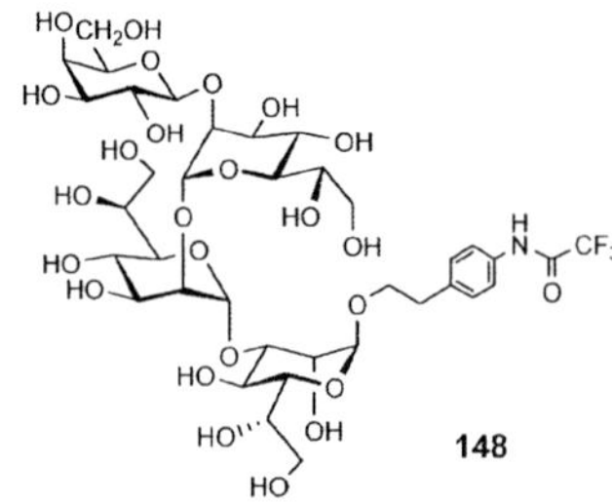

148

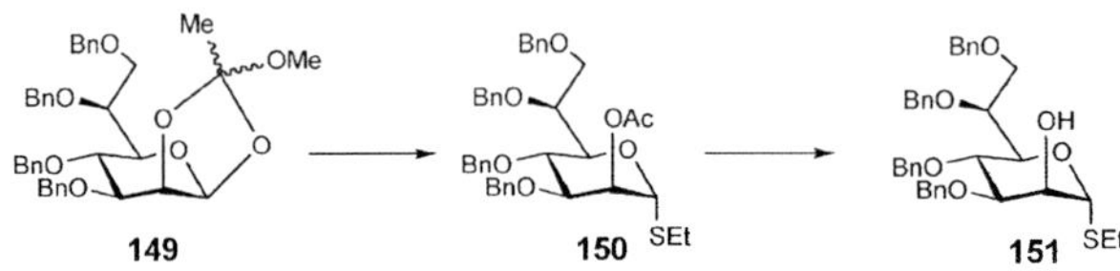

149 **150** **151**

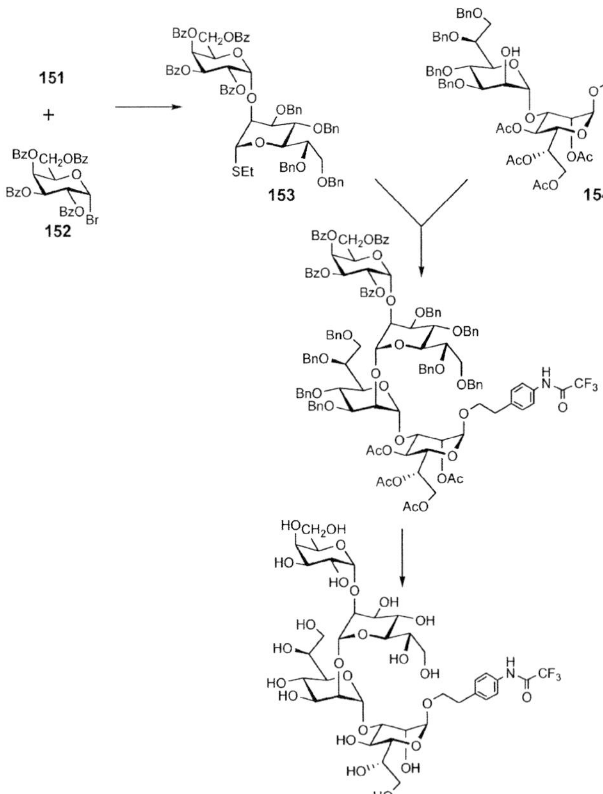

151 + **152** **153** **154** **155** **156**

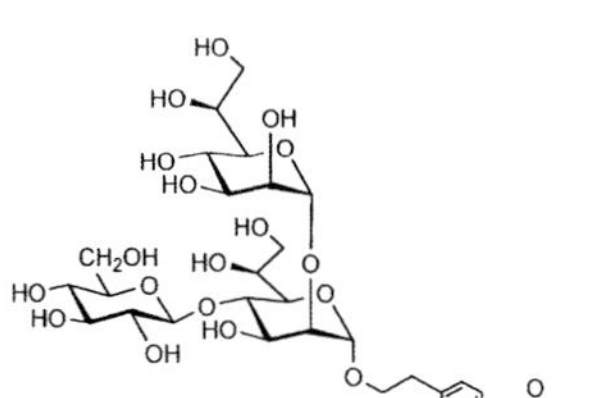

157

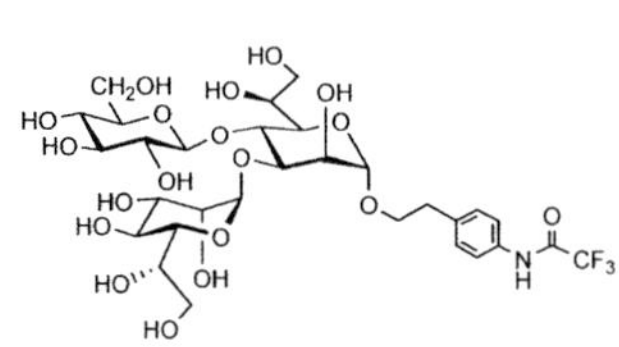

158

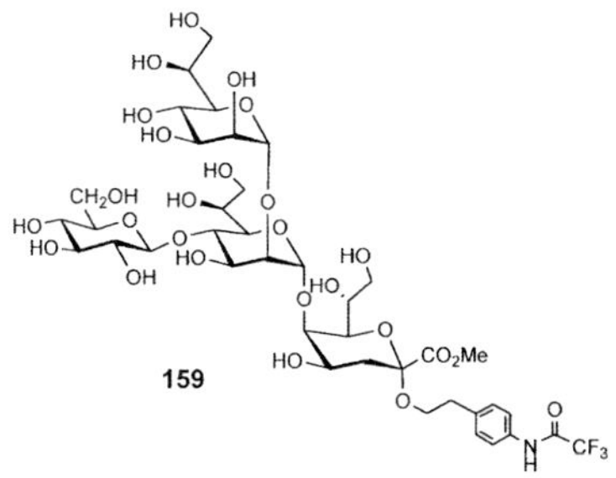

159

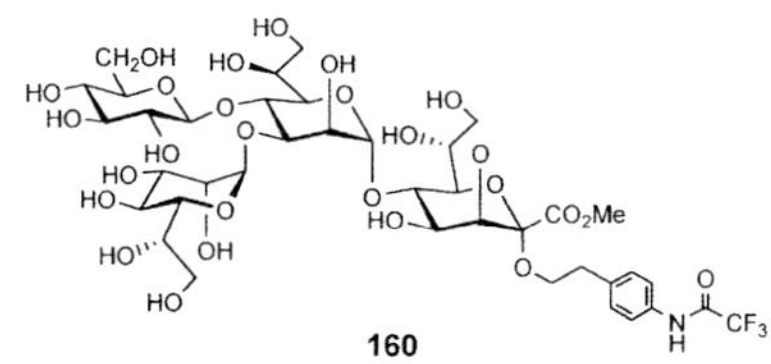

160

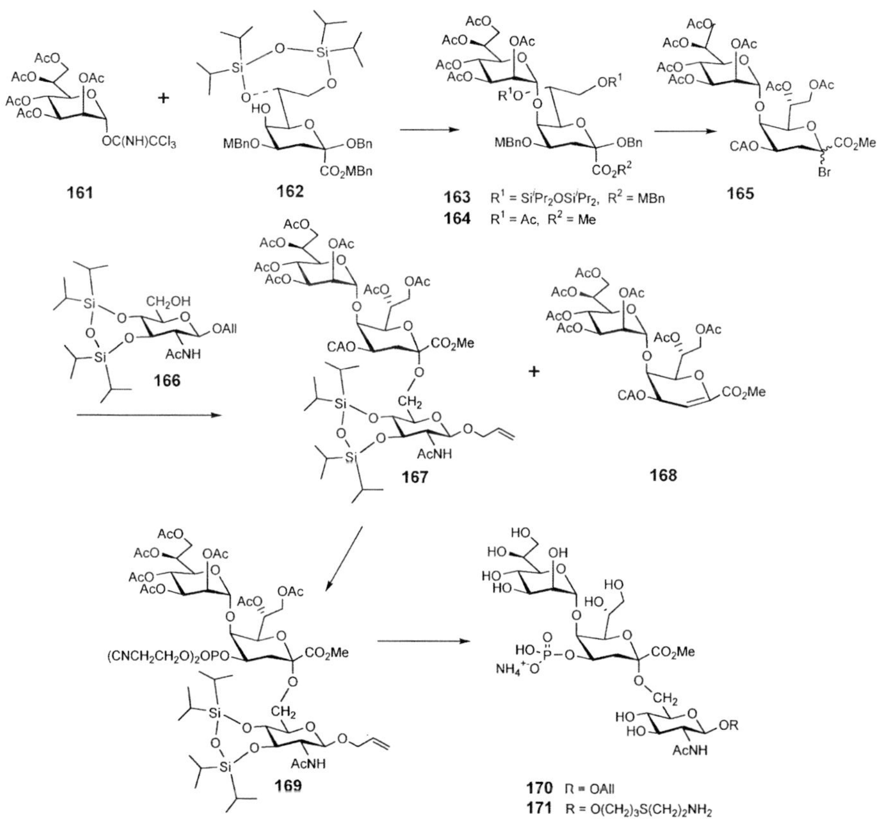

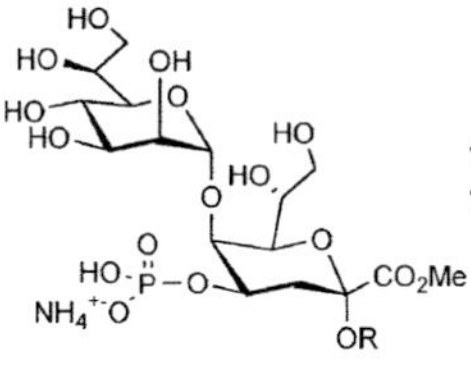

172 R = OAll
173 R = $O(CH_2)_3S(CH_2)_2NH_2$

H. influenzae Type B

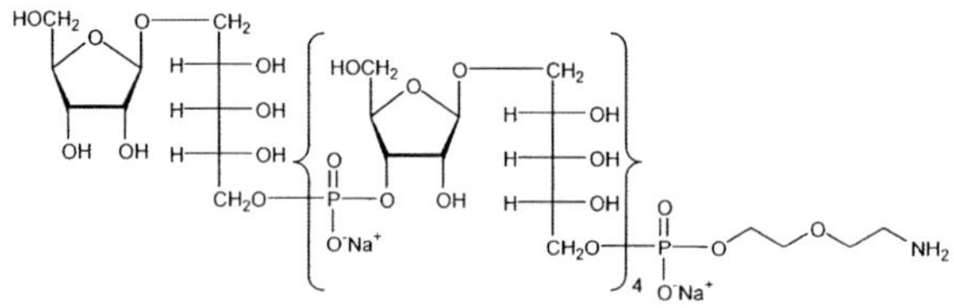

174

Verez-Bencomo reported the chemical synthesis of the pentameric fragment **174** of the native CPS in a form that can be covalently attached to proteins.[44] The synthesis was scaled-up for production purposes in a 200 g scale under GMP conditions. A covalent conjugate of this fragment with the outer membrane protein complex of *Neisseria meningitidis* elicited anti-Hib antibodies in mice. More recently, Verez-Bencomo reported that the synthetic oligosaccharide-protein conjugates are safe in humans and are undergoing Phase II clinical trials (personal communication). Preliminary results indicate that the synthetic material is as immunogenic as or even better than the commercially available licensed vaccine. At the time of the writing it is likely that this vaccine will be the first synthetic oligosaccharide-based conjugate vaccine to prevent a serious bacterial disease in humans. For another report on the design and evaluation of fully synthetic glycoconjugate vaccines for *H. influenzae* type b (Hib), see reference 45. The synthesis of a pentamer of an analog of the Hib CPS on solid support was reported by Nilsson and Norberg.[46]

H. influenzae Type C

The capsular polysaccharide of this bacterium is composed of the trisaccharide shown in formula **175** interconnected by anomerically located phosphodiester moieties. An important feature of this polysaccharide is an acetyl group at O-3 of the glucosamine residue, the presence of which severely limits the synthetic options to prepare analogs. The dimer (**187**) and trimer (**189**) corresponding to the CPS were synthesized by Oscarson's group.[47,48] Starting from azidoethyl galactoside **178** and the glucosamine thioglycoside **177**, the disaccharide **179** was prepared in 75 % yield using methyl triflate as thioglycoside activator. The anomeric selectivity was dictated by the participating nature of the phthalimido group of the donor moiety. The glycosylation step was followed by protecting group changes.

First, the phthalimido group was replaced by an acetamido by the action of ethylenediamine followed by acetic anhydride to afford compound **180**. A major side reaction was the base-promoted silyl group migration from O-4 to O-3 resulting in the formation of the byproduct **181** in 30 % yield. Next, the silyl protecting group was cleaved by treatment with $Et_3N(HF)_3$ to give the disaccharide acceptor **182** in 80 % yield. Using a similar concept except that in the acceptor moiety the azidoethyl aglycon was replaced by a temporary 2-(trimethylsilyl)ethyl group (**176**), the disaccharide **183** was prepared in 86 % under promotion with DMTST. Conversion of the phthalimido to acetamido group was accomplished as described above to yield compound **184**. Next, the aglycon was cleaved by the action of trifluoroacetic acid to produce the hemiacetal **185** in the α-form in 58 % yield. Conversion to the anomeric *H*-phosphonate derivative **186** was accomplished with a combination of phosphorous chloride and imidazole without anomerization. Next, disaccharides **182** and **186** were condensed in the presence of pivaloyl chloride in pyridine to produce the phosphodiester-bridged tetrasaccharide **187** in 71 % yield. For chain extension, the silyl group in **187** was removed by treatment with $Et_3N(HF)_3$ and the alcohol **188** was condensed with the H-phosphonate **186** as above to furnish the trimer analogue of the repeating unit **188**. Catalytic hydrogenolysis of compounds **188** and **189** afforded the partially O-acetylated dimer **190** and the trimer **191** of the repeating unit ready for coupling to proteins through the primary amino group of their linker moieties.

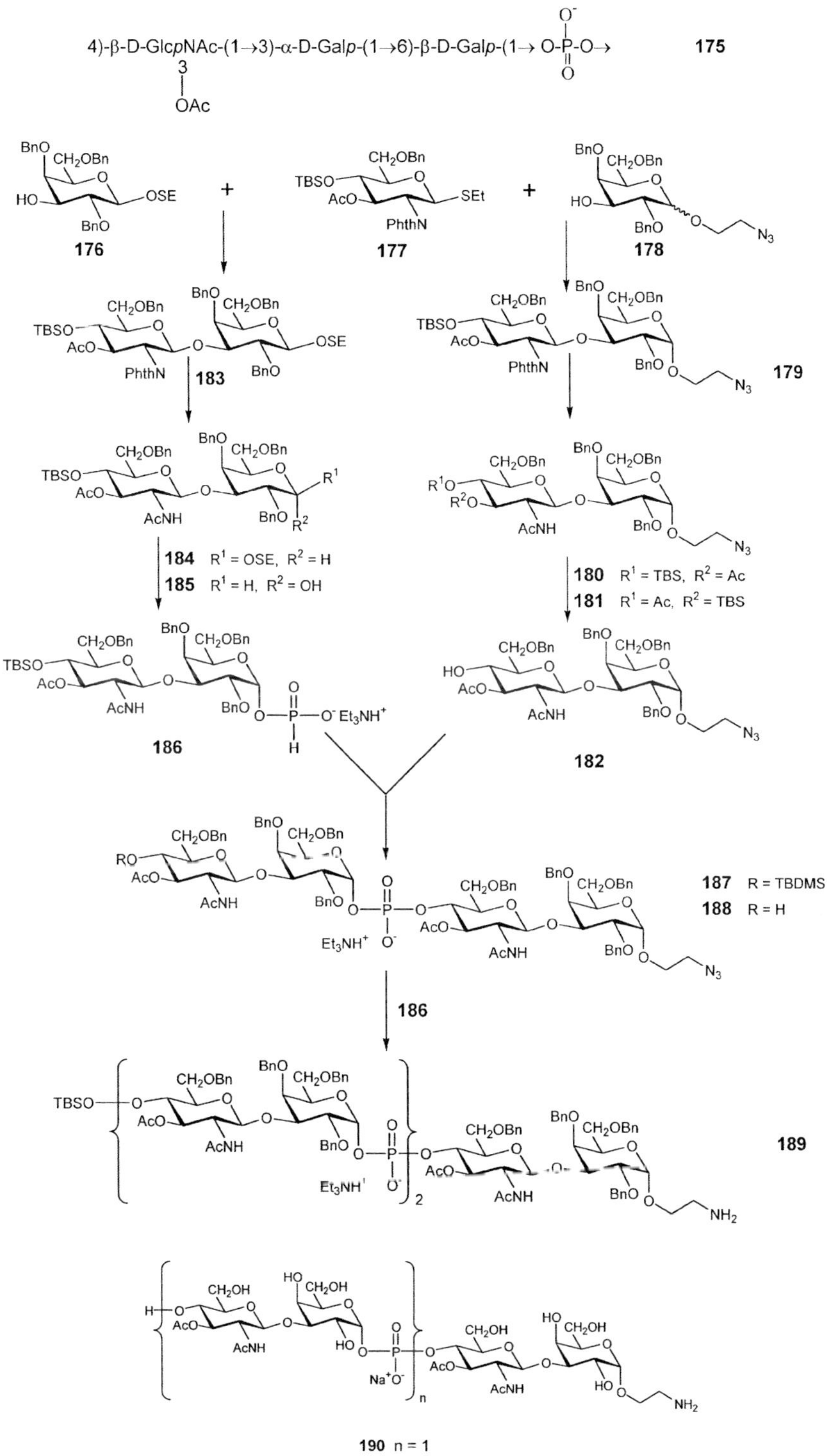
4)-β-D-GlcpNAc-(1→3)-α-D-Galp-(1→6)-β-D-Galp-(1→ O-P-O→
3
OAc
175
176
177
178
183
179
184 R¹ = OSE, R² = H
185 R¹ = H, R² = OH
180 R¹ = TBS, R² = Ac
181 R¹ = Ac, R² = TBS
186
182
187 R = TBDMS
188 R = H
186
189
190 n = 1
191 n = 2

H. influenzae Type F

Dimer (**192**) and trimer (**193**) of the CPS of this bacterium were assembled by Oscarson's group in a sequence similar to that outlined for the type c structures.[47]

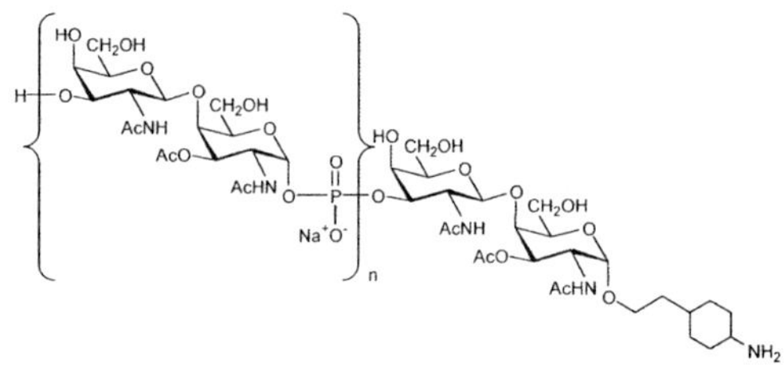

192 n = 1
193 n = 2

Hafnia alvei Strains 32 and 1192

The Gram-negative *Hafnia* is a frequent agent of nosocomial (hospital-acquired) infections and may cause enteric diseases, urinary tract infections, septicemia and meningitis. The inner core region of *Hafnia* contains the disaccharide analogue **196** in which a KDO residue is substituted at O-4 with a heptose moiety.[43] In Kosma's synthesis the carbonate-protected KDO allyl glycoside **194** was condensed with the L-*glycero*-D-*manno*-heptose trichloroacetimidate **161** using Schmidt's TMSOTf activation to produce the α-linked disaccharide **195** in 61 % yield. Base-catalyzed removal of the protecting groups (first with NaOMe, then with NaOH) afforded **196** in 93 % yield. Conversion to the allyl aglycon to the cysteamine adduct **197** was performed under radical conditions promoted by UV-irradiation. Disaccharide **197** was in situ activated by thiophosgene to afford a reactive isothiocyanate intermediate (**198**) through which the saccharide construct was linked to bovine serum albumin.

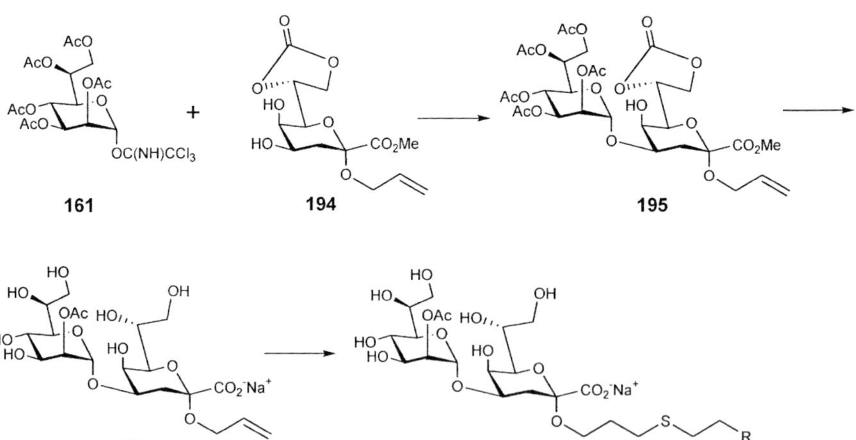

161 **194** **195** **196**

197 R = NH_2
198 R = NCS

Helicobacter pylori

H. pylori are Gram-negative bacteria associated with chronic gastritis that may cause ulcers. The Lipid A component **205** of the LPS of this organism differs from that of *E. coli* in that it has fewer fatty acids but their chains are longer. Additionally, they lack a phosphate group at the O-4' position and may contain an ethanolamine residue attached to the phosphate moiety. The Lipid A analogue **205** was synthesized by Kusumoto's group.[49] The glucosamine trichloroacetimidate donor **199** was prepared that bears a Troc amino protecting group. It was condensed with the glucosamine acceptor **200** that already contains the fatty acid moiety at its amino group, using trimethylsilyl triflate as the activator. Disaccharide **201** was obtained in 72 % yield in a stereoselective manner due to neighboring carbamate group participation.

Next, compound **201** was treated with zinc-copper alloy to remove the Troc protecting group and the amine so obtained was lipidated with the fatty acid **207** in the presence of a water-soluble carbodiimide to yield the allyl glycoside **202**. Removal of the allyl group by isomerization with an iridium complex followed by oxidation with iodine furnished the hemi-acetal **203** in 94 % yield. This was anomerically phosphorylated by treatment with a phosphoramidite in the presence of 1H-tetrazole followed by oxidation with MCPBA to afford the fully protected phosphorylated glycolipid **204** from which hydrogenolytic removal of all the protecting groups gave the target compound **205** in 25 % yield after silica gel chromatography. An analogue that lacks the ethanolamine moiety was also synthesized.

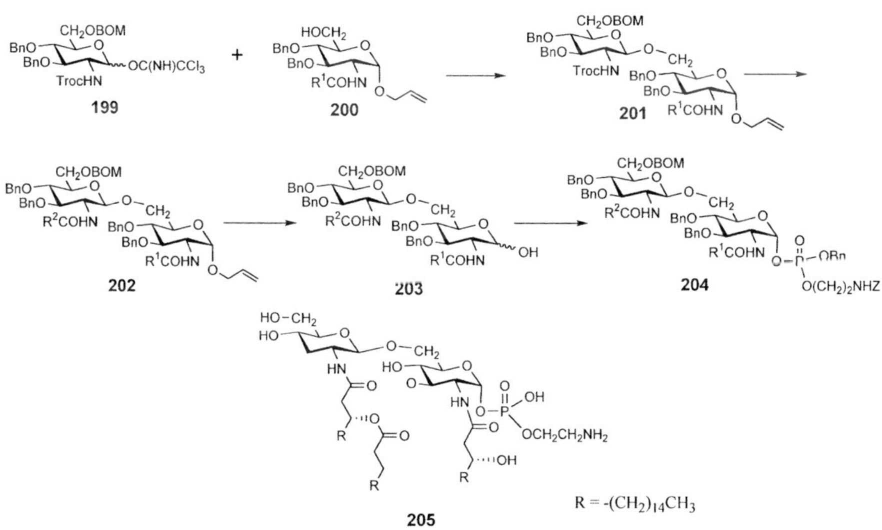

206 R^1COOH = (*R*)-3-(*p*-(trifluoromethyl)benzyloxy)octadecanoic acid
207 R^2COOH = (*R*)-3-(octadecanoyloxy)octadecanoic acid

Klebsiella

Klebsiella are Gram-negative, opportunistic pathogens that can cause a variety of diseases in humans including pneumonia, bacteremia, lesions and abscesses. The repeating unit of the O-SP of an untyped *K. pneumoniae* strain consists of the linear pentarhamnoside **208**. The synthesis of the tetrasaccharide analogues as methyl and octyl glycoside **218** and **219** was reported by Kong's group.[50] Starting from the rhamnose triol **209** and the rhamnose trichloroacetimidate **210** that were condensed in the presence of TMSOTf, the α-(1→3)-linked disaccharide **211** was obtained in a highly regioselective reaction in 64 % yield. The authors hypothesized that the good regioselectivity was due to the formation of a 3-O-linked orthoester intermediate followed by rearrangement. The disaccharide **211** was converted to the benzoylated derivative **212** by the action of benzoyl chloride and pyridine. Exposure of the allyl glycoside **212** to palladium chloride in acetate buffer caused the rearrangement of the allyl group into propenyl that was cleaved under mildly acidic conditions.

3)-α-L-Rhap-(1→3)-α-L-Rhap-(1→2)-α-L-Rhap-(1→2)-L-Rhap-(1→2)-α-L-Rhap-(1→ **208**

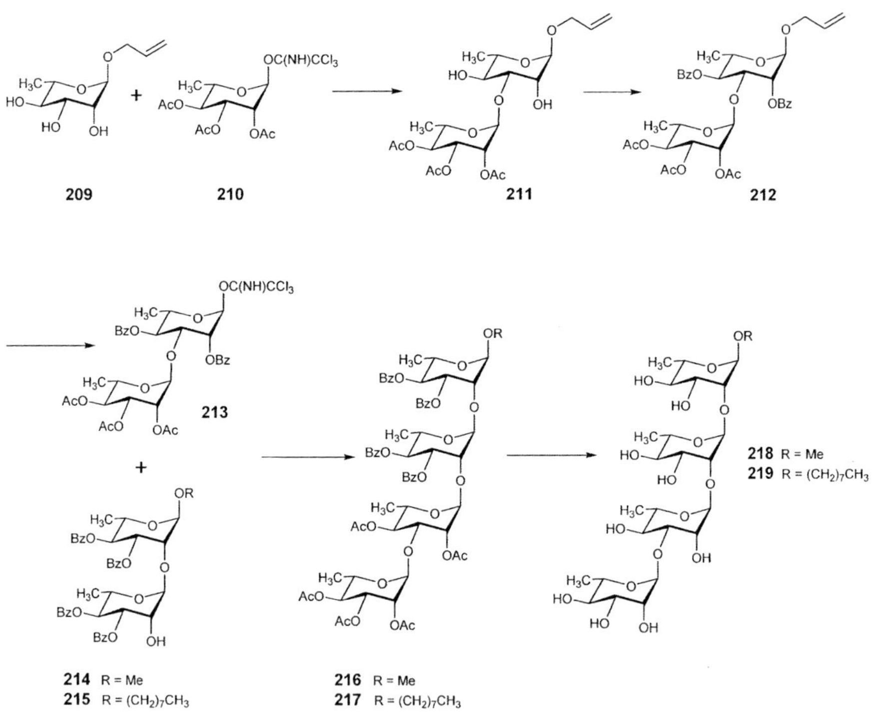

The hemiacetal so obtained was treated with trichloroacetonitrile and DBU to afford the disaccharide trichloroacetimidate **213** in 86 % combined yield. Using similar chemistry, the rhamnobiose acceptors **214** and **215** were prepared and condensed with the donor **213** to afford rhamnotetraoses **216** and **217** in 70-80 % yields. Removal of all the acyl groups by the action of ammonia in methanol afforded the target tetrasaccharides **218** and **219**. Because of the presence of the α-L-Rhap-(1→3)-α-L-Rhap motif in the O-SPs of *Shigella dysenteriae* type 1 and *Shigella flexneri* 2a, synthesis of the α-(1→ 3)-linked disaccharide element **211** also represents synthesis of the partial structures of *Shigella* strains.

Klebsiella Type 14

Di- and trisaccharide parts of its repeating units are identical to those of Type 43 described below.

Klebsiella Type 43

The CPS of this organism is composed of the branched pentasaccharide repeating unit **220** consisting of a trisaccharide backbone and an acidic disaccharide side chain. Synthesis of disaccharide **228** corresponding to the side chain of this CPS was reported by Roy's group.[51] Thus, ethylthio glucoside **221** was activated with NIS/TfOH in the presence of the allyl glucoside acceptor **222** to give the disaccharide **223** in good stereoselectivity, in 72 % yield. Removal of the acetyl group at O-2' under Zemplén conditions afforded the alcohol **224**. The required *manno* residue was created by oxidation of C-2 of the "nonreducing" glucose unit using Ac_2O-DMSO followed by reduction of the intermediate keto compound with $NaBH_4$ and subsequent O-acetylation (→ **225**).

Next, the carboxyl group was fashioned. First, the methoxybenzyl group was selectively cleaved by ceric ammonium nitrate produced the alcohol **226** in 67 % yield. Next, the hydroxymethyl group of **226** was oxidized by the action of CrO_3/H_2SO_4 followed by esterification with diazomethane to afford **227** in 64 % combined yield. Subsequent deacetylation and hydrogenolysis steps yielded the targeted trisaccharide as the propyl glycoside **228** in 69 % yield. Using the monosaccharides **123**, **229**, and **230** as precursors, similar chemistry afforded a trisaccharide analogue **231**.[51]

3)-α-D-Gal*p*-(1→3)-α-D-Man*p*-(1→2)-α-D-Man*p*-(1→
2
↑
1
β-D-Man*p*-(1→4)-β-D-Glc*p*A

220

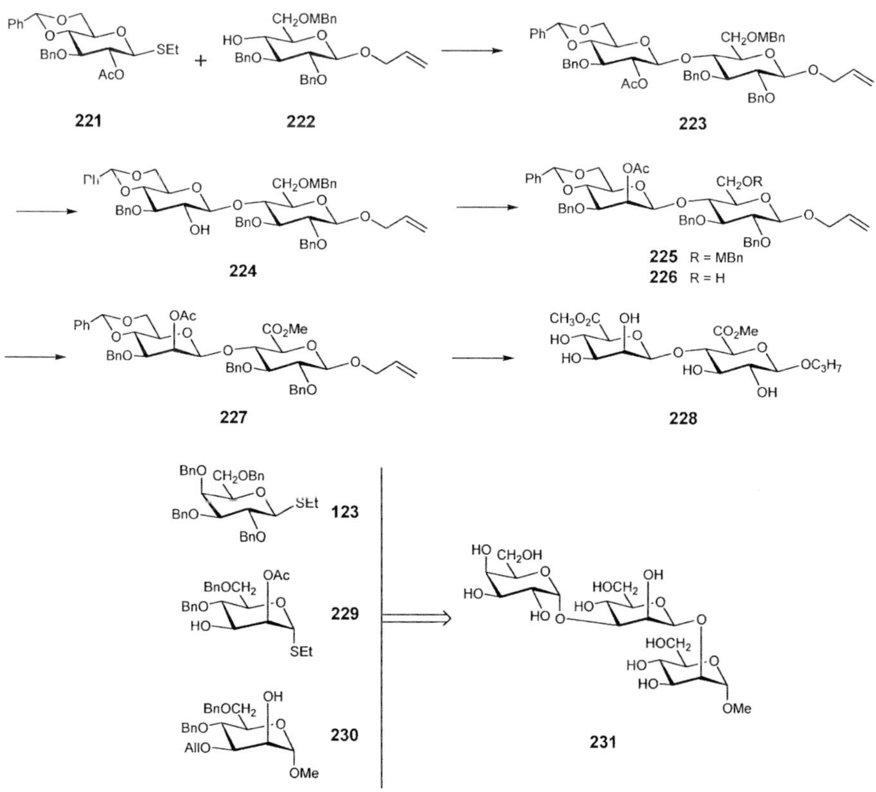

Klebsiella Type 57

The K-antigen of this organism is composed of the heterotetrasaccharide repeating unit **232** that is identical to the structure of the K-antigen from *E. coli* type 36. The tetrasaccharide analog **243** (in the form of a methyl ester) was synthesized by Roy's group starting from galactose derivatives **233** and **235** and the mannose precursors **234**.[52] Thus, the galactoside alcohol **233** was condensed with ethylthio mannoside donor **234** under NIS/TfOH activation to afford the fully protected disaccharide intermediate **236** in 95 % yield that was deacetylated to give the diol **237** (96 %). The other half of the target tetrasaccharide was prepared by condensation of the same thiomannoside **234** with the thiogalactoside **235**. In this reaction the reagent iodonium dicollidine perchlorate was used to activate the benzyl-protected, more reactive thiomannoside donor **234** and the disaccharide **238** having the ethylthio aglycon was formed in 70 % yield.

α-D-Man*p*-(1→3)-β-D-Gal*p*-(1→3)-β-D-Gal*p*A-(1→
4
↑
1
α-D-Man*p*

232

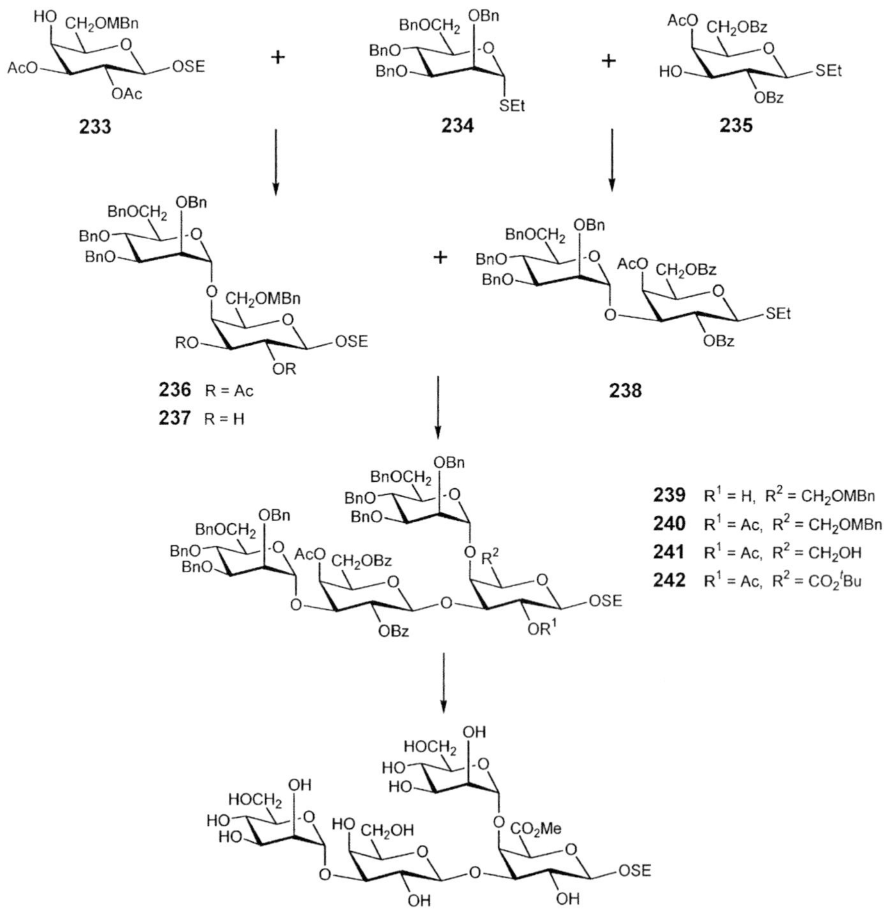

Condensation of the disaccharide diol **237** and the disaccharide thioglycoside **238** in the presence of DMTST afforded the protected tetrasaccharide **239**. The high degree of regioselectivity reflects the decreased reactivity of the HO-2 hydroxyl group in the acceptor moiety. Compound **239** was acetylated with acetic anhydride and the methoxybenzyl group in the fully protected tetrasaccharide **240** was selectively cleaved by oxidation with ceric ammonium nitrate to afford the alcohol **241** (69 %). Exposure of compound **241** the CrO_3-pyridine-Ac_2O-^{t}BuOH reagent system produced the fully protected tetrasaccharide ester **242**. In the final stages, compound **242** was treated with sodium methoxide in methanol that cleaved the O-acyl groups and fashioned the methyl ester, then the resulting intermediate was subjected to hydrogenolysis over palladium-on-charcoal to yield the target tetrasaccharide as trimethylsilylethyl glycoside **243**.

Klebsiella Type 63

The K antigen of *Klebsiella* type 63 contains the acidic trisaccharide repeating unit **244**. A synthesis of the trisaccharide analog as the methyl glycoside **250** was reported by Roy.[53] Starting from the tolylthio galactoside **245**, the thioglycoside acceptor **246** was synthesized in a series of protecting group manipulations including (i) isopropylidenation (at HO-3 and 4), (ii) O-benzylation (at HO-2 and 6), (iii) removal of the isopropylidene moiety and (iv) regioselective O-acetylation through a cyclic orthester formed on HO-3 and 4 hydroxyl groups. Acceptor **246** was coupled with the perbenzylated ethylthio fucoside **131** that was activated by the NIS/TfOH reagent to afford the 1,2-*cis*-linked disaccharide **247** in 66 % yield. This glycosylation made use of the differential reactivities of the thioglycosides **131** and **246** under thiophilic activation arising out of a combination of the different aglyconic components and protecting groups. For further chain extension, disaccharide thioglycoside **247** was activated by methyl triflate in the presence of the galacturonic acid acceptor **248** to afford a fully protected trisaccharide intermediate (**249**, 77 %) that was subjected to hydrogenolysis followed by treatment with sodium methoxide in methanol to produce the tetrasaccharide methyl ester **250**.[53]

3)-α-D-Gal*p*-(1→3)-α-D-Gal*p*A-(1→3)-α-L-Fuc*p*-(1→ **244**

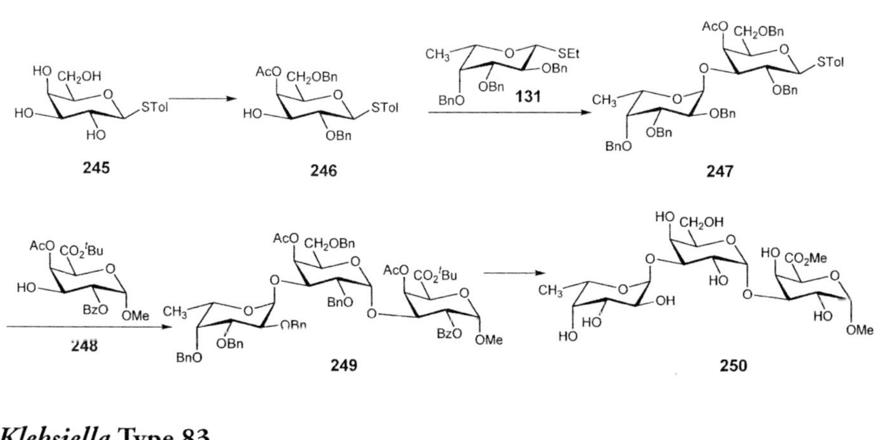

Klebsiella Type 83

Di- and trisaccharide parts of its repeating units are identical to those of Type 43 described above.

Moraxella Type A

Moraxella may cause respiratory diseases, especially in children. Their LPS is truncated: it lacks the O-specific polysaccharide part, its terminus contains a relatively short oligosaccharide of irregular structure instead. The synthesis of the octasaccharide **264** corresponding to the outer part of the LPS of this bacterium using a [3 + 5] block scheme was described by Ekelöf and Oscarson.[54] First, the lactose acceptor **251** was condensed with the galactosyl chloride donor **252** using silver triflate as the activator in diethyl ether to afford the trisaccharide block **253**. The yield of the anomerically pure trisaccharide **253** was 62 %. The pentasaccharide block **262** features a linker moiety and was constructed from mono- (**255** and **258**) and disaccharide (**254**) intermediates. Thus, disaccharide acceptor **254** was coupled with the bromosugar donor **255**, obtained in situ from the corresponding β-ethylthio glucoside by treatment with bromine, in the presence of silver triflate to afford trisaccharide **256** (86 %). Chemoselective removal of the chloroacetyl group by the action of sodium methoxide in methanol yielded the trisaccharide alcohol **257**. For further chain extension, azido-thioglucoside **258** was treated with bromine to afford the corresponding bromide that was activated with silver triflate in the presence of the trisaccharide **257** to yield the fully protected tetrasaccharide **259** in 54 % yield.

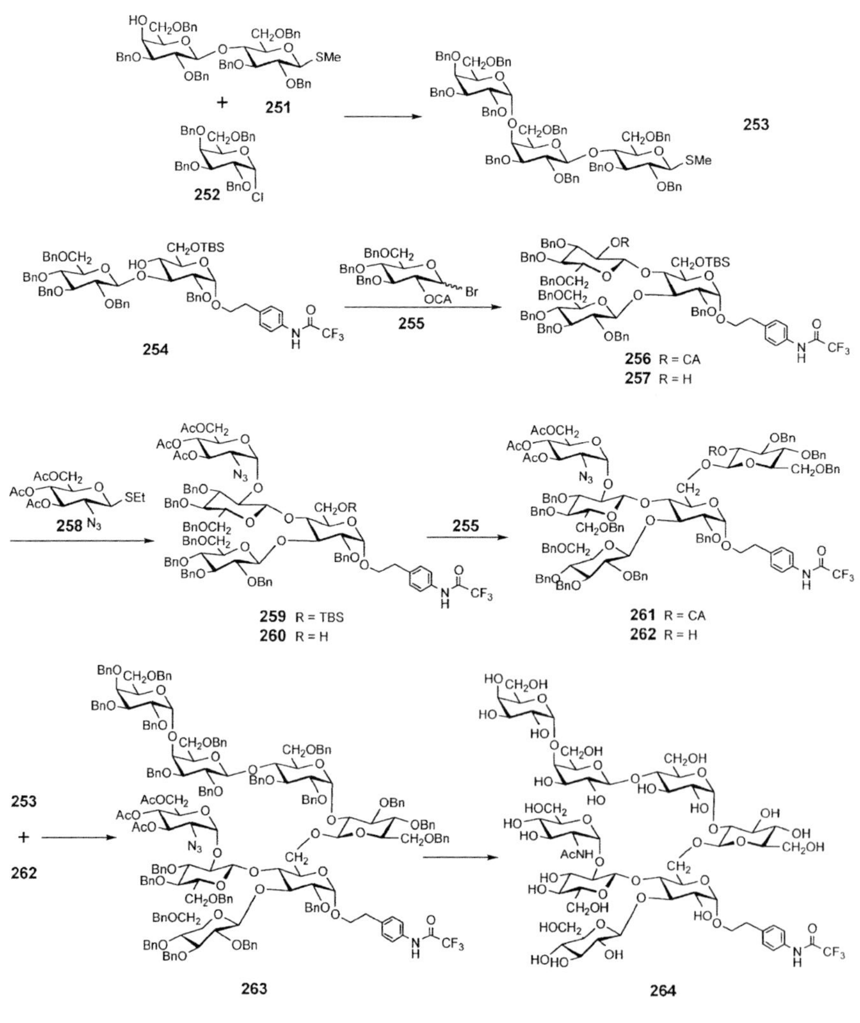

Next, tetrasaccharide **259** was desilylated using acetic acid and the alcohol **260** formed was further glycosylated with the donor **255** to afford the pentasaccharide **261** in 79 % yield. The synthesis of the pentasaccharide block was completed by treatment of **261** with hydrazine dithiocarbonate to give the acceptor **262** in 79 % yield. Next, the trisaccharide thioglycoside **253** was treated with bromine as described above then the resulting trisaccharide bromide was coupled, without isolation, with the pentasaccharide alcohol **262** in the presence of silver triflate to furnish the fully protected intermediate **263** in 63 %. Conversion of the azido to acetamido group and deprotection by sequential exposure to sodium methoxide followed by hydrogenolysis gave the target octasaccharide **264** suitable for attachment to proteins after N-deblocking.

Mycobacteria

The past decade witnessed a rise of mycobacterial infections in many parts of the world. Because of the increasing drug-resistance of these bacteria, novel approaches are sought for prevention and treatment that target the highly complex polysaccharides and glycolipids constituting the cell-wall surrounding mycobacteria. A major difference from most other bacteria is that mycobacteria lack regular, repeating structures observed in the CPSs and LPSs of many human pathogens. Two major polysaccharides of complex architecture were identified, termed arabinogalactan and lipoarabinomannan. Both the arabinose and the galactose residues occur in their rare, furanoside form. Numerous mycobacterial oligosaccharides were synthesized in the hope that they might help to understand the biosynthesis of mycobacterial polysaccharides, with the eventual goal of killing mycobacteria by arresting the biosynthesis of their surface polysaccharides.[55-69]

The synthesis of the tetrasaccharide motif **274** occupying the nonreducing termini of the above-mentioned polysaccharides was reported by Lowary's group.[55] Glycosylation of the arabinofuranoside **266** with the thioarabinoside donor **265** under NIS/AgOTf activation afforded the fully protected α-linked disaccharide **267** in 84 % yield. This was transformed to the partially protected disaccharide acceptor **268** in a four step sequence involving (i) base-catalyzed removal of the three benzoyl groups, (ii) tritylation of the primary hydroxyl group, (iii) benzylation, and (iv) detritylation in 46 % overall yield. Subsequent glycosylation with the donor **269** under activation by NIS/AgOTf led to the fully protected arabinotrioside **270** (88 % yield) containing both interglycosidic linkages in α configuration. Base-catalyzed regioselective deprotection (→ **271**, 79 %) followed by glycosylation with the fully benzylated thioglycoside donor **272** under the action of NIS/AgOTf afforded the fully protected intermediate **273** in 52 % yield. One-step deprotection by hydrogenolysis gave the tetrasaccharide methyl glycoside **274**. Similar chemistry was used for the synthesis of the branching hexasaccharide **275** and its di-pentasaccharide fragments.[55]

Mycobacterium avium Serovar 17

A particular feature of the pentasaccharide glycopeptidolipid antigen of this organism is an acylamino-dideoxy-glucopyranose residue at its terminus. The synthesis of the pentasaccharide analog **285** was reported by Lipták's group using a [3+2] block scheme.[57] Precursors for the trisaccharide part were the azidoglucose derivative **276** and the O-methylated thiorhamnoside **277** that were combined under activation by TMSOTf to afford the β-linked disaccharide **278** in 75 % yield. This was converted to the imidate **279** after treatment with NBS then with trichloroacetonitrile in 75 % overall yield. Condensation of **279** with the thiorhamnoside **280**, using TMSOTf activation afforded the trisaccharide block **281** in 55 % yield. This was combined with the disaccharide block **282** under NIS/TfOH activation that yielded the protected pentasaccharide **283** in 69 % yield.

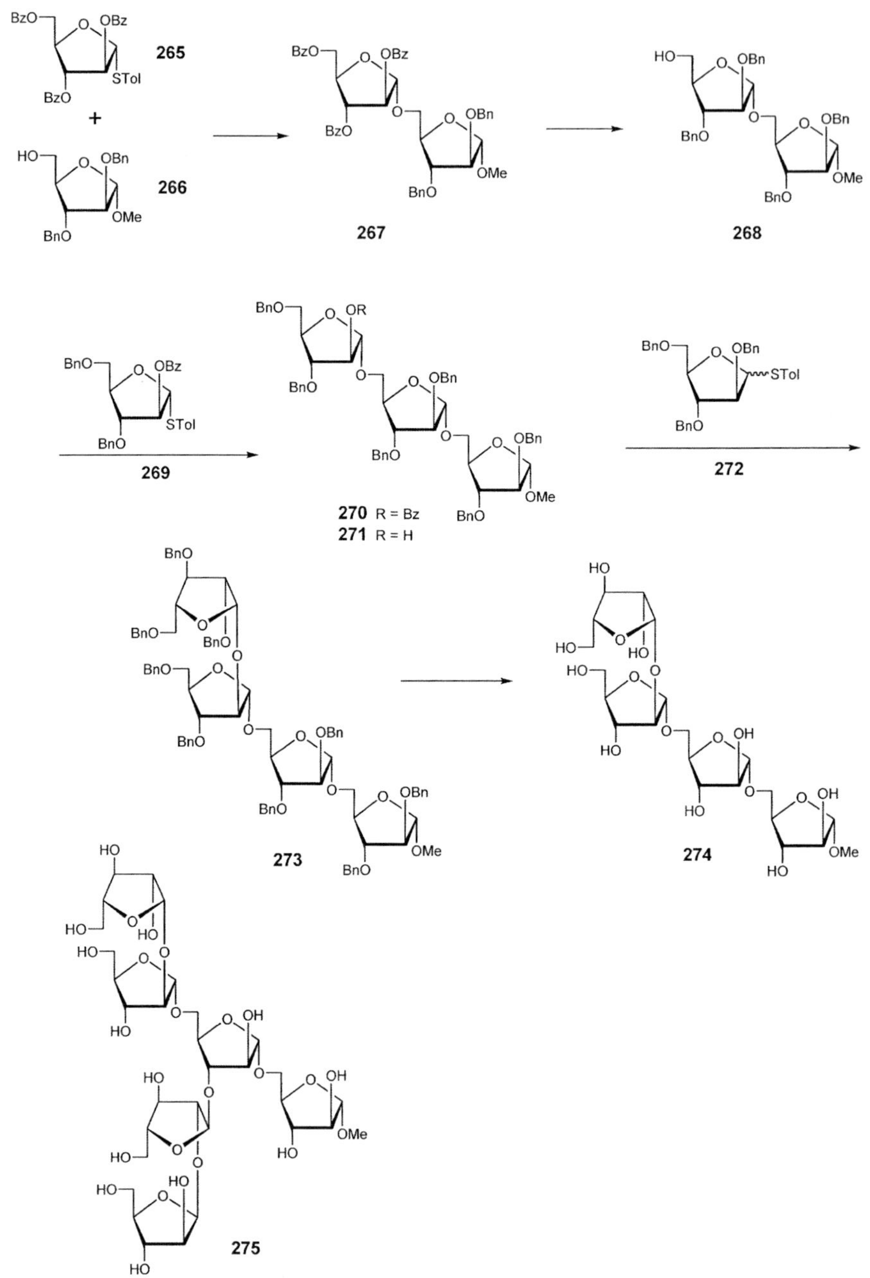
265
266
267
268
269
270 R = Bz
271 R = H
272
273
274
275

Next, the acetyl groups were removed using KO^{t}Bu (→**284**). It is remarkable that the usual NaOMe /MeOH reagent failed in this case. In the final stages the azido group was reduced using triphenylphosphine and the resulting primary amino group was acylated with (S)-3-hydroxy-(S)-2-methylbutyric acid using BOP as the coupling reagent. The synthesis was completed by the reduction of the nitro group to amino by hydrogenation over platinum followed by N-acylation with trifluoroacetic anhydride. In the final step, the benzylidene and benzyl groups were removed by hydrogenolysis over palladium-on-charcoal to afford the glycolipid analog **285**.

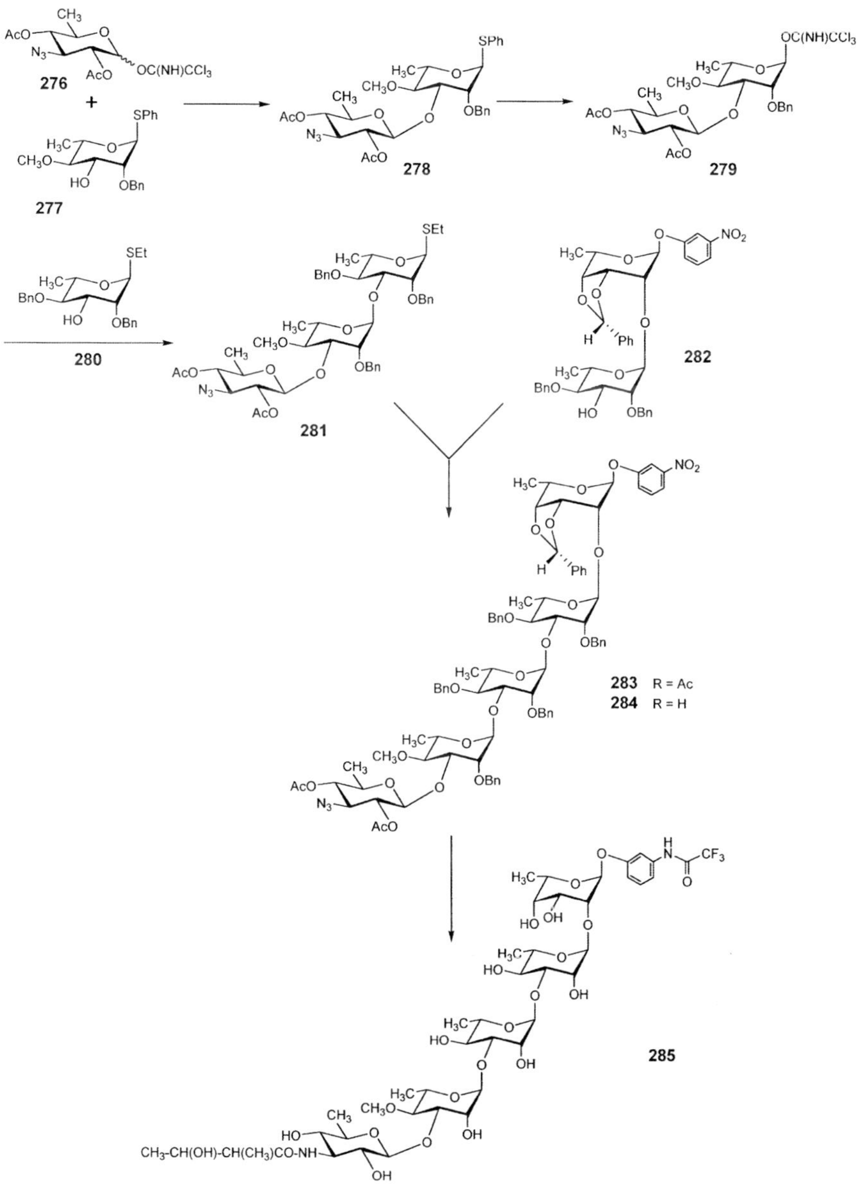

Neisseria gonorrhoea Strain 15253

The trisaccharide lactosyl-heptose **293** is part of the core region of the lipooligosaccharide of this bacterium. The synthesis of its methyl glycoside was undertaken by Yamasaki's group.[70] Thus, methoxyphenyl lactoside **286** was benzylated using benzyl bromide and sodium hydride in the presence of tetrabutylammonium iodide then its aglycon was oxidatively cleaved using cerium ammonium nitrate to afford lactose hemiacetal **287**. This was then treated with trichloroacetonitrile and potassium carbonate to produce the per-benzyl α and β-lactosyl trichloroacetimidate **288** in 66 % combined yield. Disaccharide donor **288** was coupled with the heptose derivative **290** that was obtained from mannose **289**, in the presence of TMSOTf to give the trisaccharide **291** as the major product, isolated in 59 % yield. Next the double bond was oxidatively cleaved by the reagent OsO_4-$NaIO_4$ followed by reduction with $NaBH_4$ to fashion the primary hydroxyl group in the heptose moiety (→**292**, 80 % overall yield). The protected intermediate **292** was converted to the target trisaccharide **293** by hydrogenolytic removal of all the protecting groups in one step, followed by purification through the per-O-acetylated derivative that was deprotected by transesterification with sodium methoxide in methanol.

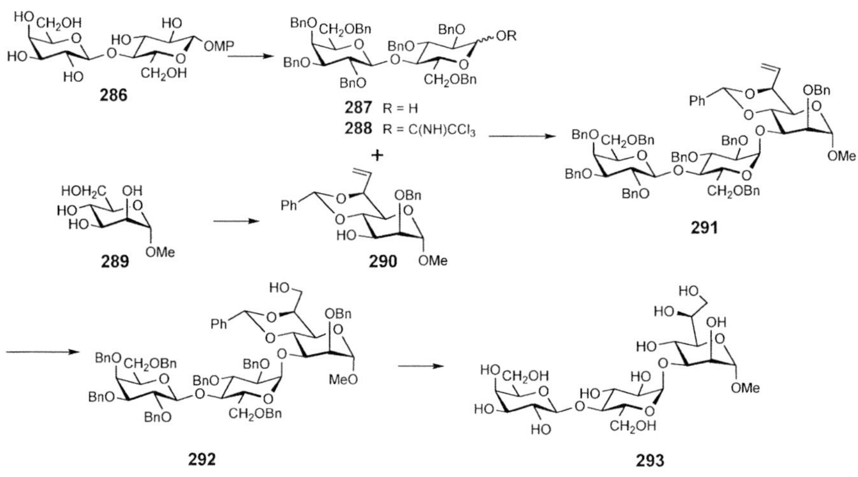

Neisseria meningitidis

N. meningitidis are Gram-negative bacteria causing endemic and epidemic septicaemia and meningitis, a potentially deadly inflammation of the brain's outer layers termed meninges, throughout the world. Currently licensed vaccines include a four component vaccine containing purified polysaccharides of serogroups A, C, W-135 and Y, targeting the population older than two years of age. Conjugate vaccines containing the group C polysaccharide covalently linked to either a mutant diphtheria toxin (CRM197) or tetanus toxoid proved efficacious in infants and toddlers and were licensed in the UK in 1999.

Lipooligosaccharide of *N. meningitidis*

The upstream terminus of the LOS of *N. meningitidis* consists of the sialylated pentasaccharide **300** and is implicated in the pathogenesis of the diseases caused by this bacterium. The synthesis of this pentasaccharide using a combined chemoenzymatic protocol was reported by Whitfield's group.[71] Starting from lactose **286**, the phthaloylated lactosamine donor **294** was prepared in a multistep sequence involving azidonitration of lactal. Condensation of **294** with

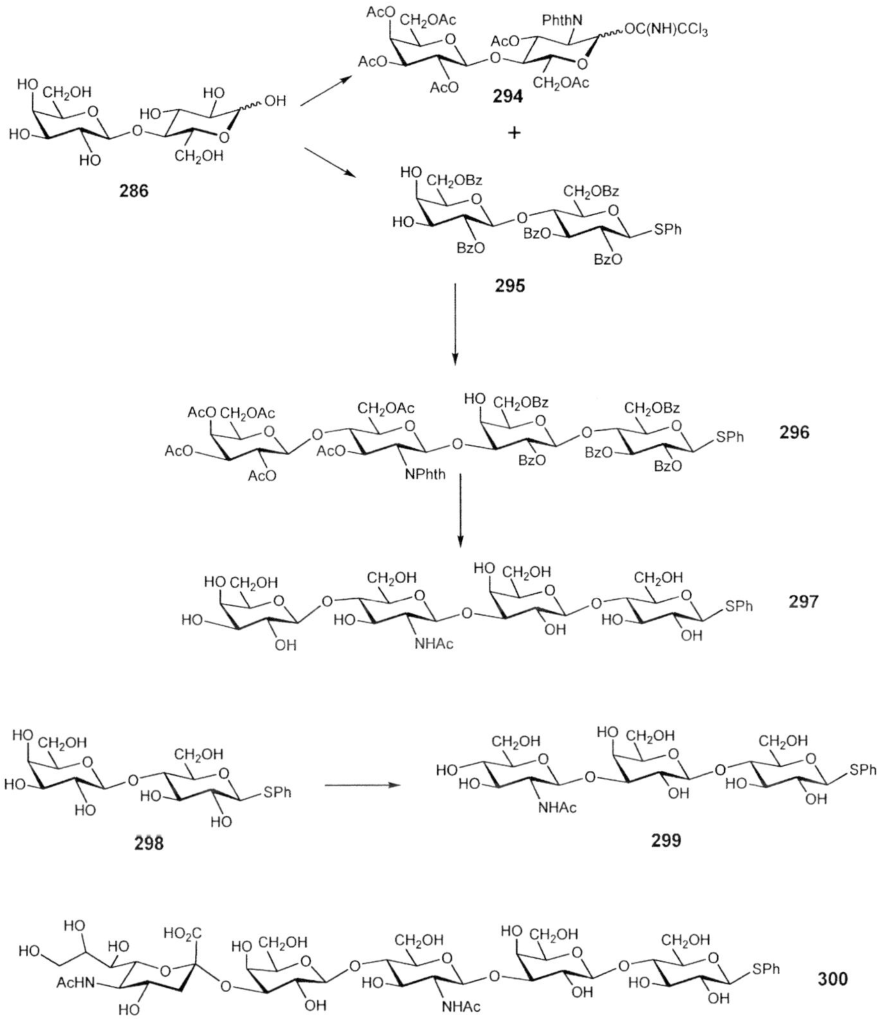

the phenylthio lactoside **295** using $BF_3.Et_2O$ as the activator under Schmidt conditions afforded the protected tetrasaccharide **296** in 40 % yield. Removal of all the protecting groups with NaOMe and hydrazine followed by O- and N-acetylation (pyridine-Ac_2O) then O-deacetylation (Zemplén) afforded the tetrasaccharide **297** in 86 % combined yield.

Chemical approaches similar to those described for the tetrasaccharide were also used to prepare the trisaccharide **299**. Whitfield and coworkers also demonstrated the power of enzymatic synthesis: in addition to the above-mentioned chemical transformations, trisaccharide **299** was also assembled from the phenylthio lactoside **298** by enzymatic transfer of N-acetylglucosamine from UDP-GlcNAc in 96% yield, using a β-(1→3)-N-acetylglucosaminyltransferase. Compound **299** was further elongated by enzymatic attachment of a galactose residue using a fusion enzyme to afford the tetrasaccharide **297** in a nearly quantitative yield. Tetrasaccharide **297** was enzymatically sialylated using CMP-Neu5Ac as NeuAc donor and the α-(2→3) sialyltrasferase from *N. meningitidis* to afford the targeted pentasaccharide **300** also in a nearly quantitative yield.

Group A *Neisseria meningitidis*

The CPS of Group A *N. meningitidis* consists of α-(1→6) -linked 2-acetamido-2-deoxy-α-D-mannopyranosyl phosphate moieties **301** in which the HO-3 group is nonstoichiometrically acetylated, up to 95 %. Synthesis of aminoethyl spacer-equipped saccharide fragments **302-305** was reported by Pozsgay's group.[72] Their approach relied on the H-phosphonate chemistry already mentioned in the section on *Haemophilus influenzae* type c. Thus, azidomannoside **306** was converted to the differentially protected derivative **307** in a sequence involving tritylation of the primary hydroxyl group followed by benzylation at O-3 and 4 in 92 % yield. Next, the trityl group and the aglycon were simultaneously cleaved by acetolysis using acetic anhydride and a catalytic amount of sulfuric acid to produce the α-acetate **308** in 65 % yield. Next, the anomeric acetyl group was selectively cleaved off under the action of dimethylamine to yield the α-hemiacetal **309** quantitatively. The hemiacetal was converted to the H-phosphonate **310** using phosphorous trichloride and imidazole, in 84 % yield, then the H-phosphonate was condensed with the spacer-linked mannosamine acceptor **311** in the presence of pivaloyl chloride to afford an intermediate that was oxidized with iodine in water.

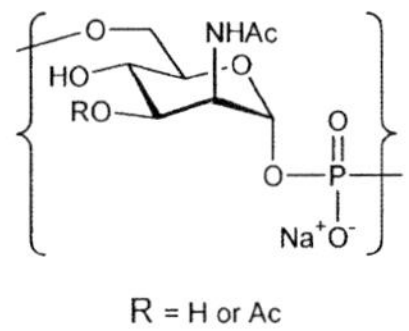

301

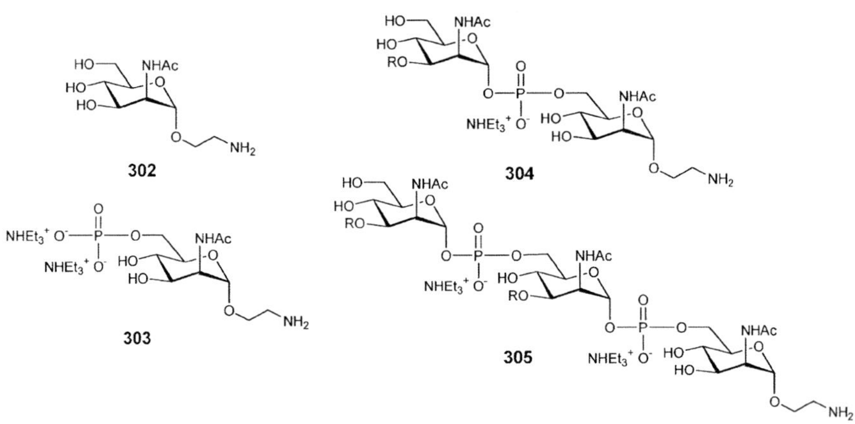

302 **303** **304** **305**

The phosphate diester-linked disaccharide **312** was obtained in 95 % yield. Conversion of the azido to acetamido group with nickel boride followed by N-acetylation and O-deacetylation afforded the dimer **313**. Iteration of the elongation step using H-phosphonate **310** (→**314**, 74%) followed by the N_3 → NHAc conversion as above gave the trisaccharide **315** in 80 % yield. Deprotection of the intermediates by deacetylation, followed by hydrogenolysis in the presence of triethylamine that was added to prevent hydrolysis of the acid-sensitive phosphodiester bonds, afforded the targeted aminoethyl saccharides **302-305**. These were covalently attached to human serum albumin using the authors' conjugation method[73] based on the Diels-Alder cycloaddition reaction. In a double immunodiffusion assay all the conjugates reacted with an antiserum raised against formaldehyde-inactivated *N. meningitidis* group A bacteria. This finding led the authors to conclude that a fragment as small as a monosaccharide is recognized by the homologous polyclonal antisera and that the O-acetyl group of the native polysaccharide is not essential for antigenicity.

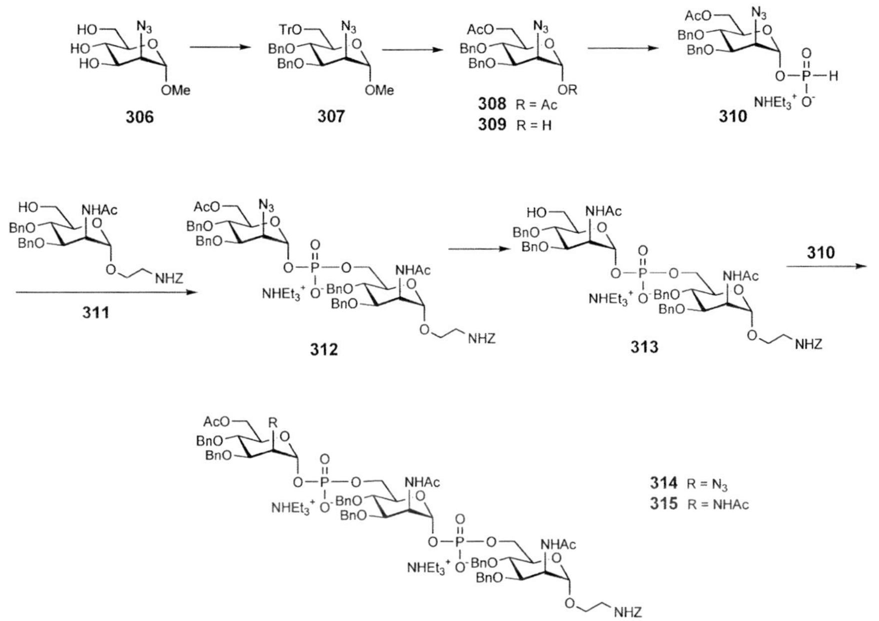

Group B *Neisseria meningitidis*

The CPS of this organism is a homopolysaccharide consisting of α-(2→8) linked N-acetylneuraminic acid (NeuAc) residues. Oligomers of this sequence also occur in numerous mammalian glycoconjugates. Syntheses of such compounds are problematic because of (i) the inherent difficulty of creating α-linkages of NeuAc residues and (ii) the low reactivity of its HO-8 hydroxyl group. An approach to this problem was provided by Schmidt's group[74,75] as shown below. Thus, NeuAc **316** was converted to the 2,3-dehydro derivative **317** that served as the key intermediate. The glycal **317** was transformed to the chlorohydrin **318** from which 3-phenylthionocarbonate **319** was prepared using chlorophenylthionocarbonate, in 89 % yield.

Compound **317** was also converted to the glycal diol **323**. This was done by deacetylation of the glycal **317** followed by isopropylidenation to afford the diol **320**. This was regioselectively acetylated at OH-4 (→**321**, 71%), followed by acid mediated acetal cleavage that exposed HO-7, 8, and 9 (→**322**, 94%). The sequence was completed by dibutyltin-acetal promoted regioselective benzylation of the triol **322** at O-9 to afford diol-glycal **323** in 70 % yield.

Condensation of **319** and **323** using silver triflate as the activator followed by reductive removal of the phenylthiono auxiliary gave exclusively the α(2→8) disaccharide **324** in 65 % combined yield.[74] The 2,3-double bond in the disaccharide **324** renders it suitable for further chain extension. In the approach of Demchenko and Boons, the di-N-acetylated glycosyl acceptor **325** was condensed with the thioglycoside donor **326** using NIS/TfOH activation to afford the disaccharide **327** in an α to β ratio of 1.7 in 50% combined yield.[76]

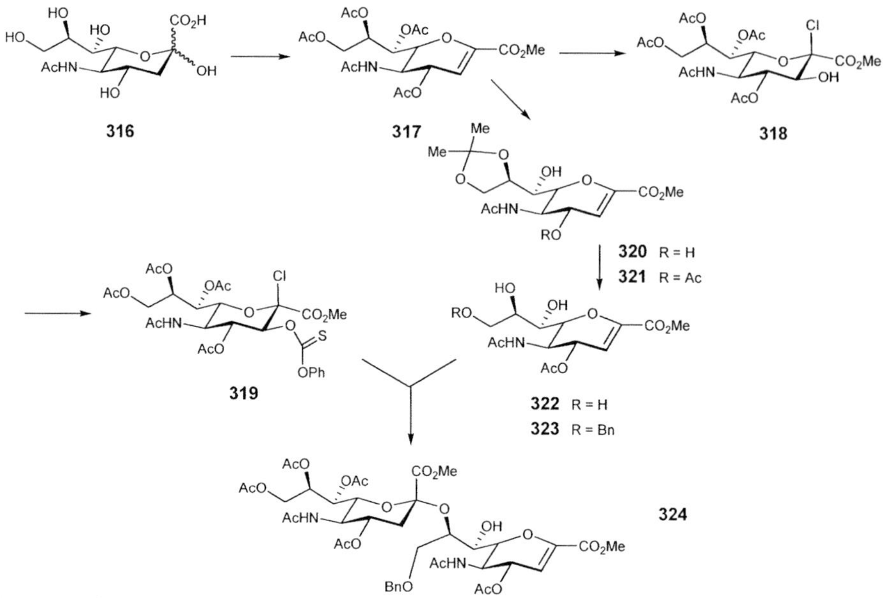

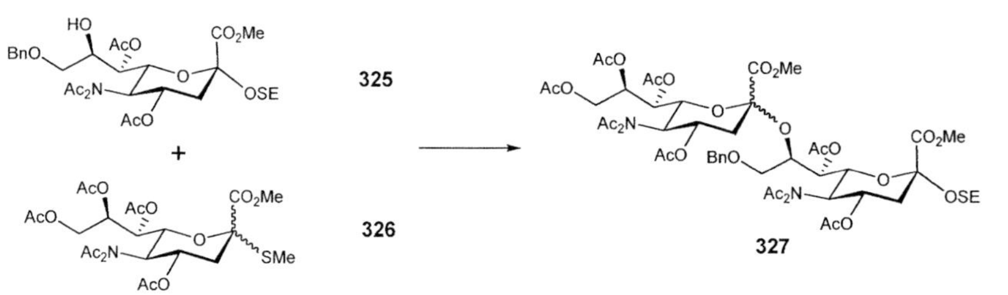

Group C *Neisseria meningitidis*

The CPS of the Group *C N. meningitidis* consists of α-(2→9)-linked NeuAc residues. Synthesis of a dimer corresponding to this CPS was reported by Demchenko and Boons in an approach closely related to that described above.[76] Thus, the 8,9-diol **328** was prepared and was condensed with the thioglycoside donor **326** using the NIS/TfOH reagent for activation. The α-(2→9)-linked dimer of NeuAc **329** was produced in 98% yield, in an α/β ratio of 2.5.

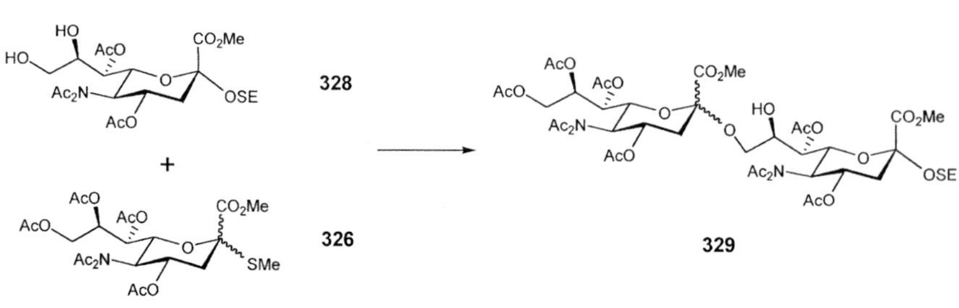

Pseudomonas aeruginosa

Pseudomonas are common nosocomial Gram-negative bacteria that are developing increasing resistance against antibiotics. It was suggested that the core region might serve as a candidate for glycoconjugate vaccine development.[77] In order to explore this possibility, Kosma and coworkers reported the synthesis of mono- and disaccharide congeners **333**, and **338** of the core oligosaccharide of *Pseudomonas aeruginosa*.[21,77] Starting compound for the monosaccharide **333** containing a 7-O-carbamoyl group was the reducing heptopyranose derivative **330** that was converted to the α-allyl glycoside **331** through the corresponding trichloroacetimidate. Deacetylation followed by reaction with trichloromethyl chloroformate installed the 6,7-O-carbonate moiety ($\rightarrow$**332**, 76% overall yield). Treatment of this intermediate with ammonia cleaved the carbonate in a regioselective manner to afford the targeted 7-O-carbamate **333** in 57% yield. The carbamoyl group in this compound is prone to migration: upon standing in an aqueous solution at room temperature for two weeks, formation of the 6-O-carbamate was noticed having the carbamoyl group at O-6.

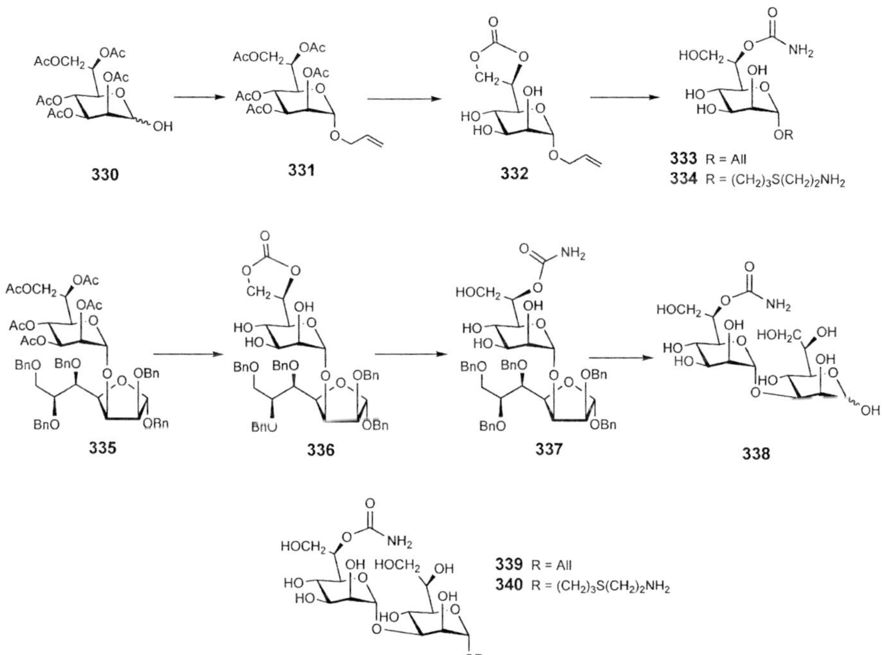

The disaccharide congeners **338** and **339** contain two α-(1$\rightarrow$3)-linked *manno*-heptopyranosyl residues.[21,77] Precursor to **338** was the α-benzyl heptofuranoside **335** that was deacetylated then was treated with the reagent trichloromethyl chloroformate as described above to produce the 6'-7'-O-carbonate-protected disaccharide derivative **336**. Treatment of this compound with ammonia afforded the 7'-O-carbamoyl derivative ($\rightarrow$**337**, 66% yield) from which removal of the benzyl protecting groups by hydrogenolysis afforded the hemiacetal **338** in 91% yield. Using similar chemistry, the disaccharide allyl glycoside **339** was also prepared. Constructs **333** and **339** were converted to the cysteamine adducts **334** and **340** under irradiation by UV light, that were activated with thiophosgene to afford the corresponding, highly reactive isothiocyanates for covalent attachment to bovine serum albumin.

The rhamnolipid **341** produced by *Pseudomonas aeruginosa* was reported to possess interesting biological properties. Its synthesis by van Boom's group utilized the cyclohexane-1,2-diacetal-protected rhamnoside **342** and the perbenzylated ethylthio rhamnoside **343**.[78] Condensation under NIS/TfOH activation afforded the dirhamnoside **344** in 75 % yield that was in situ reacted with the lipid **345** in the presence of NIS/TfOH to yield compound **346**. Deblocking of the carboxyl group with zinc in acetic acid followed by DCC-assisted condensation with a second molecule of the lipid alcohol **345** produced the protected intermediate **347** in 73 % yield from which removal of the protecting groups produced the disaccharide lipid **341**.

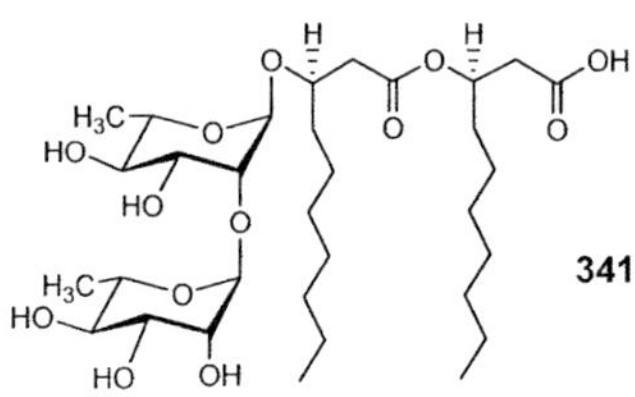

341

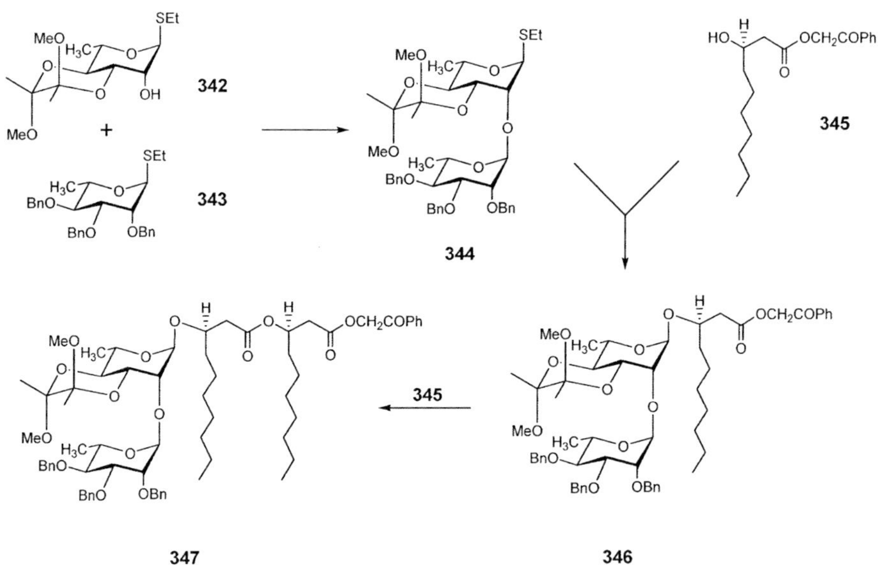

Rhizobia

Rhizobiae are Gram-negative, plant-specific bacteria that, at the outset of the infectious process, induce the formation of nodules on the roots of legumes in a symbiotic process. These organs reduce atmospheric nitrogen to ammonia that is assimilated by the plants. The signal molecules termed Nod-factors are lipopolysaccharides bearing O-specific polysaccharides.

The synthesis of the lipotetrasaccharide analogue **358** related to *rhizobial* nodulation factors was described by Robina's group.[79] The cellobiosyl donor **348** and the methoxyphenyl glucosaminide derivative **349** were synthesized and condensed under NIS/TfOH activation to afford the protected trisaccharide **350** with quantitative β-stereoselectivity. This intermediate was converted to the acetamido derivative **351** by treatment with ethylenediamine followed by acetylation in an 80 % overall yield. Next, compound **351** was deacetylated (Zemplén), then the intermediate polyol was subjected to benzaldehyde and zinc chloride to install the benzylidene acetal moiety, followed by benzylation with benzyl bromide to afford intermediate **352**. Exposure of **352** to the $NaCNBH_3/HCl$ reagent reductively opened the benzylidene acetal moiety and gave the acceptor **353** having the HO-4'' hydroxyl group available for chain extension.

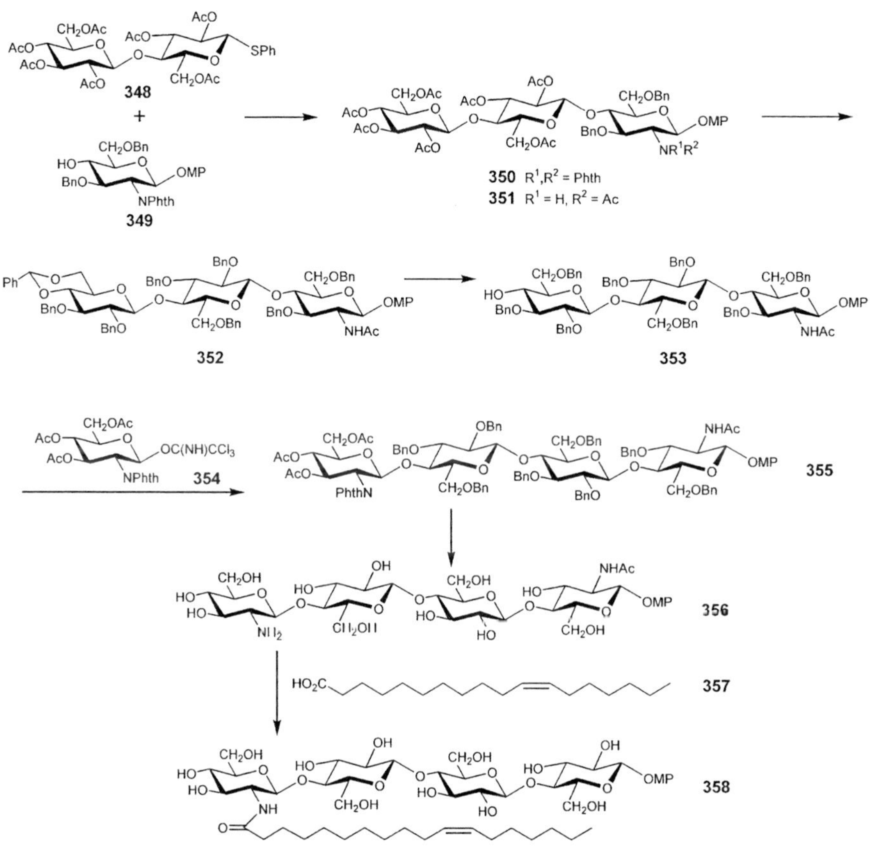

Alternatively, the benzylidene acetal moiety was removed from compound **352** by acid-hydrolysis and the benzyl group was placed at the primary hydroxyl group using the tin-acetal activation method. The two-step procedure afforded a somewhat lower overall yield (44 %) than the reductive ring opening protocol (60 %). Further chain elongation with the phthaloyl-protected glucosamine donor **354** under activation by silver triflate gave the fully protected tetrasaccharide **355** in 60 % yield. Conventional removal of the protecting groups by catalytic hydrogenolysis yielded the intermediate **356** featuring a free amino group. The synthetic sequence was completed by N-acylation with the lipid vaccenic acid **357** using 2-chloro-1-methylpyridinium iodide as the condensing agent to give the lipidated tetrasaccharide **358** in 20 % overall yield from **355**.

Rhizobium leguminosarum Biovar *Phaseoli* 127 K 87

The exopolysaccharides of several biovars of *R. leguminosarum* contain the common pyruvated tetrasaccharide motif **369**, synthesis of which was reported by Ziegler's group as the allyl glycoside for eventual conjugation purposes.[80] (The pyruvate is shown in the S configuration only, although the target compounds contained the pyruvate acetal moiety in both the S and R configuration). A key monosaccharide precursor was the pyruvated thiogalactoside **359** that was converted to the bromide **360** by exposure to bromine. The product was coupled, without isolation, with the galactoside acceptor **361** under activation by silver triflate to afford the disaccharide as benzyl glycoside **362** in 73 % yield. This was converted to the trichloroacetimidate **363** through hydrogenolysis over palladium-on charcoal and subsequent treatment with trichloroacetonitrile and potassium carbonate in 74 % yield for two steps. The other half of the tetrasaccharide was constructed from the allyl galactoside **364** and the galactosyl chloride **365**.

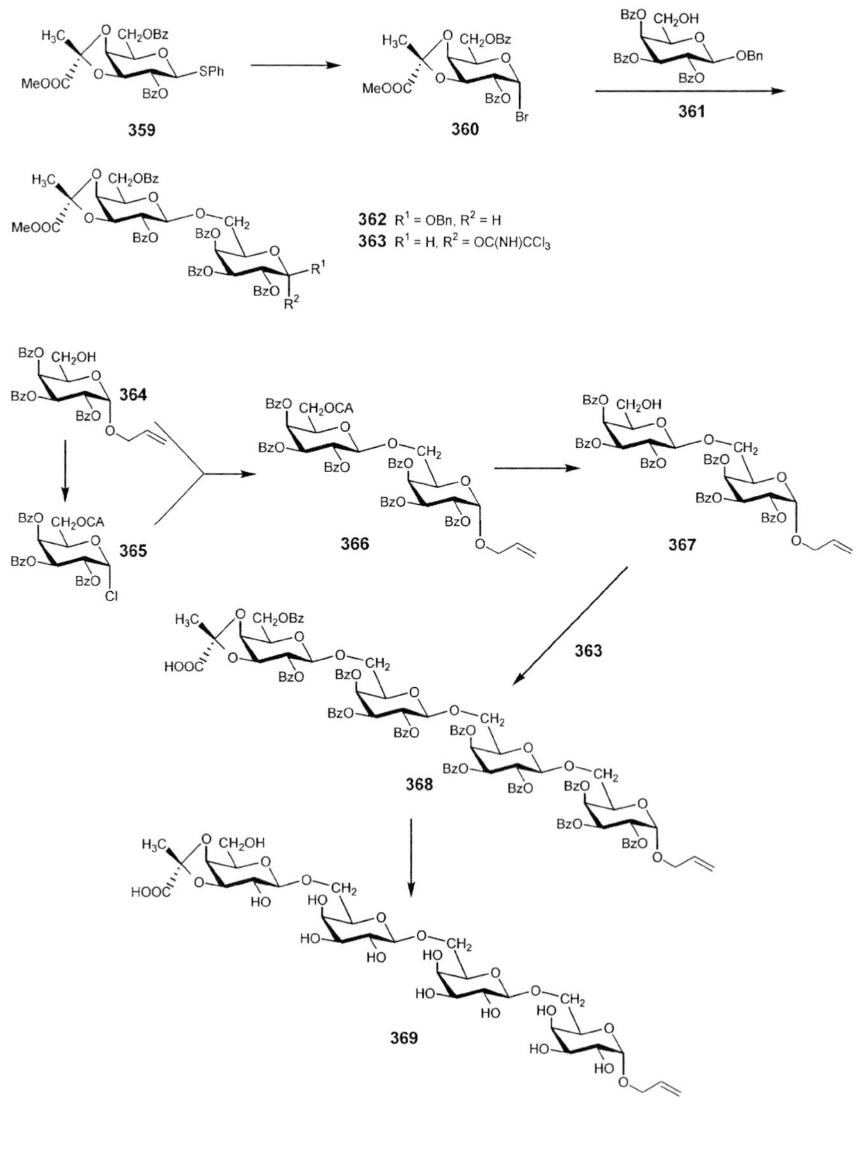

Their condensation, promoted by silver triflate, afforded the disaccharide **366** in 72% yield from which the chloroacetyl group was removed by treatment with thiourea to give the disaccharide acceptor **367** (85%). Next, the fully protected tetrasaccharide **368** was constructed by coupling the disaccharide imidate **363** and the acceptor **367** in the presence of TMSOTf, in 62 % yield. The target tetrasaccharide **369** was obtained after base-catalyzed cleavage (NaOMe/MeOH, then NaOH) of the acyl protecting groups in a quantitative yield. The allyl group in **369** makes possible its covalent attachment to macromolecular carriers after either ozonolytic formation of an aldehyde group in the aglycon or after introduction of an amino group by radical-addition of cysteamine to the allyl group.

Rhizobium leguminosarum **Biovar *Trifolii* 24**

The O-specific polysaccharide of this strain is composed of the trisaccharide repeating unit **370** containing the rare sugars 3-deoxy-D-*lyxo*-heptulosaric acid and 6-deoxy-D-talose in addition to the commonly found L-rhamnose. Synthesis of the congener **379** using compounds **371**, **372**, and **377** as key starting materials was reported by Banaszek.[81,82] Thus, 6-deoxy-L-talosyl chloride **371** and the rhamnopyranoside **372** were combined in the presence of silver triflate to furnish **373** in 72% yield. The benzyl groups were replaced by acetyls in two steps involving hydrogenolysis and acetylation to afford the disaccharide **374**. Next, the anomeric tert-butyl group was cleaved with trifluoroacetic acid and the alcohol **375** obtained was converted to the trichloroacetimidate **376** (CCl_3CN/K_2CO_3). Coupling of the disaccharide donor

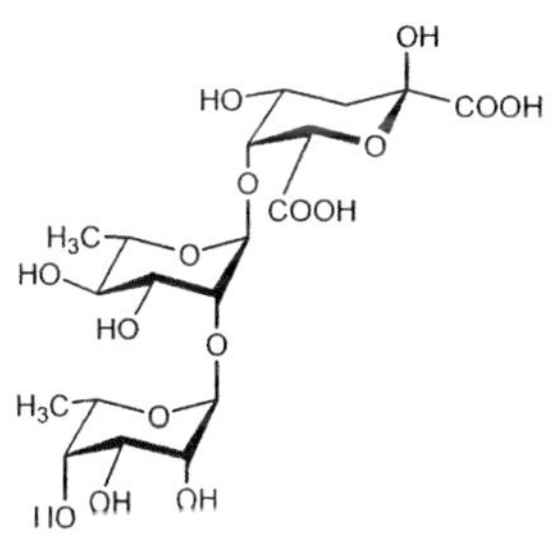

370

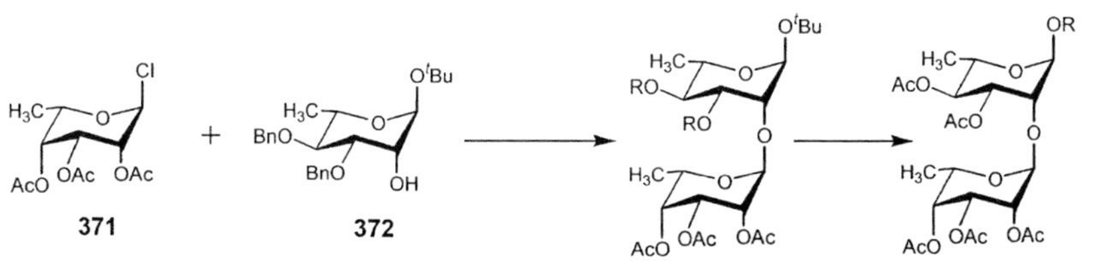

371 **372** **373** R = Bn **374** R = Ac **375** R = H **376** R = OC(NH)CCl_3

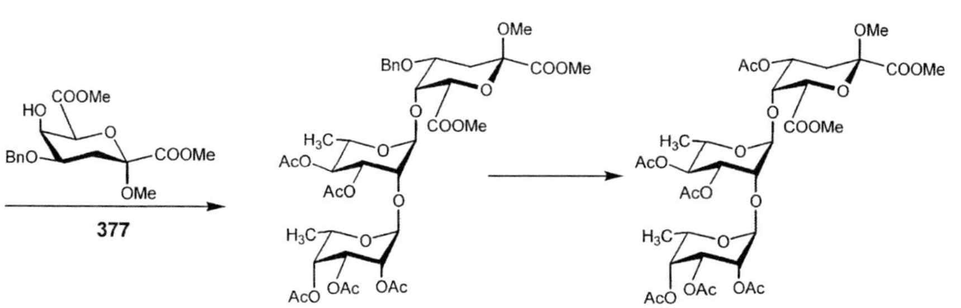

377 **378** **379**

376 with the diester acceptor **377** under promotion by boron trifluoride afforded, solely, the α-coupled trisaccharide **378** in 78% yield. This was subjected to hydrogenolysis followed by acetylation to produce the fully protected trisaccharide **379**.

Salmonella Group E_1

Zhao and Thorson reported a combined chemoenzymatic synthesis of the trisaccharide **387** representing a complete repeating unit of the O-SP of this organism.[83] The synthesis started with the condensation of acetobromo-L-rhamnose **380** and the partially silylated galactal **381** in the presence of silver triflate. The disaccharide **382** was obtained in 73% yield. In this reaction the presence of the bulky silyl group was necessary to influence the regioselectivity of the reaction as required. The glycal **382** was acetylated (→**383**) then the product was epoxidized by 3,3-dimethyldioxirane to provide the 2,3-anhydro derivative **384**. Treatment of the epoxide **384** with 4-(4-nitrophenyl)-1-butanol in the presence of one equivalent of zinc chloride produced the glycoside **385** in 43 % overall yield, in a 1:9 α/β ratio. The low yield was caused by the hydrolysis of the interglycosidic linkage and could be substantially increased (to 62 %) without changing the anomeric ratio by the use of only 0.6 equivalents of the activator. The disaccharide **385** was deprotected first with tetrabutylammonium fluoride then with sodium methoxide in 84 % overall yield (→**386**). Subsequent site-specific β-mannosylation was achieved by incubating **386** with GDP-α-β-mannose and a purified recombinant mannosyltransferase to yield the trisaccharide **387** in a > 85% turnover of **386** in less than 5 h. A purely chemical approach to the same trisaccharide was worked out by Crich and Li.[84]

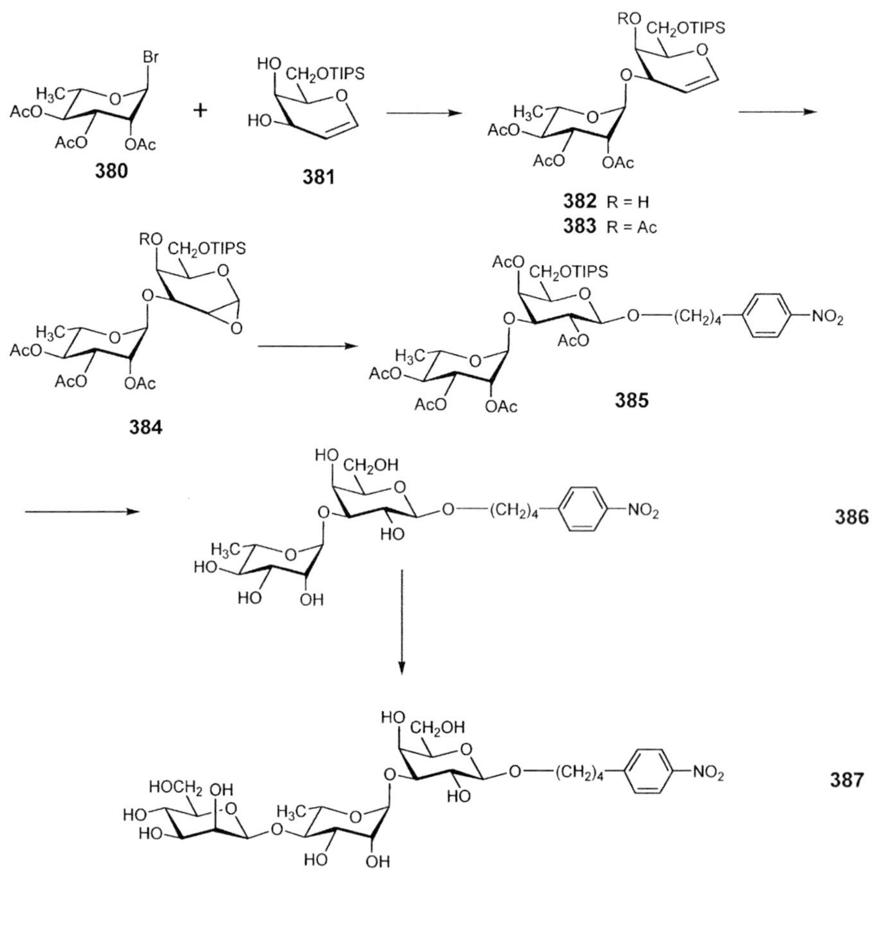

Thus, coupling of the benzylidene-galactoside acceptor **388** with the acetal-protected thiorhamnoside donor **389** using 1-(benzenesulfinyl)piperidine and triflic anhydride as the activator afforded the disaccharide **390** in 64% yield, from which removal of the silyl protecting group with tetrabutylammonium fluoride gave the disaccharide acceptor **391** in 80% yield. This was combined with the α-mannosyl donor **392** using the same activator as before to provide the fully protected trisaccharide derivative **393** in 71 % yield in addition to an α/β mixture (1:1.4, 21 % yield). The acetonide protecting group in the rhamnose moiety of **391** was essential for the highly β-selective glycosylation. The use of a carbonate protection in place of the acetal moiety gave rise to a 2:1 mixture of two isomeric trisaccharides that contain β- and the α-linked mannosyl moieties, respectively. The target **387** was obtained from **393** by (1) saponification that removed the pivaloyl group and (2) exposure to acidic conditions that cleaved all the remaining protecting groups.

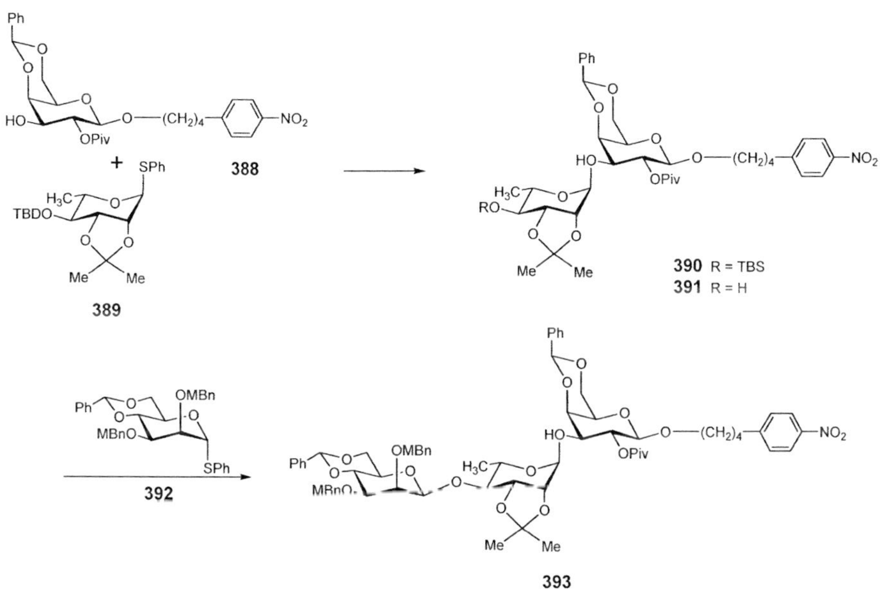

Salmonella arizona O62

A trisaccharide element of the O-SP of this organism (**109**) differs from that of *E. coli* O35 described above in that it contains a galacturonic acid residue in place of the galacturonamide moiety found in *E. coli* O35. For its synthesis, see the paragraph on *E. coli* O35.[33]

Salmonella greenside

Compound **660** in the section on *Vibrio cholera* O139 (vide infra) is also part of the O-antigen of *Salmonella greenside*.[34]

Salmonella typhi

The CPS of this bacterium is a homopolysaccharide of a-(1→4)-linked N-acetyl-D-galacturonic acid **394** having acetyl groups at O-3 in nonstoichiometric amounts. A vaccine containing the pure CPS is in clinical use (Typhim, Aventis-Pasteur). Its protein conjugate was shown to elicit superior antibody response in humans and is a candidate for a licensed vaccine.[85] Chemical synthesis of fragments of this polysaccharide was reported by Sinaÿ's group.[86] Starting from the β-galactosyl nitrate **395** the α-methyl glycoside **396** was prepared by methanolysis using sodium methoxide in 64% yield. Next, the isopropylidene moiety was removed in acetic acid followed by dibutyltin oxide-mediated regioselective methoxybenzylation to afford the alcohol **397** in 86% overall yield from **396** featuring an unprotected hydroxyl group at the site of chain elongation (O-4). The glycosyl donor ethyl xanthate **399** was obtained from the α-galactosyl nitrate **398** with EtOCSSK in 88% yield. DMTST activation of the xanthate **399** in the presence of the acceptor **397** afforded, mainly, the α-linked disaccharide **400** (53%) together with a smaller amount of the β-linked isomer (16%). Next, disaccharide **400** was subjected to the action of acetic acid to cleave the isopropylidene group followed by dibutyltin acetal-assisted regioselective methoxybenzylation of the HO-3' hydroxyl group to furnish compound **401** in 78% overall yield for two steps.

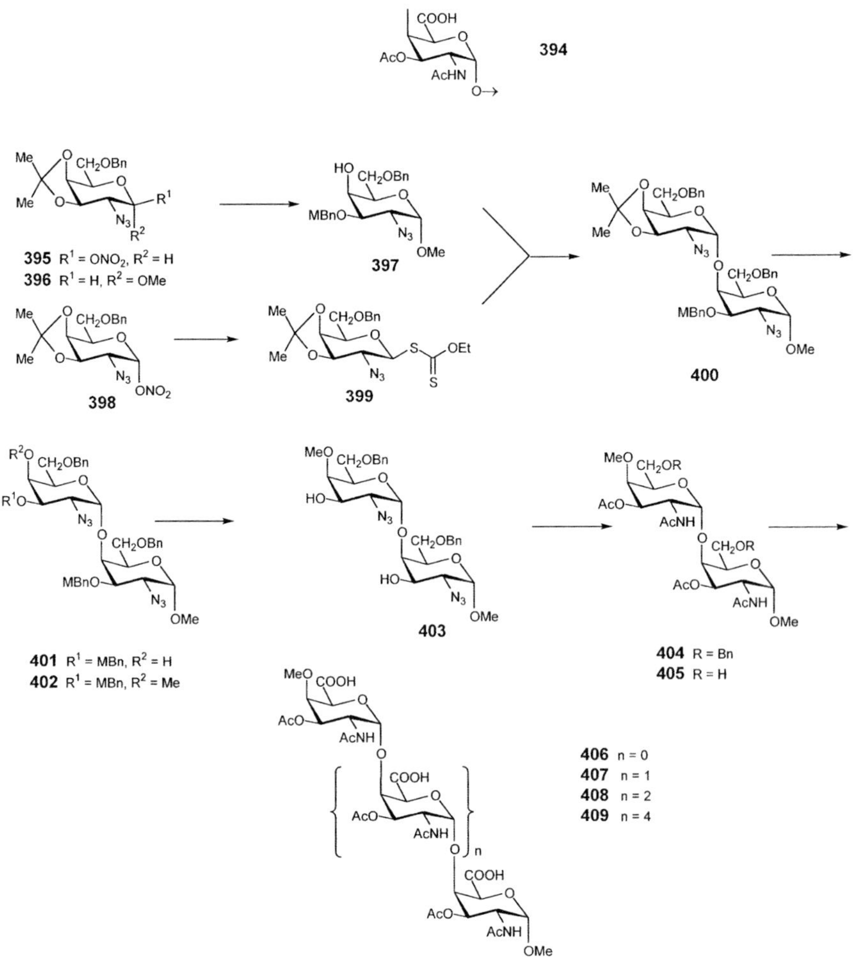

This was then converted to its 4'-O-methyl derivative **402** using methyl iodide and sodium hydride (97%) followed by acidolytic removal of the methoxybenzyl groups to yield the diol **403** in 83% yield. The methyl group was installed to block/mimic the site of the chain extension in the native polymer. Next, the two azido groups in compound **403** were converted to acetamido groups by reduction with nickel boride followed by N,O-acetylation to afford intermediate **404** (76%, two steps) from which catalytic hydrogenolysis produced the diol **405** in 93% yield. This was subjected to oxidation with $RuCl_3/NaIO_4$ to furnish the target di-acid **406** in 77% yield. A similar approach was employed for the synthesis of tri- **407**, tetra- **408**, and hexasaccharides **409**. While the disaccharide **406** and the trisaccharide **407** had no inhibitory effect on the interaction of the native PS with anti-Vi IgG antibodies obtained from mice in an ELISA assay, tetrasaccharide **408** inhibited this interaction at a concentration approximately 1,000 times higher than the native PS. The inhibitory effect of hexasaccharide **409** was approximately tenfold relative to that of tetrasaccharide **408**.

Salmonella typhimurium

A dodecasaccharide fragment **410** of the O-SP of *Salmonella* strain SH4809 corresponding to three repeating units was obtained by phage-associated endorhamnosidase-mediated site-selective cleavage of the native LPS. The fragment was then covalently attached to a peptide containing 13 amino acids termed PADRE through a scheme shown below (saccharide is represented by a pyranose ring).[87] The resulting glycoconjugate was tested for its immunogenicity in mice in admixture to complete Freund's adjuvant.

The PADRE-dodecasaccharide construct induced high titers of serum antibodies reacting with the homologous O-SP. The primary antibody pool consisted of ~ 30% IgM and ~70% IgG, while the response after the second injection was mostly of IgG type. Since Freund's adjuvant is not licensed for human use, the possible use of PADRE-based glycoconjugate vaccines in humans needs further evaluation with a medically acceptable adjuvant or without adjuvant at all.

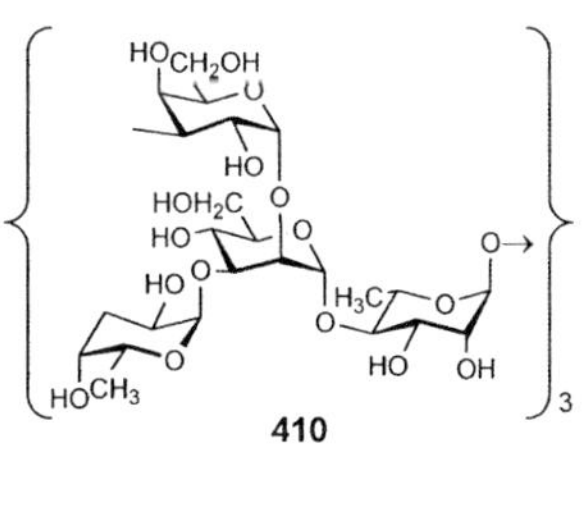

410

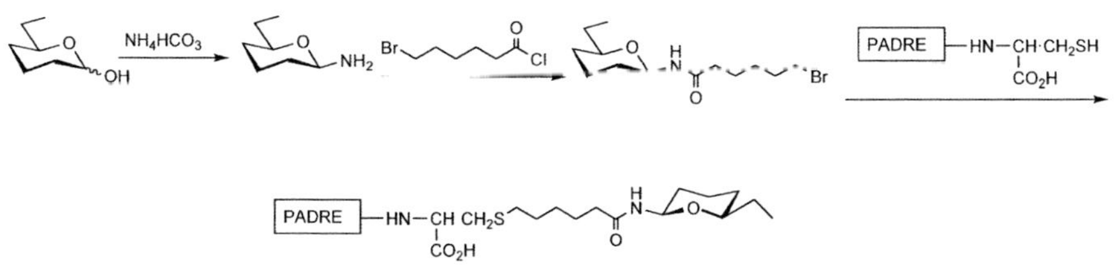

Shigellae

Shigellae are Gram-negative bacteria. Of the four groups, only four serotypes are considered major pathogens in humans. These are *Sh. dysenteriae* type 1, *Sh. flexneri* 2a, *Sh. flexneri* 5a, and *Sh. sonnei*. While they lack capsules, they express LPSs that are essential virulence factors: only strains having the O-SPs of their LPSs fully expressed are virulent. Based on the correlation between serotype-specific protection and a minimum level of serum antibodies against the O-SP, vaccines were constructed that contain the O-SPs covalently attached to an immunogenic protein.[9] These constructs elicit protective antibodies. It was surmised that protein conjugates of fragments of the *Shigellae* O-SPs might also be suitable immunogens.[88]

Shigella dysenteriae Type 1

The O-SP of this bacterium consists of the tetrasaccharide repeating unit **411** containing an N-acetyl-D-glucosamine, a D-galactose, and two L-rhamnose residues. Chemical synthesis of extended oligosaccharides corresponding to **411** was reported by Pozsgay.[88,89] In short, monosaccharide intermediates **412**, **413**, **414**, and **415** were prepared and were combined in a linear, stepwise fashion to afford the tetrasaccharide repeating unit as a thioglycoside (**416**). This unit was treated with mercury trifluoroacetate and the tetrasaccharide hemiacetal so obtained was transformed to the trichloroacetimidate **417** in 67 % overall yield.

Subsequently, **417** was condensed with the spacer **418** to afford under Schmidt conditions construct **419** in 73% yield. Next, the chloroacetyl group was removed by treatment with thiourea to yield **420** (95%) that has the site of the chain extension unprotected. Iterative chain extension involving glycosylation with the tetrasaccharide donor **417** followed by deprotection led to the free tetra- **421**, octa- **422**, dodeca- **423**, and hexadecasaccharides **424** in spacer-linked form featuring a hydrazinocarbonyl function for use in the covalent attachment to proteins. The synthetic oligosaccharide hydrazides were coupled to the secondary spacer **425** containing an activated carboxyl group and a masked aldehydo function. For example, exposure of the hexadecasaccharide **424** to the active ester **425** afforded the intermediate **426** containing the masked aldehydo derivative that was deprotected by mild acid treatment to produce the aldehyde **427**. Coupling of the aldehyde with the amino groups of the protein human serum albumin (HSA) by reductive amination using $NaCNBH_3$ as the reducing agent produced neoglycoconjugate **428**. The immunogenicity of the tetra-, octa-, dodeca-, and hexadeca-saccharide/protein conjugates were tested in mice.[90]

The maximum immune response was obtained with conjugates having approximately 10 dodeca- and hexadecasaccharide chains. Interestingly, high loading (an average of ~ 20 saccharide chains per HSA molecule) had a detrimental effect on immunogenicity. The synthesis of analogues containing up to 24 monosaccharide residues was also published.[14] Mulard and Glaudemans have reported the synthesis of deoxygenated (**433**, **435**) and fluorinated (**434**, **436**) tri- and tetrasaccharides, using L-rhamnose (**429**), 3-fluoro-3-deoxy-D-galactose (**430**, **431**), and D-glucosamine (**432**) building blocks.[91] Data on the carbohydrate binding specificities of antibodies directed against the native polysaccharides and related conformational studies have been reported.[92-95]

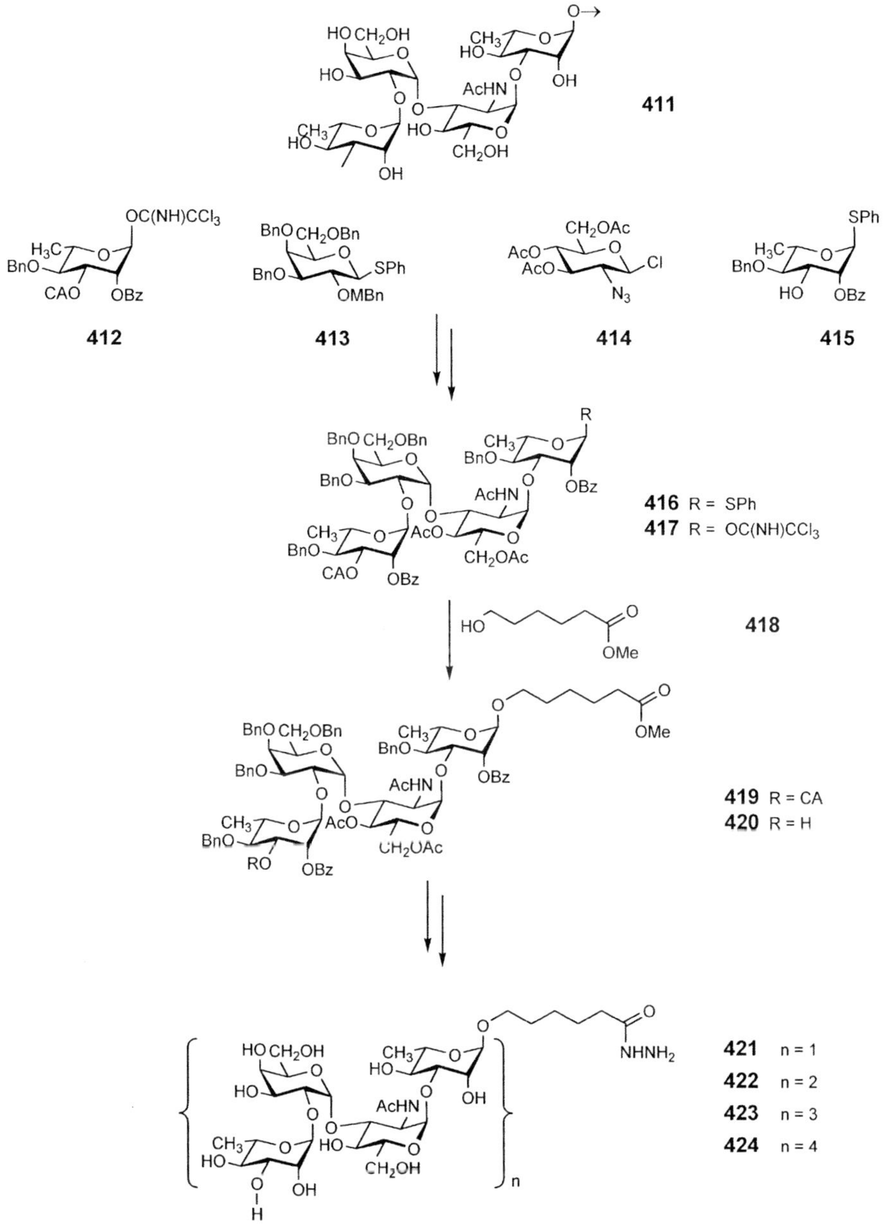
411
412
413
414
415
416 R = SPh
417 R = OC(NH)CCl3
418
419 R = CA
420 R = H
421 n = 1
422 n = 2
423 n = 3
424 n = 4

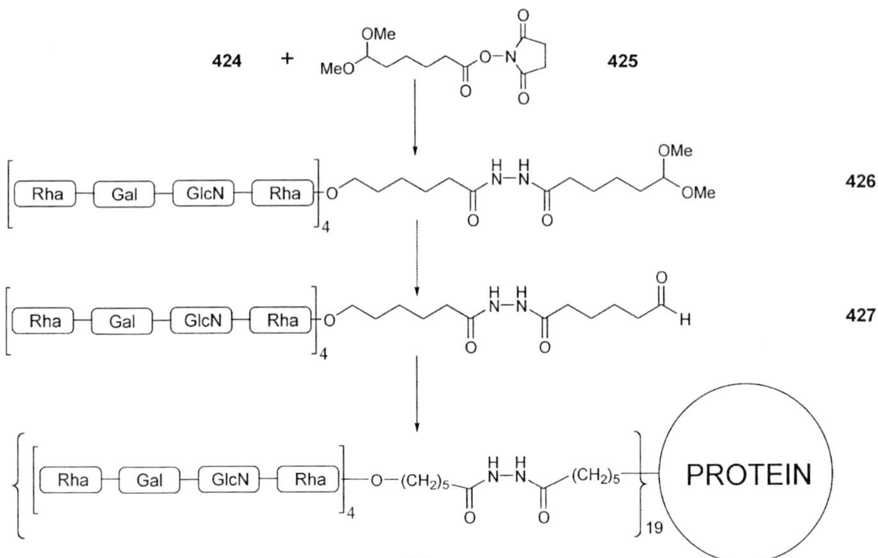
424
+
OMe
MeO
O
O–N
O
O
425
Rha
Gal
GlcN
Rha
O
4
H H
N–N
O
O
OMe
OMe
426
O
H
427
O–(CH2)5
(CH2)5
19
PROTEIN

428

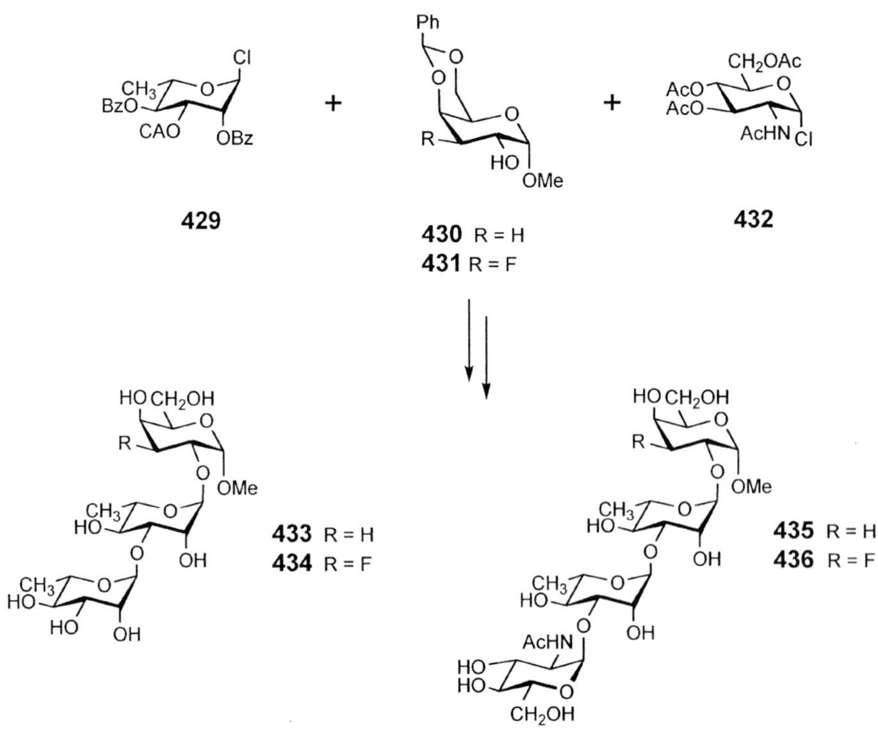
Cl
CH3
BzO
CAO
O
OBz
429
+
Ph
O
O
R
HO
OMe
430 R = H
431 R = F
CH2OAc
AcO
AcO
AcHN
Cl
432
HO CH2OH
R
OMe
CH3
HO
OH
433 R = H
434 R = F
435 R = H
436 R = F
AcHN
HO
HO
CH2OH

Shigella dysenteriae Type 5

The repeating unit of the O-SP consists of the partially O-acetylated tetrasaccharide **437** containing a (R)-lactic acid moiety at its side-chain L-rhamnose residue. Synthesis of this structure using thioglycoside chemistry was reported.[96] Starting from thiorhamnoside **280** and (S)-2-bromopropionic acid, the lactic acid derivative **438** was synthesized. Selective activation of the latter with iodonium dicollidine perchlorate in the presence of the thiomannoside acceptor **439** produced the disaccharide **440** in 71% yield. The other half of the target was assembled from the thiomannoside donor **441** and the glucosamine derivative **442** as the acceptor, using the NIS/TfOH reagent to afford the disaccharide **443** in 90% yield. Treatment of **443** with $NaCNBH_3$ and hydrogen chloride led to regioselective reductive acetal ring opening to furnish the disaccharide acceptor **444**. Condensation of the latter with the thioglycoside **440** using NIS/TfOH as the activator gave the fully protected tetrasaccharide **445** in 79% yield, from which conventional protecting group manipulations yielded the tetrasaccharide methyl ester as trimethylsilylethyl glycoside **446**.

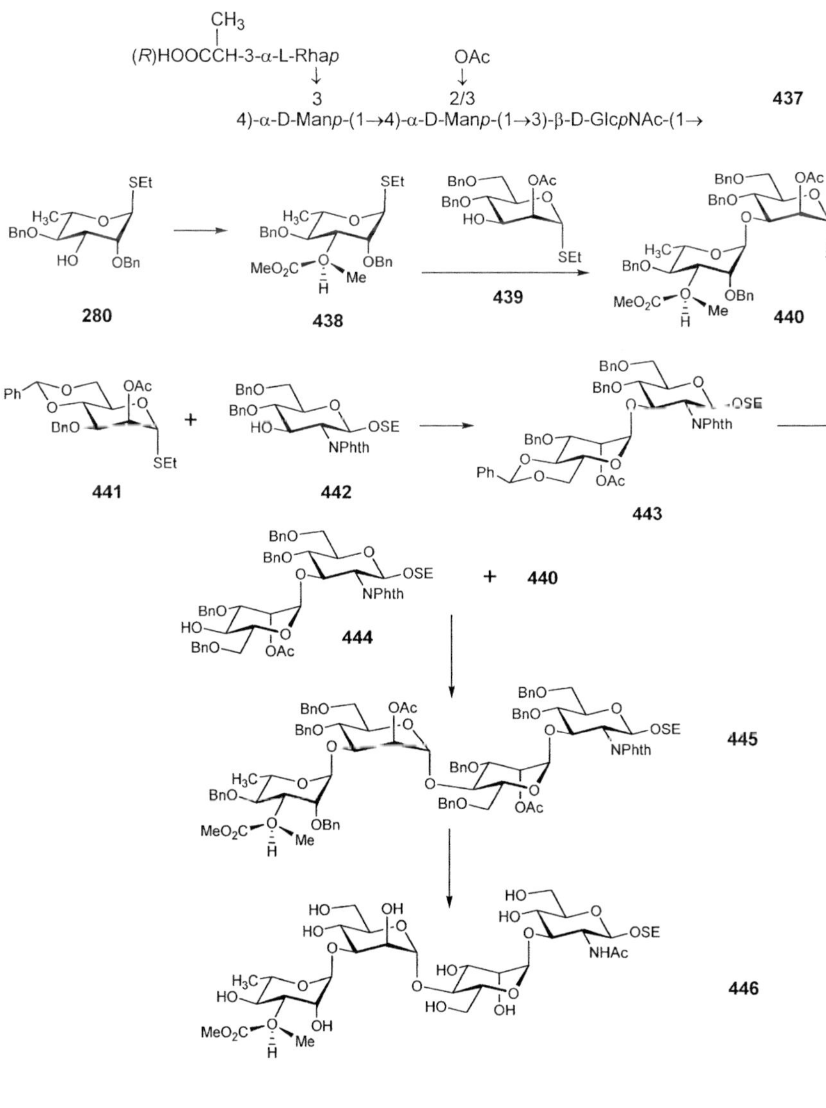

Shigella flexneri Type 2a

The O-SP of the LPS of this organism is composed of the branching pentasaccharide repeating unit **447**. Disaccharides **448** and **449** and linear **450** and branching trisaccharide **451** fragments of the repeating unit were synthesized as their methyl glycosides by Mulard's group.[97] In a typical sequence, the silyl-protected allyl rhamnoside **452** was used as the key starting material in which the role of the trimethylsilyl group was to increase the nucleophilicity of O-4. Condensation of **452** and the fluoride donor **453** under promotion by triflic anhydride as the activator yielded the disaccharide **454** in 55% yield. The β-linked isomer was also isolated in a 23% yield. Compound **454** was subjected to acid hydrolysis to cleave the acetal moiety followed by benzoylation of the hydroxyl groups (→**455**, 86% overall yield). Next, the allyl group was cleaved by isomerization with a cationic iridinium complex followed by hydrolysis with mercury salts to yield a hemiacetal from which the trichloroacetimidate **456** was prepared by the action of trichloroacetonitrile and a catalytic amount of DBU, in 90 % yield.

This was condensed with the glucosaminyl acceptor **457** in the presence of TMSOTf in acetonitrile to furnish the protected trisaccharide derivative **458** in 74% yield. Using boron trifluoride as the activator, the yield could be improved to 81%. A sequence of deprotecting steps including Zemplén-type debenzoylation, hydrolytic removal of the acetal protection followed by debenzylation using catalytic hydrogenolysis afforded the target trisaccharide **450** in 90% yield. Syntheses of the branched tetra- **459** and the penta-saccharide **460** as their methyl glycosides were also reported.[98]

α-D-Glc*p*
1
↓
4
2)-α-L-Rha*p*-(1→2)-α-L-Rha*p*-(1→3)–α-L-Rha*p*-(1→3)-β-D-Glc*p*NAc-(1→ **447**

α-D-Glc*p*-(1→4)-α-L-Rha*p*-(1→OMe **448**

β-D-Glc*p*-(1→4)-α-L-Rha*p*-(1→OMe **449**

α-D-Glc*p*-(1→4)–α-L-Rha*p*-(1→3)-β-D-Glc*p*NAc-(1→OMe **450**

α-L-Rha*p*-(1→3)-[α-D-Glc*p*-(1→4)]–α-L-Rha*p*-(1→OMe **451**

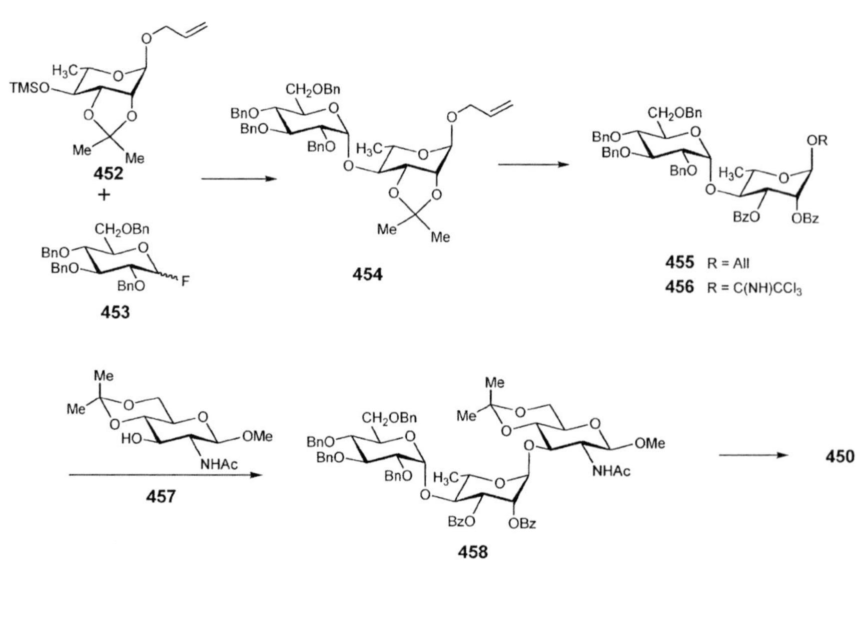

```
                        α-D-Glcp
                            1
                            ↓
                            4
α-L-Rha-(1→2)-α-L-Rhap-(1→3)-α-L-Rhap-(1→OMe
```
459

```
                                            α-D-Glcp
                                                1
                                                ↓
                                                4
β-D-GlcpNAc-(1→2)-α-L-Rhap-(1→2)-α-L-Rhap-(1→3)-α-L-Rhap-(1→OMe
```
460

Shigella flexneri Type 5a

The O-SP of this organism shares an identical backbone with the O-SP of *Sh. flexneri* serotype 2a. The difference between the two polysaccharides is the location of the α-D-glucopyranosyl side-chain residue that, in serotype 5a (**461**) is at the rhamnose unit that is flanked by Rha residues at each side.

Mulard's group reported the synthesis of a panel of overlapping oligosaccharides as their methyl glycosides including the tri- (**462**, **463**), tetra- (**464**, **465**), and penta-saccharide fragments (**466**, **467**).[99-101] As a typical example, the assembly of the pentasaccharide **466** was performed in a stepwise manner based on the Schmidt glycosylation protocol.

```
                  α-D-Glcp
                      1
                      ↓
                      4
2)-α-L-Rhap-(1→2)-α-L-Rhap-(1→3)-α-L-Rhap-(1→3)-β-D-GlcpNAc-(1→
```
461

```
α-D-Glcp-(1→3)-α-L-Rhap-(1→3)-α-L-Rhap-(1→OMe
```
462

```
       α-D-Glcp
           1
           ↓
           3
α-L-Rhap-(1→2)-α-L-Rhap-(1→OMe
```
463

```
       α-D-Glcp
           1
           ↓
           3
α-L-Rhap-(1→2)-α-L-Rhap-(1→3)-α-L-Rhap-(1→OMe
```
464

```
                               α-D-Glcp
                                   1
                                   ↓
                                   3
β-D-GlcpNAc-(1→2)-α-L-Rhap-(1→2)-α-L-Rhap-(1→OMe
```
465

```
                               α-D-Glcp
                                   1
                                   ↓
                                   3
β-D-GlcpNAc-(1→2)-α-L-Rhap-(1→2)-α-L-Rhap-(1→3)-α-L-Rhap-(1→OMe
```
466

```
                                                     α-D-Glcp
                                                         1
                                                         ↓
                                                         3
α-L-Rhap-(1→3)-β-D-GlcpNAc-(1→2)-α-L-Rhap-(1→2)-α-L-Rhap-(1→OMe
```
467

The synthetic sequence started with the condensation of the methyl rhamnoside acceptor **119** and the rhamnosyl trichloroacetimidate **468** in the presence of TMSOTf to give a fully protected disaccharide **469** in 70% yield. Deacetylation (Zemplén) gave the diol **470** that was regioselectively acetylated at O-3 through a cyclic orthoester formed on the HO-2 and 3 hydroxyl groups followed by hydrolytic ring opening in aqueous acetic acid to afford the disaccharide acceptor **471** in a quantitative yield.[99] Next, the disaccharide **471** was condensed with the glucosyl fluoride **453** using titanium tetrafluoride as the activator to furnish the trisaccharide **472**α,β as an inseparable anomeric mixture in an 85:15 α,β ratio.

This mixture was exposed to sodium methoxide that was needed in a stoichiometric amount for complete removal of the acetyl group. Repeated chromatography of the deacetylated mixture afforded the anomerically pure trisaccharide **473** in 45% yield. The corresponding isomer containing β-linked glucopyranosyl moiety was isolated in 9% yield. Condensation of the trisaccharide acceptor **473** with rhamnosyl trichloroacetimidate **474** under TMSOTf activation afforded the fully protected tetrasaccharide **475** in 94% yield. Next, the tetrasaccharide was deacetylated to produce the tetrasaccharide alcohol **476**. Glycosylation of the latter with the protected glucosamine trichloroacetimidate **477** in the presence of a catalytic amount of TMSOTf gave the protected pentasaccharide **478** in 81 % yield that was sequentially deprotected. First, compound **478** was exposed to an iridinium complex then to mercury salts to remove the allyl groups. The intermediate so obtained was treated with sodium methoxide followed by catalytic hydrogenolysis then hydrogenation to produce the pentasaccharide as the methyl glycoside **466**.

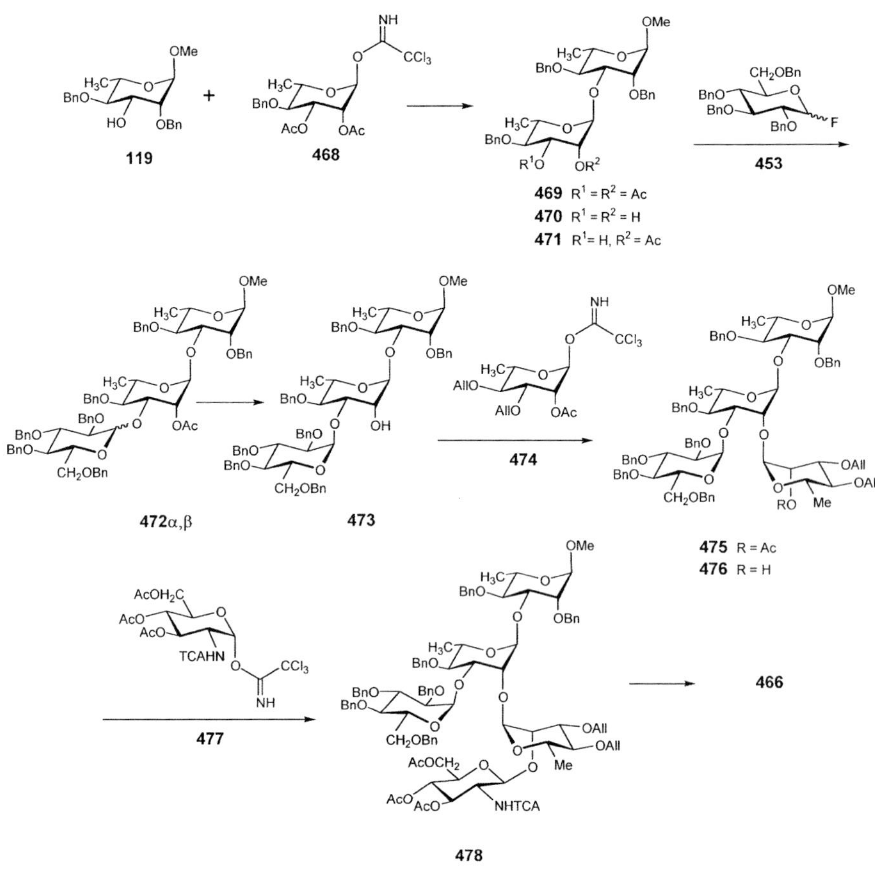

Shigella sonnei

Sh. sonnei contains a unique O-SP that is composed of the disaccharide repeating unit **479** containing the rare acetamido-altruronic acid **480** and the 6-deoxygalactose congener **481** shown as methyl glycosides. The synthesis of these saccharides was reported by Lipták' group.[102,103] The route to the altruronic acid derivative **480** started from methyl α-L-glucopyranoside **482** that was converted to the di-tosylate **483** through benzylidene acetal formation and tosylation (47 % yield for two steps) from which the epoxide **484** was prepared in the presence of sodium methoxide in methanol in 77% yield. Opening of the epoxide ring with sodium azide established the L-*altro* stereochemistry in a highly regioselective reaction (→ **485** 66%). Next, the azido-alcohol **485** was O-benzylated and the acetal protecting group was removed to afford the diol **486** (62% overall yield). Exposure of the diol **486** to the TEMPO/NaOCl reagent followed by esterification with methyl iodide produced the altruronic acid ester **487** in 62% yield for two steps. Simultaneous reduction of the azido group and hydrogenolytic cleavage of the benzyl protecting group in a hydrogen atmosphere in the presence of palladium hydroxide followed by N-acetylation yielded the target uronic acid derivative **488** in 90% overall yield.

The synthesis of the disaccharide repeating unit as the methyl glycoside **493** was also reported.[103] Starting from the trideoxy-galactose donor **489** and the altruronic acid acceptor **490**, the fully protected disaccharide **491** was synthesized in the presence of NIS/TfOH, in

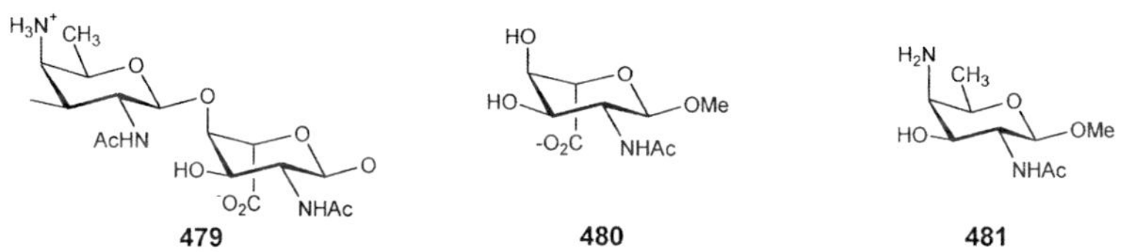

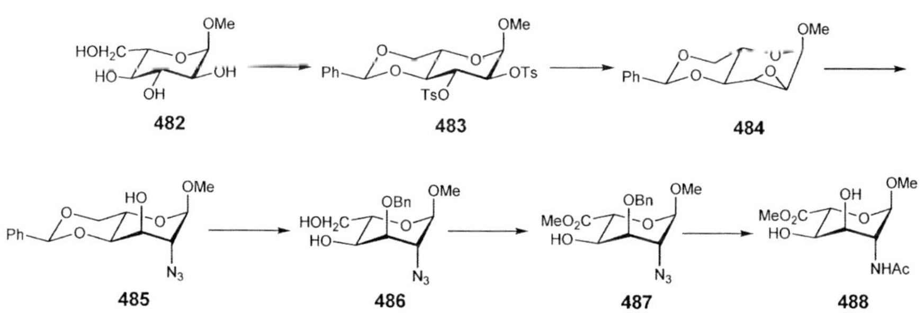

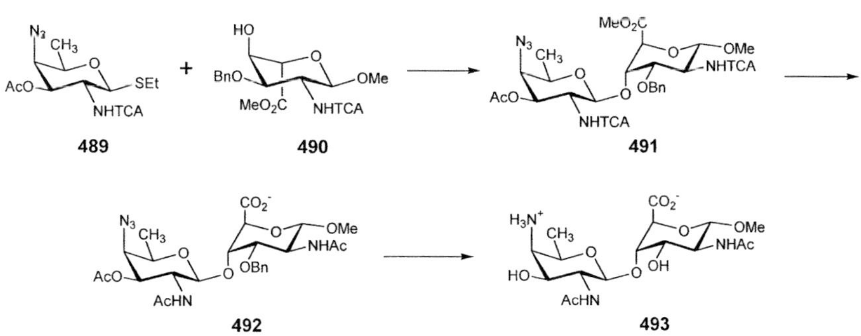

85% yield. The *N*-trichloroacetyl group in the donor residue **489** assisted β-selective glycosylation. Intermediate **491** was subjected to sodium hydroxide to remove the *N*-trichloroacetyl groups and cleave the ester, then the diamine product was *N*-acetylated to furnish compound **492**. One-step hydrogenation/hydrogenolysis over palladium produced the target zwitterionic structure **493**. All three saccharides **480**, **481**, and **493** were recognized by rabbit antibodies raised against *Sh. sonnei* giving support for the proposed structure of the O-SP. The synthesis of related oligosaccharides has also been described.[104]

Stenotrophomonas maltophilia JCC 18, 2623 1999

The O-SP of the serogroup O2 of this organism contains the branched trisaccharide repeating unit **494** consisting of the L-rhamnosyl-D-mannosyl backbone carrying L-xylose side chains at the rhamnose moieties. In an approach to the repeating unit, Wang and Kong reacted acetobromo-rhamnose **380** with the partially *O*-benzoylated allyl mannoside **495** in the presence of silver triflate to afford the orthoester **496** as a sole product in a nearly quantitative yield.[105] Deacylation with sodium methoxide in methanol followed by benzoylation afforded the perbenzoylated intermediate **497** that was subjected to acid-catalyzed rearrangement of the orthoester portion to furnish disaccharide **498**. This was deacetylated with hydrogen chloride in methanol to produce the alcohol **499** in 96% yield. Subsequent condensation with acetobromo-L-xylose **500** using silver triflate as the activator provided an orthoester intermediate from which acid-catalyzed rearrangement led to the fully protected allyl glycoside **501**. The target trisaccharide was obtained by deacylation (Zemplén) as the allyl glycoside **502** in 96% yield.

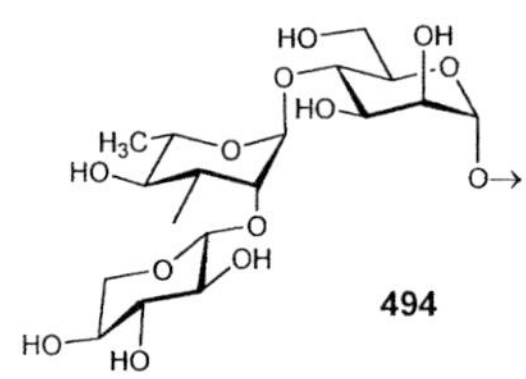

494

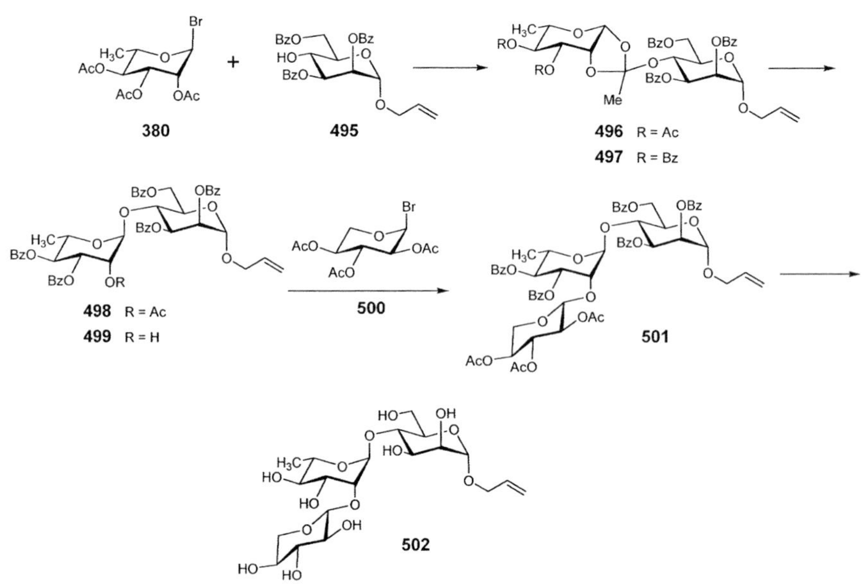

380, 495, 496 R = Ac, 497 R = Bz, 498 R = Ac, 499 R = H, 500, 501, 502

Streptococci

Because of their clinical importance, numerous synthetic endeavors target oligosaccharide portions of *Streptococcal* CPSs. In fact, a disaccharide fragment of type 3 *Streptococcal* CPS was already synthesized in 1938. For a catalogue of the synthetic fragments up to 1999 the reader is referred to Kamerling's monograph on *Streptococcal* carbohydrates from a chemist's view.[2]

Group A *Streptococcus*

This organism (in popular parlance "flesh-eating bacteria") can cause pharyngitis, pneumonia, toxic shock, and severe wounds. Pinto's group reported[106] the synthesis of a number of oligosaccharide fragments of its cell wall polysaccharide e.g., the hexasaccharide **503** and fragments thereof. Conjugation of these oligosaccharides to proteins and solid supports was achieved using Tietze's squarate method.[107]

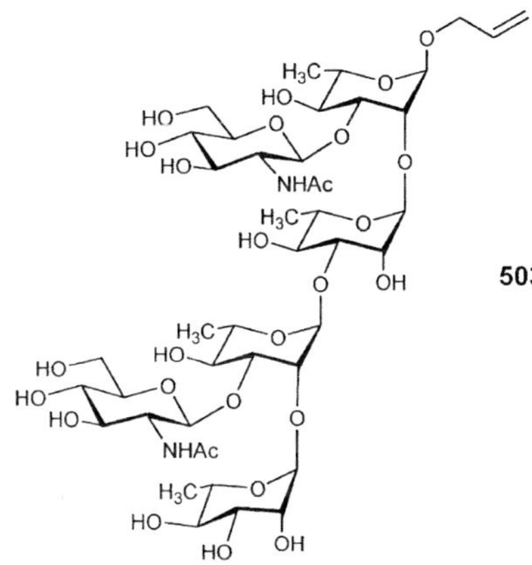

503

In this two-stage procedure, an aminoalkyl glycoside **504** is reacted with diethyl squarate **505** at neutral pH to afford the intermediate **506**. In the second phase the half-ester **506** is condensed with a protein (**507**) at elevated pH (10 to 10.5) to furnish the conjugate 1,2-bisamide **508** for use as an immunogen or immunoadsorbent. Using synthetic oligosaccharides the key components of the carbohydrate epitope of the cell-wall polysaccharide of Group A *Streptococcus* recognized by an anti-PS antibody were identified.[108]

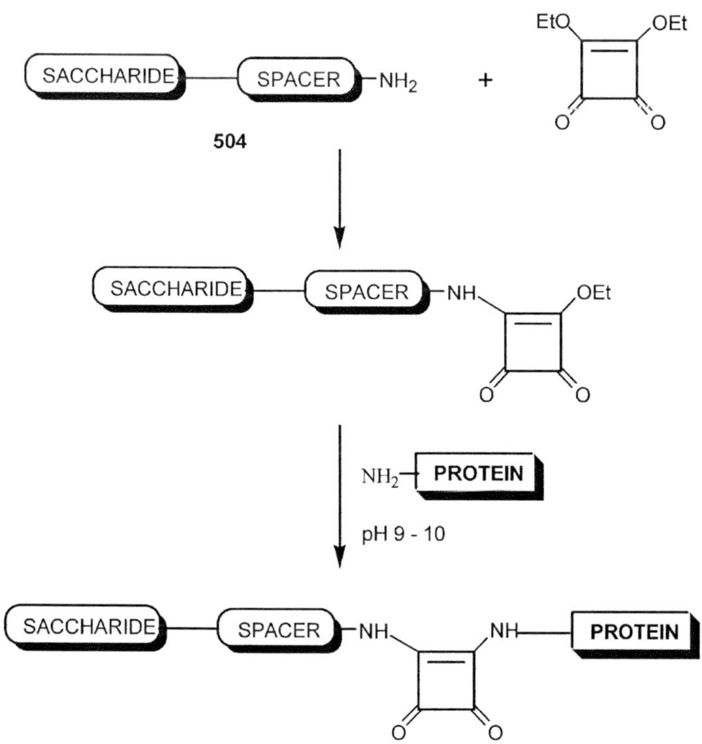

504 **505** **506** **507** **508**

Group B *Streptococcus*

Group B streptococci (GBS) are a major cause of neonatal sepsis. The organisms belonging to this group share a common polysaccharide that has a complex, branching structure containing, among others, several α-(1→2)- and α-(1→3)-linked L-rhamnose residues. In addition to GBS, such linkages occur in a large number of CPSs and O-SPs and attracted the interests of several authors. The most recent synthesis of the dirhamnoside **509** is that of Widmalm's group who prepared a deuterated derivative for NMR studies.[101] The synthesis of the rhamnotriose **510** was reported by Verez-Bencomo's group in a spacer-linked form that allows its covalent connection to proteins.[109]

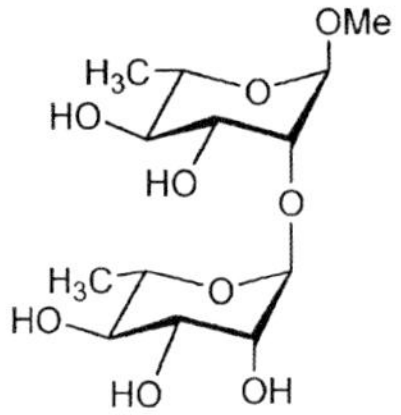

509

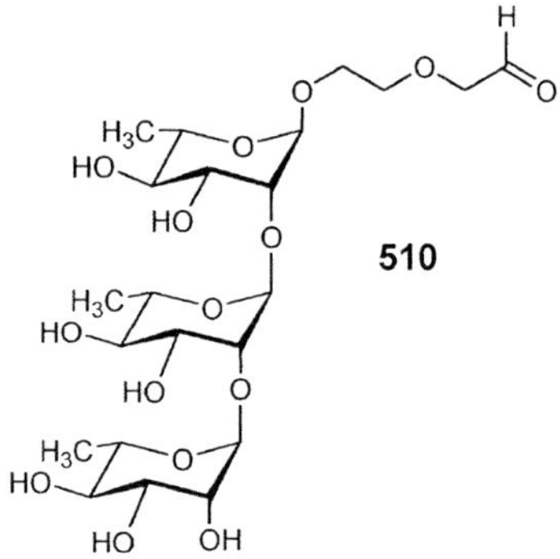

510

Type IA GBS

The CPS of this organism is composed of the branching pentasaccharide repeating unit **511** having a disaccharide backbone and a trisaccharide side-chain. Synthesis of the branching trisaccharide portion **520** of the repeating unit was reported by Mehta and Whitfield.[110] Thus, the differentially protected galactose derivative **512** was attached to the soluble monomethyl ether of polyethylene glycol **513** featuring a dioxyxylene linker using the NIS/AgOTf reagent for anomeric activation. Under these reaction conditions, in addition to the desired glycosylation reaction, iodination of the methoxybenzyl protecting group also occurred giving rise to the unexpected iodinated derivative **514** in 97% yield. The acid-stability of the iodinated methoxybenzyl group is elevated relative to that of the methoxybenzyl group. Removal of the levulinoyl group with a solution of hydrazine hydrate in pyridine and acetic acid afforded the alcohol **515** in 90% yield. Next, the acceptor **515** was glucosylated with the trichloroacetimidate donor **516** under activation by triethylsilyl triflate to produce the polymer-supported disaccharide **517** in a nearly quantitative yield. Treatment of this compound with 10 % TFA in CH_2Cl_2 removed the 3-iodo-4-methoxybenzyl group giving rise to the disaccharide acceptor **518**. In the final stages of the synthesis the disaccharide **518** was aminoglucosylated with the phthalimido-derivative **519** using NIS/AgOTf as the activator to afford a support-linked trisaccharide. The targeted trisaccharide **520** was obtained from the polymer support by cleavage with the scandium triflate/acetic anhydride reagent.

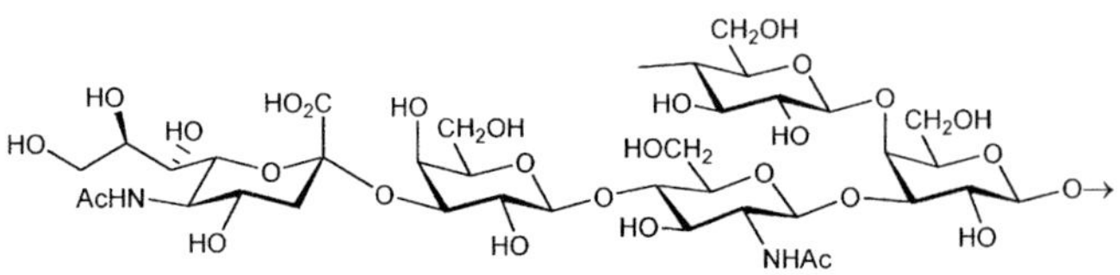

511

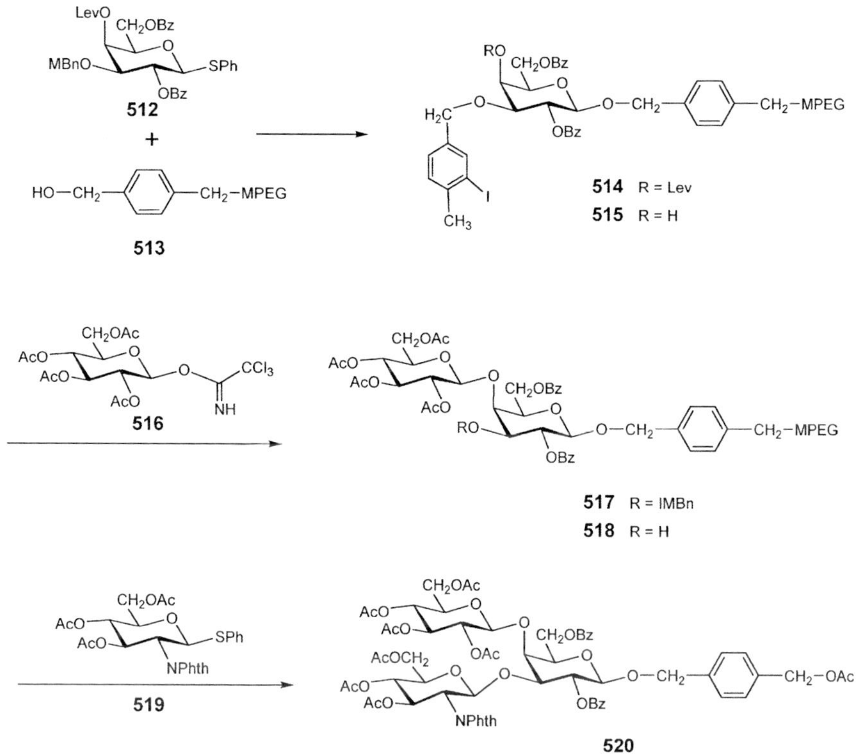

Type 3 GBS

The repeating unit of Type 3 Group B streptococcal polysaccharide (GBS) consists of the branching pentasaccharide **521** in which a sialyl-galactose disaccharide side-chain is attached to the trisaccharide backbone. Synthetic studies[111-114] culminated in a chemoenzymatic synthesis of a decasaccharide[115] (**522**) corresponding to two contiguous repeating units where the two sialic acid residues were introduced enzymatically to a synthetically prepared octasaccharide. Jennings' group also reported the synthesis of a frame-shifted decasaccharide and its C-13-labeled and N-propionyl-substituted sialic acid analogues.[116] The syntheses of a hexasaccharide[117] **523** and its sialylated derivative[118] **524** were described by Demchenko and Boons. The latter construct contains an aminopropyl aglycon for eventual coupling to proteins. Jenning's group also used a combination of controlled polysaccharide degradation, chemical modification and enzymatic sialylation to prepare higher-membered oligosaccharides up to a 30-mer (**525**) containing five contiguous repeating units.[119]

In 1997, a [5 + 3] block synthesis of octasaccharide **538** was reported that contains (S)-1-carboxyethyl groups in place of the sialic acid residues.[120] This synthesis employed two disaccharide building blocks, namely the lactosyl trichloroacetimidate **526** and the lactosamine diol **527** that were combined in the presence of TMSOTf. The HO-3 hydroxyl group of the glucosamine residue in **527** appears to be unreactive under the conditions of glycosylation. The resulting intermediate **528** (44%) was deallylated with palladium chloride to give the tetrasaccharide hemiacetal **529** that was converted to the imidate **530** by the action of trichloroacetonitrile and potassium carbonate. Subsequent condensation with the galactoside acceptor **531** in the presence of trimethylsilyl trifluoromethanesulfonate produced the pentasaccharide **532** in 66% yield. Next, the isopropylidene protecting group was removed

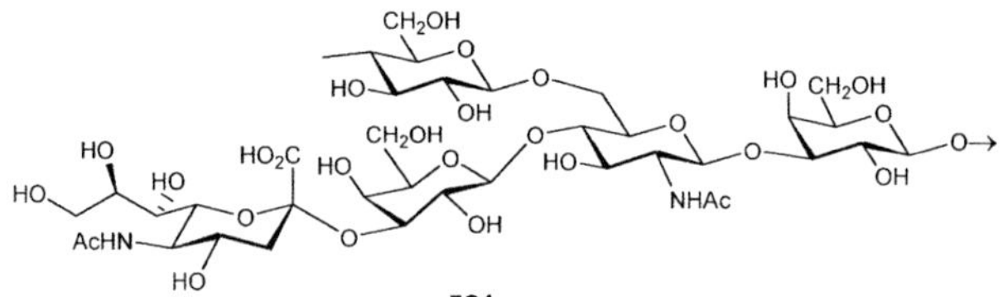

521

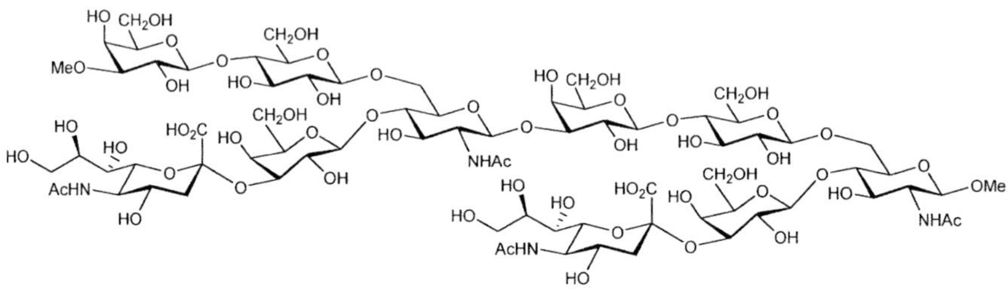

522

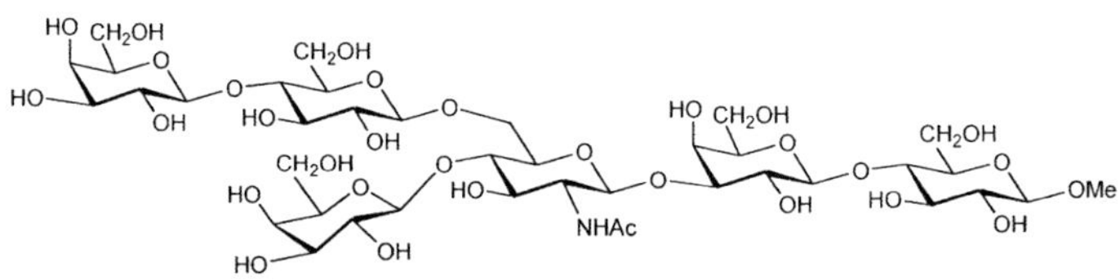

523

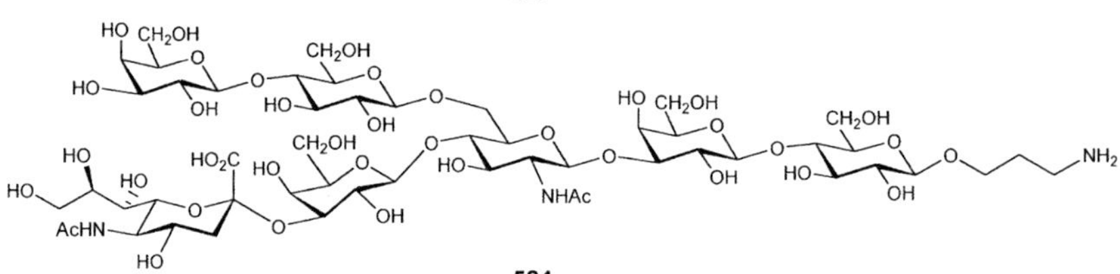

524

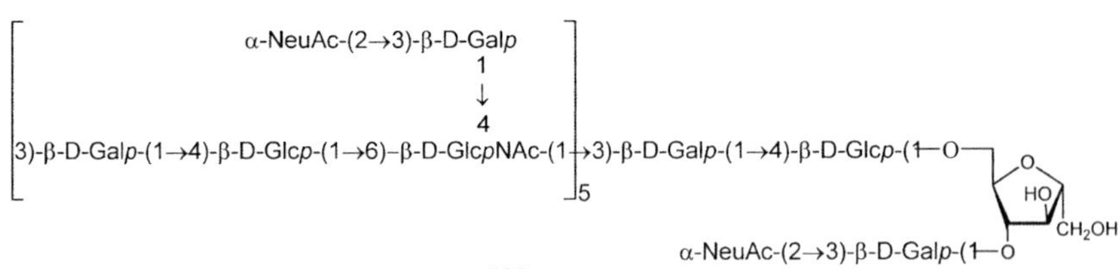

525

using trifluoroacetic acid to afford the pentasaccharide triol **533** (97 %) for further chain extension. Compound **533** features three hydroxyl groups that are potential glycosylation sites. The HO-3 hydroxyl group proved to be unreactive in the previous glycosylation step. In general in galactose derivatives having both HO-3 and HO-4 free, the former is more reactive in glycosylation reactions than the latter. Therefore, subsequent glycosylation was expected to occur at HO-3 of the nonreducing terminal galactose moiety. The trisaccharide block was prepared from the disaccharide diol **527** and acetobromo glucose **534**. Their condensation was mediated by mercuric cyanide to afford the trisaccharide **535** in 88% yield.

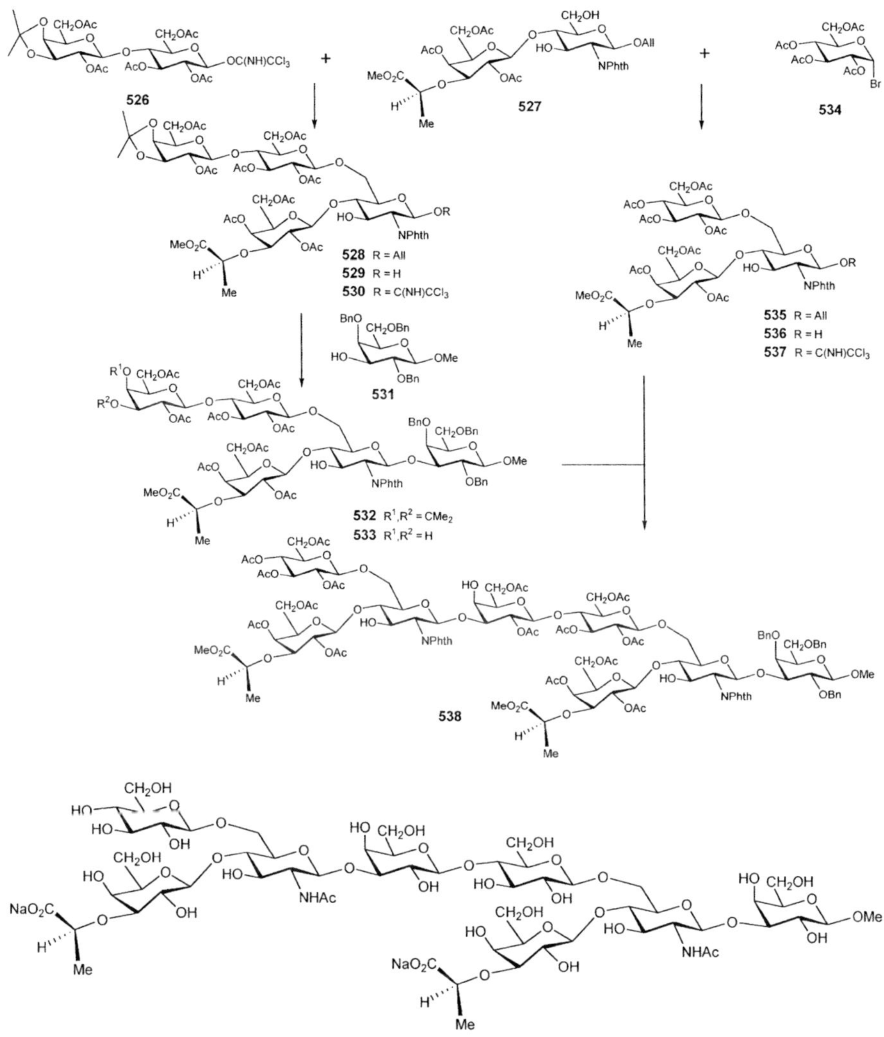

Removal of the allyl group as above followed by treatment of the hemiacetal **536** so obtained with CCl_3CN gave the imidate **537** (47% yield from **535**). In the final stages the pentasaccharide acceptor **533** and the trisaccharide donor **537** were combined in the presence of TMSOTf to produce the protected octasaccharide **538** in a highly regioselective fashion in 62% yield. The usual deprotection steps yielded the decasaccharide mimetic **539** in 30% overall yield. Using a number of oligosaccharide probes, the groups of Jennings and Kasper confirmed that two contiguous repeating units is the minimum binding epitope with anti-GBS CPS monoclonal antibodies. Increased affinity of binding was observed with longer saccharides consisting of up to 20 repeating units.

Type 8 GBS

A single repeating unit of type 8 GBS CPS that corresponds to the frame shown in formula **540** was synthesized as a trimethylsilylethyl glycoside by Jennings' group.[121] The same group reported the synthesis of the penta- **550** and the hexa-saccharide **551** fragments of polysaccharide using a combined chemical and enzymatic approach.[122] Starting from the galactosyl trichloroacetimidate donor **541** and the β-rhamnosyl-glucose disaccharide acceptor **542** the trisaccharide **543** was assembled using triethylsilyl trifluoromethanesulfonate as the activator, in 67% yield. Next, the trisaccharide thioglycoside **543** was activated with NIS/TfOH in the presence of the sialyl-galactoside **544**. The latter has two potential sites for glycosylation. Because of the higher reactivity of HO-4 relative to HO-2 of the galactose residue, the branched pentasaccharide **545** was formed predominantly with high stereoselectivity and was isolated in 34% yield. An isomeric pentasaccharide in which the trisaccharide **543** is linked to HO-2 of the galactose residue in **544** was also formed and was isolated as an impure fraction in approximately 24% yield. Next, the pentasaccharide **545** was subjected to hydrogenolysis followed by O-acetylation to afford the intermediate **546**.

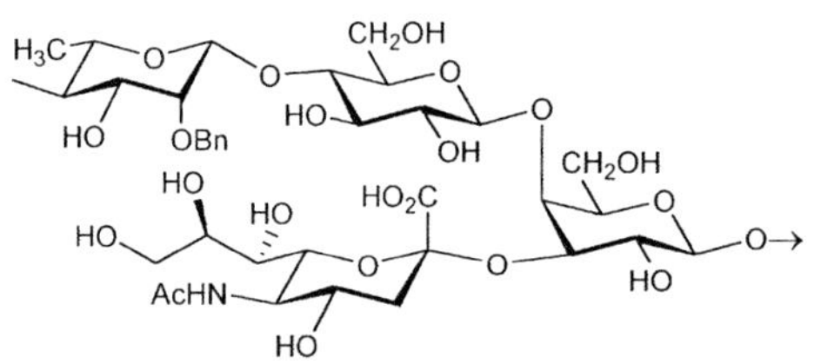

540

This was converted to the trichloroacetimidate **548**. First, compound **546** was subjected to the action of trifluoroacetic acid that cleaved the trimethylsilylethyl aglycon (→ **547** 89%). Next the hemiacetal **547** was treated with trichloroacetonitrile and DBU to afford the imidate **548** that was coupled with the linker 3-azido-propan-1-ol to furnish the pentasaccharide glycoside **549** in 77% yield. Treatment of the latter with sodium methoxide followed by sodium hydroxide afforded the unprotected target pentasaccharide **550** in 81% yield. In order to attach the second sialic acid residue, compound **550** was enzymatically sialylated using an α-(2→3)-sialyltransferase from *Campylobacter jejuni* and CMP-β-Neu5Ac as the sialic acid donor. Hexasaccharide **551** containing two sialic acid residues was isolated in 85% yield. Remarkably, the α-(2→3)-Neu5Ac transferase from *Neisseria meningitidis* could not catalyze this condensation.

Type 3 *Streptococcus pneumoniae*

The repeating unit of type 3 *S. pneumoniae* consists of a β-(1→3)-linked cellobiuronic acid **552**. Synthesis of the disaccharide analogue **557** was reported by Garegg's group.[123] Thus, the benzylidene-protected thioglycoside donor **553** was combined with the anhydroglucose acceptor **554** in a DMTST-promoted reaction to give the disaccharide **555** in 86% yield.

Removal of the benzylidene-acetal protecting group in acetic acid followed by oxidation of the primary C-6 carbon atom by the TEMPO/NaOCl reagent afforded the uronic acid-containing intermediate **556**. Subsequent acid-catalyzed esterification and O-acetylation yielded the desired cellobiuronic acid analogue **557** in 61% yield. Thiolysis of the anhydro ring of the closely related benzoyl derivative **558** using cyclohexylthiotrimethylsilane and the Lewis acid zinc iodide followed by benzoylation afforded the cellobiuronic acid thioglycoside **559** (68%). This was used as a donor for coupling with the heterobifunctional spacer **560** in a DMTST-assisted reaction that yielded the fully protected spacer-linked cellobiuronic

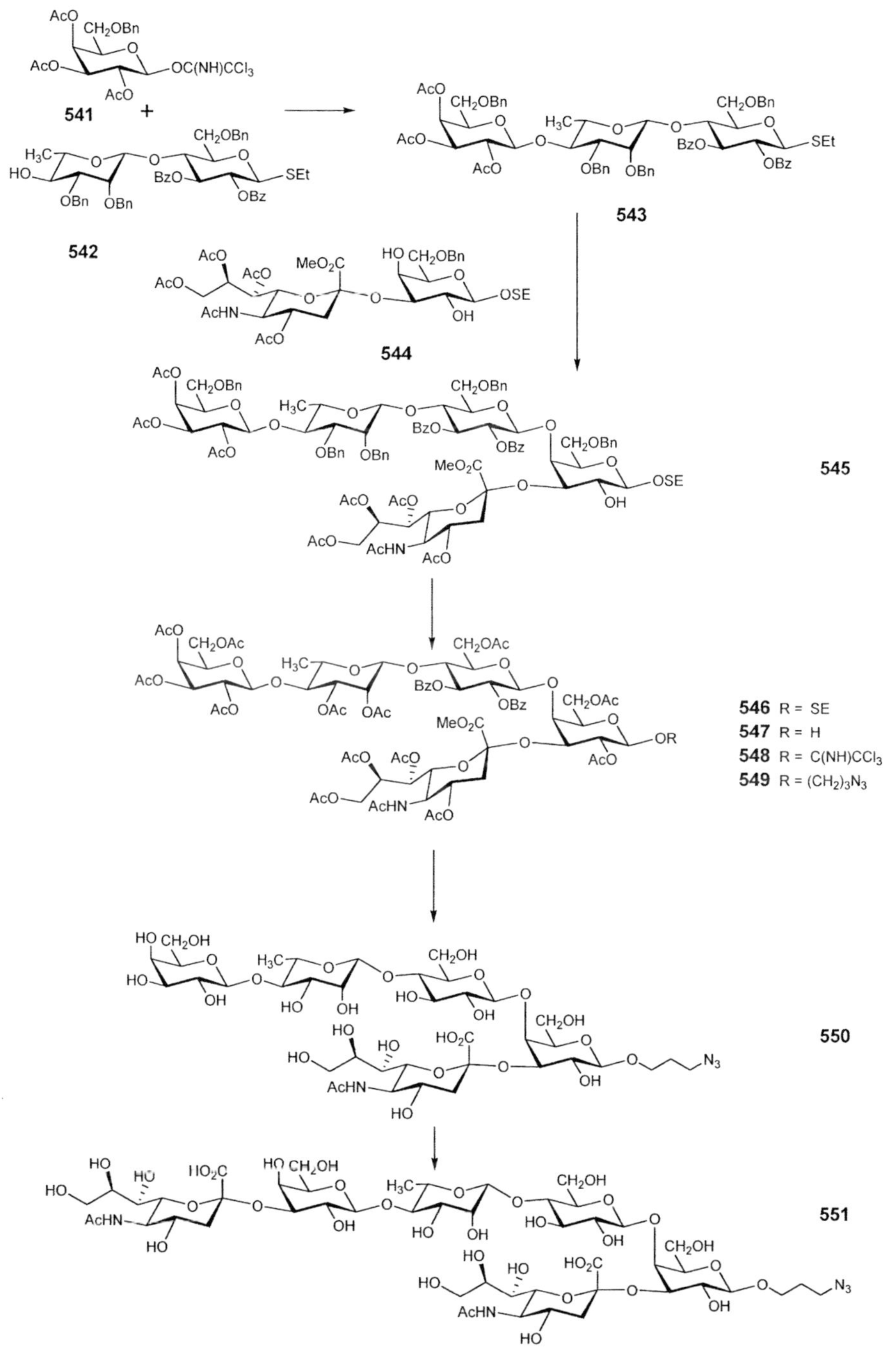
541
542
543
544
545
546 R = SE
547 R = H
548 R = C(NH)CCl3
549 R = (CH2)3N3
550
551

acid analogue **561** in 84% yield. Removal of the benzyl group by hydrogenolysis gave compound **562** that was suitable for the preparation of higher oligomers.[123]

Vliegenthart's group described the synthesis of the di- (**563**), tri- (**564**) and tetra-saccharide (**565**) fragments as their 3-aminopropyl glycosides for attachment to proteins.[124] In the synthesis of disaccharide **563** D-glucose acceptor **567** containing the azidopropyl spacer was combined with the acetal-protected D-glucosyl trichloroacetimidate **566** using 8% TMSOTf as the imidate activator, to afford the fully protected disaccharide **568** in 78% yield. This was converted to the triol **570** by sequential removal of the monochloroacetyl group with DABCO (→ **569**, 98%) followed by trifluoroacetic acid-mediated acetal cleavage (90%). Regioselective oxidation using TEMPO and aqueous sodium hypochlorite produced the protected cellobiuronic acid derivative **571** in 85% yield from which the unprotected cellobiuronic acid glycoside **563** was prepared in a two-step sequence. First, the benzoyl groups were removed under Zemplén conditions. Subsequently the azido group was reduced by adding a solution of the azido derivative in 0.1M sodium hydroxide to a suspension of palladium-on-charcoal and sodium borohydride to afford the target **563**.

The key intermediate for longer homologs was the disaccharide alcohol **573** that was extended either by the thioglucoside **572** or by the disaccharide imidate **575** to afford the protected tri- **574** and tetramers **576**. Compounds **564** and **565** were obtained by a combination of deprotection and oxidation steps applied to **574** and **576** as outlined for the disaccharide **563**. A range of protein conjugates of di-, tri- and tetra-saccharide **563**, **564**, and **565** with three unrelated proteins including a mutant of diphtheria toxin (CRM_{197}), tetanus toxoid, and

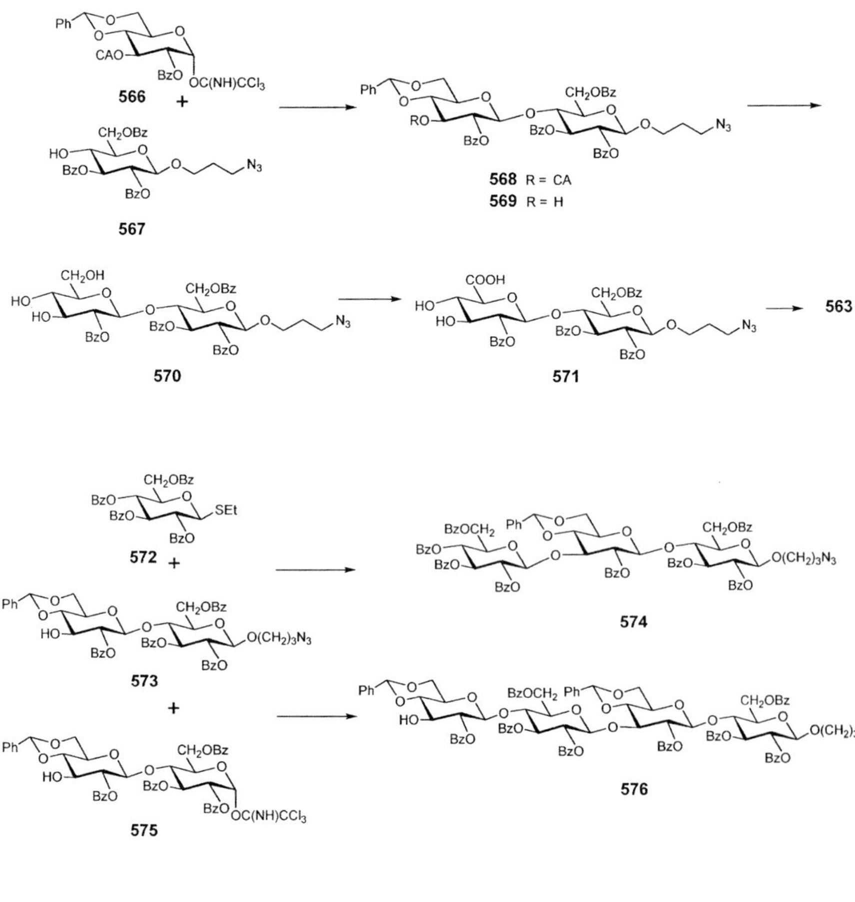

keyhole limpet hemocyanin were prepared using the squarate method of conjugation.[107] The average incorporations were 2.9-12 oligosaccharide chains per CRM_{197} and 13 to 25 chains per KLH. The oligosaccharide-CRM_{197} conjugates elicited anti-polysaccharide specific IgG antibodies in mice that lasted for at least 7 weeks after booster injections.[125] These experiments provide further examples that synthetic oligosaccharide-protein conjugates may, indeed, have a potential as vaccines against bacterial infections. Vliegenthart's group also reported the synthesis of a spacer-linked hexasaccharide fragment comprising three contiguous repeating units **552**.[126]

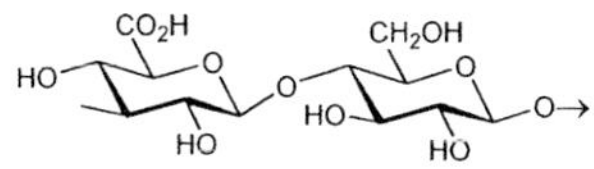

552

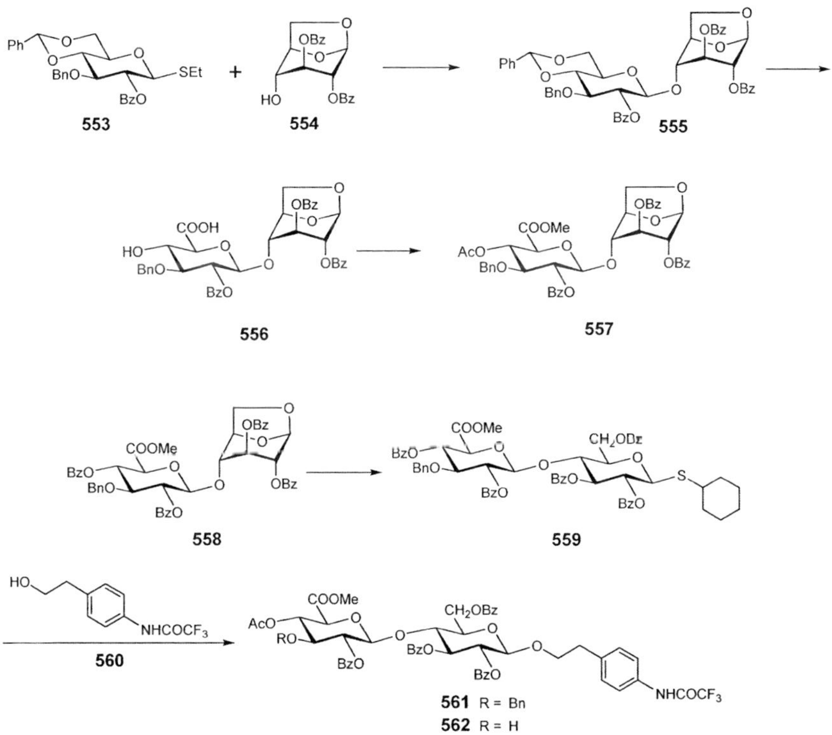

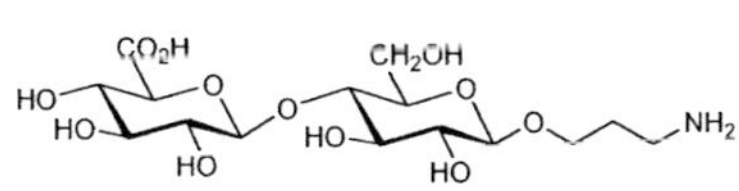

563

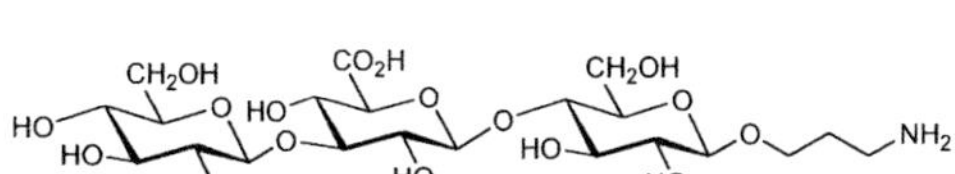

564

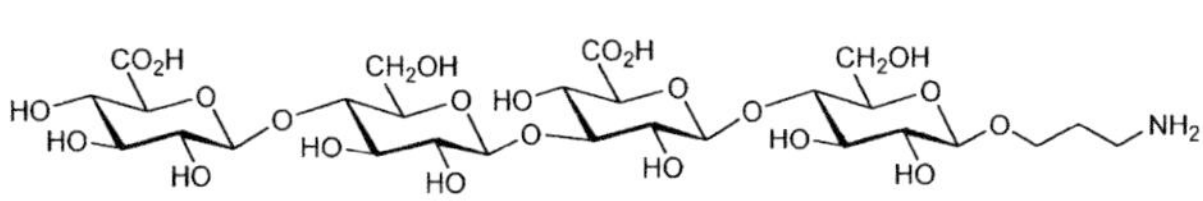

565

Type 6b *Streptococcus pneumoniae*

The structure of the repeating unit of the CPS of this bacterium (**577**) features a phosphodiester linkage interconnecting ribitol and galactopyranose residues. Chemical approaches to such structures require not only the stereoselective formation of interglycosidic connections but also the installment of the phosphodiester linkages that are usually highly sensitive to acids. In two successive papers Kamerling's group reported the synthesis of a series of di- to tetrasaccharides (**578-585**) related to **577** featuring an aminopropyl aglycon that allows their covalent attachment to protein carriers.[127,128]

Precursors in their syntheses were the ethylthio rhamnoside **586** and the partially protected ribitol **587** that were condensed using NIS/TfOH to afford the dimer **588** in 78% yield that was deallylated using Wilkinson's catalyst followed by treatment with mercury salts (→ **589**,

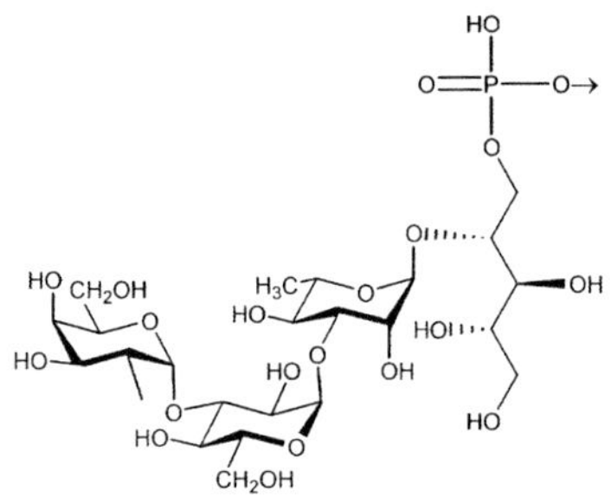

577

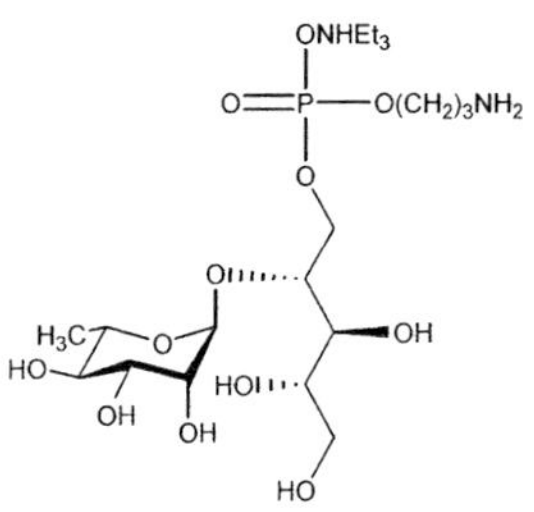

578

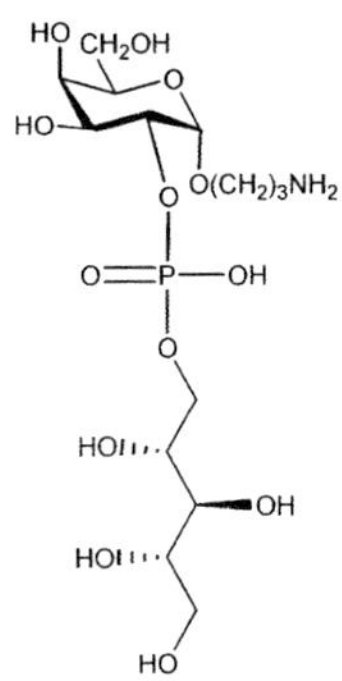

579

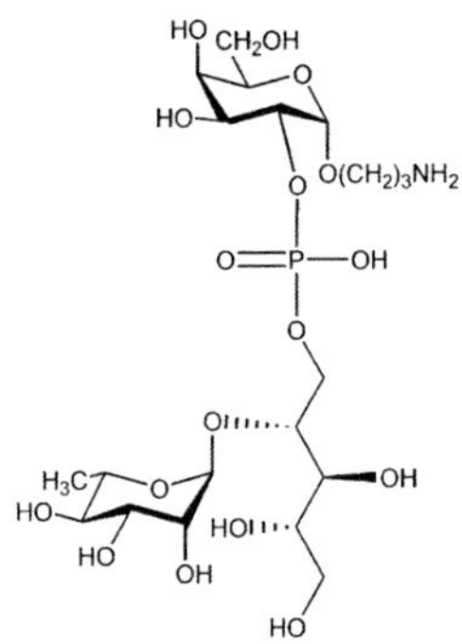

580

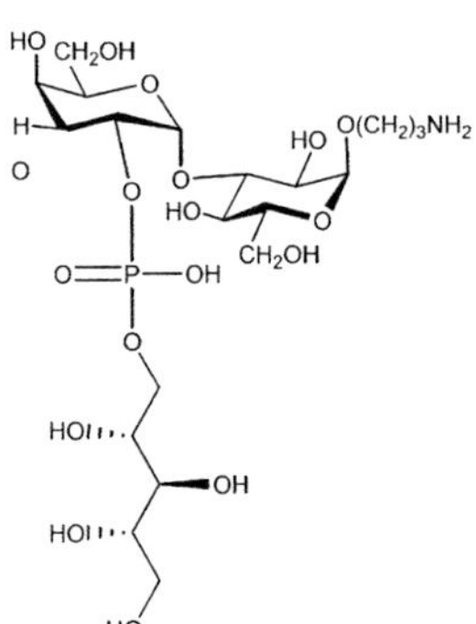

581

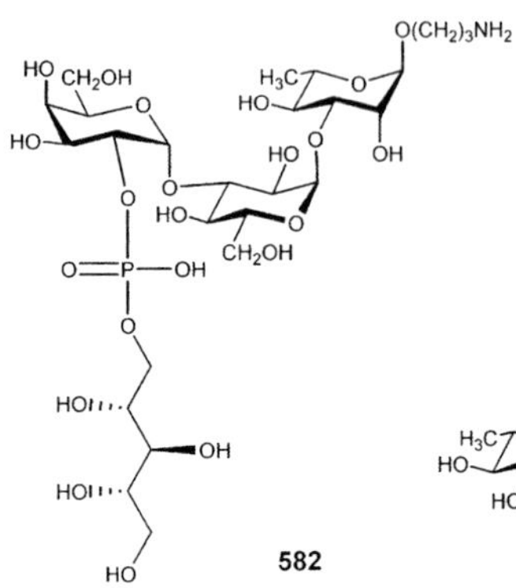

582

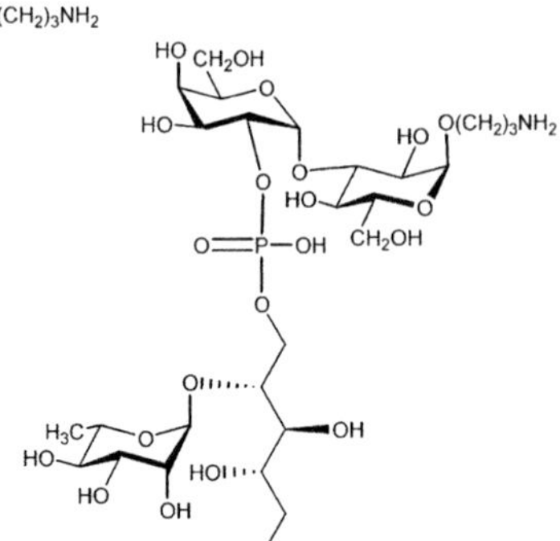

583

68%). Next, the phosphate function was introduced using the *H*-phosphonate method.[129-136] Thus, the alcohol **589** was reacted with the phosphitylating reagent salicylchlorophosphite **590** developed by van Boom[133] to afford the *H*-phosphonate ester **591**.

Phosphonylation of the spacer alcohol **592** with the *H*-phosphonate **591** in the presence of pivaloyl chloride afforded an intermediate that was in situ oxidized with iodine in pyridine-water mixture to furnish the phosphodiester derivative **593** (43% for two steps). Deprotection of **593** was performed first with ammonia followed by hydrogenolytic removal of the benzyl and benzyloxycarbonyl groups to yield the spacer-linked rhamnosyl-ribitol-phosphate as the triethylammonium salt **578**. The phosphodiester-linked fragments **579-585** were obtained using a similar strategy. Keyhole limpet hemocyanin conjugates of disaccharide **578**, trisaccharide **581** and tetrasaccharide **583** elicited high levels of antibodies directed against the native Type 6b CPS.[137] Antisera raised by the conjugates of the tri- and tetra-saccharides also reacted with Type 6a CPS. It was demonstrated that conjugates of both the disaccharide and the tetrasaccharide are capable of inducing Type 6b specific protective antibodies in both rabbits and mice.

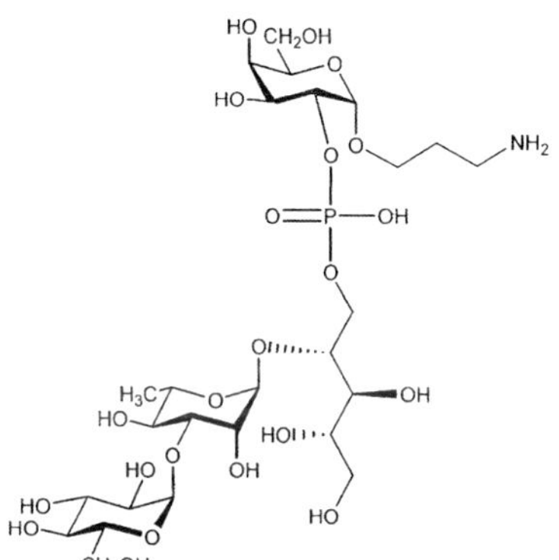

584

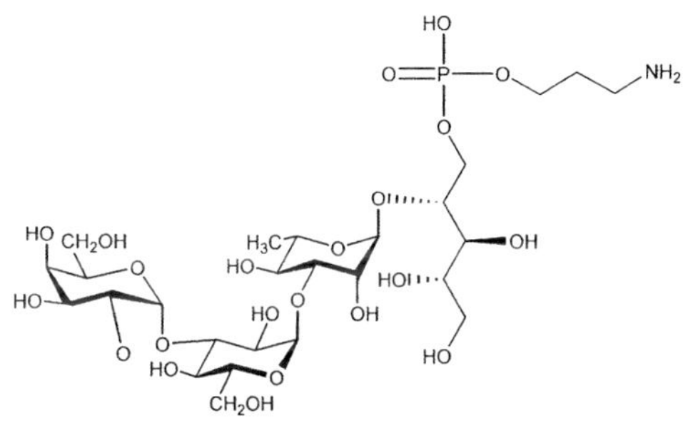

585

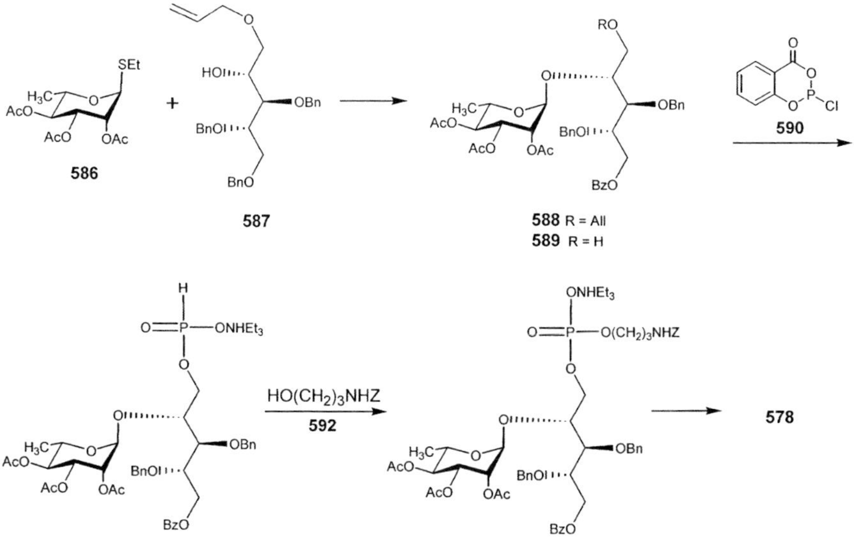

Type 14 *Streptococcus pneumoniae*

The repeating unit of the CPS of this organism consists of the branched tetrasaccharide **594** having a trisaccharide backbone. A combined chemoenzymatic approach to the branched tetrasaccharide **599** was reported by the Utrecht group.[138] The synthesis started with the condensation of the peracetylated lactosyl donor **595** with the glucosamine derivative **596** in the presence of TMSOTf to produce the linear trisaccharide **597** in 63% yield. Next, the O-acyl groups were removed by transesterification in the presence of sodium methoxide followed by removal of the phthaloyl protecting group. A cycle of per-O and N-acetylation followed by de-O-acetylation (Zemplén) afforded the trisaccharide **598**. Attempted one-pot deacylation and dephthaloylation with ethylenediamine followed by re-N- and O-acetylation also produced the N-*p*-methoxybenzoylated congener as a by-product in 32% yield. Next, the glucosamine moiety in **598** was galactosylated using UDP-galactose as the galactose donor and β-1,4-galactosyltrasferase isolated from bovine milk. The resulting allyl glycoside **599** was obtained in 70% yield and was converted to the cysteamine adduct **600** under UV irradiation for eventual attachment of proteins. A similar chemoenzymatic strategy was used to prepare the hexa **601**- and octa-saccharide **602** mimics of **594**.[139]

β-D-Gal*p*
1
↓
3
3)-β-D-Gal*p*-(1→4)-β-D-Glc*p*-(1→6)-β-D-Glc*p*NAc-(1→ **594**

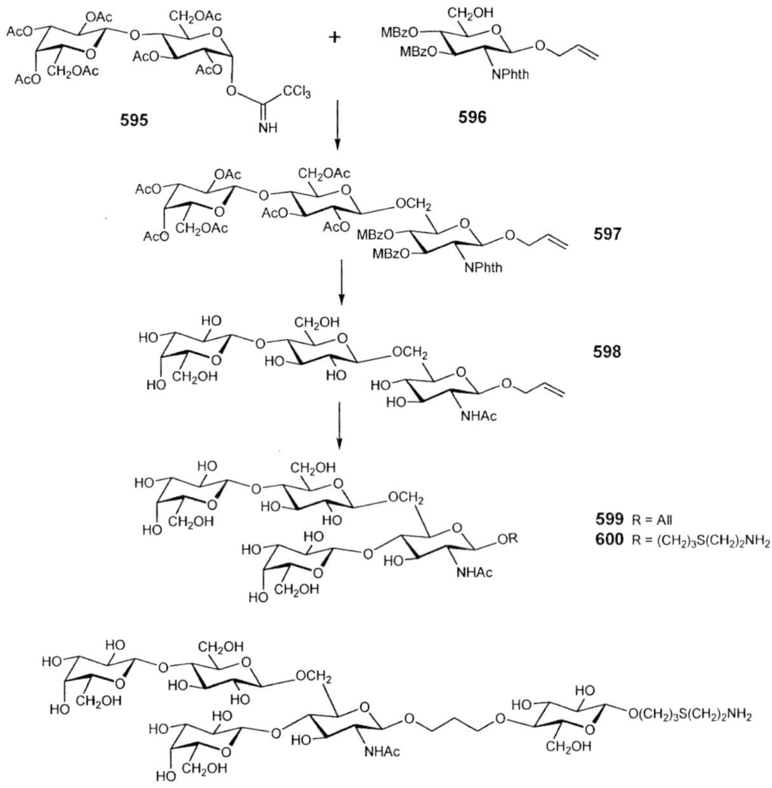

601

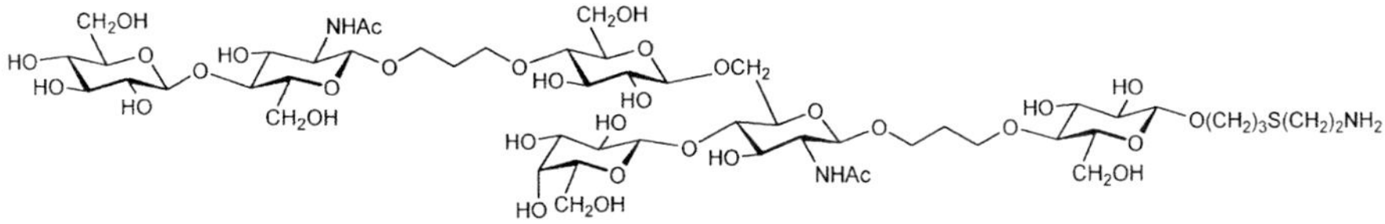

602

Type 19F *Streptococcus pneumoniae*

The CPS of this organism consists of the phosphodiester-containing repeating unit **603**. Synthesis of the trisaccharide fragment **613** utilizing the sulfoxide-glycosylation method was published by Russo's group in 1998.[140] Thus, the glucose sulfoxide **604** was prepared from the corresponding phenylthio glucoside and was stereoselectively condensed with the allyl rhamnoside **605** using triflic anhydride as the promoter and 2,6-di-*tert*-butyl-4-methylpyridine as the acid scavenger to provide the α-linked disaccharide **606** as the predominant product in 80% yield. Removal of the benzylidene ring by hydrolysis with aqueous trifluoroacetic acid followed by dibutyltin-assisted regioselective benzylation afforded the alcohol **607** (80%). Chain elongation was performed with glucosyl sulfoxide **608** under conditions described above to yield the trisaccharide **609** (77%). The exclusive, β-stereoselectivity was achieved by the presence of the participating acetyl protecting group at O-2 of the donor. In the final stage of the synthesis the nonreducing-end *gluco* residue was converted to *manno* azide **612**. First, the acetyl group was removed in the presence of sodium methoxide (→**610**) then an imidazolylsulfonyl group was introduced at O-2 to yield the sulfonate **611**. Treatment of compound **611** with tetrabutylammonium azide introduced the azide group and thus created the *manno* configuration in a nearly quantitative yield.

Compound **612** was then exposed to the action of zinc activated with copper sulfate in a THF-AcOH-Ac_2O mixture that not only reduced the azido group to amino but also resulted in in situ N-acetylation to produce the target trisaccharide **613** in 84% yield.

Synthesis of a nonasaccharide **614**, representing three consecutive repeating units of the CPS, in a form that is amenable to covalent attachment to proteins was achieved by Nilsson and Norberg.[141] Thus, coupling of benzylidene acetal-protected glucosyl bromide **615** and the thioglucoside acceptor **616** afforded the disaccharide **617** in the presence of silver triflate, in 66% yield. In order to fashion the β-linked mannose moiety from the nonreducing end *gluco*-configured residue, compound **617** was debenzoylated with sodium methoxide (→**618**) followed by reaction with triflic anhydride. Treatment of the resulting sulfonate **619** with sodium azide gave the *manno* azide **620** in 75% yield from **618**. This was subjected to regioselective opening of the acetal group using sodium cyanoborohydride in combination with tetrafluoroboric acid that formed a benzyl group at O-6. Subsequent O-acetylation at O-4 yielded intermediate **621** in 76% overall yield. The thioglycoside **621** so prepared was then used to glycosylate allyl rhamnoside **622** in the presence of methyl triflate.

The resulting trisaccharide was obtained as an inseparable α/β mixture **623**α/β in 87% combined yield. Sequential reduction of the azido group in a Staudinger reaction using triphenyl phosphine followed by N-acetylation and chromatographic purification afforded the anomerically homogeneous trisaccharide **624** in 58% yield. The corresponding β-anomerically linked trisaccharide was also isolated, in 28% yield. Compound **624** was converted to the hemiacetal **625** by isomerization with Wilkinson's catalyst and hydrolysis of the resulting propenyl glycoside with acetic acid (80%). Subsequent treatment of the hemiacetal **625** with phosphorous acid and the dioxaphosphorinane **626** afforded the key α-*H*-phosphonate intermediate **627** in 61% yield, that was accompanied by only very little β product. Compound **627** was first coupled with the spacer N-benzyloxycarbonyl-ethanolamine using the standard *H*-phosphonate protocol involving condensation by pivaloyl chloride followed by oxidation with iodine in aqueous pyridine to afford the spacer-linked repeating unit analogue **628**. This was deacetylated using methanolic sodium methoxide to give the acceptor **629** in 91% yield.

Application of the *H*-phosphonate coupling/deprotection in two additional cycles followed by protecting group removal by deacetylation and hydrogenolysis provided the target nonasaccharide **614** in a form that is suitable for conjugation with proteins.

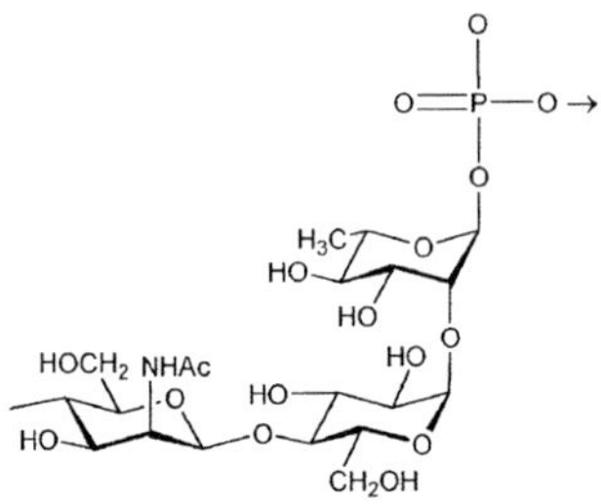

603

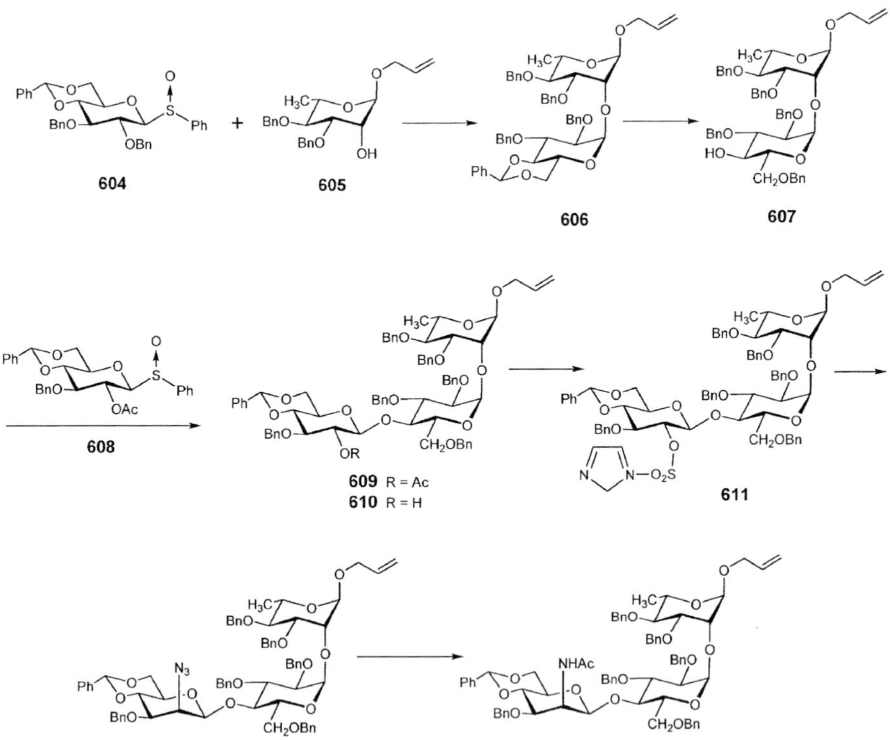

604 **605** **606** **607**

608 **609** R = Ac **610** R = H **611**

612 **613**

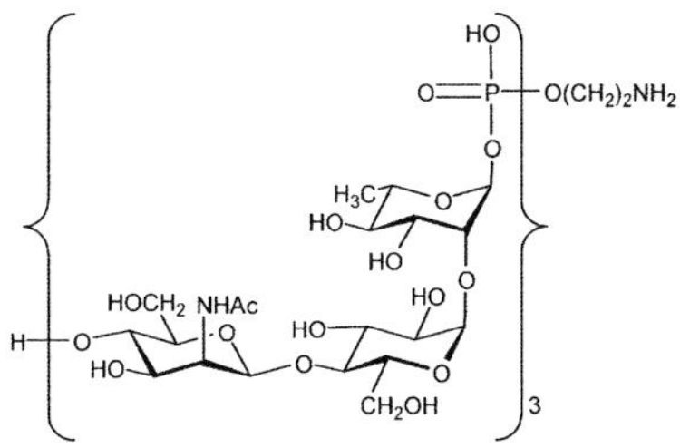
HO
O=P—O(CH2)2NH2
O
H3C
O
HO
HO
O
HOCH2 NHAc
HO
H—O
O
HO
HO
O
O
CH2OH
3

614

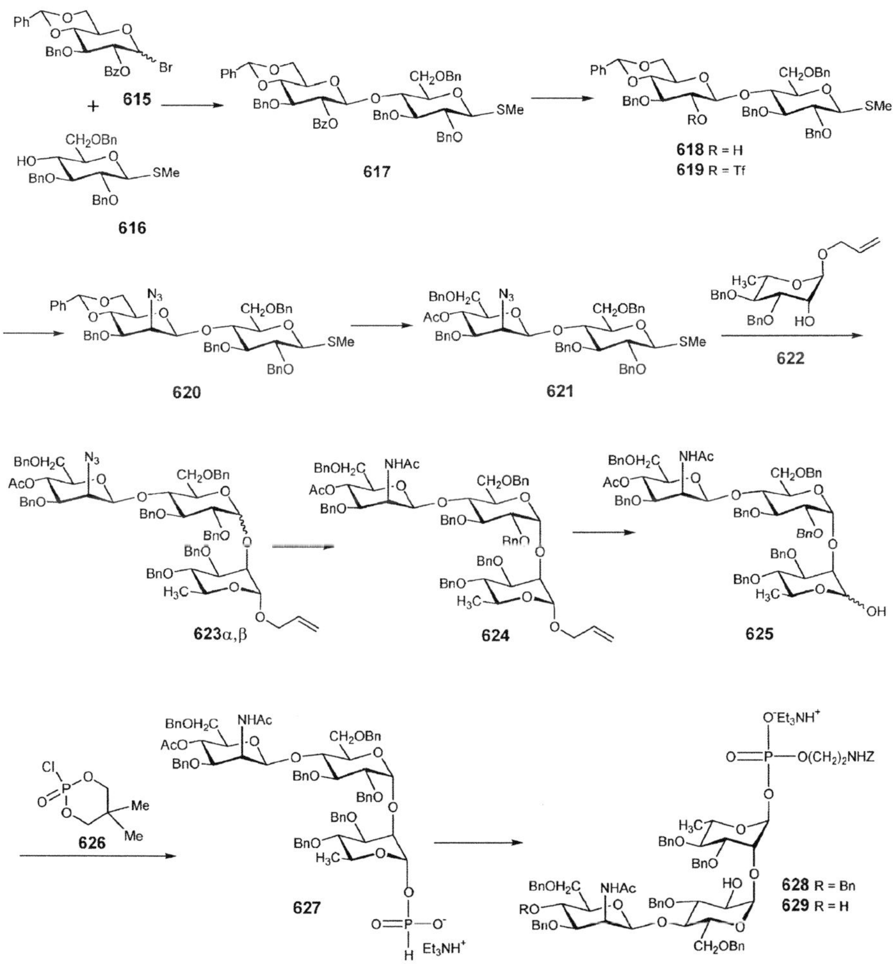
615
616
617
618 R = H
619 R = Tf
620
621
622
623α,β
624
625
626
627
628 R = Bn
629 R = H

Type 29 *Streptococcus pneumoniae*

The repeating unit of the CPS of this organism consists of the phosphodiester-linked pentasaccharide **630** containing an *N*-acetylgalactosamine, a galactopyranose, a ribitol, and two galactofuranose residues. Synthesis of the trigalactoside motif **636** of this pentasaccharide was reported by Roy's group.[142] The synthetic scheme is based on reactivity differences between two thioglycosides, namely thiogalactofuranoside **631** and tolylthio galactoside **632**. Thus, activation of the more reactive ethylthio glycoside **631** with NIS/TfOH in the presence of the galactoside acceptor **632** produced the disaccharide **633** as *p*-tolyl thioglycoside in 80% yield. Subsequent treatment of the thioglycoside disaccharide **633** with the same activator in the presence of the galactofuranoside acceptor **634** yielded trisaccharide **635** in 75% yield. This was deblocked in one step using sodium methoxide in methanol to give the desired galactotriose as the methyl glycoside **636** (68%).

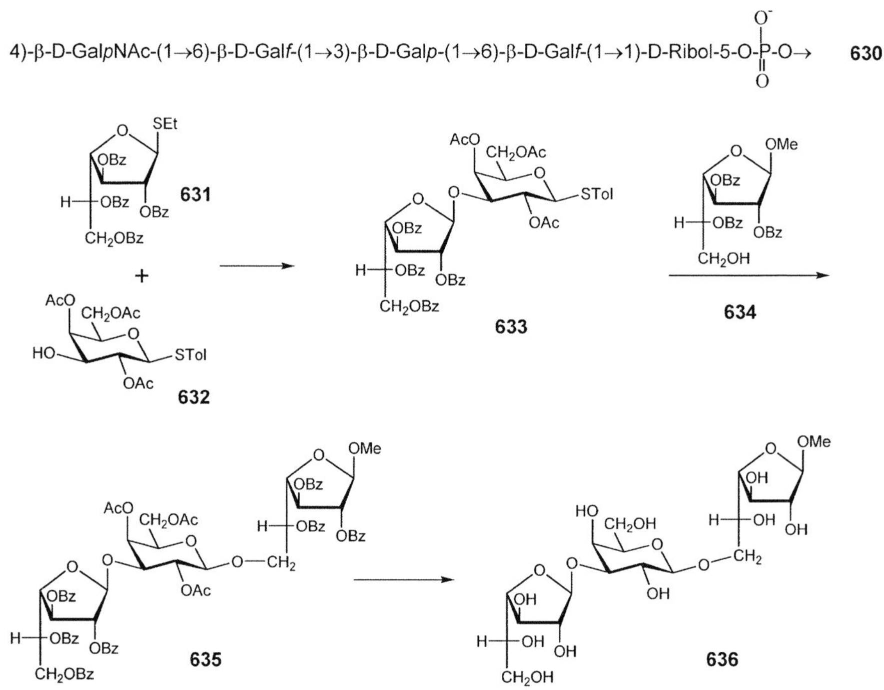

Vibrio cholera O1 Serotype Inaba

The O-SP of this organism is composed of approximately fifteen α-(1→2)-linked residues of 4-amino-4,6-dideoxy-D-mannose *N*-acylated with 3-deoxy-L-*glycero*-tetronic acid **637**. Verez-Bencomo's group reported the synthesis of the disaccharide **644**[143] Starting compounds were the azido glycoside **638** and the dideoxymannosyl chloride **639** that were condensed in a silver triflate-promoted reaction to furnish the α-linked disaccharide **640** in 60% yield. Next, the acetyl group was removed under Zemplén conditions and the azido groups were converted to amino groups under hydrogen atmosphere in the presence of a palladium catalyst (→**641**). The O-benzyl protecting groups were not cleaved under these conditions. Next the disaccharide **641** was subjected to N-acylation with the tetronic acid derivative **642** in the presence of the condensing reagent EEDQ (2-ethoxy-1-ethoxycarbonyl-1,2-dihydroquinoline) to produce the amide **643** in 57% yield. Hydrogenolytic cleavage of the benzyl groups using palladium-on-charcoal as the catalyst followed by exposure to the Zemplén conditions to cleave the acetyl groups of the side chain produced the disaccharide **644**.

Similar approaches afforded the disaccharide fragment **645** equipped with a dioxolane-type spacer that was used for covalent attachment to the proteins bovine serum albumin and meningococcal outer membrane protein complex.[144] The synthesis of the trisaccharide **647** is also described.[145] The molecular basis of the serotype specificity was investigated by binding studies with monoclonal antibodies[146] and X-ray crystallography.[147] Using similar building blocks, the chemical synthesis of a hexamer **649** was reported by Kovac's group.[148]

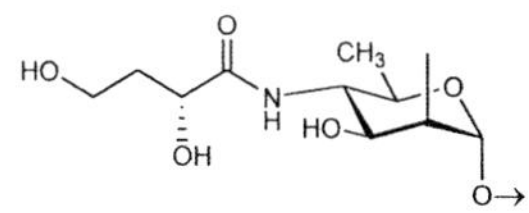

637

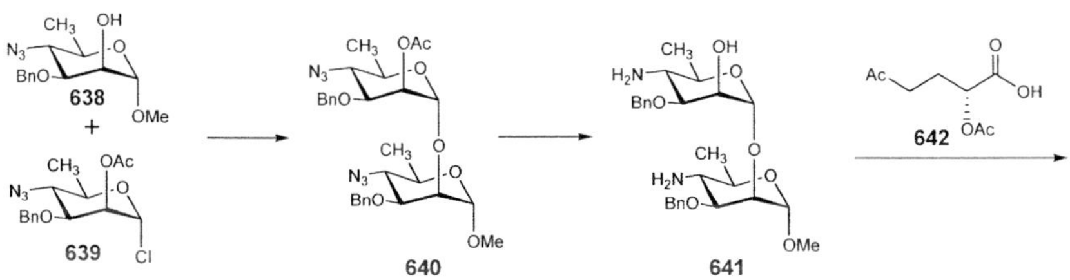

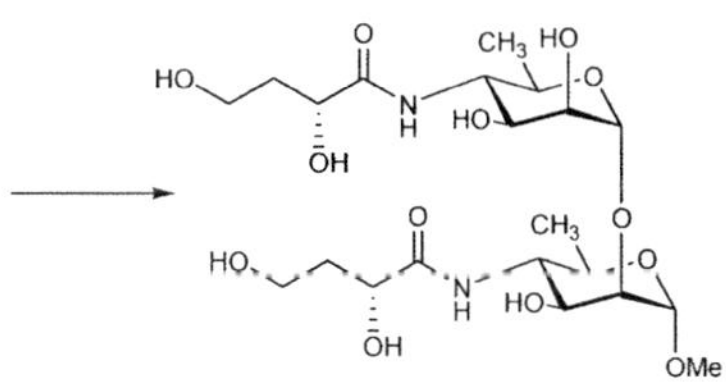

643 R = Bn
644 R = H

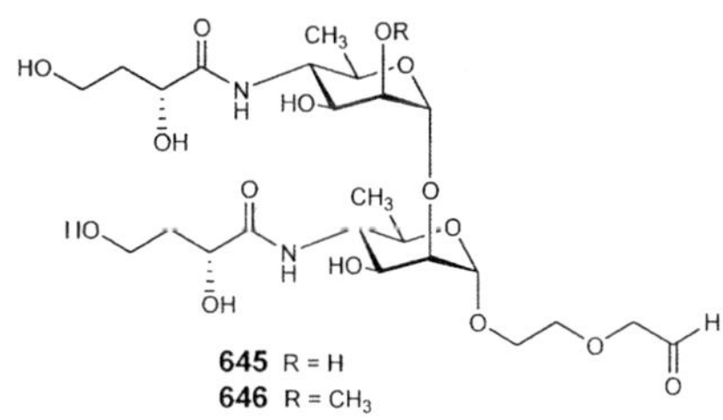

645 R = H
646 R = CH_3

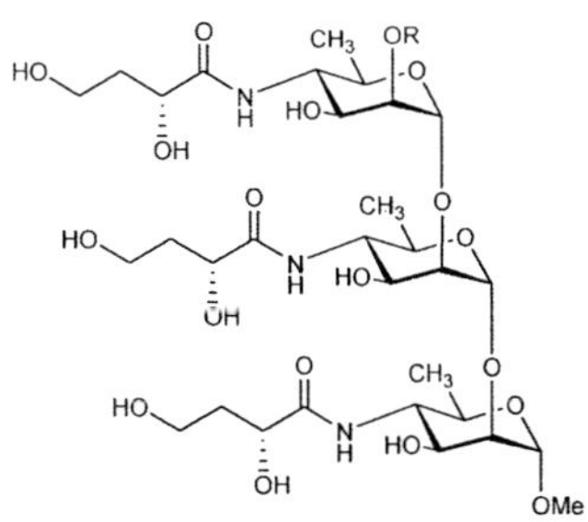

647 R = H
648 R = CH_3

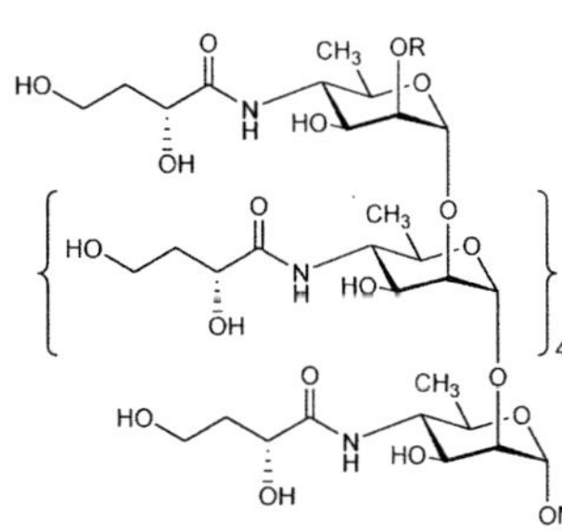

649 R = H
650 R = CH_3

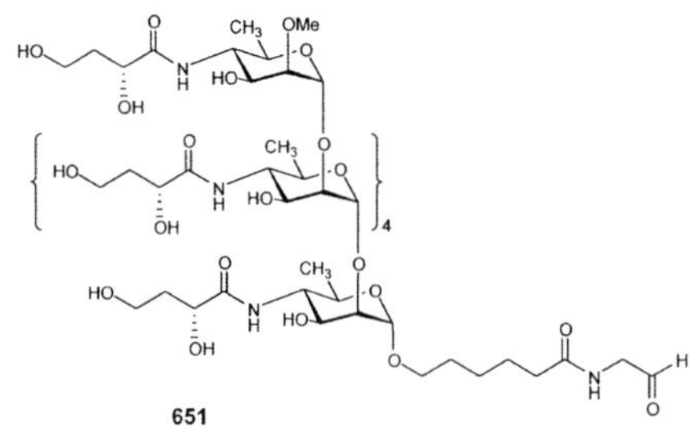

651

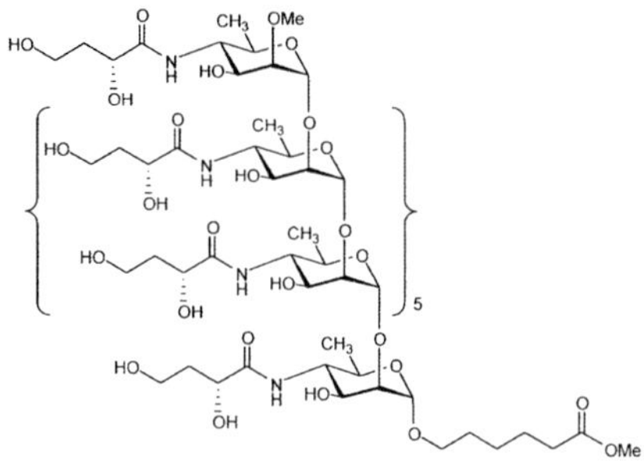

652

Vibrio cholera O1 Serotype Ogawa

The O-SP of this organism differs from that of *V. cholera* serotype Inaba only in that it bears a methyl group at O-2 of its "nonreducing" terminal residue. Verez-Bencomo reported the synthesis of the O-methylated analogs **646** and **648**[144] in a manner similar to the synthesis of the nonmethylated congener **647** and demonstrated that there were no significant differences in their major conformations.[145] Compound **648** is a good inhibitor of the interaction between the Ogawa LPS and antibodies directed to the common antigen against both the Inaba and the Ogawa serotypes.[143] It was concluded that **646** represents the carbohydrate epitope recognized by most of the anti-Ogawa polyclonal antibody pool. The synthesis of the hexameric fragment **650** is described in the above-referenced paper of Zhang and Kovac.[148] For the synthesis of deoxygenated,[149] fluorinated,[150] and other analogues,[148,151] the reader is referred to the original literature. Neoglycoconjugates from the mono-[152] and hexasaccharide fragments[153] **651** using the diethyl squarate method were also reported and the synthesis of a dodecasaccharide fragment **652** was described.[154]

Vibrio cholera O139

The CPS of this organism contains a complex oligosaccharide that is also found in its LPS. Tri- (**660** and **663**) and tetrasaccharide fragments (**666**) of these structures were assembled by Oscarson's group.[34] A key precursor was the colitose-derived thioglycoside donor **657** that was prepared from thiofucoside **653**. Thus, triol **653** was first converted to the *endo*-acetal **654** using α,α-dimethoxytoluene (benzaldehyde dimethyl acetal) and toluenesulfonic acid that was benzylated to yield the *endo* intermediate **655** in 34% overall yield for two steps.

Reductive ring opening of the benzylidene acetal **655** with $NaCNBH_3$/HCl afforded the dibenzyl-alcohol derivative **656** in 85% yield. Next this compound was deoxygenated using a two-step protocol. Thus, alcohol **656** was treated with thiocarbonyldiimidazole and the thiocarbonyl intermediate formed was reduced with tributyltin hydride to yield the key colitose donor **657** in 66% yield. This protocol was accompanied by the well-documented formation of the starting alcohol **656** that was recovered in 25% yield.

Condensation of the thiocolitoside **657** with the disaccharides **658**, **661**, and **664** using DMTST as the activator led to fully protected tri- **659** and **662** and tetrasaccharide **665** intermediates, respectively, in high yields. Optimizing the reaction time was critical for obtaining high yields. Attempts to use the NIS/AgOTf combination produced the *N*-succinimide glycoside of the donor instead of the expected glycoside. Deprotection was accomplished by catalytic hydrogenolysis. This was performed in the presence of a basic ion-exchange resin to avoid hydrolytic cleavage of the acid-sensitive colitose residue to yield the targeted saccharides **660**, **663**, and **666** in spacer-linked form.

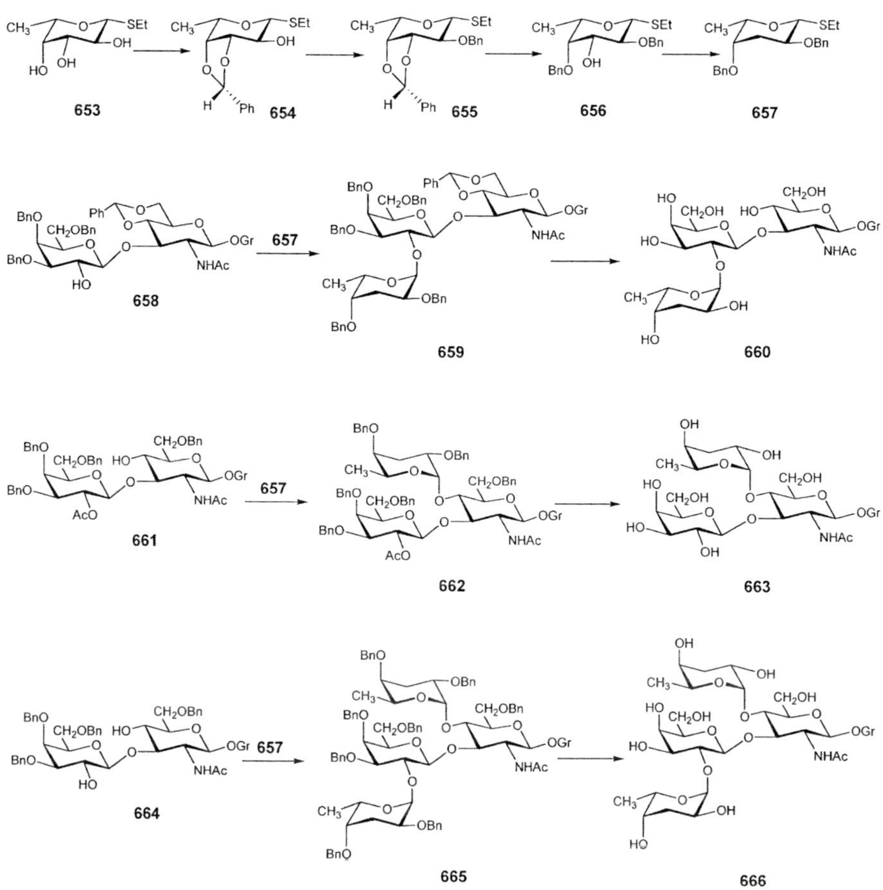

Vibrio parahaemolyticus

The linear acidic trisaccharide fragment **682** of the inner core LPS region of this bacterium was synthesized by van Boom's group.[155] The key intermediate in the synthesis was compound **674** that was prepared from the isopropylidene-protected mannoside **667** having HO-6 unprotected. Thus, Swern oxidation of **667** using oxalyl chloride afforded the aldehyde **668** that was transformed to the olefin **669** in a Wittig reaction with methyltriphenylphosphonium bromide in 71% yield for two steps. Subsequent *syn*-dihydroxylation using potassium osmate afforded a diastereomeric mixture. The major isomer was identified as the D-*glycero*-D-*manno*-heptopyranose **670** and was isolated in 79% yield. Its conversion to the acceptor **671** (69% overall yield) involved (i) benzylation of the hydroxyl groups, (ii) acid-catalyzed removal of the acetal protecting group and (iii) phase transfer-catalyzed benzylation of the HO-2 hydroxyl group. The L-*glycero*-D-*manno*-heptopyranoside intermediate **674** was obtained from the thiomannoside derivative **672**. Thus, reaction of **672** with the Grignard reagent (phenyldimethylsilyl)methyl magnesium chloride produced compound **673** from which protecting group changes yielded the thioglycoside donor **674**.

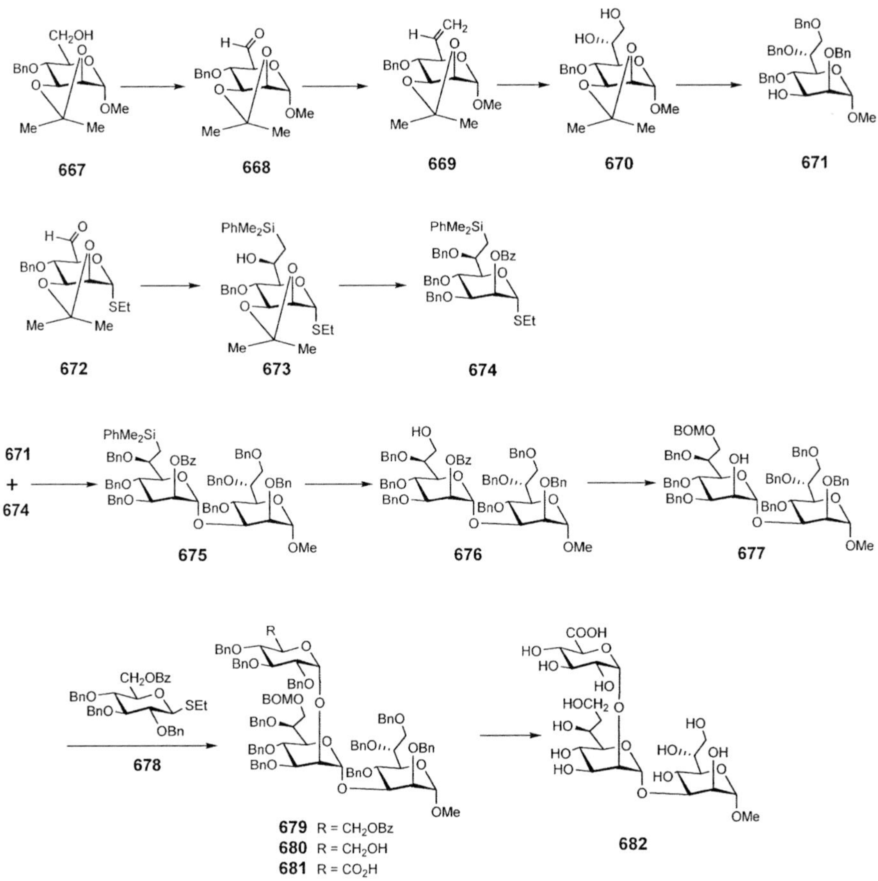

Condensation of the monosaccharide precursors **671** and **674** in the presence of NIS/TfOH produced the fully protected disaccharide **675** in 76% yield. Oxidative removal of the silyl group using peracetic acid gave the alcohol **676** (85%). Next, the primary hydroxyl group in compound **676** was protected with a benzyloxymethyl group and the benzoate group was removed by treatment with KO^{t}Bu in methanol to yield the disaccharide acceptor **677** (56% for two steps) ready for chain extension. The glucuronyl moiety was introduced in a two-step protocol. First, the disaccharide acceptor **677** and the thioglucoside donor **678** were condensed in the presence of iodonium dicollidine triflate to yield the fully protected trisaccharide **679** in a highly α-stereoselective fashion in 73% yield. Next, Zemplén-type debenzoylation furnished the partially protected trisaccharide **680**. This was subjected to a two-step oxidation protocol involving Swern oxidation and treatment with buffered sodium chlorite to yield acidic trisaccharide **681**. Finally, hydrogenolysis of the benzyloxymethyl and benzyl groups over palladium-on-charcoal yielded the unprotected acidic trisaccharide **682**.

References

1. Heidelberger M, Avery OT. The soluble specific substance of pneumococcus. J Exp Med 1923; 38:73-79.
2. Kamerling JP.; In: Tomasz A, ed. Streptococcus pneumoniae. Larchmont: Mary Ann Liebert, Inc., 2000:81-114.
3. Croakert F, Lismont MJ, van der Linden MP et al. Determination of complement-mediated serum bactericidal activity against Gram-negative bacteria. Rev Med Microbiol 1992; 3:241-248.
4. In: Brade H, Opal SM, Vogel S et al, eds. Endotoxin in Health and Disease. New York, Basel: Marcel Dekker, Inc., 1999.
5. Giardina PC, Apicella MA, Gibson B et al. In: Brade H, Opal SM, Vogel SN et al, eds. Endotoxin in Health and Disease. New York, Basel: Marcel Dekker, Inc., 1999:55-65.
6. Francis T Jr, Tillet WS. Cutaneous reactions in pneumonia: the development of antibodies folowing intrademal injection of type specific polysaccharide. J Exp Med 1930; 52:573-585.
7. Heidelberger M, Dilapi MM, Siegel M et al. Presence of antibodies in human subjets injected with pneumococcal polysaccharides. J Immunol 1950; 65:535-541.
8. Avery OT, Goebel WF. Chemo-immunological studies on conjugated carbohydrate proteins. V. Immunological specificity of an antigen prepared by combining the capsular polysaccharide of type III penumococcus with foreign antigen. J Exp Med 1931; 54:437-447.
9. Passwell JH, Harlev E, Ashkenazi S et al. Safety and immunogenicity of improved Shigella O-specific polysaccharide-protein conjugate vaccines in adults in Israel. Infect Immun 2001; 69:1351-1357.
10. Pavliakova D, Moncrief JS, Lyerly DM et al. Clostridium difficile recombinant toxin A repeating units as a carrier protein for conjugate vaccines: Studies of pneumococcal type 14, Escherichia coli K1, and Shigella flexneri type 2a polysaccharides in mice. Infect Immun 2000; 68:2161-2166.
11. Svenson SB, Lindberg AA. Immunochemistry of Salmonellea O-antigens: Preparation of an octasaccharide-bovine serum albumin immunogen representetive of Salmonella serogroup B O-antigen and characterization of the antibody response. J Immunol 1978; 120:1750-1757.
12. Konadu EY, Lin FYC, Ho VA et al. Phase 1 and phase 2 studies of Salmonella enterica serovar Paratyphi A O-specific polysaccharide-tetanus toroid conjugates in adults, teenagers, and 2-to 4-year-old children in Vietnam. Infect Immun 2000; 68:1529-1534.
13. Pozsgay V. Oligosaccharide-protein conjugates as vaccine candidates against bacteria. Adv Carbohydr Chem Biochem 2001; 56:153-199.
14. Pozsgay V. A new strategy in oligosaccharide synthesis using lipophilic protecting groups: synthesis of a tetracosasaccharide. Tetrahedron-Asymmetry. 2000; 11:151-172.
15. Bundle DR. Synthesis of oligosaccharides related to bacterial O-antigens. Top Curr Chem 1990; 154:1-37.
16. Kosma P. In: Brade H, Opal SM, Vogel SN et al, eds. Endotoxin in Health and Disease. New York, Basel: Marcel Dekker, Inc., 1999:257-281.
17. Kusumoto S, Fukase K, Oikawa M. In: Brade H, Opal SM, Vogel SN et al, eds. Endotoxin in Health and Disease. New York, Basel: Marcel Dekker, Inc., 1999:243-256.
18. Lipták A, Borbás A, Bajza I. Synthesis of carbohydrate-containing surface antigens of Mycobacteria. Med Res Rev 1994; 14:307-352.
19. Hansson J, Oscarson S. Complex bacterial carbohydrate surface antigen structures: Syntheses of Kdo- and heptose-containing lipopolysaccharide core structures and anomerically phosphodiester-linked oligosaccharide structures. Curr Org Chem 2000; 4:535-564.
20. Oscarson S. Synthesis of oligosaccharides of bacterial origin containing heptoses, uronic acids and fructofuranoses as synthetic challenges. Top Curr Chem 1997; 186:171-202.
21. Kosma P, Reiter A, Zamyatina A et al. Synthesis of inner core antigens related to Chlamydia, Pseudomonas and Acinetobacter LPS. J Endotoxin Res 1999; 5:157-163.
22. Wimmer N, Brade H, Kosma P. Synthesis of neoglycoproteins containing D-glycero-D-talo-oct-2-ulopyranosylonic acid (Ko) ligands corresponding to core units from Burkholderia and Acinetobacter lipopolysaccharide. Carbohydr Res 2000; 329:549-560.
23. Lemanski G, Ziegler T. Prearranged glycosides, 9. Chemical synthesis of a tetrasaccharide fragment related to the exopolysaccharide of Arthrobacter sp. CE-17. Eur J Org Chem 2000:181-186.
24. Pozsgay V. In: Ogura H, Hasegawa A, Suami T, eds. Carbohydrates, Synthetic Methods and Applications in Medicinal Chemistry. Tokyo-Weinheim: Kodansha-Verlag Chemie, 1992:188-227.
25. Nilsson M, Norberg T. Synthesis of a spacer-containing disaccharide fragment of Bordetella pertussis lipopolysaccharide. Carbohydr Res 2000; 327:261-267.
26. Muller R, Brade H, Kosma P. Synthesis of deoxy analogues of (2→8)-linked 3-deoxy-alpha-D-manno-oct-2-ulopyranosylonic acid (Kdo) disaccharides for binding studies with Chlamydia specific monoclonal antibodies. J Endotoxin Res 1997; 4:347-355.

27. Brade L, Zych K, Rozalski A et al. Structural requirements of synthetic oligosaccharides to bind monoclonal antibodies against Chlamydia lipopolysaccharide. Glycobiology 1997, 7, 819-827.
28. Kosma P, Reiter A, Hofinger A et al. Synthesis of neoglycoproteins containing Kdo epitopes specific for Chlamydophila psittaci lipopolysaccharide. J Endotoxin Res 2000; 6:57-69.
29. Hirooka M, Mori Y, Sasaki A et al. Synthesis of beta-D-ribofuranosyl-(1→3)-alpha-L-rhamnopyranosyl-(1→3)-L-rhamnopyranose by in situ activating glycosylation using 1-OH sugar derivative and Me_3SiBr-$CoBr_2$-Bu_4NBr-molecular sieves 4A system. Bull Chem Soc Jpn 2001; 74:1679-1694.
30. Garegg PJ, Olsson L, Oscarson S. Synthesis of oligosaccharides corresponding to structures found in capsular polysaccharides of Cryptococcus neoformans II. Bioorg Med Chem 1996; 4:1867-1871.
31. Liu WC, Oikawa M, Fukase K et al. A divergent synthesis of lipid A and its chemically stable unnatural analogues. Bull Chem Soc Jpn 1999; 72:1377-1385.
32. Antonov KV, Backinowsky LV, Grzeszczyk B et al. Synthesis and serological characterization of L-glycero-alpha-D-manno-heptopyranose-containing di- and tri-saccharides of the non-reducing terminus of the Escherichia coli K-12 LPS core oligosaccharide. Carbohydr Res 1998; 314:85-93.
33. Soderman P, Widmalm G. Synthesis of trisaccharides related to the branching region of the O-antigen polysaccharides from Escherichia coli O35 and Salmonella arizona O62. Eur J Org Chem 2001:3453-3456.
34. Oscarson S, Tedebark U, Turek D. Synthesis of colitose-containing oligosaccharide structures found in polysaccharides from Vibrio cholerae O139 synonym Bengal using thioglycoside donors. Carbohydr Res 1997; 299:159-164.
35. Misra AK, Das SK, Roy N. Preparation of trans diequatorial 2,3-pyruvate acetals in a di- and a trisaccharide related to the K-antigen from E. coli O101 : K103 : H-. Synthetic Communications 1998; 28:1471-1480.
36. Bajza I, Kerékgyártó J, Hajko J et al. Chemical synthesis of the pyruvic acetal-containing trisaccharide unit of the species-specific glycopeptidolipid from Mycobacterium avium serovariant-8. Carbohydr Res 1994; 253:111-120.
37. Sengupta P, Sarbajna S, Basu S. Synthesis of disaccharides related to the O-specific side chains from E. coli O126 & O128 lipopolysaccharides. J Carbohydr Chem 1999; 18:87-97.
38. Misra AK, Roy N. Synthesis of the tetrasaccharide repeating unit of the antigen from Escherichia coli O126 as its methyl glycoside. J Carbohydr Chem 1998: 17:1047-1056.
39. Bernlind C, Bennett S, Oscarson S. Synthesis of a D,D- and L,D-heptose-containing hexasaccharide corresponding to a structure from Haemophilus ducreyi lipopolysaccharides Tetrahedron-Asymmetry. 2000; 11:481-492.
40. Robbins JB, Schneerson R, Anderson P et al. Prevention of systemic infections, especially meningitis, caused by Haemophilus influenzae type b. J Am Med Assoc 1996; 276:1181-1185.
41. Bernlind C, Oscarson S. Synthesis of L-glycero-D-manno-heptopyranose-containing oligosaccharide structures found in lipopolysaccharides from Haemophilus influenzae. Carbohydr Res 1997; 297:251-260.
42. Bernlind C, Oscarson S. Synthesis of a branched heptose- and Kdo-containing common tetrasaccharide cores structure of Haemophilus influenzae lipopolysaccharides via a 1,6-anhydro-L-glycero-β-manno-heptopyranose intermediate. J Org Chem 1998; 63:7780-7788.
43. Sekljic H, Wimmer N, Hofinger A et al. Synthesis of neoglycoproteins containing L-glycero-alpha-D-manno-heptopyranosyl-(1→4)- and -(1→5)-linked 3-deoxy-alpha-D-manno-oct-2-ulopyranosylonic acid (Kdo) phosphate determinants. J Chem Soc Perkin 1 1997:1973-1982.
44. Verez Bencomo V, Fernandez Santana V, Figueroa Perez I et al. Development of a Haemophilus influenzae type b conjugate vaccine based on synthetic oligosaccharide as antigen XIXth. Int Carbohydr Symp. San Diego: Abstract BP 1998:141.
45. Chong P, Chan N, Kandil A et al. A strategy for rational design of fully synthetic glycopeptide conjugate vaccines. Infect Immun 1997; 65:4918-4925.
46. Nilsson M, Norberg T. Solid-phase synthesis of an analog of Haemophilus influenzae type B capsular polysaccharide. J Carbohydr Chem 1998; 17:305-316.
47. Hansson J, Garegg PJ, Oscarson S. Syntheses of anomerically phosphodiester-linked oligomers of the repeating units of the Haemophilus influenzae types c and f capsular polysaccharides. J Org Chem 2001; 66:6234-6243.
48. Garegg PJ, Hansson J, Helland AC et al. Formation of anomeric phosphodiester linkages using H-phosphonate acceptors. Tetrahedron Lett 1999; 40:3049-3052.
49. Sakai Y, Oikawa M, Yoshizaki H et al. Synthesis of Helicobacter pylori lipid A and its analogue using p-(trifluoromethyl)benzyl protecting group. Tetrahedron Lett 2000; 41:6843-6847.

50. Zhang J, Zhu Y, Kong F. Synthesis of an L-rhamnose tetrasaccharide. Carbohydr Res 2001; 336:229-235.
51. Sarbanja S, Misra AK, Roy N. Synthesis of a di- and a trisaccharide related to the antigen from Klebsiella type 43. Synth Commun 1998; 28:2559-2570.
52. Sarbajana S, Roy N. Synthesis of the tetrasaccharide repeating unit of the K-antigen from Klebsiella type 57. Carbohydr Res 1998; 306:401-407.
53. Choudhury AK, Roy N. Synthesis of the trisaccharide repeating unit of the K-antigen from Klebsiella type-63. J Carbohydr Chem 1997; 16:1363-1371.
54. Ekelöf K, Oscarson S. Synthesis of oligosaccharide structures from the lipopolysaccharide of Moraxella catarrhalis. J Org Chem 1996; 61:7711-7718.
55. Yin HF, D'Souza FW, Lowary TL. Arabinofuranosides from Mycobacteria: Synthesis of a highly branched hexasaccharide and related fragments containing beta- axabinofuranosyl residues. J Org Chem 2002; 67:892-903.
56. Yin H, Lowary TL. Synthesis of arabinofuranosides via low-temperature activation of thioglycosides. Tetrahedron Lett 2001; 42:5829-5832.
57. Varga Z, Bajza I, Batta G et al. Synthesis of the pentasaccharide hapten from the glycopeptidolipid antigen of Mycobacterium avium serovar 17. Tetrahedron Lett 2001; 42:5283-5286.
58. Gurjar MK, Reddy LK, Hotha S. Synthesis of oligosaccharides of motifs D and E of arabinogalactan present in Mycobacterium tuberculosis. J Org Chem 2001; 66:4657-4660.
59. Gurjar MK, Reddy LK, Hotha S. Synthesis of ethyl 5-O-(alpha-D-arabinofuranosyl)-6-O-(beta-D-galactofuranosyl)-beta-D-galactofuranoside present in motif E of the Mycobacterium tuberculosis cell wall. Organic Letters 2001; 3:321-323.
60. Du YG, Pan QF, Kong FZ. An efficient and concise regioselective synthesis of alpha-(1→5)-linked L-arabinofuranosyl oligosaccharides. Carbohydr Res 2000; 329:17-24.
61. D'Souza FW, Lowary TL. The first total synthesis of a highly branched arabinofuranosyl hexasaccharide found at the nonreducing termini of mycobacterial arabinogalactan and lipoarabinomannan Organic Letters 2000, 2, 1493-1495.
62. D'Souza FW, Ayers JD, McCarren PR et al. Arabinofuranosyl oligosaccharides from Mycobacteria: Synthesis and effect of glycosylation on ring conformation and hydroxymethyl group rotamer populations. J Am Chem Soc 2000; 122:1251-1260.
63. Subramaniam V, Lowary TL. Synthesis of oligosaccharide fragments of mannosylated lipoarabinomannan from Mycobacterium tuberculosis. Tetrahedron 1999; 55:5965-5976.
64. Wu XM, Marino-Albernas JR, Auzanneau FI et al. Synthesis and NMR analysis of C-13-labeled oligosaccharides corresponding to the major glycolipid from Mycobacterium leprae. Carbohydr Res 1998; 306:493-503.
65. Bajza I, Kövér KE, Lipták A. Synthesis of p-trifluoroacetamidophenyl (4,6-dideoxy-4-formamido-3-C-methyl-2-O-methyl-alpha-L-mannopyranosyl)-(1→3)-(2-O-methyl-alpha-D-rhamnopyranosyl)-(1→3)-(2-O-methyl-alpha-L-fucopyranosyl)-(1→3)-(alpha-L-rhamnopyranosyl)-(1→2)-6-deoxy-alpha-L-talopyranoside: a spacer-armed pentasaccharide glycopeptidolipid antigen of Mycobacterium avium serovar 14. Carbohydr Res 1998; 308:247-258.
66. Severn WB, Furneaux RH, Falshaw R et al. Chemical and spectroscopic characterisation of the phosphatidylinositol manno-oligosaccharides from Mycobacterium bovis AN5 and WAg201 and Mycobacterium smegmatis mc(2) 155. Carbohydr Res 1998; 308:397-408.
67. Mereyala HB, Hotha S, Gurjar MK. Synthesis of pentaarabinofuranosyl structure motif a of Mycobacterium tuberculosis. J Chem Soc Chem Commun 1998:685-686.
68. Ayers JD, Lowary TL, Morehouse CB et al. Synthetic arabinofuranosyl oligosaccharides as mycobacterial arabinosyltransferase substrates. Bioorg Med Chem Lett 1998; 8:437-442.
69. Gurjar MK, Adhikari S. Synthesis of the unique terminal branched tetrasaccharide of Mycobacterium gordonae strain 990. Tetrahedron 1997; 53:8629-8634.
70. Ishii K, Kubo H, Yamasaki R. Synthesis of alpha-lactosyl-(1→3)-L-glycero-alpha-D-manno-heptopyranoside, a partial oligosaccharide structure expressed within the lipooligosaccharide produced by Neisseria gonorrhoeae strain 15253. Carbohydr Res 2002; 337:11-20.
71. Yan FY, Wakarchuk WW, Gilbert M et al. Polymer-supported and chemoenzymatic synthesis of the Neisseria meningitidis pentasaccharide: a methodological comparison. Carbohydr Res 2000; 328:3-16.
72. Berkin A, Coxon B, Pozsgay V. Towards a synthetic glycoconjugate vaccine against Neisseria meningitidis A. Chem Eur J 2002; 8:4424-4433.
73. Pozsgay V, Vieira NE, Yergey A. A method for bioconjugation of carbohydrates using Diels-Alder cycloaddition. Org Lett 2002; 4:3191-3194.
74. Castro-Palomino JC, Tsvetkov YE, Schneider R et al. 8-O-Sialylation of neuraminic acid acceptor reactivity and anomeric stereocontrol. Tetrahedron Lett 1997; 38:6837-6840.

75. Castro-Palomino JC, Tsvetkov YE, Schmidt RR. 8-O-Sialylation of neuraminic acid. J Am Chem Soc 1998; 120:5434-5440.
76. Demchenko AV, Boons GJ. A novel direct glycosylation approach for the synthesis of dimers of N-acetylneuraminic acid. Chem Eur J 1999; 5:1278-1283.
77. Reiter A, Zamyatina A, Schindl H et al. Synthesis of Pseudomonas aeruginosa lipopolysaccharide core antigens containing 7-O-carbamoyl-L-glycero-alpha-D-manno-heptopyranosyl residues. Carbohydr Res 1999; 317:39-52.
78. Duynstee HI, van Vliet MJ, van der Marel GA et al. An efficient synthesis of (R)-3-{(R)-3-[2-O-(alpha-L- rhamnopyranosyl)-alpha-L- rhamnopyranosyl]oxydecanoyl}oxydecanoic acid, a rhamnolipid from Pseudomonas aeruginosa. Eur J Org Chem 1998:303-307.
79. Robina I, LopezBarba E, JimenezBarbero J et al. Synthesis and conformational analysis of a lipotetrasaccharide related to the nodulation factor of Rhizobium bacteria.Tetrahedron-Asymmetry 1997; 8:1207-1224.
80. Zou W, Jennings HJ. The conformational epitope or type III group B Streptococcus capsular polysaccharide. Adv Exp Med Biol 2001; 491:473-484.
81. Banaszek A. Synthesis of the unique trisaccharide repeating unit, isolated from lipopolysaccharides Rhizobium leguminosarum bv. trifolii 24, and its analogue. Carbohydr Res 1998; 306:379-385.
82. Banaszek A, Ciunik Z. Synthesis of the unique trisaccharide: 6-d-L-Talp-alpha(1→2)-L-Rhaf-beta(1→5)-DHA. Tetrahedron Lett 1997; 38:273-276.
83. Zhao YX, Thorson JS. Chemoenzymatic synthesis of the Salmonella group E-1 core trisaccharide using a recombinant beta-(1→4)- mannosyltransferase. Carbohydr Res 1999; 319:184-191.
84. Crich D, Li H. Synthesis of the Salmonella type E-1 core trisaccharide as a probe for the generality of 1-(benzenesulfinyl)piperidine/triflic anhydride combination for glycosidic bond formation from thioglycosides. J Org Chem 2002; 67:4640-4646.
85. Lin FYC, Ho VA, Khiem HB et al. The efficacy of a Salmonella typhi Vi conjugate vaccine in two-to-five-year-old children. N Engl J Med 2001; 344:1263-1269.
86. Shi-Shun LK, Mallet JM, Moreau M et al. Synthesis of oligomers of the capsular polysaccharide of Salmonella typhi, etiological agent of typhoid fever. Tetrahedron 1999; 55:14043-14068.
87. Alexander J, del Guercio MF, Maewal A et al. Linear PADRE T helper epitope and carbohydrate B cell epitope conjugates induce specific high titer IgG antibody responses. J Immunol 2000; 164:1625-1633.
88. Pozsgay VJ. Org. Chem. Synthesis of glycoconjugate vaccines against Shigella dysenteriae type 1 1998; 63:5983-5999.
89. Pozsgay V. Synthetic Shigella Vaccines: A carbohydrate-protein conjugate with totally synthetic hexadecasaccharide haptens. Angew Chem Int Ed 1998; 37:138-142.
90. Pozsgay V, Chu CY, Pannell L et al. Protein conjugates of synthetic saccharides elicit higher levels of serum IgG lipopolysaccharide antibodies in mice than do those of the O-specific polysaccharide from Shigella dysenteriae type 1. Proc Natl Acad Sci USA 1999; 96:5194-5197.
91. Mulard LA, Glaudemans CPJ. Synthesis of ligands related to the O-specific antigen of Shigella dysenteriae type 1. Part 10—Synthesis of tri- and tetrasaccharide fragments of the Shigella dysenteriae type 1 O- antigen deoxygenated and fluorinated at position 3 of the methyl alpha-D-galactopyranoside terminus. Carbohydr Res 1998; 311:121-133.
92. Miller CE, Karpas A, Schneerson R et al. Of four murine, anti-Shigella dysenteriae type 1 O-polysaccharide antibodies, three employ V-genes that differ extensively from those of the fourth. Molec Immun 1996; 33:1217-1222.
93. Nyholm PG, Mulard LA, Miller CE et al. Conformation ofa the O-specific polysaccharide of Shigella dysenteriae type 1. Molecular modeling shows a helical structure with efficient exposure of the antigenic determinant α-L-Rhap-(1→2)-α-D-Galp. Glycobiology 2001; 11:945-955.
94. Miller CE, Mulard LA, Padlan EA et al. Binding of modified fragments of the Shigella dysenteriae type 1 O-specific polysaccharide to monoclonal IgM 3707 E9 and docking of the immunodeterminant to its modeled Fv. Carbohydr Res 1998; 309:219-226.
95. Pozsgay V, Coxon B, Glaudemans CPJ et al. Towards an oligosaccharide-based glycoconjugate vaccine against Shigella dysenteriae type 1. Synlett 2003:743-767.
96. Mukherjee I, Das SK, Mukherjee A et al. Synthesis of the tetrasaccharide related to the repeating unit of the antigen from Shigella dysenteriae type 5. Carbohydr Res 2000; 325:245-252.
97. Mulard LA, Costachel C, Sansonetti PJ. Synthesis of the methyl glycosides of a di- and two trisaccharide fragments specific for the Shigella flexneri serotype 2a O-antigen. J Carbohydr Chem 2000; 19:849-877.
98. Costachel C, Sansonetti PJ, Mulard LA. Synthesis of ligands related to the O-specific polysaccharides of Shigella flexneri serotype 2a and Shigella flexneri serotype 5a part 5—Linear synthesis of the methyl glycosides of tetra and pentasaccharide fragments specific for the Shigella flexneri serotype 2a O-antigen J. Carbohydr. Chem. 2000, 19, 1131-1150.

99. Mulard LA, Ughetto-Monfrin J. Linear synthesis of the methyl glycosides of tri-, tetra-, and pentasaccharide fragments of the Shigella flexneri serotype 5a O-antigen. J Carbohydr Chem 2000; 19:503-526.
100. Mulard LA, Ughetto-Monfrin J. First synthesis of a branched pentasaccharide representative of the repeating unit of the Shigella flexneri serotype 5a O- antigen. J Carbohydr Chem 2000; 19:193-220.
101. Mulard LA, Ughetto-Monfrin J. Synthesis of a tri- and tetrasaccharide fragment specific for the Shigella flexneri serotype 5a O-antigen. A reinvestigation. J Carbohydr Chem 1999: 18:721-753.
102. Medgyes A, Farkas E, Lipták A et al. Synthesis of the monosaccharide units of the O-specific polysaccharide of Shigella sonnei. Tetrahedron 1997; 53:4159-4178.
103. Toth A, Medgyes A, Bajza I et al. Synthesis of the repeating unit of the O-specific polysaccharide of Shigella sonnei and quantitation of its serologic activity. Bioorg Med Chem Lett 2000; 10:19-21.
104. Medgyes A, Bajza I, Farkas E et al. Synthetic studies towards the O-specific polysaccharide of Shigella sonnei. J Carbohydr Chem 2000; 19:285-310.
105. Wang W, Kong FZ. Synthesis of an L-rhamnose tetrasaccharide. J Carbohydr Chem 1999; 18:263-273.
106. Auzanneau FI, Forooghian F, Pinto BM. Efficient, convergent syntheses of oligosaccharide allyl glycosides corresponding to the Streptococcus Group A cell- wall polysaccharide. Carbohydr Res 1996; 291:21-41.
107. Tietze LF, Schröter C, Gabius S et al. Conjugation of p-aminophenyl glycosides with squaric acid diester to a carrier protein and the use of neoglycoprotein in the histochemical detection of lectins. Bioconj Chem 1991; 2:148-153.
108. Pitner JB, Beyer WF, Venetta TM et al. Bivalency and epitope specificity of a high-affinity IgG3 monoclonal antibody to the Streptococcus Group A carbohydrate antigen. Molecular modeling of a Fv fragment. Carbohydr Res 2000; 324:17-29.
109. Palomino JCC, Rensoli MH, Verez Bencomo V. Synthesis of the trisaccharide alpha-L-Rha-(1-2)-alpha-L-Rha-(1-2)-alpha-L-Rha with a dioxolane-type spacer arm. J Carbohydr Chem 1996; 15:137-146.
110. Mehta S, Whitfield DM. Polymer-supported synthesis of a branched trisaccharide of the type IA group B Streptococcus capsular polysaccharide: 3-iodo- 4-methoxybenzyl as a new O-protecting group. Tetrahedron 2000; 56:6415-6425.
111. Pozsgay V, Brisson JR, Jennings HJ et al. Synthetic oligosaccharides related to group-B streptococcal polysaccharides .5. Combined chemical and enzymatic-synthesis of a pentasaccharide repeating unit of the capsular polysaccharide of type-iii group-B Streptococcus and one- dimensional and 2-dimensional nmr spectroscopic studies. J Org Chem 1991; 56:3377-3385.
112. Pozsgay V, Brisson JR, Jennings HJ. Synthetic oligosaccharides related to group-B streptococcal polysaccharides .4. Synthesis of a trisaccharide and a tetrasaccharide fragment of the capsular polysaccharide of type-iii group-B Streptococcus. Carbohydr Res 1990; 205:133-146.
113. Pozsgay V, Jennings HJ. Synthesis of oligosaccharides corresponding to the common antigen of Group B Streptococci. J Org Chem 1988; 53:4042-4052.
114. Pozsgay V, Jennings HJ. Synthetic oligosaccharides related to group-B streptococcal polysaccharides.2. Synthesis of a di-saccharide, tri- saccharide, and tetra-saccharide unit of the group-B streptococcal common antigen. Carbohydr Res 1988; 179:61-75.
115. Pozsgay V, Gaudino J, Paulson JC et al. SYNTHETIC oligosaccharides related to group-b streptococcal polysaccharides. 6. Chemoenzymatic synthesis of a branching decasaccharide fragment of the capsular polysaccharide of type- iii group-B Streptococcus. Bioorg Med Chem Lett 1991; 1:391-394.
116. Zou W, Brisson JR, Yang QL et al. Synthesis and NMR assignment of two repeating units (decasaccharide) of the type III group B Streptococcus capsular polysaccharide and its C-13-labeled and N-propionyl substituted sialic acid analogues. Carbohydr Res 1996; 295:209-228.
117. Demchenko A, Boons GJ. A highly convergent synthesis of a hexasaccharide derived from the oligosaccharide of group B type III Streptococcus. Tetrahedron Lett 1997; 38:1629-1632.
118. Demchenko AV, Boons GJ. A highly convergent synthesis of a complex oligosaccharide derived from group B type III Streptococcus. J Org Chem 2001; 66:2547-2554.
119. Zou W, Laferriere CA, Jennings HJ. Oligosaccharide fragments of the type III group B streptococcal polysaccharide derived from S. pneumoniae type 14 capsular polysaccharide by a chemoenzymatic method. Carbohydr Res 1998; 309:297-301.
120. Zou W, Jennings HJ. Mimics of the structural elements of type III group B Streptococcus capsular polysaccharide .3. Two repeating units (octasaccharide) with (S)-1-carboxyethyl groups replacing sialic acids. Bioorg Med Chem Lett 1997; 7:647-650.
121. Eichler E, Jennings HJ, Whitfield DM. Synthesis of a single repeat unit of type VIII group B Streptococcus capsular polysaccharide. J Carbohydr Chem 1997; 16:385-411.

122. Eichler E, Jennings HJ, Gilbert M et al. Synthesis of a disialylated hexasaccharide of Type VIII Group B Streptococcus capsular polysaccharide. Carbohydr Res 1999; 319:1-16.
123. Garegg PJ, Oscarson S, Tedebark U. Synthesis of the repeating unit of the capsular polysaccharide of Streptococcus pneumoniae type 3 as a building block suitable for formation of oligomers. J Carbohydr Chem 1998; 17:587-594.
124. Lefeber DJ, Kamerling JP, Vliegenthart JFG. Synthesis of Streptococcus pneumoniae type 3 neoglycoproteins varying in oligosaccharide chain length, loading and carrier protein. Chem Eur J 2001; 7:4411-4421.
125. Benaissa-Trouw B, Lefeber DJ, Kamerling JP et al. Synthetic polysaccharide type 3-related di-, tri-, and tetrasaccharide-CRM197 conjugates induce protection against Streptococcus pneumoniae type 3 in mice. Infect Immun 2001; 69:4698-4701.
126. Lefeber DJ, Arevalo EA, Kamerling JP et al. Synthesis of a hexasaccharide fragment of the capsular polysaccharide of Streptococcus pneumoniae type 3. Can J Chem 2002; 80:76-81.
127. Thijssen MJL, van Rijswijk MN, Kamerling JP et al. Synthesis of spacer-containing di- and tri-saccharides that represent parts of the capsular polysaccharide of Streptococcus pneumoniae type 6B. Carbohydr Res 1998; 306:93-109.
128. Thijssen MJL, Bijkerk MHG, Kamerling JP et al. Synthesis of four spacer-containing 'tetrasaccharides' that represent four possible repeating units of the capsular polysaccharide of Streptococcus pneumoniae type 6B. Carbohydr Res 1998; 306:111-125.
129. Corby NS, Kenner GW, Todd AR. Nucleotides. Part XVI. Ribonucleoside-5' phospites. J Chem Soc 1952:3669.
130. Stawinsi J, Kraszewski A. How to get the most out of two phosphorous chemistries. Studies on H-phosphonates. Acc Chem Res 2002; 35:952-960.
131. Froehler BC, Matteucci MD. Nucleoside H-phosphonates—valuable intermediates in the synthesis of deoxyoligonucleotides. Tetrahedron Lett 1986; 27:469-472.
132. Garegg PJ, Regberg T, Strawinski J et al. Formation of internucleotidic bond via phosphonate intermediate. Chemica Scripta 1986; 26:59-62.
133. Marugg JE, Tromp M, Kuyl-Yeheskiely E et al. A convenient and general approach to the synthesis of properly protected d-nucleoside-3'-hydrogenphosphonates via phosphite intermediates. Tetrahedron Lett 1986; 27:2661-2664.
134. Marugg JE, Burik A, Tromp M et al. A new and versatile approach to the preparation of valuable deoxynucleoside 3'-phosphite intermediates. Tetrahedron Lett 1986; 27:2271-2274.
135. Westerduin P, Veeneman GH, van der Marel GA et al. Synthesis of the fragment GlcNAc-α(1→p→6)-GlcNAc of the cell wall polymer of staphylococcus lactis having repeating N-acetyl-D-glucosamine phosphate units. Tetrahedron Lett 1986; 27:6271-6274.
136. Lindberg M, Norberg T. Synthesis of sucrose 4'-(L-arabinose-2-yl phosphate) (agrocinopine A) using an arabinose 2-H-phosphonate intermediate. J Carbohydr Chem 1988; 7:749-755.
137. Jansen WTM, Hogenboom S, Thijssen MJL et al. Synthetic 6B di-, tri-, and tetrasaccharide-protein conjugates contain pneumococcal type 6A and 6B common and 6B-specific epitopes that elicit protective antibodies in mice. Infect Immun 2001; 69:787-793.
138. Niggemann J, Kamerling JP, Vliegenthart JFG. Beta-1,4-galactosyltransferase-catalyzed synthesis of the branched tetrasaccharide repeating unit of Streptococcus pneumoniae type 14. Bioorg Med Chem 1998; 6:1605-1612.
139. Niggemann J, Kamerling JP, Vliegenthart JFG. Application of beta-1,4-galactosyltransferase in the synthesis of complex branched-chain oligosaccharide mimics of fragments of the capsular polysaccharide of Streptococcus pneumoniae type 14. J Chem Soc Perkin 1998; 1:3011-3020.
140. Bousquet E, Khitri M, Lay L et al. Capsular polysaccharide of Streptococcus pneumoniae type 19F: synthesis of the repeating unit. Carbohydr Res 1998; 311:171-181.
141. Nilsson M, Norberg T. Synthesis of a spacer-containing nonasaccharide fragment of Streptococcus pneumoniae 19F capsular polysaccharide. J Chem Soc Perkin Trans 1998; 1:1699-1704.
142. Choudhury AK, Mukherjee I, Roy N. Synthesis of a trisaccharide related to the K-antigen from Streptococcus pneumoniae type 29. Synth Commun 1998; 28:3115-3120.
143. Arencibia-Mohar A, Ariosa-Alvarez A, Madrazo-Alonso O et al. Synthesis of terminal disaccharide elements corresponding to the Ogawa and Inaba antigenic determinant from Vibrio cholerae O1. Carbohydr Res 1998; 306:163-170.
144. Ariosa-Alvarez A, Arencibia-Mohar A, Madrazo-Alonso O et al. Synthesis of the Vibrio cholerae O1 Ogawa and Inaba terminal disaccharides with dioxolane-type spacers and their coupling to proteins. J Carbohydr Chem 1998; 17:1307-1320.
145. Gonzalez L, Asensio JL, Ariosa-Alvarez A et al. Solution conformation and dynamics of the trisaccharide fragments of the O-antigen of Vibrio cholerae O1, serotypes Inaba and Ogawa. Carbohydr Res 1999; 321:88-95.

146. Wang J, Villeneuve S, Zhang J et al. On the antigenic determinants of the lipopolysaccharides of Vibrio cholerae O : 1, serotypes Ogawa and Inaba. J Biol Chem 1998; 273:2777-2783.
147. Villeneuve S, Souchon H, Riottot MM et al. Crystal structure of an anti-carbohydrate antibody directed against Vibrio cholerae 01 in complex with antigen: Molecular basis for serotype specificity. Proc Natl Acad Sci USA 2000; 97:8433-8438.
148. Zhang J, Kovac P. Synthesis of methyl α-glycosides of some higher oligosaccharide fragments of the O-antigen of Vibrio cholerae O1, serotype Inaba and Ogawa. Carbohydr Res 1997; 300:329-339.
149. Zhang J, Kovac P. Synthesis of some analogs of the methyl alpha-glycoside of the presumed antigenic determinant of the O-specific polysaccharide of Vibrio cholerae O : 1, serotype Ogawa. J Carbohydr Chem 1998; 17:341-357.
150. Chang AHC, Horton D, Kovac P. Syntheses of fluorinated ligands to probe binding of antigenic determinants of Vibrio cholerae O : 1, serotypes Inaba and Ogawa, to antibodies. Tetrahedron-Asymmetry 2000; 11:595-606.
151. Poirot E, Zhang XD, Whittaker NF et al. Syntheses of the L-manno and some other analogs of the terminal determinants of the O-PS of Vibrio cholerae O : 1. Carbohydr Res 2001; 330:7-20.
152. Zhang J, Yergey A, Kowalak J et al. Studies towards neoglycoconjugates from the monosaccharide determinant of Vibrio cholerae O : 1, serotype Ogawa using the diethyl squarate reagent. Carbohydr Res 1998; 313:15-20.
153. Zhang JA, Kovac P. Studies on vaccines against cholera. Synthesis of neoglycoconjugates from the hexasaccharide determinant of Vibrio cholerae O : 1, serotype Ogawa, by single-point attachment or by attachment of the hapten in the form of clusters. Carbohydr Res 1999; 321:157-167.
154. Ogawa Y, Kovac P. Synthesis of the dodecasaccharide fragment representing the O-polysaccharide of Vibrio cholerae O:1, serotype Ogawa, bearing an aglycon offering flexibility for chemical linking to proteins. Glycoconj J 1997: 14:433-438.
155. vanStraten NCR, Kriek NMAJ, Timmers CM et al. Synthesis of a trisaccharide fragment corresponding to the lipopolysaccharide region of Vibrio parahaemolyticus. J Carbohydr Chem 1997; 16:947-966.

CHAPTER 13

Challenges and Opportunities in the Development of New Conjugate Vaccines against Infectious Diseases

P. Moingeon, M. Moreau and A.A. Lindberg

Abstract

Chemical conjugation between capsular polysaccharides (CPS) or lipopolysaccharides (LPS) and carrier proteins represents a powerful means to create vaccines targeting bacterial carbohydrate antigens with increased immunogenicity. Using such conjugate vaccines, infections against *Haemophilus influenza* type b, *Streptococcus pneumoniae* and *Neisseria meningitidis* type C can be successfully controlled. Additional conjugate vaccines targeting multiple serotypes of *N. meningitidis, Salmonella typhi, and Staphylococcus aureus* have entered clinical efficacy trials. In this chapter, we discuss recent developments in targeting other infectious pathogens, searching for new carrier proteins, and improving conjugation methods. We also review some of the challenges associated with the development of conjugate vaccines, including issues linked with product characterization, and the need to incorporate multiple polysaccharides (PS) to cover against several serotypes for a given pathogen. In addition, both the phenomenon of epitopic suppression as well as the theoretical risk that long-term vaccination may facilitate capsular switching or serotype replacements have to be taken into consideration.

The Success of Vaccines Based on Glycoconjugates

It was recognized in the first half of the 20th century that surface capsular polysaccharides (CPS) extracted from bacteria can be used as immunogens, in order to elicit protective antibody responses against a number of infectious diseases.[1-4] The first *N.meningitidis* CPS vaccine was licensed in 1972, followed by vaccines against *Streptococcus pneumoniae* (7-valent in 1977, and via a 14-valent to a 23-valent in 1983), *Haemophilus influenzae* type b (1985), and *Salmonella typhi* (1993).[1,2] However, polysaccharides (PS) are often poorly immunogenic in infants where some of these diseases are prevalent. In addition, polysaccharides do not elicit memory functions, and hence the immune responses cannot be boosted.[1-4] Consequently, these vaccines gained little success, except in epidemics where meningococcal outbreaks could be controlled. In addition, the appearance of antimicrobial compounds slowed, or even stopped, vaccine development. Interest was renewed in the 1960s, most particularly when immunogenicity problems were overcome by linking the CPS to an immunogenic carrier protein. Such a chemical conjugation was found to enhance immune responses, both in terms of magnitude and duration, particularly in young infants.[1-5] This strategy of conjugation of PS to carrier proteins has materialized successfully into commercially available safe and cost-effective vaccines against three major infectious pathogens: *N. meningitidis* type C, *S. pneumoniae* (7serotypes) and *H. influenzae* type b (Table 1).[6-18]

Immunobiology of Carbohydrates, edited by Simon Y.C. Wong and Gemma Arsequell.
©2003 Eurekah.com and Kluwer Academic / Plenum Publishers.

Table 1. From polysaccharides to glycoconjugate vaccines

Infectious Pathogen	Comments	Ref.
Haemophilus influenzae type b (Hib)	A PS-based vaccine against Hib was registered in the US in 1985. Eventually, four different conjugate vaccines were subsequently developed and shown to be highly efficacious (> 90 %) in developed countries in preventing Hib-associated meningitis. Conjugate vaccines were also shown to have a dramatic impact on lower respiratory tract infections in developing countries.	7-10
N. meningitidis	PS vaccines against serogroups A & C were registered as early as 1972. A conjugate vaccine against group C was commerciali-zed in the UK in 2001. Additional conjugate vaccines are under development, with the aim to protect against multiple serotypes (*e.g.* A,C, W135, Y).	6, 11-13
S .pneumoniae	There are pneumococcal polysaccharide vaccines which are composed of the 23 most common pneumococcal serotypes. They represent 85- 90 % of the serotypes that cause invasive infections in adults in industrialized countries. In addition, a 7-valent conjugate vaccine (Prevnar™) was licensed in the US in 2000. Nine- and eleven- valent conjugate vaccines are currently in development.	1, 14-18

Conjugate Vaccines against H.Influenzae *Type b (Hib)*

Prior to the introduction of a conjugate Hib vaccine, invasive disease associated with capsulated *Haemophilus influenzae* type b in Europe ranged from 30 to 60 cases per 100,000 children under 5 years old.[1] In the US, corresponding figures were 50-100 per 100,000 children. Hib infections were less common in adults, with an annual incidence of 0-22 per 100 000 being reported. The peak incidence of Hib disease in the prevaccine era was observed in children, most particularly between the ages of 5 and 12 months, coinciding with the waning of maternal antibodies, and prior to the appearance of anti-capsular antibodies. Four different PRP conjugate vaccines were commercialized in the early 1990s. These have differences in carrier proteins, polysaccharide size, linkage type with or without spacers, and links at either the two PS ends or along the chains.[1,2] Irrespective of conjugation characteristics, the vaccines elicited solid and protective immune responses. The introduction of Hib conjugate vaccines has had a major effect on the overall burden of Hib disease: a 97% reduction in disease burden has been observed in Finland, UK, Israel, Australia and USA where these vaccines are part of the routine infant immunization programmes.[7-10,19] Despite this very high efficacy, the involvement of herd immunity was questioned.[9] Importantly, a randomized efficacy trial in Gambia demonstrated an efficacy of more than 90 % against Hib meningitis. In addition, in this study a dramatic impact on acute lower respiratory tract infections was also documented.[20] The efficacy of Hib conjugate vaccines has now been further confirmed in a number of developing countries.[19]

Conjugate Vaccine against N. Meningitidis *Group C*

Prompted by the high effectiveness of the Hib conjugate vaccines, there has been a strong impetus to develop conjugates of other capsular polysaccharide vaccines.[1,2] The technology was applied to the development of vaccines against meningitis caused by *N.meningitidis*. Although the total number of cases is small, it is a disease that attracts much attention because of its mortality. Here as for Hib, a polysaccharide capsule is an important virulence factor, and prospects for prophylactic immunization are good. *N. meningitidis* is an exclusive human pathogen, and transmission is by droplets from colonized upper respiratory mucosal membranes. There are 12 serogroups based on the structure of the capsular polysaccharide: A, B, C, 29E, H, I, K, L, W135, X, Y and Z. Approximately 90% of cases of meningitis are caused by strains belonging to serotypes A, B and C. Meningococcal meningitis is spread over all age groups: one third occurs from 0 to 4 years of age, one third among 5 to 9 years old, and the final one third at age 20 and above.

Research efforts culminated in the successful licensing of a *N. meningitidis* serogroup C CPS-carrier protein conjugate in the UK. This vaccine was shown to be safe, immunogenic and to prime for immunological memory.[11] As such, it was introduced into the routine infant schedule and was in parallel offered to all children less than 18 years-old, in a phased program started in November 1999.[13] An efficacy study conducted over a 9-month period following the introduction of the meningococcal serogroup C conjugate vaccine in the UK established that the short-term efficacy of the vaccine was 97 and 92% for teenagers and toddlers, respectively.[13] Importantly, carriage of serogroup C meningococci 1 year after the introduction of the vaccine was found to be reduced by 66% in 15-17 years old adolescents.[12]

Conjugate Vaccine against S. Pneumoniae

Infections caused by *Streptococcus pneumoniae* are still a major cause of morbidity and mortality in adults and children despite the availability of effective antimicrobial therapy.[15] An important virulence factor is the CPS, and there are now 90 different capsular types described.[1] Invasiveness is dependent on the capsular structure rather than on the amount being produced. Pneumococci cause a wide variety of infections ranging from relatively mild mucosal infections like acute otitis media, to more serious bronchopneumonia and potentially life-threatening meningitis. Clinical illness is a result of the spread of the pneumococci to tissues from the oropharynx.

Pneumococci colonize the respiratory mucosa of both healthy and sick individuals. The carrier rates are higher in children than in adults, even higher in children attending day-care centers and still higher in those with respiratory infections compared with healthy children. In developing countries, pneumococcal infections cause a significant disease burden in the form of acute lower respiratory infections (with more than 4 million deaths annually, most of them in children <5 years of age). The immunological defence against pneumococcal infection depends on type-specific anticapsular polysaccharide antibodies interacting with complement. This opsonizes the bacteria for phagocytosis and clearance. Of the 90 capsular types identified, only a fraction are common causes of invasive pneumococcal disease. Seven- and nine-valent (1, 3, 4, 5, 6B, 7F, 9V, 14, 18C, 19F and 23F) conjugates are expected to cover from 60 to 85% of invasive pneumococcal infections. However, variations can be expected dependent on local differences in serotype prevalences in industrialized as well as developing countries.

The structure of each polysaccharide serotype differs from that of other serotypes and from that of PRP. Hence different coupling strategies have to be used for the different polysaccharides.

Several conjugates have been through phase I, II and III trials and a 7-valent conjugate was licensed in 2000 in the US and 2001 in France and the UK.[14-18] To date, the following conclusions can be drawn: (i) in adults the conjugates are well tolerated although local reactions at the injection side appear to be more common than after vaccination with capsular polysaccharides, (ii) in toddlers all conjugates tested have been immunogenic, (iii) in infants all conjugates tested have been well tolerated and no immediate, serious vaccine-related adverse events have been reported.

The seven-valent vaccine was found to elicit protection against invasive disease associated with all seven included serotypes (with an efficacy of up to 98%).[15] Importantly, no major epitopic suppression nor immune interference were observed as a consequence of the inclusion of multiple PS in the conjugate vaccine, although it cannot be excluded that it may occur with a vaccine designed to cover more serotypes.

Similarly to the group C meningococci conjugate vaccine, a 9-valent conjugate vaccine in development was shown to reduce carriage in toddlers.[18] Such a carriage reduction is critically important given that humans are the only reservoir of such pathogenic bacteria: as a consequence, breaking human transmission might likely lead to disease eradication, most particularly as a consequence of herd immunity. That the licensed pneumococcal conjugate is a 7-valent vaccine has raised concerns that there could be replacement by nonvaccine serotypes, following the widespread use of such a vaccine. As a matter of fact, introduction of the 7-valent pneumococcal conjugate in Finland was associated with an increased incidence of otitis media caused by nonvaccine serotypes.[21] Conversely, nasopharyngeal carriage in children receiving three doses of a nine-valent conjugate vaccine was barely biased toward nonvaccine serotypes.[22] It is evident that the circulation of different pneumococcal serotypes in society should be carefully monitored in the coming years.

Conjugate Vaccines in Development

Given the efficacy of currently licensed CPS-carrier protein conjugate vaccines, it is not surprising that most candidate vaccines targeting polysaccharides surface-expressed by infectious pathogens are being developed directly as conjugates.[1,2,23] Interestingly, the capsulated pathogens for which a vaccine is available (i.e., *H. influenzae*, *S. pneumoniae*, and *N.meningitidis*, see Table 1) all cause bacteremia at some stage in the infection. Consequently, elicited antibodies have a large probability of eliminating bacteria from the bloodstream. Nevertheless, a number of CPS-carrier protein conjugate vaccines are currently being tested against pathogens that infect predominantly mucosal surfaces, although bacteremia may occur during the disease (e.g., *Staphylococcus aureus* and *Shigella* spp). The likelihood of success of the glycoconjugate approach is, however, bolstered by the experiences of remarkable clinical effectiveness of the Hib, *N. meningitidis* serogroup C, and *S pneumoniae* CPS-carrier protein conjugate vaccines in reducing colonization of the nasopharynx.[1,2,24] This is thought to be a consequence of effusion of IgG antibodies over the mucosal membranes.

The idea of using chemically or synthetically produced oligosaccharides (as opposed to purified PS) linked to carrier proteins was pioneered by Avery and Goebel in the 1920s.[25] It was subsequently used in the 1980s to characterize epitope and antibody combining sites in Salmonella and Shigella.[1] Recently, such studies with the aim of making human vaccines have been extended into human trials.[26-28]

Other infectious pathogens being considered as targets for conjugate vaccines include, for example, bacteria associated with nosocomial infections (such as *Staphylococcus aureus*,[29,30] *Pseudomonas aeruginosa*[39,41] as well as the group A beta-hemolytic[47] and group B streptococci[35-38] (GBS). Information regarding the nature and status of these experimental conjugate vaccines is summarized in (Table 2). We will focus below predominantly on three promising conjugate vaccines that have reached late stage clinical development, and for which some efficacy data has been generated in large-scale clinical studies. These candidate vaccines target *S. aureus*, *S. typhi* and group B streptococci, respectively.

Staphylococcus Aureus

A double-blind trial was conducted with a conjugate vaccine in patients with end-stage renal disease receiving hemodialysis. This vaccine consisted of *S. aureus*-derived type 5 and 8 capsular polysaccharides conjugated to nontoxic recombinant exotoxin A from *P. aeruginosa*.[29] Overall, the vaccine was well tolerated, and elicited antibody responses at a level of at least 80 µg/ml (i.e., the estimated minimal level conferring protection) in 80% of patients for type 5, and 75% of patients for type 8.[30] The efficacy evaluated over a period of 40 weeks was about 57 %. Subsequently, however, the protection conferred by the vaccine was found to wane as antibody levels decreased.[30]

Table 2. Main conjugate vaccines against infectious pathogens currently in development

Infectious Pathogen	Status	Comments	Ref.
Staphylococcus aureus	Phase III	A candidate conjugate vaccine against types 5 and 8 capsular PS was shown to be partially efficacious in patients undergoing hemodialysis (see main text).	29, 30
Salmonella typhi	Phase III	A Vi polysaccharide vaccine against typhoid fever is being commercialized. A Vi PS-exoprotein A (from *P. aeruginosa*) conjugate was found to be safe and im munogenic in both adults and children (see main text).	31-33
Neisseria meningitidis serogroups A C, W135, Y	Phase II/III	To complement the licensed conjugate vaccine against meningococcal C, multivalent conjugate vaccines pro tecting against other major serogroups are being devel oped. Three doses of a *Neisseria meningitidis* A+C dip htheria conjugate vaccine were found to be immunoge nic in Nigerian infants.	1, 6, 34
Streptococcus Group B (S. agalactiae)	Phase I/II	See main text. Vaccines based on PS alone were found to be poorly immunogenic. Conjugate vaccines based on PS types Ia, Ib, II, III and V have been tested in humans, and found to be immunogenic.	35-38
Shigella spp. (*S. flexneri, S. sonnei, S. dysenteriae*)	Phase I/II	Associated with infectious diarrheas, most particularly in chilren less than 5 years of age, in developing countries. Due to the existence of multiple serotypes, a tetra or even pentavalent conjugate vaccine is needed to protect against 80% of infections. A 74% efficacy rate was obtained in adults for a conjugate vaccine against *S. sonnei*. Other candidate vaccines against *S. sonnei* and *S. flexneri* are being tested in adults and infants.	23, 26,27
Pseudomonas aeruginosa	Phase I/II	Many serotypes. An eight-valent O-polysaccharide toxin A conjugate vaccine has been developed for patients with acute trauma or cystic fibrosis, and appears to delay colonization.	39-41
Vibrio cholera	Phase I	Two serogroups, O1 and O139, are associated with epidemics. A conjugate vaccine based on polysacch arides from LPS of V. cholerae O1 was tested in humans. Similar candidate vaccines against *V. cholerae* O139 are in preclinical development. Vaccines based on synthetic polysaccharides are also considered.	28, 42- 44
Neisseria meningitidis serogroup B	Preclinical/ phase I	Tolerance issue because of homology between Neu NAc α28 polymer with human tissue antigens. A N-propionylated group B meningococcal vaccine was found to bypass tolerance and still elicit bactericidal antibodies against a bacterial capsular epitope. Safety is of concern.	45, 46
Streptococcus Group A	Preclinical	Conjugate vaccines are being tested as a means to prevent rheumatic fever	47
Mycobacterium tuberculosis	Preclinical	Synthetic oligosaccharide-protein conjugates are being used to elicit immune responses against cell envelope lipoarabinomannans.	48

continued on next page

Table 2. Continued

Infectious Pathogen	Status	Comments	Ref.
Moraxella catharrhalis	Preclinical	Detoxified lipooligosaccharide conjugated to proteins enhances clearance of bacteria from murine lungs.	49, 50
Haemophilus influenzae NT	Preclinical	Detoxified lipooligosaccharide conjugated to proteins was found to confer protection against otitis media in chinchillas	51
E. coli O157	Preclinical	Parenteral immunization with a glycoconjugate vaccine containing the O157 antigen of *E.coli* O157:H7 linked to horse serum albumin, was found to elicit a systemic antibody response in mice. In this model, colonization by the pathogen was not prevented.	52

Salmonella typhi

Licensed typhoid vaccines (killed whole cell, live attenuated or Vi polysaccharide) confer only about 70% immunity, and do not efficiently protect young children. Thus, they are not used for routine vaccination. A conjugate vaccine based on the capsular polysaccharide of *S. typhi*, Vi, bound to the nontoxic recombinant exotoxin A protein from *P. aeruginosa* was tested in both adults, 5 -14 years as well as 2-4 years old children.[31-33] The vaccine was found to be safe and immunogenic in all populations, and exhibited more than 90% efficacy in 2-4 year old children in an efficacy trial in Vietnam.[33]

Group B Streptococci

Infections with Group B streptococci (type I, II and III) represent a major threat in neonates and young infants since these pathogens can be transferred from the mother to the baby through the vaginal tract. Thus, a strategy being pursued by several groups is to vaccinate the mother for a passive transfer of antibodies to the fetus.[35] Clinical trials have been initially performed in healthy women, using an unconjugated tetravalent mixture of polysaccharides covering type Ia, Ib, II and III. In such studies,antibodies were induced against all polysaccharides, albeit at low titers. Subsequently, conjugate vaccines against these polysaccharides have been developed and tested successively in humans. As expected, conjugation to a carrier protein was shown to improve the immunogenicity of the polysaccharides.[36-38] In all phase I and phase II clinical trials evaluating such conjugate vaccines, the vaccine was well tolerated and induced a dose-dependent CPS-specific IgG response in healthy non pregnant adults.[36-37] Attempts to increase the immunogenicity using alum as adjuvant were not confirmed as successful in adults, despite the fact that encouraging results had been obtained in baboons.

New approaches in this field include the development of a vaccine against type V polysaccharides, and intranasal immunization against polysaccharides types Ia and II. To this end, polysaccharides are being conjugated to the B subunit of cholera toxin, to target complexes on GM1 gangliosides expressed on epithelial cells at mucosal surfaces.[37] The ultimate goal is to examine if colonization of the female genital tract can be prevented, thereby reducing or eliminating the risk of infecting the baby when passing through the birth canal.

New Directions

Immune Thoughts

Diphtheria or tetanus toxoid-based conjugates against Hib, *N. meningitidis*, *S. pneumoniae*, were designed to elicit high serum antibody titers, in order to neutralize bacteria in the bloodstream. As a benefit, high IgG titers appear to effuse to membrane surfaces effectively, thereby preventing colonization that is a prerequisite for infection. Low IgA levels are, however, elicited with these conjugate vaccines at mucosal surfaces. In this context, new conjugates are rather designed to affect colonization directly, as opposed to targeting bacteriemia. For example, a vaccine against GBS should ideally reduce vaginal carriage in the mother, in that passively transferred antibodies can protect the baby, but active immunization of the baby will not ensure timely protection. Similarly, protective immunity against Shigella should be elicited at intestinal mucosal surfaces.

Thus, although the route of injection for conjugate vaccines was traditionally subcutaneous or intramuscular, attempts to move towards new administration routes such as the intranasal route are being made in order to achieve mucosal immunization.[53,55] Preclinical experiments, conducted with pneumococcal conjugates associated with a nontoxic mutant of heat-labile enterotoxin, appear to yield promising results both in terms of immunogenicity and protection.[54] Also, in order to enhance the immunogenicity of conjugate vaccines, new formulations and adjuvants besides Alum are being developed.[56] For example the heat-labile enterotoxin from *E. coli* gives quick, protective immune responses in animals when associated with polysaccharides.[54,57] Similarly, the non toxic B subunit of cholera, well known for its mucosal binding properties, has been used successfully as a carrier and an adjuvant with group B streptococci polysaccharide.[58,59] A number of preclinical experiments have also been conducted, with some success using CpG oligonucleotides,[60] genetically detoxified mutants of enterotoxins[54,61] and cytokines such as IL12[62] as adjuvants. In some of these experiments, the adjuvant was found to significantly enhance serum IgG responses against both the carbohydrate antigen and the carrier protein, thereby resulting in a better protection against bacterial challenge. Pneumococcal polysaccharide conjugate vaccines, formulated as microparticles[63] or alginate microspheres,[55]s were shown to elicit mucosal immunity in murine models. However, to the best of our knowledge, studies in man have not yet been conducted.

Which Carrier Protein?

The carrier which allows the T-dependent immune response against the carbohydrate moiety of conjugates should contain at least one T-cell epitope motif to induce cellular immunity. Ideally however, it is more likely that the presence of several T-cell epitopes is beneficial. At present, commercialized conjugates have been prepared using either genetically modified toxin proteins without toxin activity (e.g., CRM 197),[61] chemically detoxified toxins (diphtheria or tetanus toxoid)[7,57] or an outer membrane protein complex from *Neisseria meningitidis*,[64] and consequently several T-cell epitopes are present.[64] Clinically, they have all proven to be efficacious and very successful. New carriers are however still needed to improve the immunogenicity of some conjugate vaccines, and to minimize the potential problem of carrier-induced epitopic suppression.[65]

Some candidate carriers, most of them chosen for their potential usefulness in vaccination, are currently being tested in newly developed conjugate vaccines: a nonexhaustive list of new carriers at various stages of development, is presented in (Table 3). As a general conclusion, even if most of these proteins have been shown to function as carriers, for characterization purposes (see below), it should be emphasized that small proteins or recombinant proteins may be preferred. Interestingly, rationally designed strings of promiscuous CD4+ T-cell epitopes from various pathogen-derived antigens, including tetanus toxoid (TT) and proteins from *P. falciparum*, influenza virus and hepatitis B virus, were shown to provide help to Hib oligosaccharides following conjugation.[67] This holds promise for the design and development of synthetic and immunologically defined carrier proteins. Obviously, the introduction of new carrier proteins requires significant efforts, both in terms of process development and of clinical evaluation in order to document safety in large human populations.

In addition to its function as a carrier, a protein can potentially represent an immunogenic protein eliciting protective antibodies. Carriers such as tetanus or diphtheria toxoids induce protective antibodies, as demonstrated in a number of preclinical tests. Some clinical studies

Table 3. Potential carrier proteins and their utilization

Protein	Pre Clinical Vaccine	Phase1-3 Vaccine	Commercialized Vaccine	Refs.
Tetanus toxoid	x	x	x	1, 2, 65
Diphtheria toxoid	x	x	x	1, 2, 61
Recombinant exotoxin A (rEPA) from *P. aeruginosa*	x	x	x	68
Cross-reacting molecule CRM (non toxic mutant of diphtheria toxin)	x	x	x	61
Protein D from non typeable *Haemophilus influenzae*	x	x		69
Pneumolysin	x	x		70
T-cell epitopes from heat shock protein 60 Kd	x			71
Meningococcal porin 3	x	x		72
OMP complex from *N. meningitidis* B	x	x	x	64,73
PspA (*S. pneumoniae* surface protein A)	x	x		74
P 40 from *K. pneumoniae*	x	x	x	75
Cholera toxin (CT and CTB)	x	x		54, 57, 76
T-cell epitopes	x			67, 77
100-mer polypeptide from meningococcal IgA protease	x			78
Bordetella pertussis fimbriae	x			79

are being conducted in order to document the possibility of eliminating the tetanus toxoid valence from combination vaccines. If shown to be possible, this would avoid protein over-loading and would potentially enhance antibody responses against polysaccharide antigens in TT conjugates.

Chemical Conjugation

CPS or LPS as Polysaccharidic Antigens

Polysaccharide conjugation needs to be supported by a good knowledge of the polysaccharide moiety: structure, lability of the glycosidic linkages or substituents. Carbohydrate antigen candidates include both capsular polysaccharides, saccharide core region or O-antigen from detoxified lipooligosaccharides (LOS) or LPS.[1,2,74] Having some information regarding the main immunogenic determinants is critical, but unfortunately, it is lacking for most of the bacterial polysaccharides. It is nevertheless known that both *O*-acetyl groups of *Salmonella typhi* Vi antigen and pneumococcal serotype 1 polysaccharide represent powerful antigenic determinants.[80,81] It is, however, not always a given conclusion that antibodies against O-acetyls are required for vaccine efficacy. Recent studies established that de-*O*-acetylated pneumococcal type 9V or meningococcal group C polysaccharides are still immunogenic.[82] Since little is known on carbohydrate structure-immunogenicity relationships and animal models are poorly predictive, the conjugation technologies are based on nondegradative coupling chemistry in order to maintain the native structure of the polysaccharide as much as possible.[74] Molecular weights for polysaccharides may range from 2,000 daltons (e.g., for LPS) to several hundred thousands daltons (for repeating units constituting CPS or LPS). For very high molecular weight CPS, depolymerization may be advantageous prior to coupling with the carrier protein. This, however, has to be adapted to the structural characteristics of the polysaccharides, since removal of labile substituents happen as a consequence of chemical treatments.

Altogether, it appears that for most of the PS, the depolymerization process should preserve the initial structure containing all the labile groups such as phosphodiester groups or *O*-acetyl groups, as well as branch substituents. Gentle depolymerization reactions, mild acid hydrolysis, controlled periodic oxidation, enzymatic hydrolysis and more recently, oxidative-reductive depolymerization techniques have all been used successfully to preserve labile groups.[83]

Lastly, the size of PS[84] as well as the supramolecular structure of the glycolipid may be of importance for conformationally determined epitopes and hence there is a need to use a suitable conjugation method.

Coupling Technologies

The choice of the coupling method depends on the chemical functions available on the polysaccharide (e.g., amino, carbonyl, carboxyl or hydroxyl groups) as well as on the nature of the protein, and its content of amino, sulfhydryl, and carboxyl groups.[74] A variety of chemical reactions can be used, but the most appropriate coupling approach is the one avoiding degradation of constituents of the immunogen. Thus, gentle chemistry is required. Some examples of the reagents or reactions used in the conjugation reaction, depending on the functional groups available, are shown in Table 4.

To overcome potential steric hindrance issues during the coupling step and to optimize antigen presentation, spacer molecules can also be used during the conjugation reaction.[74] Spacers may in addition improve the yield during the conjugation step: as a consequence, some coupling reagents can be used as activators of the reaction but also as spacers. In all cases the chemistry and the length of the spacer will depend on the choice of the conjugation process.

Surprisingly however, during the last 10 years few innovative coupling processes have been developed. It has been suggested that PS activation with CDAP may advantageously replace the toxic cyanogen bromide.[85,86] This approach can be used at neutral pH, while preserving polysaccharide coupling ratios. Other coupling reagents, such as vinyl sulfone[87] or squarate derivatives,[88] have recently been tested for oligosaccharide conjugation, and seem to provide a good yield in terms of carbohydrate integration, even if no data on large PS have been obtained yet. Importantly, any new activator or spacer introducing a residual motif in the conjugate molecule should be proven to be safe, on the basis of dedicated toxicological studies conducted in animals, prior to being incorporated in vaccine candidates administered to humans.

Specific Challenges in the Development of Conjugate Vaccines

Product Characterization

Conjugates are dual and complex molecules, and thus difficult to characterize. Depending on the conjugation chemistry, molecules with different conformations are obtained (Fig. 1). As indicated above, reagents and technologies can be used in various combinations to produce the conjugate vaccines. Because large PS are frequently linked to large protein carriers, a level of random coupling is generally observed, affecting the number of polysaccharide chains linked to each carrier, as well as the attachment sites.[74] The end result is a variety of either protein-polysaccharide (Fig. 1A), protein-protein–polysaccharide (Fig. 1B, with polysaccharides being linked with polymeric proteins cross-linked during the conjugation process) or complexes resulting from multipoint attachments between polysaccharide and protein (Fig. 1C). As the conjugate is a polydispersed mixture, standardization of the conjugate reaction is difficult to achieve, and the various parameters of the reaction have to be customized, depending on each semi-synthesized molecule. The resulting conjugates are nevertheless released on a set of predefined criteria.

Table 4. Chemical reactions/reagents involved in coupling[74,85-88]

Carbohydrate Antigens	Protein Carrier	Reagents	Reaction	Spacer
CHO		NH_2	Reductive amination	Diamino alkyl Carbonyldiimidazol
OH CHOH-CHOH		periodate	Cyanogen bromide Periodate oxidation	None
COOH	COOH	Carbodiimide	Esterification	Di amino hydrazide diamino-alkyl
NH_2	NH_2	N-OH succinimidyl	Esterification	None
	SH	Vinyl sulfone Maleimide Haloacetyl	Oxidation	Alkyl di-acid

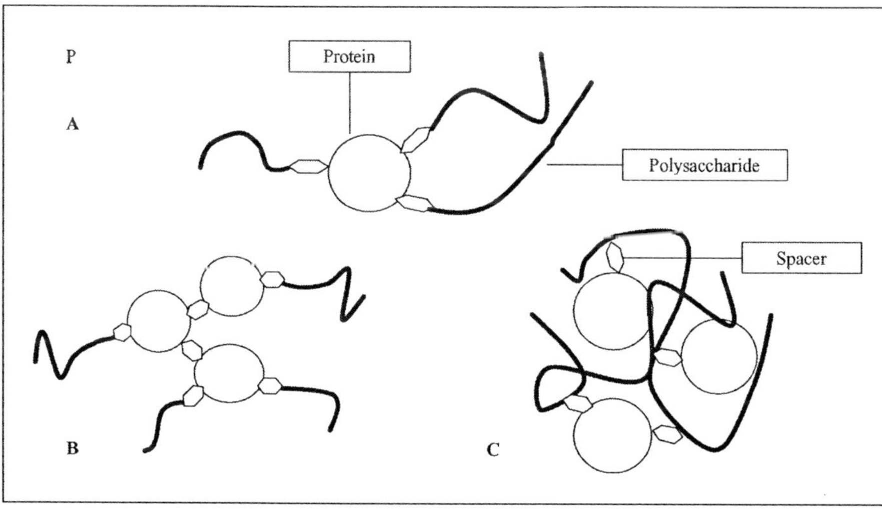

Figure 1. Schematic representation of PS-carrier protein conjugates.
Three main conformations are usually obtained when using different chemical processes to make chemical conjugates between polysaccharides (dark lines) and carrier proteins (circle), depending upon the number and nature of attachment sites:
A: Mono-activated polysaccharide conjugated to proteins. Mono-activation is generally obtained through terminal reduction of sugar.
B: Mono-activated polysaccharide conjugated to a cross-linked protein. When relying upon EDAC activation (see main text on chemical conjugation), a linkage between carboxyl and amino groups of multiple proteins is frequently observed.
C: Multi-activated polysaccharide conjugated to protein: this conformation is obtained following activation of multiple functions in the polysaccharide repeating units.

It is often difficult to characterize precisely conjugate molecules. Quality controls have been initially based on the determination of size, content in polysaccharide and protein, residual unconjugated polysaccharides, and polysaccharide/protein ratio, the latter being characteristic of the conjugate molecule. In order to further improve product characterization, new analytical technologies are developed, which are more specific, more sensitive, and require less material. Nuclear magnetic resonance spectroscopy (NMR) is a powerful tool to analyze, before and after conjugation, the structural integrity of the antigen (Fig. 2) When conjugates are made with oligosaccharides consisting of 10-30 repeating units, NMR can be used to assess the state of depolymerization and activation of the polysaccharide.[89] It is likely that NMR will be used as a routine control method for product characterization.

The determination of the size of the molecule is also critical to control both the quality of the conjugate synthesis and the consistency of the conjugation process, as well as to follow-up the stability of the vaccine antigen. In this regard, low-pressure size exclusion chromatography was initially used, but more recently, HPLC combined with light scattering detection emerged as a preferred method to determine the size of the conjugate. Polysaccharide content is also now documented by assessing monosaccharide composition in a quantitative manner, using high performance anion exchange chromatography.[90] In this method, carbohydrates are detected following ionization by amperometry (Fig. 3). This technology allows detection and quantification of even very low amounts of polysaccharide. This represents a critical technological advance, given that commonly used doses for conjugate vaccines are usually in the range of 1 to 10 µg. It should be mentioned that it is difficult to quantify the different polysaccharides in a multivalent conjugate vaccine, such as a 7, 9 or 11-valent pneumococcal conjugate. To date only immuno-chemical assays have been developed to this aim, including for example rate nephelometry relying upon PS-specific antibodies.[91]

Since the choice of the vaccine dose is made on the basis of polysaccharidic content, the amount of unconjugated antigen has also to be determined. This is difficult however, because of the complexity and heterogeneity of the conjugate molecules, in terms of conformation, hydrophobicity and size. Therefore, new reliable and standardized physicochemical characterization methods still remain to be developed to fulfill this need. The quantification of free protein is not usually performed. Protein content should ideally be documented, for example using capillary electrophoresis,[92] since the polysaccharide/protein ratio is a main characteristic to control the conjugate molecule.

Documenting the Immunogenicity of Conjugate Vaccines

Experience with conjugate vaccines focuses the need for a predictive animal potency test that would correlate with immunogenicity in human infants. For example, when comparing the antibody responses obtained with four tetanus toxoid-based conjugates in mice and in infants, we observed that an unadjuvanted pneumococcal 23F conjugate induced very few anti-polysaccharide antibodies in mice following two injections, and no further antibody increase was observed after a third injection. The same antigen, however, was found to induce significant anti-polysaccharide antibody titers in infants. Therefore, human studies are essential. In addition, immunogenicity varies from one polysaccharide serotype to another. Pneumococcal types 3 and 4 polysaccharides elicit significant antibody responses in infants after a single injection and these are further increased after the second and third injections. For pneumococcal serotypes 6B, 9V, 18C and 23F however, significant increases in antibody levels are observed after the third injection only (M. Moreau, unpublished results).

The most likely explanation for such variability in terms of immunogenicity between the various polysaccharidic antigens is linked with differences in polysaccharide structures. It seems that 6B conjugates are always among the weakest immunogens in infants, independently of the carrier protein and the coupling technology being used; thus, in this particular case, the kinetics of the response seems linked to the structure of the polysaccharide.[9]

In fact, our knowledge in conjugate processing by the immune system is very limited, despite the well-documented efficacy of the first conjugate vaccines against *Haemophilus influenzae*

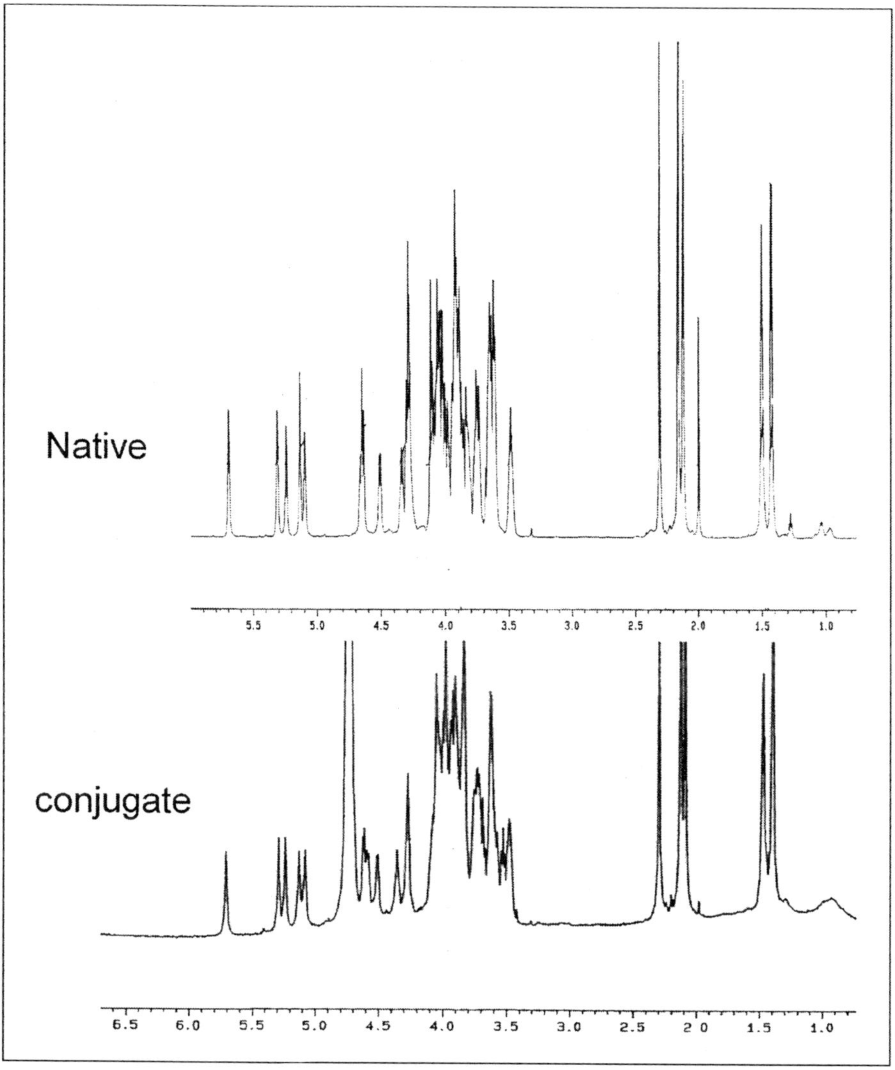

Figure 2. 600Mhz H NMR spectra of native and conjugated serotype 7F pneumococcal polysaccharide.

type b in infants. Multi-center research programs are now being undertaken to better understand the mechanisms through which conjugate vaccines are processed by the immune system, and to find a correlation between physicochemical characteristics and immune potency of conjugate molecules.

The Interference Issue

In order to simplify vaccination schedules, a number of pediatric vaccines are being administered as combinations or associations, leading to the risk of a potential interference and competition in terms of immune responses directed at the various antigens.[65,66] It was for example established that infants receiving simultaneously a tetravalent pneumococcal vaccine conjugated to TT and a diphtheria-tetanus-pertussis-poliovirus-Hib-tetanus conjugate vaccine

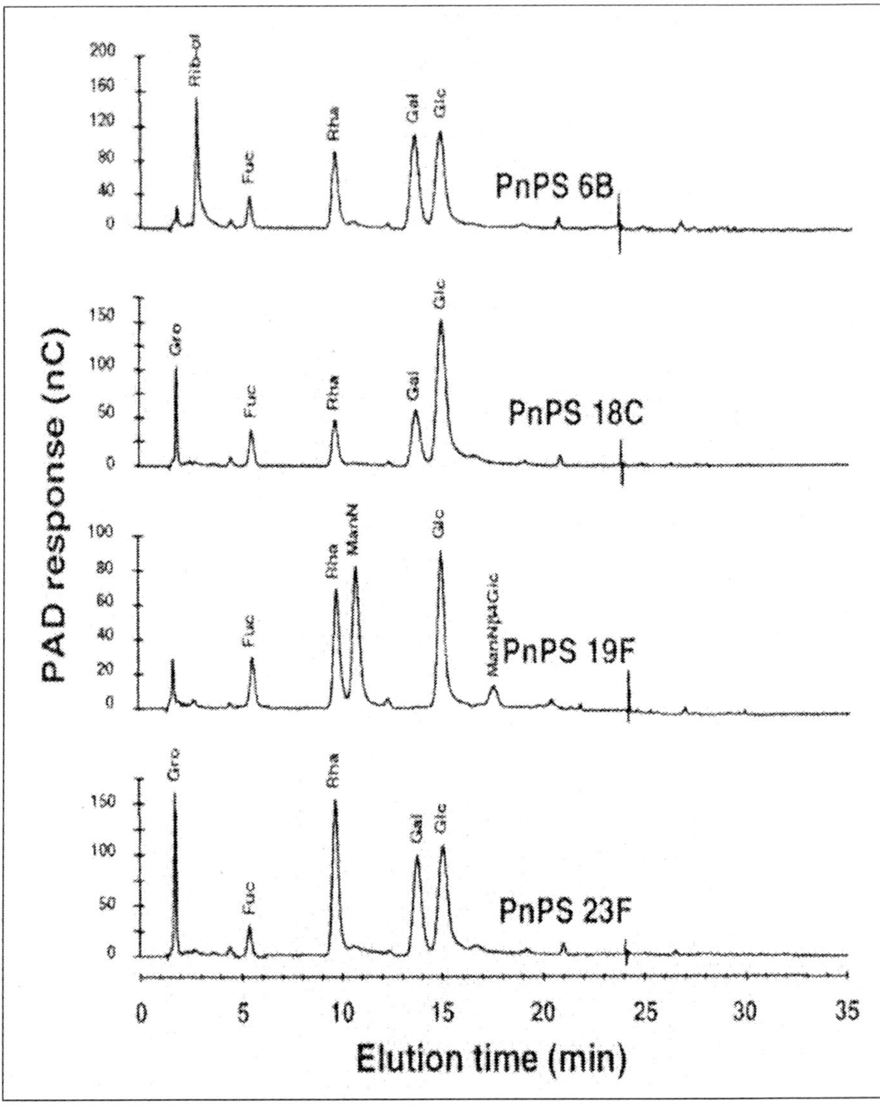

Figure 3. Carbohydrate analysis by HPEAC-PAD.
A quantitative analysis of carbohydrate content is performed following monosaccharide release by hydrolysis with both hydrofluoric acid and trifluoroacetic acid, using monosaccharide solutions as standards. This very sensitive detection method allows quantitation of 1-10 μg of injected PS following hydrolysis. The ion exchange chromatography profile is characteristic of the PS composition, and can be used for polysaccharide identification.

elicited significantly lower anti Hib PRP antibody responses, when compared with infants receiving either a tetravalent pneumococcal vaccine conjugated to diphtheria toxoid or placebo. It was further shown that the anti PRP antibody concentrations were inversely correlated to the TT content of the conjugate vaccine.[66] Noteworthy, these lower antibody responses were only observed when the Hib conjugate was included in combination vaccines containing acellular, as opposed to whole-cell pertussis. Similarly, immunological interference was also detected in infants receiving the 7-valent pneumococcal conjugate vaccine, together with vaccine combinations containing acellular pertussis antigens.[93] Even if the lower antibody responses are not associated with impaired function of the antibodies induced, nor with a loss in immune memory, these results raise the risk of epitopic suppression when immunizing simultaneously with multiple conjugate vaccines, most particularly if based on a common carrier. Therefore, for multivalent vaccines such as pneumococcal conjugate vaccines, efforts towards minimizing carrier protein loads in the vaccine doses are being sought by all the manufacturers. Using multiple as opposed to a single carrier protein(s) may also be appropriate in order to avoid immune interference, when conjugates are included in combination vaccines.

Conclusions

The development of glycoconjugates based on linking a polysaccharide moiety to an immunogenic protein has resulted in vaccines which elicit protective immune responses in both infants and adults, and which create immunological memory for long-lasting immunity. In this category belong glycoconjugates targeted against CPS in *H.influenzae* type b, *N. meningitidis* type C and 7 serotypes of *S. pneumoniae*, which all have given protective efficacies in the 90 to > 97 % ranges.

New glycoconjugate vaccines are currently being developed against additional serotypes of *S. pneumoniae* and *N.meningitidis,* against CPS of *S.typhi* and *S. aureus*, as well as polysaccharides from the LPS of *S. dysenteriae*, *S.flexneri*, *S. sonnei*, *P. aeruginosa*, *E. coli*, nontypeable *H.influenzae* and *Moraxella*. These vaccine candidates are presently either in clinical or preclinical development.

One of the main advantages of bacterial conjugate vaccines is their capacity to reduce carriage and transmission, leading consequently to herd immunity. In spite of the success with pneumococci, meningococci and Hib, where antibodies neutralize bacteria in the bloodstream or reduce colonization, the challenge remains when the predominant target is at a mucosal surface, as for example for group B streptococci and Shigella. Here, new routes of administration as well as innovative adjuvants may be needed.

Lastly, many of today's conjugate vaccines are polydisperse and difficult to characterize. Attempts are ongoing to develop fully synthetic conjugates, i.e., both carrier and immune epitopes, in order to facilitate manufacturing consistency and product characterization in the future.

Acknowledgements

The authors wish to thank Dr. Mark Fletcher for fruitful discussions on new potential targets for conjugate vaccines.

References

1. Lindberg AA. Glycoprotein conjugate vaccines. Vaccine 1999; 17:528-536.
2. Mulard L, Alonso JM, Fournier JM. Vaccins polyosidiques. Ann Inst Pasteur 2002; 12:37-54.
3. Jennings HJ. Capsular polysaccharides as vaccine candidates. Curr Top Microbiol Immunol 1990; 150:97-127.
4. MacLeod CM, Hodge SRG, Heidelberger M et al. Prevention of pneumococcal pneumonia by immunization with specific capsular polysaccharides. J Exp Med 1945; 82:445-465.
5. Schneerson R, Robbins JB, Parke JC. Quantitative and qualitative analyses of serum antibodies elicited in adults by Haemophilus influenzae type b and pneumococcus type 6A capsular polysaccharide-tetanus toxoid conjugates. Infect Immun 1986; 502:519-528.
6. Jodar L, Feavers I, Salisbury D et al. Development of vaccines against meningococcal disease. Lancet 2002; 359:1499-1508.
7. Schneerson R, Barrera A, Sutton A et al. Preparation, characterization and immunogenicity of Haemophilus influenza type b polysaccharide-protein conjugate. J Exp Med 1980; 152:361-376.
8. Dagan R, Fraser D, Roitman M et al. Effectiveness of a nationwide infant immunization program against Haemophilus influenzae b. Vaccine 1999; 17:134-141.

9. Van Alphen L, Spanjaard L, Van der Erde A et al. Effect of a nationwide vaccination of 3-month-old infants in The Netherlands with conjugate Haemophilus influenzae type b vaccine: High efficacy and lack of herd immunity. J Pediatr 1997; 131:869-873.
10. Moxon ER, Heath PT, Booy R et al. The impact of Hib conjugate vaccines in preventing invasive H. influenzae diseases in the UK. Vaccine 1999; 17:S11-S13.
11. MacLennan JM, Shackley F, Heath PF et al. Safety, immunogenicity and induction of immunologic memory by a serogroup C meningococcal conjugate vaccine in infants. JAMA 2000; 283:2795-2801.
12. Maiden MCJ, Stuart JM. Carriage of serogroup C meningococci 1 year after meningococcal C conjugate polysaccharide vaccination. Lancet 2002; 359:1829.
13. Ramsay ME, Andrews N, Kaczmarski EB et al. Efficacy of meningococcal serogroup C conjugate vaccine in teenagers and toddlers in England. Lancet 2001; 357:195.
14. Watson L, Wilson BJ, Waugh N. Pneumococcal polysaccharide vaccine: A systematic review of clinical effectiveness in adults. Vaccine 2002; 20:2166-2173.
15. Black S, Shinefield H, Fireman B et al. Efficacy, safety and immunogenicity of heptavalent pneumococcal conjugate vaccine in children. Pediatr Infect Dis J 2000; 19:187-195.
16. Eskola J, Kilpi T, Palmu A et al. Efficacy of a pneumococcal conjugate vaccine against acute otitis media. N Engl J Med 2001; 344:403-409.
17. Puumalainen T, Zeta-Capeding MR, Kaythy et al. Antibody response to an eleven valent diphtheria- and tetanus –conjugated pneumococcal conjugate vaccine in Filipino infants. Ped Infect Dis J 2002; 21:309-314.
18. Obaro SK, Adegbola R. The pneumococcus: Carriage, disease and conjugate vaccines. J Med Microbiol 2002; 51:98-104.
19. Steinhoff MC. Haemophilus influenzae type b infections are preventable everywhere. Lancet 1997; 349:1186-1187.
20. Mulholland K, Hilton S, Adegbola R et al. Randomised trial of Haemophilus influenzae type b tetanus protein conjugate vaccine for prevention of pneumonia and meningitis in Gambian infants. Lancet 1997; 349:1191-1197.
21. Obaro SK. Protein conjugate vaccine – how much is enough? Trends in Microbiol 2001; 9:364-365.
22. Obaro SK. Safety and immunogenicity of a nonavalent pneumococcal vaccine conjugated to CRM197 administered in Gambian infants. Pediatr Infect Dis J 2000; 19:463-469.
23. Levine MM, Noriega F. Vaccines to prevent bacterial enteric infections in children. Ped Ann 1993; 22:719-725.
24. Robbins JB, Schneerson R, Szu SC. Perspective: Hypothesis: Serum IgG antibody is sufficient to confer protection against infectious diseases by inactivating the inoculum. J Infec Dis 1995; 171:1387-1398.
25. Avery OT, Goebel WF. Chemo-immunological studies on conjugated carbohydrate-proteins. Part II: Immunological specificity of synthetic sugar-proteins. J Exp Med 1929; 50:521-533.
26. Cohen D, Ashkenazi S, Green MS. Double-blind vaccine controlled randomised efficacy trial of an investigational Shigella sonnei conjugate vaccine in young adults. Lancet 1997; 349:155-159.
27. Taylor DN, Trofa AC, Sadoff J et al. Synthesis, characterization and clinical evaluation of conjugate vaccines composed of the O-specific polysaccharides of Shigella dysenteriae type 1, Shigella flexneri type 2a, and Shigella sonnei (Plesiomonas shigelloides) bound to bacterial toxoids. Infect Immun 1993; 61:3678-3687.
28. Zhang J, Kovac P. Studies on vaccines against cholera. Synthesis of neoglycoconjugates from the hexasaccharide determinant of Vibrio cholerae O:1, serotype Ogawa, by single-point attachment or by attachment of the hapten in the form of clusters. Carbohydrate Res 1999; 321:157-167.
29. Mckkenney D, Pouliot KL, Wang Y et al. Broadly protective vaccine for Staphylococcus aureus based on an in vivo-expressed antigen. Science 1999; 284:1523-1527.
30. Shinefield H, Black S, Fattom A et al. Use of a Staphylococcus aureus conjugate vaccine in patients receiving hemodialysis. N Engl J Med 2002; 346:491-493.
31. Szu SC, Stone AL, Robbins JB et al. Vi Capsular polysaccharide-protein conjugates for prevention of typhoid fever. J Exp Med 1987; 166:1510-1524.
32. Szu SC, Taylor DN, Trofa AC et al. Laboratory and preliminary clinical characterization of Vi capsular polysaccharide-protein conjugate vaccines. Infect Immun 1994; 62:4440-4444.
33. Lin FYC, Ho VA, Khiem HB et al. The efficacy of a Salmonella typhi Vi conjugate vaccine in two-to-five-year-old children. N Engl J Med 2001; 344:1263-1267.
34. Campagne G, Garba A, Fabre P et al. Safety and immunogenicity of three doses of a Neisseria meningitidis A+C diphtheria conjugate vaccine in infants from Niger. Ped Infect Dis J 2000; 19:144-150.

35. Michon F, Fusco PC, D'ambra AJ et al. Combination conjugate vaccines against multiple serotypes of group B streptococci. Adv Exp Med Biol 1997; 418:847-850.
36. Wessels MR, Paoletti LC, Rodewald AK. Stimulation of protective antibodies against type Ia and Ib group B streptococci by a type Ia polysaccharide-tetanus toxoid conjugate vaccine. Infect Immun 1993; 61:4760-4766.
37. Baker CJ, Paoletti LC, Rench MA et al. Use of capsular polysaccharide -tetanus toxoid conjugate for type II group B streptococcus in healthy women. J Infect Dis 2001; 182:1129-1138.
38. Paoletti LC. Potency of clinical group B streptococcal conjugate vaccines. Vaccine 2001; 19:2118-2126.
39. Edelman R, Taylor DN, Wasserman SS et al. Phase 1 trial of a 24-valent Klebsiella capsular polysaccharide vaccine and an eight-valent Pseudomonas O-polysaccharide conjugate vaccine administered simultaneously. Vaccine 1994; 12:1288-1294.
40. Campbell WN, Hendrix E, Cryz S et al. Immunogenicity of a 24-valent Klebsiella capsular polysaccharide vaccine and an eight-valent Pseudomonas O-polysaccharide conjugate vaccine administered to victims of acute trauma. Clin Infect Dis 1996; 23:179-181.
41. Cryz SJ, Lang A, Rudeberg A et al. Immunization of cystic fibrosis patients with a Pseudomonas aeruginosa O-polysaccharide-toxin A conjugate vaccine. Behring Institute Mitteilungen 1997; 345-349.
42. Fournier JM, Villeneuve S. Cholera update and vaccination problems. Med Trop 1998; 58:32-35.
43. Johnson JA, Joseph A, Morris JR. Capsular polysaccharide-protein conjugate vaccines against Vibrio cholerae O139 Bengal. Bull Inst Pasteur 1995; 93:285-290.
44. Kossaczka Z, Shiloach J, Johnson V et al. Vibrio cholerae O139 conjugate vaccines: Synthesis and immunogenicity of V. cholerae O139 capsular polysaccharide conjugates with recombinant diphtheria toxin mutant in mice. Infect Immun 2000; 68:5037-5043.
45. Jennings HJ. The capsular polysaccharide of group B Neisseria meningitidis as a vehicle for vaccine development. Contr Microbiol Immunol 1989; 10:151-165.
46. Pon RA, Lussier M, Yang QL et al. N-propionylated group B meningococcal polysaccharide mimics a unique bactericidal capsular epitope in group B Neisseria meningitidis. J Exp Med 1997; 185:1929-1938.
47. Brandt ER, Good MF. Vaccine strategies to prevent rheumatic fever. Immunol Res 1999; 19:89-103.
48. Hamasur B, Kallenius G, Svenson SB. Synthesis and immunologic characterisation of Mycobacterium tuberculosis lipoarabinomannan specific oligosaccharide-protein conjugates. Vaccine 1999; 17:2853-2861.
49. Gu X, Chen J, Barenkamp SJ et al. Synthesis and characterization of lipooligosaccharide-based conjugates as vaccine candidates for Moraxella (Branhamella) catarrhalis. Infect Immun 1998; 66:1891-1897.
50. Hu WG, Chen J, Battey JF et al. Enhancement of clearance of bacteria from murine lungs by immunization with detoxified lipooligosaccharide from Moraxella catarrhalis conjugated to proteins. Infect Immun 2000; 68:4980-4985.
51. Gu X, Sun J, Jin S et al. Detoxified lipooligosaccharide from nontypeable Haemophilus influenzae conjugated to proteins confers protection against otitis media in chinchillas. Infect Immun 1997; 65:4488-4493.
52. Conlan JW, Cox AD, Kuolee R et al. Parenteral immunization with a glycoconjugate vaccine containing the O157 antigen of Escherichia coli O157:H7 elicits a systemic humoral immune response in mice, but fails to prevent colonization by the pathogen. Can J Microbiol 1999; 45:279-286.
53. Shen X, Lagergard T, Yang Y et al. Group B streptococci polysaccharide–cholera toxin b subunit conjugate prepared by different methods for intranasal immunization. Infect Immun 2001; 69:297-306.
54. Jakobsen H, Bjarnarson S, del Giudice G et al. Intranasal immunization with pneumococcal conjugate vaccines with LT-K63, a non toxic mutant of heat-labile enterotoxin, as adjuvant rapidly induces protective immunity against lethal pneumococcal infections in neonatal mice. Infect Immun 2002; 70:1443-1452.
55. Cho NH, Seong SY, Chun KH et al. Novel mucosal immunization with polysaccharide-protein conjugates entrapped in alginate microspheres. J Control Releas 1998; 53:215-224.
56. Paoletti LC, Rench MA, Kasper DL et al. Effects of alum adjuvant or a booster dose on immunogenicity during clinical trials of group B streptococcal type III conjugate vaccines. Infect Immun 2001; 69:6696-6701.
57. Rappuoli R, Pizza M, Douce G et al. Structure and mucosal adjuvanticity of cholera and Escherichia coli heat-labile enterotoxins. Immunol Today 1999; 20:493-500.

58. Mariotti S, Teloni R, Von Hunolstein C et al. Immunogenicity of anti-Haemophilus influenzae type b CRM197 conjugate following mucosal vaccination with oligodeoxynucleotide containing immunostimulatory sequences as adjuvant. Vaccine 2002; 20:2229-2239.
59. Gravekamp C, Kasper DL, Paoletti LC et al. Alpha C protein as a carrier for type III capsular polysaccharide and as a protective protein in group B streptococcal vaccines. Infect Immun 1999; 67:2491-2496.
60. Mariotti S, Teloni R, Von Hunolstein C et al. Immunogenicity of anti-Haemophilus influenzae type b CM197 conjugate following mucosal vaccination with oligodeoxynucleotide containing immunostimulatory sequences as adjuvant. Vaccine 2002; 20:2229-2239.
61. Anderson PW, Pichichero ME, Insel RA. Immunogens consisting of oligosaccharides from the capsule of Haemophilus influenzae type b coupled to diphtheria toxoid or the toxin protein CRM_{197}. J Clin Invest 1985; 76:52-59.
62. Buchanan RM, Briles DE, Arulanandam BP et al. IL-12-mediated increases in protection elicited by pneumococcal and meningococcal conjugate vaccines. Vaccine 2001; 19:2020-2028.
63. Flanagan MP, Michael JG. Oral immunization with a Streptococcal pneumoniae polysaccharide conjugate vaccine in enterocoated microparticles induces serum antibodies against type specific polysaccharides. Vaccine 1999; 17:72-81.
64. Marburg S, Jorn D, Tolman RL et al. Bimolecular chemistry of macromolecules: Synthesis of bacterial polysaccharide conjugates with Neisseria meningitidis membrane protein. J Am Chem Soc 1986; 108:5282-5287.
65. Schutze MP, Leclerc C, Jolivet M et al. Carrier-induced epitopic suppression, a major issue for future synthetic vaccine. J Immunol 1985; 135:2319-2322.
66. Dagan R, Eskola J, Leclerc C et al. Reduced response to multiple vaccines sharing common protein epitopes that are administered simultaneously to infants. Infect Immun 1998; 66:2093-2098.
67. Falugi F, Petracca R, Mariani M et al. Rationally designed strings of promiscuous CD4+ T cell epitopes provide help to Haemophilus influenzae type b oligosaccharide: A model for new conjugate vaccines. Eur J Immunol 2001; 31:3816-3824.
68. Szu SC, Taylor DN, Trofa JD et al. Laboratory and preliminary clinical characterization of Vi capsular polysaccharide-protein conjugate vaccines. Infect Immun 1994; 62:440-4444.
69. Akkoyunlu M, Melhus A, Capiau C et al. The acylated form of protein D of Haemophilus influenzae is more immunogenic than the nonacylated form and elicits an adjuvant effect when it is used as a carrier conjugated to polyribosyl ribitol phosphate. Infect Immun 1997; 12:5010-5016.
70. Kuo J, Douglas M, Ree HK et al. Characterization of a recombinant pneumolysin and its use as protein carrier for pneumococcal type 18C conjugate vaccines. Infect Immun 1995; 63:2706-2713.
71. Konen-Waisman S, Fridkin M, Cohen I. Self and foreign 60-KD heat shock protein T cell epitope peptides serve as immunogenic carriers for a T cell independent sugar antigen. Am Assoc Immunol 1995; 59:77-85.
72. Fusco PC, Michon F, Laude-Sharp M et al. Preclinical studies on a recombinant group B meningococcal porin as a carrier for a novel Haemophilus influenzae type b conjugate vaccine. Vaccine 1998; 19:1842-1849.
73. Bhattacharjee AK, Opal SM, Taylor R et al. Noncovalent complex vaccine prepared with detoxified Escherichia coli J5 (Rc chemotype) lipopolysaccharide and Neisseria meningitidis group B outer membrane protein produces protective antibodies against gram-negative bacteremia. J Infect Dis 1996; 173:1157-1163.
74. Moreau M. Conjugation technologies in vaccinia, vaccination and vaccinology: Jenner, Pasteur and their successors. In: Plotkin S, Fantini B, eds. 1996:145-149.
75. Libon C, Haeuw JF, Crouzet F et al. Streptococcus pneumoniae polysaccharide conjugated to outer membrane protein A from Klebsiella pneumoniae elicit protective antibodies. Vaccine 2002; 20:2174-2180.
76. Gupta RK, Taylor DN, Bryla DA et al. Phase I evaluation in adult volunteers of Vibrio cholera O:1 serotype Inaba, polysaccharide-cholera toxin conjugates. Infect Immun 1998; 66:3095-3099.
77. Chong P, Neville C, Kandil A et al. A strategy for rational design of fully synthetic glycoconjugate vaccine. Infect Immun 1997; 65:4918-4925.
78. Achtman M, Moreau M. IgA1 protease fragment as carrier peptide. 1998 patent WO 98/31791.
79. Reddin KM, Crowley-Luke A, Clark SO et al. Bordetella pertussis fimbriae are effective carrier proteins in Neisseria meningitidis serogroup C conjugate vaccines. FEMS Immun Med Microbiol 2001; 31:153-162.
80. Landy M, Johnson AG, Webster ME. Studies on Vi antigen. VII: Role of acetyl in antigenic activity. Am J Hyg 1961; 73:55-65.

81. Heidelberger M, Goebel WF, Avery OT. Chemo-immunological studies on the soluble specific substance of pneumococcus. I:The isolation and the properties of the acetylated polysaccharide of pneumococcus type 1. J Exp Med 1933; 58:731-755.
82. McNeely TB, Staub JM, Rush CM et al. Antibody response to capsular polysaccharide backbone and O-acetyl side groups of Streptococcus pneumoniae type 9 in humans and rhesus macaques. Infect Immun 1998; 66:3705-3710.
83. Moreau M. Oligosaccharide derived from antigenic polysaccharide obtained from a pathogenic agent. US Patent 1999; 6,007,818.
84. Costantino P, Norelli F, Giannozzi A et al. Size fractionation of bacterial capsular polysaccharides for their use in conjugate vaccines. Vaccine 1999; 17:1251-1263.
85. Bystricky S, Machova E, Bartek et al. Conjugation of yeast mannans with protein employing cyanopyridinium agent (CDAP) - an effective route of antifungal vaccine preparation. Glycoconjugate J 2000; 17:677-680.
86. Lees A, Nelson BL, Mond JJ. Activation of soluble polysaccharides with 1- cyano - 4-dimethylaminopyridinium tetrafluoroborate for use in protein-polysaccharide conjugate vaccines and immunological reagents. Vaccine 1996; 14:190-198.
87. Lees A. Protein-polysaccharide conjugate vaccines and other immunological reagents prepared using homobifunctional vinylsulfones and process for preparing the conjugates. Patent 2001 US 6,309,646 B1.
88. Kamath VP, Diedrich P, Hindsgaul O. Use of diethyl squarate for coupling of oligosaccharide amines to carrier proteins and characterization of the resulting neoglycoproteins by MALDI-TOF. Glycoconjugate J 1996; 13:315-319.
89. Ravencroft N, d'Ascenzi, Projetti D et al. Physicochemical characterization of the oligosaccharide component of vaccines. In: Brown F, Corbel M, Griffiths E, eds. Physicochemical procedures for characterization of vaccines. Dev Biol Basel Karger 2001:103:35-47.
90. Talaga P, Vialle S, Moreau M. Development of a high-performance anion-exchange chromatography with pulsed-amperometric detection based quantification assay for pneumococcal polysaccharide and conjugates. Vaccine 2002; 20:2474-2484.
91. Lee CJ. The quantitative immunochemical determination of pneumococcal and meningococcal capsular polysaccharides by light scattering rate nephelometry. J Biol Stand 1983; 11:55-64.
92. Lamb DH, Summa L, Lei QP et al. Determination of free carrier protein in protein-polysaccharide conjugate vaccines by micellar electrokinic chromatography. J Chromatography 2000; 894:311-318.
93. Eskola J, Ward J, Dagan R et al. Combined vaccination of Haemophilus influenzae type b conjugate and diphtheria-tetanus-pertussis containing acellular pertussis. Lancet1999; 354:2063-2068.

CHAPTER 14

Carbohydrate-Based Targets and Vehicles for Cancer and Infectious Diseases Vaccines

Vasso Apostolopoulos, Magdalena Plebanski and Ian McKenzie

Abstract

The last decade has seen carbohydrates used not only as targets for effective vaccines against bacteria, but also developed as adjuvants and vaccine carriers for protein antigens for immunotherapy. This chapter focuses on carbohydrate targeting in bacterial and parasitic models for vaccine development, and in the current leading edge technologies for inducing T cells specific to tumor antigens for cancer therapy. This is made possible by the fact that the cells of the immune system, and specifically antigen presenting cells, express carbohydrate receptors which provide both a danger signal to the cell, and deliver protein attached to the carbohydrate for effective processing and presentation to T cells. Examples of such carbohydrate receptors include the mannose and scavenger receptors. The mechanisms by which they lead to effective antigen processing and stimulation of the immune system are also considered.

Bacterial Carbohydrates As Danger Signals

The immune system has evolved a set of receptors on phagocytic and antigen presenting cells (APCs) to recognize specifically and rapidly bacterial carbohydrates, especially in the form of liposaccharides. The best characterized are bacterial lipopolysaccharides (LPS) found abundantly on all Gram negative bacteria such as *Escherichia coli*. Interaction of LPS with receptors, such as CD14, on dendritic cells (DCs), monocytes and macrophages promotes phagocytosis of the bacteria, and provides an activation or 'danger' signal to the cell.[1-3] LPS-stimulated DCs express potent costimulatory molecules for T cell activation, such as CD40, CD80 and CD86, which makes them uniquely potent at priming both CD8 and CD4 T cell responses.[4] LPS-activated monocytes and macrophages do not only become more effective at destroying the bacteria, but also secrete large amounts of proinflammatory molecules such as tumor necrosis factor alpha (TNFα), which cause general reactions to infection, such as fever.[1] Although these responses help to eliminate the bacteria, excessive proinflammatory reactions can be deleterious to the host. Indeed, toxic shock induced by LPS injection is lethal in many species.[1] The diverse biological activities of LPS may be further dissociated into chemical moieties with the lipid A portion being the main determinant of toxicity.[5]

Although unmodified LPS may therefore not be a safe adjuvant, cell-wall skeleton (CWS) fractions of mycobacterial cells (present in Freund's complete adjuvant, the widely used adjuvant for experimental animals) have also been found to contain carbohydrate-lipid or protein conjugates with immunostimulatory activity.[6] It has been suggested that DCs and macrophages express both Toll-like receptors, TLR-2 and TLR-4, and a receptor for mycobacterial CWS whose signaling pathways promote an activation state of the immune system. Synthetic adjuvants such as muramyldipeptide (MDP) derivatives and trehalose-dimycolates (TDM) have

Immunobiology of Carbohydrates, edited by Simon Y.C. Wong and Gemma Arsequell.
©2003 Eurekah.com and Kluwer Academic / Plenum Publishers.

been developed to activate similar signaling pathways.[6] Mycobacterial lipoarabinomannan (LAM), mannosylated LAM (ManLAM) and LAM lacking the terminal mannosyl units (AraLAM) induce distinct responses in human polymorphonuclear (PMNs) and mononuclear phagocytes. Thus, AraLAM and ManLAM affect mononuclear, but not PMN, phagocyte functions, but both forms are chemotactic for monocytes and monocyte-derived macrophages (MDMs).[7]

Bacterial Carbohydrates As Human Vaccines

Bacterial carbohydrates, particularly those of the bacterial capsules of *Haemophilus influenzae, Streptococcus pneumoniae* and *Neisseria meningitidis* can be the targets of protective antibodies. These antibodies may have the dual function of opsonizing bacteria for phagocytosis and targeting them for destruction by complement. In addition, their toxin neutralizing effect on the soluble LPS released during bacterial destruction may limit the extent of potentially harmful proinflammatory reactions. Purified capsules can often elicit antibody responses in adults and children older than 2 years, but these thymus-independent responses only promote the generation of low-affinity antibodies and fail to generate long-term memory. The use of immunogenic carrier proteins containing helper T cell epitopes linked to the polysaccharide creates more powerful immunogens able to induce high affinity antibodies and prime for boosting either with the glycoconjugate or the polysaccharide itself.

Prior to the implementation of vaccination programs, the weighted worldwide incidence of *H. influenzae* type B (Hib) diseases except nonbacteremic pneumonia was 71/100,000 in patients younger than 5 years. Vaccination against Hib using a polysaccharide vaccine was initiated over 25 years ago. However, the initial formulation suffered from poor immunogenicity, particularly in infants. A dramatic improvement in efficacy was observed upon the introduction of four Hib glycoconjugate vaccines in the 1990s.[8-10] Indeed, since the immunization programs began, the disease has been almost eliminated in countries with high immunization coverage with an estimated 38,000 cases prevented each year. However, Hib vaccination still has had only limited impact globally. It is hoped that the development of vaccines that are practical for administration in the Third World, probably using glycoconjugates in a reduced number of doses and in combination with other vaccines, may extend the full benefit of vaccination to less privileged countries, where most Hib disease occurs.

The weighted worldwide incidence of meningitis in patients younger than 5 years prior to the implementation of vaccination programs was estimated at 57/100,000.[11] Human trials in the 1970s and 1980s showed that polysaccharide vaccines prevent meningococcal meningitis. Most of these vaccines have focused on group A meningococcus. A study elegantly summarizing the results of eight such trials suggested that the protective effect within the first year was consistent across all trials, with an overall vaccine efficacy of 95% (Exact 95% CI 87%, 99%). Protection extended into the second (in two studies) and third (in one study) years after vaccination, but the results were not statistically significant, with variations in the level and duration of protection, particularly among young children.[8,10] Recently, a meningococcal serogroup C glycoconjugate vaccine was successfully implemented nationally in the UK.[12] The glycoconjugate approach to increase immunogenicity and priming ability has now been extended with A, C, W 135 and Y, but not B serotypes in registration phases.[10] The potential inclusion of new virulence factors such as outer membrane proteins (OMPs) and lipopolysaccharides (LPS) may further the development of effective meningococcal immunogens, currently focused on providing a vaccine for serogroup B.[12]

S. pneumoniae (pneumococcus) is a major cause of morbidity and mortality worldwide, causing over 1 million of the 4 million annual deaths from acute lower respiratory infections in children under 5 years of age. The currently licensed pneumococcal vaccine comprises 23 capsular pneumococcal polysaccharides. Unfortunately, many of these are poorly immunogenic in young children. Similarly to Hib, clinical trials of protein-polysaccharide conjugates have shown promising results in safety and immunogenicity studies.[11] However, the development of a conjugate vaccine against pneumococcal disease is further complicated by the existence a very large number of serotypes. Since nonvaccine serotypes are already present in the community as confirmed in studies of the etiology of acute purulent otitis media, these nonvaccine serotypes may become more common in vaccination areas.[11] Indeed, a shift in *S. pneumoniae* serotypes colonizing the nasopharynx in children receiving the vaccine can be observed.[11] The identification of protective antigens (proteins or carbohydrates) common to clinically important strains could provide alternative immunogens for inclusion in a generic pneumococcal vaccine. Serotype diversity may be anticipated to be a problem for other capsular bacterial vaccines, for example although a group B streptococcus (GBS) polysaccharide-protein conjugate vaccine given to women of reproductive age was well tolerated and highly immunogenic, new capsular serotypes are now causing an important proportion of clinical infections.[13]

Bacterial Carbohydrates As the Basis of Novel Vaccine Approaches

The use of glycoconjugates of capsular polysaccharides with carrier proteins has proven highly effective in generating serotype-specific protective immunity. The problem of targeting multiple serotypes is a serious one. In addition, there is a need in generating vaccines that are not only effective but affordable and easy to administer in poor countries where the burden of bacterial disease is highest.[14] The identification of conserved antigenic regions in carbohydrates as well as proteins, in order to provide cross-strain protection, is a direct approach to tackle this problem. Two more unusual generic approaches are exemplified below.

Staphylococcus aureus and *S. epidermidis* are common causes of nosocomial infection and are major pathogens for domesticated animals. *S. aureus* is frequently found in community-acquired infections. Poly-N-succinyl β-1-6 glucosamine (PNSG), a chemical form of the *S. epidermidis* capsular polysaccharide/adhesin (PS/A) is a target for protective antibodies.[15] *S. aureus* cells in infected human sputa and lung also elaborate PNSG and immunization of mice with PNSG protects them against metastatic kidney infections. However, PNSG is not usually expressed by the bacteria in vitro, and indeed the challenge strains can be initially PNSG negative. Antigens like PNSG, present only under specific in vitro and in vivo conditions, offer exciting potentials as new targets for the development of protective bacterial immunity.

N. meningitidis is a major cause of meningitis and sepsis. Despite nearly 25 years of work, there is currently no vaccine for meningococcal B strains. The capsular polysaccharide of this organism is conserved and antibodies to it confer protection against disease. The immunogenicity of meningococcal B polysaccharide-based vaccines is poor. Although the use of immunogenic protein carriers is expected to enhance immunogenicity of the polysaccharide, this is not generally found to be case for group B capsular polysaccharides. There is also a concern that a portion of the antibody elicited by the capsule has autoimmune activity towards brain glycolipids and NCAM molecules on specialized cells. To avoid this problem, a panel of murine monoclonal antibodies (Mabs) to capsular polysaccharide epitopes on meningococcal B that are distinct from host polysialic acid was identified. These antibodies confer passive protection in animal models. The Mabs were then used to identify molecular mimetics from phage display peptide libraries. Although the resulting mimetic peptides have thus far failed to induce high levels of anti-capsular antibodies, this approach when optimized may be an exciting alternative line of research in vaccine development targeting bacterial polysaccharide structures.[16]

Parasite Carbohydrates As Targets of Protective Immunity

Bacterial carbohydrates have been known for many years to be the target of protective immunity, with effective human vaccines developed on this solid basis. Parasite carbohydrates, in contrast, are still largely understudied as targets of protective immunity or as vaccine components.

Evidence to show that protective antibodies can target carbohydrates is growing for a variety of parasitic diseases. *Trypanosoma cruzi* is the causative pathogen of Chagas disease. Antibodies that lyse trypomastigotes in a complement-mediated reaction are thought to mediate protection against virulent *T. cruzi*. Titers of antibodies to α-galactosyl determinants markedly increase in Chagas's disease.[17] Binding of these antibodies to *T. cruzi* causes complement-mediated lysis of trypomastigotes. Anti-gal antibodies from human serum may similarly inhibit the growth of *Plasmodium falciparum*, the parasite responsible for lethal malaria.[18] Most of the lytic power of the serum anti-Gal induced by Chagas's disease is removed by absorption with Galα1,3Galβ1,4GlcNAc. Lytic antibodies are also partly absorbed by *Serratia marcescens* but not by *E. coli* O111.[17] However, crossreactivity between some bacterial polysaccharides and *T. cruzi* may also occur. Indeed, rabbits and human volunteers immunized respectively with purified meningococcal polysaccharide C and the AC-polysaccharide vaccine produced antibodies crossreactive to *T. cruzi* infective forms. Furthermore, these crossreactive antibodies were able to target trypomastigotes for complement-mediated lysis. Nonsialylated epitopes expressed on infective forms of the parasite are the target of these antibodies, and could be considered as new immunogens for the development of *T. cruzi* vaccines.[19]

A radiation-attenuated *Shistosoma mansoni* vaccine in chimpanzees induced specific IgM and IgG to glycans on antigens released by cercariae. These antibodies were crossreactive to soluble antigens from larvae, adult worms, and eggs. Egg deposition was the major antigenic stimulus after challenge. Glycan epitopes recognized included GalNAcβ1-4GlcNAc-(LacdiNAc), fucosylated LacdiNAc, Lewis X (weakly), and those on keyhole limpet hemocyanin. Antibodies to peptide epitopes became prominent only during the chronic phase of infection, as glycan-specific IgM and IgG decreased. It is unclear whether these anti-glycan responses are protective or a "smoke screen" to divert the immune system away from more vulnerable larval peptide epitopes.[20] C57BL/6J and CBA/J mice vaccinated with irradiated cercariae have protective antibodies recognizing carbohydrate epitopes on schistosomal glutathione S-transferase. Lacto-N-fucopentaose III, a carbohydrate structure relevant for cell trafficking, is recognized predominantly. In contrast to the primate studies, there is no binding to its nonfucosylated homologue, lacto-N-neotetraose, or to oligosaccharides present on keyhole limpet hemocyanin.[21] It may be that the fine specificity of the anti-carbohydrate response determines its role in protective immunity.

As well as carbohydrates on parasites, carbohydrates of the host can play a role in parasitic infection by serving as receptors for the attachment and entry of parasites into cells. A clear example is provided by *P. falciparum's* use of chondroitin sulphate to attach itself to vascular endothelium, particularly of the placenta. Attachment of infected red blood cells allows this parasite to sequester itself and multiply to high density.[22] The inhibition of these carbohydrate-protein interactions leads to protective immunity, as demonstrated by the protective effect of anti-pfEMP1 antibodies, which target the parasite lectin[23] or by interfering with this interaction directly using competing carbohydrates in vitro.[24]

Parasite Carbohydrates As Immunomodulators

Galactosyl residues and mannan have been described to be immunostimulators or as carriers to target specific receptors on APCs. These moieties, found on many bacteria, yeasts and parasites, are also involved as classical exogeneous "danger" signals that usually activate APCs to promote a potent Th1 response. In contrast, a phosphorylcholine-containing glycoprotein (ES-62) secreted by the filarial nematode, *Acanthocheilonema viteae*, promotes the maturation of DCs with the preferential capacity to induce Th2 responses.[3] Other glycoconjugates with immunomodulatory roles include lipophosphoglycan (LPG), a major surface glycoconjugate of *Leishmania promastigotes. Leishmania* parasites use LPG to impair the normal activity of phagocytes, and thus protect the parasites within phagolysosomes.[25] The LPG-mediated escape mechanisms of promastigotes from human phagocyte responses have been shown to include impairment of oxidative burst and chemotactic activity.

Similar to some bacterial toxins, a dominant glycolipid from *P. falciparum* may contribute to the fever response associated with malaria infection. Parasite-derived glycosylphosphatidylinositol (GPI), free or associated with protein, induces TNFα and interleukin-1 production by macrophages and regulates glucose metabolism in adipocytes. Deacylation with specific phospholipases abolishes cytokine induction, as do inhibitors of protein kinase C. When administered to mice in vivo GPI from the parasite induces cytokine release, a transient pyrexia, and hypoglycemia. When administered with sensitizing agents it can elicit a profound and lethal cachexia. Antibody to the GPI inhibits these toxic activities, suggesting that GPI could be a useful target for inclusion in vaccines against malarial disease.[26]

Targeting the Mannose Receptor for Vaccine Development

The mannose receptor (MR) is primarily present on DCs and macrophages. It recognizes carbohydrates (mannose, fucose, glucose, GlcNAc, maltose) on the cell walls of infectious agents (mainly bacteria and yeast). Upon binding, there is aggregation and receptor mediated endocytosis and phagocytosis. The MR is prototypical member of the multilectin receptor family and provides a link between innate and adaptive immunity.

Human DCs and macrophages bind agalactosyl IgG, found in several autoimmune diseases, which has the terminal galactose residues removed, thus exposing N-acetylglucosamine and providing a binding site for the MR.[27] Antibodies or antigen-antibody complexes are taken into DCs or macrophages and generate Ig-derived peptides that bind MHC class II molecules and activate T cells.[27] The expression and functional state of the MR is governed by various cytokines, immunoglobulin receptors and pathogens. It is downregulated during IFNγ mediated macrophage activation even though the affinity of ligand binding is not affected, however the functional properties are modified.[28] IFNγ treatment of MDMs increases their capacity to kill *Candida albicans* in an MR-dependent manner.[28] The addition of IL-4 acts synergistically with IFNγ to enhance MR-dependent uptake.[28] In addition, IL-4 increases cell surface expression of MR and MR-mediated endocytosis whereas IFNγ decreases these effects.[29] However, both IL-4 and IFNγ either alone or together increased MR-mediated phagocytosis; IL-13 exerted similar effects to IL-4.[29] This demonstrates that phagocytosis of microorganisms could be enhanced in the presence of T1- and T2-type cytokines at sites of inflammation. Furthermore, when the MR binds to microorganisms, a variety of intracellular responses are triggered such as cytokine secretion,[30-32] lysosomal enzyme secretion,[33] and modulation of other cell surface receptors. In addition, glycosylated viral envelope proteins (HIV, HSV) stimulate IFNα production by DCs.[34] This stimulation can be inhibited by sugars specific for the MR implying that the MR is an important receptor for the recognition of enveloped viruses by DCs. Thus, the MR has more functions than just phagocytosis of pathogens.

Targeting the Mannose Receptor for Drug Therapy

The biology of MR has given new insights of its use as a target for delivering drugs to macrophages that have internalized bacteria or other infectious organisms. One study targeted the macrophage MR with a norfloxacin antibiotic, which is active against intracellular bacteria.[35] The antibiotic was conjugated to mannose with a poly(L-lysine citramide imide) carrier which successfully targeted the MR of macrophages infected by intracellular bacteria. Thus, targeting the MR with mannose linked to drugs to kill ingested microorganisms or viruses, is a new and exciting approach for therapeutic drug delivery. In addition, mannosylated poly(L-lysine) (Man-PLL) was synthesized as a carrier molecule and mixed with a plasmid DNA encoding chloramphenicol acetyltransferase (CAT) to form a DNA-Man-PLL complex.[36] The complex bound specifically to the MR in the liver after intravenous injection, indicating that a cell-specific gene delivery system can be developed by regulating the biodistribution of DNA-carrier complex.[36] Furthermore, the MR has been investigated for MR-mediated gene transfer into macrophages using mannosylated cationic liposomes and high transfection activity due to recognition by the MR was demonstrated.[37]

Targeting the Mannose Receptor for Antigen Delivery

The high expression of MR on DCs and macrophages indicates that the MR is a key molecule in antigen recognition.[30,38] The MR on macrophages and immature DCs is involved in endocytosis and phagocytosis[39] and is an important pathway for antigen uptake and delivery to MHC class II molecules.[40,41] Recently, we demonstrated that the MR is also involved in antigen uptake and delivery to MHC class I molecules,[30,42] particularly with mannan modified by oxidation.

The MR is found in endosomes where it transits to the cell surface to bind ligands. It recycles from the cell surface through the endosomal pathway. Dissociation of the bound ligands occurs at the lower pH found in endosomes.[43] Endocytosis by DCs via the MR takes place in small coated vesicles, shortly after internalization the MR and its ligand appear in larger vesicles, followed by colocalization with MHC class II molecules in lysosomes.[41] Mannosylated peptides and proteins are able to stimulate MHC class II restricted peptide specific T cells with 200-10,000 fold higher efficiency than peptides or proteins which have not been mannosylated.[41] Furthermore, uptake by DCs via the MR results in 100-fold enhanced presentation of soluble antigens to T cells than antigens internalized via fluid phase pinocytosis.[40] Recently, a DC receptor for endocytosis, DEC-205, was found to mediate a 100-fold increase in antigen presentation via the MHC class II pathway to $CD4^+$ T cells.[44]

After internalization, the MR transports antigens to MHC class II-containing compartments in immature DCs for antigen processing and presentation to T cells.[44] Human peripheral blood mononuclear cells cultured in GM-CSF and IL-4 for 5 days, develop into DCs which are able to efficiently present antigens to T cells. It has been demonstrated that DCs can endocytose antigens via the MR and deliver processed peptides to MHC class II molecules.[44] In addition, the antigen presenting function of the DCs has been shown to be associated with high level expression of the DEC-205 MR in mice.[45]

CD1 proteins have been implicated to have antigen presentation function.[46,47] Human CD1b can present nonprotein antigens from mycobacteria to T cells, including lipid mycolic acid and LAM. The antigen presentation pathway for LAM has been characterized and the macrophage MR is clearly responsible for uptake.[48] MR is abundant in early endosomes and the MHC class II loading and presentation pathway. LAM is taken into early endosomes via the MR, transported to late endosomes and then loaded onto CD1b molecules for T cell presentation.[48] This study links the MR to presentation of glycolipids via CD1 and indicates that the MR may play a critical role in processing of carbohydrate antigens.

A fusion protein containing the cysteine-rich domain of the murine MR and the Fc portion of human IgG1 was able to bind cells which were MHC class II, sialoadhesin and CD11c positive, and negative for other markers such as F4/80, FDC-M2, CD11b, B220 and CD4.[49] These cells have been found to localize to B cell follicles and initiate humoral immune responses and activation of T cells.

The use of mannan to aid in the induction of T1-mediated immune responses and cytotoxic T cells (CTLs) has also been investigated. Cationic liposomes, containing HIV-1 DNA and coated with mannan, can significantly enhance HIV-specific CTL responses, T1-type cytokine IFNγ, IgG2a and IgA antibodies and delayed-type hypersensitivity responses.[50] Furthermore, HER2 protein conjugated to either mannan or to polysaccharides having cholesteryl groups was able to induce $CD8^+$ CTLs which rejected HER2+tumors.[51]

We have demonstrated that mannan conjugated to the tumor associated antigen MUC1 can induce strong T1 or T2 type immune responses, depending on the mode of conjugation. MUC1 conjugated to mannan under reducing conditions (where the mannan contains no aldehydes and no Schiff bases) induces strong T2-type immune responses with high IgG1 antibodies, IL-4 cytokine production, low CTL precursor frequency and no protection in mice against a tumor challenge.[52-57] However, conjugation of MUC1 to mannan under oxidizing conditions (where it contains aldehydes and Schiff bases) generates a T1-type response, with $CD8^+$ CTL responses, low IgG2a antibodies, and IL-12 and IFNγ production.[52-57] Both conjugate formulations bound equally to the MR and were taken up into early endosomes.[30,42] However, reduced mannan-MUC1 was preferentially presented by the MHC class II pathway whereas oxidized mannan-MUC1 was preferentially presented by the MHC class I pathway.[42] This was the first demonstration that the MR could aid in introducing antigens into the MHC class I pathway. Furthermore, studies using oxidized mannan-MUC1 to target ex vivo macrophages or DCs via the MR showed that these pulsed APCs, when transferred into naïve mice, could induce strong CTL responses and protect mice against MUC1 tumor challenge.[30]

Our laboratory has immunized cynomologous monkeys with MUC1 fusion protein conjugated to oxidized mannan. Immunized monkeys generated strong anti-MUC1 antibodies, MUC1-specific $CD4^+$ and $CD8^+$ T cell proliferative responses and specific CTL precursor cells.[58,59] In a phase I clinical trial 25 patients with advanced metastatic adenocarcinoma were injected with increasing doses of mannan-MUC1.[60] High titers of IgG1 anti-MUC1 antibodies were produced in 13/25 patients (with antibody titers by ELISA of 1/320-1/20,480). In addition, T cell proliferation was found in 4/15 patients, and CTL responses were seen in 2/10 patients.[60,61] Recently it was demonstrated, by flow cytometric analysis of peripheral blood mononuclear cells of patients immunized with MUC1-mannan conjugates, that intracellular cytokines IL-2, IL-4, IFNγ, and TNFα were produced by $CD4^+CD69^+$ and $CD8^+CD69^+$ activated T cells upon MUC1 antigen stimulation.[62] Taken together these studies suggest that patients can successfully be immunized for the generation of both humoral and cellular responses using mannan-antigen conjugates. We are currently performing clinical trials where MR on macrophage/DC is targeted ex vivo with mannan-MUC1.

Targeting the Scavenger Receptor for Vaccine Development

The scavenger receptor is primarily present on macrophages and can internalize endotoxins, oxidized low density lipoproteins and other negatively charged proteins. Maleylated ovalbumin has been demonstrated to bind to the scavenger receptor, this enhances its presentation to ovalbumin-specific MHC class I restricted CTL by macrophages and B cells.[63] Maleylated diphtheria toxoid has also been demonstrated to be more immunogenic than nonmaleylated diphtheria toxoid which generated enhanced antibody and T cell proliferative responses.[64] In chickens, immunization with maleylated-bovine serum albumin (BSA) specifically bound to the scavenger receptor and modulated the Th1 immune response with weak antibodies. In addition, splenocytes expressed high levels of mRNA for IFNγ. Non maleylated BSA induced Th2 immune responses.[65]

Alcohol metabolites malondialdehyde and acetaldehyde when combined form stable adducts (oxidative product). These adducts when conjugated to proteins, such as hen egg lysozyme (HEL), induced a strong antibody response and T cell proliferation. Studies have suggested that the immune responses may be mediated by scavenger receptors that recognize malondialdehyde and acetaldehyde adducted proteins.[66]

Future Prospects

Carbohydrates have been known to be the target of protective immune responses against bacterial diseases for over two decades. Lately, conjugation of carbohydrates to proteins to provide a helper T cell response has dramatically enhanced the efficacy of polysaccharide vaccines aiming to induce bacterial immunity. It is anticipated parasitic diseases may soon follow suit and be the focus of studies investigating the role of carbohydrates in generating protective immune responses. Recently a number of different novel immunization strategies based on the use of glycoconjugates have been tested in animal models and humans. These have been shown to induce cellular and humoral responses, and may provide a simple, safe and effective approach to the development of vaccines. DCs have emerged as the main stimulating APCs of the immune system. Delivering antigens (e.g., proteins or antigenic peptides) to DCs has proven to be a successful approach for the induction of powerful immune responses, and protection in cancer models. Initial glycoconjugate strategies have focused on targeting the mannose receptor, expressed abundantly on both macrophages and DCs. Recently, scavenger receptors have shown similar promise, and we may anticipate that this field will continue to provide exciting developments.

Acknowledgements

This work was supported in part by a NH&MRC of Australia CJ Martin Fellowship (VA), Howard Hughes Fellowship (MP) and The Austin Research Institute (VA, MP, IFCM).

References

1. Alexander C, Rietschel ET. Bacterial lipopolysaccharides and innate immunity. J Endotoxin Res 2001; 7:167-202.
2. Koski GK, Lyakh LA, Rice NR. Rapid lipopolysaccharide-induced differentiation of CD14(+) monocytes into CD83(+) dendritic cells is modulated under serum-free conditions by exogenously added IFN-gamma and endogenously produced IL-10. Eur J Immunol 2001; 31:3773-3781.
3. Whelan M, Harnett MM, Houston KM et al. A filarial nematode-secreted product signals dendritic cells to acquire a phenotype that drives development of Th2 cells. J Immunol 2000; 164:6453-6460.
4. Ridge JP, Di Rosa F, Matzinger P. A conditioned dendritic cell can be a temporal bridge between a CD4+ T- helper and a T-killer cell [see comments]. Nature 1998; 393:474-478.
5. David SA. Towards a rational development of anti-endotoxin agents: Novel approaches to sequestration of bacterial endotoxins with small molecules. J Mol Recognit 2001; 14:370-87.
6. Azuma I, Seya T. Development of immunoadjuvants for immunotherapy of cancer. Int Immunopharmacol 2001; 1:1249-1259.
7. Fietta A, Francioli C, Gialdroni Grassi G. Mycobacterial lipoarabinomannan affects human polymorphonuclear and mononuclear phagocyte functions differently. Haematologica 2000; 85:11-18.
8. Patel M. Polysaccharide vaccines for preventing serogroup a meningococcal meningitis. Cochrane Database Syst Rev 2000; 2:CD001093.
9. Peltola H. Worldwide haemophilus influenzae type b disease at the beginning of the 21st century: Global analysis of the disease burden 25 years after the use of the polysaccharide vaccine and a decade after the advent of conjugates. Clin Microbiol Rev 2000; 13:302-317.
10. Lindberg AA. Glycoprotein conjugate vaccines. Vaccine 1999; 17 Suppl 2:S28-36.
11. Pelton SI. Acute otitis media in the era of effective pneumococcal conjugate vaccine: Will new pathogens emerge? Vaccine 2000; 19 Suppl 1:S96-99.
12. Riddell A, Buttery J. Vaccines against meningococcal disease: Current and future technologies. Expert Opin Biol Ther 2001; 1:385-399.
13. Schuchat A. Group B streptococcus. Lancet 1999; 353:51-56.
14. Obaro SK. Prospects for pneumococcal vaccination in African children. Acta Trop 2000; 75:141-153.
15. McKenney D, Pouliot K, Wang Y et al. Vaccine potential of poly-1-6 beta-D-N-succinylglucosamine, an immunoprotective surface polysaccharide of Staphylococcus aureus and Staphylococcus epidermidis. J Biotechnol 2000; 83:37-44.
16. Moe GR, Tan S, Granoff DM. Molecular mimetics of polysaccharide epitopes as vaccine candidates for prevention of Neisseria meningitidis serogroup B disease. FEMS Immunol Med Microbiol 1999; 26:209-226.
17. Almeida IC, Milani SR, Gorin PA et al. Complement-mediated lysis of Trypanosoma cruzi trypomastigotes by human anti-alpha-galactosyl antibodies. J Immunol 1991; 146:2394-2400.
18. Ramasamy R, Rajakaruna R. Association of malaria with inactivation of alpha1,3-galactosyl transferase in catarrhines. Biochim Biophys Acta 1997; 1360:241-246.
19. Oliveira TG, Milani SR, Travassos LR. Polyclonal B-cell activation by Neisseria meningitidis capsular polysaccharides elicit antibodies protective against Trypanosoma cruzi infection in vitro. J Clin Lab Anal 1996; 10:220-228.
20. Eberl M, Langermans JA, Vervenne RA et al. Antibodies to glycans dominate the host response to schistosome larvae and eggs: Is their role protective or subversive? J Infect Dis 2001; 183:1238-1247.
21. Richter D, Incani RN, Harn DA. Lacto-N-fucopentaose III (Lewis x), a target of the antibody response in mice vaccinated with irradiated cercariae of Schistosoma mansoni. Infect Immun 1996; 64:1826-1831.
22. Duffy PE, Fried M. Malaria during pregnancy: Parasites, antibodies and chondroitin sulphate A. Biochem Soc Trans 1999; 27:478-482.
23. Bull PC, Lowe BS, Kortok M et al. Parasite antigens on the infected red cell surface are targets for naturally acquired immunity to malaria. Nat Med 1998; 4:358-360.

24. Beeson JG, Chai W, Rogerson SJ et al. Inhibition of binding of malaria-infected erythrocytes by a tetradecasaccharide fraction from chondroitin sulfate A. Infect Immun 1998; 66:3397-3402.
25. Panaro MA, Panunzio M, Jirillo E et al. Parasite escape mechanisms: The role of Leishmania lipophosphoglycan on the human phagocyte functions. Immunopharmacol Immunotoxicol 1995; 17:595-605.
26. Schofield L, Hackett F. Signal transduction in host cells by a glycosylphosphatidylinositol toxin of malaria parasites. J Exp Med 1993; 177:145-153.
27. Dong X, Storkus WJ, Salter RD. Binding and uptake of agalactosyl IgG by mannose receptor on macrophages and dendritic cells. J Immunol 1999; 163:5427-5434.
28. Stahl PD, Ezekowitz RA. The mannose receptor is a pattern recognition receptor involved in host defense. Curr Opin Immunol 1998; 10:50-55.
29. Raveh D, Kruskal BA, Farland J et al. Th1 and Th2 cytokines cooperate to stimulate mannose-receptor-mediated phagocytosis. J Leukoc Biol 1998; 64:108-113.
30. Apostolopoulos V, Barnes N, Pietersz GA et al. Ex vivo targeting of the macrophage mannose receptor generates anti- tumor CTL responses. Vaccine 2000; 18:3174-3184.
31. Shibata Y, Metzger WJ, Myrvik QN. Chitin particle-induced cell-mediated immunity is inhibited by soluble mannan: Mannose receptor-mediated phagocytosis initiates IL-12 production. J Immunol 1997; 159:2462-2467.
32. Yamamoto Y, Klein TW, Friedman H. Involvement of mannose receptor in cytokine interleukin-1beta (IL- 1beta), IL-6, and granulocyte-macrophage colony-stimulating factor responses, but not in chemokine macrophage inflammatory protein 1beta (MIP-1beta), MIP-2, and KC responses, caused by attachment of Candida albicans to macrophages. Infect Immun 1997; 65:1077-1082.
33. Ohsumi Y, Lee YC. Mannose-receptor ligands stimulate secretion of lysosomal enzymes from rabbit alveolar macrophages. J Biol Chem 1987; 262:7955-7962.
34. Milone MC, Fitzgerald-Bocarsly P. The mannose receptor mediates induction of IFN-alpha in peripheral blood dendritic cells by enveloped RNA and DNA viruses. J Immunol 1998; 161:2391-2399.
35. Gac S, Coudane J, Boustta M et al. Synthesis, characterisation and in vivo behaviour of a norfloxacin-poly(L-lysine citramide imide) conjugate bearing mannosyl residues. J Drug Target 2000; 7:393-406.
36. Nishikawa M, Takemura S, Yamashita F et al. Pharmacokinetics and in vivo gene transfer of plasmid DNA complexed with mannosylated poly(L-lysine) in mice. J Drug Target 2000; 8:29-38.
37. Kawakami S, Sato A, Nishikawa M et al. Mannose receptor-mediated gene transfer into macrophages using novel mannosylated cationic liposomes. Gene Ther 2000; 7:292-299.
38. Avrameas A, McIlroy D, Hosmalin A et al. Expression of a mannose/fucose membrane lectin on human dendritic cells. Eur J Immunol 1996; 26:394-400.
39. Ezekowitz RA, Sastry K, Bailly P et al. Molecular characterization of the human macrophage mannose receptor: Demonstration of multiple carbohydrate recognition-like domains and phagocytosis of yeasts in Cos-1 cells. J Exp Med 1990; 172:1785-1794.
40. Engering AJ, Cella M, Fluitsma D et al. The mannose receptor functions as a high capacity and broad specificity antigen receptor in human dendritic cells. Eur J Immunol 1997; 27:2417-2425.
41. Tan MC, Mommaas AM, Drijfhout JW et al. Mannose receptor-mediated uptake of antigens strongly enhances HLA class II-restricted antigen presentation by cultured dendritic cells. Eur J Immunol 1997; 27:2426-2435.
42. Apostolopoulos V, Pietersz GA, Gordon S et al. Aldehyde-mannan antigen complexes target the MHC class I antigen- presentation pathway. Eur J Immunol 2000; 30:1714-1723.
43. Tietze C, Schlesinger P, Stahl P. Mannose-specific endocytosis receptor of alveolar macrophages: Demonstration of two functionally distinct intracellular pools of receptor and their roles in receptor recycling. J Cell Biol 1982; 92:417-424.
44. Mahnke K, Guo M, Lee S et al. The dendritic cell receptor for endocytosis, DEC-205, can recycle and enhance antigen presentation via major histocompatibility complex class II-positive lysosomal compartments. J Cell Biol 2000; 151:673-684.
45. Jiang W, Swiggard WJ, Heufler C et al. The receptor DEC-205 expressed by dendritic cells and thymic epithelial cells is involved in antigen processing. Nature 1995; 375:151-155.
46. Porcelli SA, Modlin RL. CD1 and the expanding universe of T cell antigens. J Immunol 1995; 155:3709-3710.
47. Porcelli SA. The CD1 family: A third lineage of antigen-presenting molecules. Adv Immunol 1995; 59:1-98.
48. Prigozy TI, Sieling PA, Clemens D et al. The mannose receptor delivers lipoglycan antigens to endosomes for presentation to T cells by CD1b molecules. Immunity 1997; 6:187-197.

49. Berney C, Herren S, Power CA et al. A member of the dendritic cell family that enters B cell follicles and stimulates primary antibody responses identified by a mannose receptor fusion protein. J Exp Med 1999; 190:851-860.
50. Toda S, Ishii N, Okada E et al. HIV-1-specific cell-mediated immune responses induced by DNA vaccination were enhanced by mannan-coated liposomes and inhibited by anti-interferon-gamma antibody. Immunology 1997; 92:111-117.
51. Shiku H, Wang L, Ikuta Y et al. Development of a cancer vaccine: Peptides, proteins, and DNA. Cancer Chemother Pharmacol 2000; 46:S77-82.
52. Lees CJ, Apostolopoulos V, Acres B et al. The effect of T1 and T2 cytokines on the cytotoxic T cell response to mannan-MUC1. Cancer Immunol Immunother 2000; 48:644-652.
53. Lees CJ, Apostolopoulos V, McKenzie IF. Cytokine production from murine CD4 and CD8 cells after mannan-MUC1 immunization. J Interferon Cytokine Res 1999; 19:1373-1379.
54. McKenzie IF, Apostolopoulos V, Lees C et al. Oxidised mannan antigen conjugates preferentially stimulate T1 type immune responses. Vet Immunol Immunopathol 1998; 63:185-190.
55. Lofthouse SA, Apostolopoulos V, Pietersz GA et al. Induction of T1 (cytotoxic lymphocyte) and/ or T2 (antibody) responses to a mucin-1 tumour antigen. Vaccine 1997; 15:1586-1593.
56. Apostolopoulos V, Pietersz GA, McKenzie IF. Cell-mediated immune responses to MUC1 fusion protein coupled to mannan. Vaccine 1996; 14:930-938.
57. Apostolopoulos V, Pietersz GA, Loveland BE et al. Oxidative/reductive conjugation of mannan to antigen selects for T1 or T2 immune responses. Proc Natl Acad Sci USA 1995; 92:10128-10132.
58. Vaughan HA, Ho DW, Karanikas V et al. The immune response of mice and cynomolgus monkeys to macaque mucin 1- mannan. Vaccine 2000; 18:3297-3309.
59. Vaughan HA, Ho DW, Karanikas VA et al. Induction of humoral and cellular responses in cynomolgus monkeys immunised with mannan-human MUC1 conjugates. Vaccine 1999; 17:2740-2752.
60. Karanikas V, Hwang LA, Pearson J et al. Antibody and T cell responses of patients with adenocarcinoma immunized with mannan-MUC1 fusion protein. J Clin Invest 1997; 100:2783-2792.
61. Karanikas V, Thynne G, Mitchell P et al. Mannan Mucin-1 Peptide Immunization: Influence of Cyclophosphamide and the Route of Injection. J Immunother 2001; 24:172-183.
62. Karanikas V, Lodding J, Maino VC et al. Flow cytometric measurement of intracellular cytokines detects immune responses in MUC1 immunotherapy. Clin Cancer Res 2000; 6:829-837.
63. Bansal P, Mukherjee P, Basu SK et al. MHC class I-restricted presentation of maleylated protein binding to scavenger receptors. J Immunol 1999; 162:4430-4437.
64. Abraham R, Singh N, Mukhopadhyay A et al. Modulation of immunogenicity and antigenicity of proteins by maleylation to target scavenger receptors on macrophages. J Immunol 1995; 154:1-8.
65. Vandaveer SS, Erf GF, Durdik JM. Avian T helper one/two immune response balance can be shifted toward inflammation by antigen delivery to scavenger receptors. Poult Sci 2001;80:172-181.
66. Willis MS, Klassen LW, Tuma DJ et al. Adduction of soluble proteins with malondialdehyde-acetaldehyde (MAA) induces antibody production and enhances T-cell proliferation. Alcohol Clin Exp Res 2002; 26:94-106.

Chapter 15

The Interaction between Anti-Gal and the α-Gal Epitope As an Immunologic Barrier to Xenotransplantation

Uri Galili

Abstract

One of the areas in medicine in which anti-carbohydrate immune response plays the major pathophysiologic role is xenotransplantation (i.e., transplantation of pig or other mammalian organs or cells into humans). The binding of the natural anti-Gal antibody to the α-gal epitope (Galα1-3Galβ1-4GlcNAc-R) on pig cells results in a rapid rejection of pig organs and cells in human and monkey recipients. Xenotransplantation is of increasing interest because of the current shortage in organ donors for allotransplantation. More than 75% of patients in need of heart, kidney or liver transplants fail to receive the needed organ because of an insufficient number of available allografts. The similarity in size of pig and human organs has made pigs the target species as an alternative source for organs. However, pig organs transplanted into monkeys are rejected within minutes to hours, in a process designated as hyper acute rejection. Numerous studies have shown that this rejection is primarily the result of the interaction between anti-Gal and the α-gal epitope. The unique characteristics of this antibody and of its carbohydrate ligand, their role in xenotransplantation and the potential solutions for overcoming this immunological barrier, are discussed in this chapter.

Introduction

Anti-Gal is the most prevalent antibody in humans, comprising approximately 1% of circulating immunoglobulins.[1] It was originally identified by its isolation from normal human serum using Sepharose columns with α-galactosyl structures, such as melibiose (Galα1-6Glc) or α-methyl galactoside.[1] The identification of the exact specificity of anti-Gal was achieved by analysis of its binding to glycolipids with known carbohydrate structures on thin layer chromatography plates.[2,3] These studies indicated that anti-Gal interacts specifically with the carbohydrate structure Galα1-3Galβ1-4GlcNAc-R (designated as the α-gal epitope) on glycolipids of various sizes that were extracted from rabbit red blood cells (RBCs). Moreover, >80% of anti-blood group B activity in blood group A and O individuals was found to be due to anti-Gal antibodies that are capable of binding to α-gal epitopes despite the branching fucose, as in blood group B antigen (i.e., Galα1 - 3(Fucα1-2)Galβ1-4GlcNAc-R).[3] However, in blood group B and AB individuals, anti-Gal exclusively interacts with the α-gal epitope and not with other carbohydrate structures. The anti-Gal antibody was further found to interact with α-gal epitopes on a variety of glycoproteins from various nonprimate mammalian species. These include laminin, thyroglobulin, fibrinogen and immunoglobulins.[4,5]

Immunobiology of Carbohydrates, edited by Simon Y.C. Wong and Gemma Arsequell.
©2003 Eurekah.com and Kluwer Academic / Plenum Publishers.

The affinity of anti-Gal to the α-gal epitope could be determined by equilibrium dialysis studies using a free α-gal epitope as the radiolabeled trisaccharide [^{3}H]Galα1-3Galβ1-4GlcNAc. The affinity of this antibody was found to be highly variable in different individuals and to range between $2x10^5$ and $6x10^6$ M^{-1}.[6] The affinity decreased by ten fold when the disaccharide [^{3}H]Galα1-3Gal was used instead of the trisaccharide, implying that the putative "pocket" of the combining site of anti-Gal accommodates at least a trisaccharide structure. The disaccharide displays a lower affinity because of a lower number of interaction points between the carbohydrate and the antibody. Interestingly, even the presence of the N-acetyl group on the glucose at the reducing end was found to affect the binding. Thus, the trisaccharide [^{3}H]Galα1-3Galβ1-4Glc displayed an affinity to anti-Gal which was lower by 30% than that of [^{3}H]Galα1-3Galβ1-4GlcNAc.[6]

Anti-Gal is produced throughout life in humans as a result of continuous antigenic stimulation by gastrointestinal bacteria that have carbohydrate structures similar to the α-gal epitope.[7] As many as 1% of circulating B lymphocytes in humans are capable of producing anti-Gal (designated anti-Gal B cells). This is indicated by the finding that one in a hundred human B cells immortalized by Epstein Barr virus can secrete this antibody in vitro.[8] Most of these B cells are in a quiescent state as memory B cells, and only those along the gastrointestinal tract continuously produce this natural antibody. Analysis of the immunoglobulin genes of anti-Gal B cells indicated that this is a polyclonal population. However, the immunoglobulin heavy chain genes in most clones cluster in the VH3 family.[9]

The Reciprocal Expression of the α-Gal Epitope and Anti-Gal in Mammals

The identification of the α-gal epitope on glycolipids and glycoproteins as the ligand for anti-Gal raised the question of whether there is a pattern in the expression of the α-gal epitope and of anti-Gal in mammals. The epitope was originally identified in glycolipids extracted from rabbit RBC membranes,[10,11] where the α-gal epitope was identified in ceramide pentahexoside (CPH- Galα1-3Galβ1-4GlcNAcβ1-3Galβ1-4Glc-Cer). The α-gal epitope was subsequently shown to be present on rabbit RBC glycolipids with carbohydrate chains of various lengths, including 7, 10, 15 and up to 32 carbohydrate units in length,[12,13] and was also found on glycolipids of bovine RBCs.[14-16] By using *Griffonia (Bandeiraea) simplicifolia* IB4, a lectin which interacts specifically with α-gal epitopes,[17] α gal epitope expression was detected on mouse Ehrlich ascites cells.[18] Other mouse cells reported to carry this epitope were mouse lymphoma cells[19] and 3T3 fibroblasts.[20] Variable expression of α-gal epitope was identified on thyroglobulin molecules from various nonprimate mammalian species but notably not on human thyroglobulin.[21]

In order to determine whether there is a distinct pattern of distribution of α-gal epitopes in mammals, we analyzed expression of this epitope on RBCs[22] and nucleated cells from a large variety of species.[23] No α-gal epitope expression was detected on cells from fish, amphibians, reptiles, or birds; however, we found an abundance of this epitope ($1x10^6$–$30x10^6$ epitopes/cell) on cells of nonprimate marsupial and placental mammals such as kangaroo, opossum, mouse, rat, rabbit, bat, pig, cow, horse, cat, dog, and dolphin.[23] A similar expression of this epitope was detected on cells of prosimians (e.g., lemurs), and New World monkeys (i.e., monkeys of South America), but not on cells of Old World monkeys (monkeys of Asia and Africa), apes (e.g., chimpanzee, gorilla and orangutan), and humans.[22,23] In contrast, humans, apes and Old World monkeys all produce large amounts of the natural anti-Gal antibody against the α-gal epitope.[22]

The α1,3galactosyltransferase and Its Evolution in Mammals

The α-gal epitope is synthesized on glycoproteins and glycolipids within the Golgi apparatus of mammalian cells by the glycosylation enzyme α1,3galactosyltransferase (α1,3GT). This enzyme transfers galactose from the sugar donor UDP-Gal to the sugar acceptor N-acetyllactosamine residues to form the α-gal epitope in the following reaction:

Galβ1-4GlcNAc-R +	UDP-Gal	α1,3GT	Gal α1-3Galb1-4GlcNAc-R	+ UDP
N-acetyllactosamine	Uridine		α-gal epitope on cell	Uridine
on carbohydrate	diphosphate		surface carbohydrate	diphosphate
chains	galactose		chains	

This enzyme was first reported in rabbit cells[24,25] and subsequently was found in mouse and bovine cells.[26,27] As expected from the distribution pattern of α-gal epitopes, we found α1,3GT to be active in cells of nonprimate mammals and New World monkeys but not in Old World monkeys or humans.[23,28]

The α1,3GT gene was simultaneously cloned in mouse[29] and cow.[30] This led to the subsequent demonstration of the presence of this gene as a pseudogene in humans, apes, and Old World monkeys.[31,32] This gene, although found in humans and in Old World monkeys, is expressed in these species at a much lower level than in other mammals.[30,33] In addition, frame shift mutations in the open reading frame of the α1,3GT pseudogene, which result in premature stop codons, were reported in the human, ape and Old World monkey pseudogene.[31-33] Comparison of the sequence of this pseudogene in humans and in other primates led us to suggest that the α1,3GT gene was inactivated in ancestral Old World primates, after apes and monkeys diverged from each other, 20-25 million years ago.[32]

We have further suggested that this inactivation of the α1,3GT gene in ancestral Old World monkeys and apes could have been the result of an extensive evolutionary pressure exerted by a pathogen. This pathogen was detrimental to primates and was endemic only in the Old World (i.e., Africa, Asia and Europe). It did not spread to the New World because of geographic barriers and thus did not affect the evolution of New World monkeys.[23,32,34] Pathogens expressing α-gal epitopes are common among enveloped viruses propagated in cells containing active α1,3GT gene,[35–37] in various bacteria[7] and in protozoa.[38,39] An alternative scenario for the evolutionary inactivation of the α1,3GT gene could be the appearance of a pathogen that used the α-gal epitope as a receptor for affecting host cells. A present example of such a pathogen is *Clostridium difficile*. The enterotoxin A produced by this bacterium binds to α-gal epitopes on various cells.[40] Either scenario could have resulted in selective pressure for the evolutionary suppression of α-gal epitope expression by the inactivation of the α1,3GT gene, followed by production of anti-Gal (as a protective means in the first scenario) because of the loss of immune tolerance to the α-gal epitope.

The Interaction between Anti-Gal and α-Gal Epitopes on Xenografts

The production of anti-Gal in humans, apes and Old World monkeys, resulting from the evolutionary inactivation of the α1,3GT gene, has formed an immunologic barrier against transplantation of mammalian organs or cells that express the α-gal epitope.[41] It has been shown that incubation of mammalian cells expressing α-gal epitopes with human serum results in cell lysis because of the binding of anti-Gal IgM molecules to α-gal epitopes, followed by activation of the complement system.[42,43] Specific adsorption of the antibody on affinity columns with synthetic α-gal epitopes resulted in the elimination of this cytotoxic activity.[42] Moreover, transplantation of pig or New World monkey xenografts into Old World monkeys was followed by in vivo binding of anti-Gal to α-gal epitopes on endothelial cells of grafts, resulting in complement mediated lysis of these cells, collapse of the vascular bed and hyper acute rejection of the xenograft.[44] Accordingly, in vivo neutralization of anti-Gal by oligosaccharides,[45] or removal of anti-Gal by adsorption on affinity columns delayed xenograft rejection by hours to days.[46,47]

The hyper acute rejection of pig xenografts in monkeys could be prevented by transplantation of organs from pigs transgenic for complement regulatory proteins (i.e., human proteins that inactivate complement), which prevent complement activation by anti-Gal IgM molecules.[48,49] However, such organs were rejected as a result of the detrimental effect of the IgG isotype of anti-Gal on the xenografts. After the binding of anti-Gal IgG molecules to α-gal epitopes on xenograft cells, the Fc portion of the antibody readily interacts with Fcγ receptors on granulocytes, macrophages, monocytes and NK cells. This interaction directs these effector cells to destroy the xenograft cells by the antibody-dependent cell-mediated cytotoxicity (ADCC) mechanism.[41,50,51] This destruction of xenograft cells by ADCC is much slower than that mediated by complement, but this type of destruction mediates the delayed or chronic xenograft rejection phenomenon which occurs within several days post transplantation, if hyper acute rejection is avoided.

Elicited Anti-Gal Response in Xenograft Recipients

In addition to the continuous production of anti-Gal in response to antigenic stimulation by gastrointestinal bacteria, the many quiescent anti-Gal B cells can be effectively activated by α-gal epitopes on xenograft cells, following transplantation of mammalian cells expressing this epitope into humans. As a result of this immune response, anti-Gal IgG titer increases within a short period of two weeks by 30–300 fold. This immune response is effective in inducing rejection of xenografts by continuous destruction of cells expressing α-gal epitopes via the ADCC mechanism.

This anti-Gal immune response to α-gal epitopes on xenografts was first demonstrated in diabetic patients who were transplanted with allogeneic kidney and with pig fetal islet cell clusters.[52,53] These pig islet cell clusters were produced by in vitro culturing of fetal pig pancreatic tissues, which results in proliferation and clustering of islet cells.[54] The volume of transplanted islet cell clusters corresponded to 2–6 ml packed cells.[54] Follow up of anti-Gal activity in the serum of these patients has indicated that it increased by 20–100 fold within the period of 25–50 days post transplantation.[52,53] This was the result of an increase in titer of anti-Gal IgG, and to a lesser extent of the IgM and IgA isotypes of this antibody.[53] The observed increase in anti-Gal production occurred despite the immunosuppressive treatment, that was effective in preventing the rejection of the kidney allografts.[54] The increase in the titer of anti-Gal in these patients was found to be due to an increase in the concentration of this antibody as well as an increase in its affinity,[53] probably as a result of affinity maturation by the process of somatic mutations within anti-Gal B cell clones.

A detailed follow up of anti-Gal production in response to exposure of the immune system to α-gal epitopes on xenograft cells was achieved in studies on the immune response in ovarian carcinoma patients. These patients were treated (at the Human Gene Therapy Research Institute in Des Moines, Iowa) by three intraperitoneal infusions, each of $6x10^9$ mouse fibroblasts that released a replication-defective retrovirus containing the thymidine kinase gene. Tumor cells infected in situ by the virus are killed by subsequent administration of ganciclovir.[55] The amount of infused mouse fibroblasts expressing α-gal epitopes corresponds to transplantation of an approximately 50g xenograft. These patients could be regarded, therefore, as xenograft recipients of mouse cells expressing α-gal epitopes. Within one week post transplantation, the titer of anti-Gal increased by approximately ten fold, and two weeks post transplantation by approximately 100 fold.[56] The antibody activity remained at that high level after the second and third infusions. This response was the result of activation of the many anti-Gal B cell clones that engage the α-gal epitopes on the glycoproteins released from the mouse fibroblasts. The ten fold increase in anti-Gal titer within the first week post transplantation was the result of an increase in the concentration of anti-Gal in the serum, and the additional ten fold increase within the second week was found to be the result of a corresponding increase in the affinity of this antibody.[56] A similar increase in anti-Gal titer was observed in patients with impaired liver function, who were treated by temporary extracorporeal perfusion of their blood through a pig liver.[57] Thus, the release of xenoglycoproteins from the pig liver, perfused for several hours, was sufficient to induce the activation of the many quiescent anti-Gal B cells for production of anti-Gal.

We could also study the long term anti-Gal response to xenograft in a primate model of cynomolgus monkeys transplanted with pig meniscus or articular cartilage.[58,59] The use of these xenografts enables the long term follow up of the immune response because cartilage is avascular, therefore, neither hyper acute rejection nor delayed rejection occur. Within two weeks post transplantation anti-Gal IgG activity increased by at least ten fold. Four weeks post transplantation, anti-Gal activity in the various monkeys increased by 30 to 300 fold and remained at that high level as long as the cartilage xenografts were present within the monkeys.[58,59] All these studies imply that anti-Gal B cells are present in large numbers in humans and Old World monkeys. It is probable that many of these cells are activated upon engaging α-gal epitopes on xenoglycoproteins. As a result of this activation, anti-Gal B cell clones expand by proliferation and undergo isotype switch to produce large amounts of high affinity anti-Gal IgG. Without prevention of this anti-xenograft immune response it will be difficult to progress in transplantation of xenografts in humans.

The high affinity anti-Gal produced in xenograft recipients is very detrimental to xenografts. High affinity anti-Gal can very effectively mediate ADCC by binding to α-gal epitopes on xenograft cells, even if the number of these epitopes is very low. It is probable that the binding of few thousand such IgG molecules is sufficient for targeting various monocytes, macrophages, NK cells, or granulocytes to exert their cytotoxic potential on xenograft cells. This would result in chronic destruction of xenograft cells. In addition, this high affinity anti-Gal was found to induce continuous activation of the endothelial cells in the blood vessels of the xenograft. A large proportion of the cell surface molecules on endothelial cells of pig or mouse are glycoproteins expressing α-gal epitopes. Strong binding of anti-Gal to glycoproteins that are cell surface receptors may mimic the stimulatory effect of the ligands for such receptors. Indeed, incubation of pig endothelial cells with sera containing high affinity anti-Gal was found to result in the activation of the endothelial cells, as indicated by the induced expression of E-selectin.[60,61] Such sera were obtained from monkeys transplanted with pig cartilage, thus containing high affinity anti-Gal. The specific removal of anti-Gal from these sera abolished their stimulatory activity in inducing E-selectin expression. Pretransplantation sera containing low affinity anti-Gal lacked this stimulatory activity.[60,61] These findings imply that high affinity anti-Gal binding to and activation of receptors on pig endothelial cells of the xenograft can induce de novo expression of E-selectin and other adhesion molecules on these cells. This is likely to result in a chronic inflammatory response that will further contribute to xenograft rejection beyond the ADCC detrimental effect.

T Cell Association with Anti-Gal Response in Xenograft Recipients

Studies on the mechanism of anti-Gal response in xenograft recipients could be performed in the experimental animal model of α1,3GT KO mice. These mice can be induced to produce high titers of anti-Gal IgG by repeated immunization with pig kidney membranes (PKM). This immune response was found to be the result of combined activation of anti-Gal producing B cells and of helper T cells.[62] Anti-Gal B cells bind α-gal epitopes on xenoglycoproteins and internalize them. These B cells, acting as antigen presenting cells, process and present immunogenic peptides of these xenoglycoproteins for the activation of adjacent helper T cells with receptors to these peptides. The activated T cells express activation molecules such as CD40L which interact with CD40 on the B cells. Together with lymphokines secreted by the activated T cells, this interaction induces activation of the anti-Gal-producing B cell to proliferate, undergo isotype switch and somatic mutations, leading to production of large amounts of high affinity anti-Gal IgG.[62] The activation of helper T cells by these B cells is a very effective process since most pig glycoproteins differ from homologous human glycoproteins in their amino acid sequence. Therefore, many of the peptides produced from pig glycoproteins processed and presented by anti-Gal B cell are likely to activate helper T cells with many different specificities. This wide range of helper T cell activation is likely to be one of the main reasons for the inability of immunosuppressive regimens to prevent elicited anti-Gal production in the recipients of pig fetal islet cells described above.

It should be stressed that the α-gal epitope cannot directly activate T cells. Crystallography studies on the expression of mammalian carbohydrate chains on MHC molecules have indicated that the carbohydrate chain protrudes to a considerable distance from the MHC groove.[63] This protrusion prevents any subsequent interaction between T cell receptor molecules and the corresponding ligands on the antigen presenting cell. Indeed, analysis of T cell responses in PKM immunized mice indicated that T cells in the immunized mice do not recognize the α-gal epitope.[62] Since many cell surface glycoconjugates have core structures that are similar to those carrying the α-gal epitope, it is probable that other glycoconjugates with long carbohydrate chains lack the ability of direct activation of T cells via the T cell receptor.

Overcoming Anti-Gal Mediated Rejection of Xenografts

Prevention of pig organ rejection in humans is an endeavor that faces many immunologic challenges. At present the most obvious obstacles, which also seem to be the most important, are the prevention of the recipient's anti-Gal interaction with α-gal epitopes on the xenograft and the prevention of production of high affinity anti-Gal in xenograft recipients. The research approaches for overcoming these obstacles can be divided into two major groups:

i) Elimination of α-gal epitopes on pig cells by glycosyltransferase transgene, or by knockout (KO) of the α1,3GT gene.
ii) Prevention of anti-Gal production in xenograft recipients.

Elimination of α-Gal Epitopes

Glycosyltransferase Transgenes

It has been suggested that generation of transgenic pigs for α1,2fucosyltransferase (α1,2FT) would result in competition between α1,3GT and α1,2FT for capping the common N-acetyllactosamine sugar acceptor resulting in a significant decrease in the expression of α-gal epitopes.[64] In vitro studies on transfected pig cells[64] and in vivo studies on transgenic pigs for α1,2FT[65,66] demonstrated an approximately 70% decrease in α-gal epitope expression on pig cells. A similar approach was used with recombinant sialyltransferase[67] and recombinant GlcNAc-transferase III,[68,69] as enzymes competing with endogenous α1,3GT. This type of competition also demonstrated a decrease of approximately 70% in α-gal epitope expression. It seems that such a decrease would be far from having a significant effect of the prevention of anti-Gal mediated rejection. This is because pig endothelial cells (and probably many other pig cells) express $10x10^6$ to $30x10^6$ α-gal epitopes per cell. Because of the large number of anti-Gal B cells in humans, it is possible that the expression of even as few as $1x10^5$ epitopes per cell may be sufficient for inducing production of high affinity anti-Gal IgG. A higher elimination of α-gal epitopes on pig cells was observed by transfection both with α1,2FT and with the human α-galactosidase gene.[70] The combination of trans-Golgi targeted α-galactosidase, removing terminal α-galactosyl units from α-gal epitopes, and of α1,2FT competing with α1,3GT, seems to be more effective than the activity of α1,2FT alone in decreasing α-gal epitope expression. It remains to be determined whether transgenic pigs for these two genes display sufficiently low α-gal epitope expression to prevent production of high affinity anti-Gal.

Knockout Pigs for α1,3GT

It is evident that complete elimination of α1,3GT activity in pig cells (i.e., the generation of α1,3GT knockout pigs) would be the most effective way for preventing α-gal epitope expression. Such knockout of the α1,3GT gene may be achieved by homologous recombination with the gene that is disrupted by a selection marker (e.g., neomycin resistance gene). Thall et al[71] and Tearle et al[72] demonstrated the feasibility of this approach in mice. Both groups disrupted the exon containing most of the catalytic domain of α1,3GT (termed exon 9[73]) in mouse embryonic stem cells and succeeded in generating homozygous mice for the disrupted α1,3GT gene. These mice lack α-gal epitopes[71,72] and are capable of producing anti-Gal, in particular when immunized with xenogeneic cells.[62,74] A similar disruption of the α1,3GT gene in pig cells and cloning of a heterozygous pig for this gene by nuclear transfer has been reported recently.[75] It remains to be determined whether mating of such heterozygous pigs for the disrupted α1,3GT gene can result in the development of pigs that are homozygous for the disrupted gene, i.e., homozygous α1,3GT KO pigs which completely lack the α-gal epitope. Interestingly, it was recently further reported[76] on the generation of a homozygous knockout pig for the two disrupted α1,3GT alleles, by performing in vitro double knockout of the gene prior to the nuclear transfer step. This study implies that pigs may develop in the absence of α-gal epitopes. Nevertheless, it remains to be determined whether fertilization can occur when both sperm and egg lack α-gal epitopes.

Prevention of Anti-Gal Response in Xenograft Recipients

An alternative to knockout pigs for elimination of the anti-Gal/α-gal epitope barrier may be the manipulation of the immune system of xenograft recipients in order to induce long-term immune tolerance to the α-gal epitope. In view of the strong immune response toward this epitope on xenografts, despite immunosuppressive regimens,[53] it is unlikely that anti-Gal production can be prevented by nonspecific immunosuppressive drugs, unless these drugs are used at a dose that completely paralyzes the immune system. Such an approach is impractical since it leaves the xenograft recipient completely defenseless against opportunistic infections.

Studies on tolerance induction to the α-gal epitope have demonstrated that such tolerance can be achieved by the exposure of maturing lymphocytes to α-gal epitopes as a self-antigen on bone marrow cells.[77,78] For this purpose α1,3GT KO mice were irradiated and received bone marrow from wild type mice i.e., bone marrow cells expressing the α-gal epitope. Such mice were found to lack the ability to produce anti-Gal. Similarly, tolerance has been observed in mice that received bone marrow cells from α1,3GT KO mouse which were transfected in vitro with a retrovirus containing the α1,3GT gene.[79] It has been postulated that this tolerance is induced by a mechanism similar to that preventing maturation of immature B cells that interact with self-antigens, by the induction of programmed cell death (i.e., apoptosis).[79] Recent studies in our laboratory have indicated, however, that tolerance to the α-gal epitope can also be achieved by exposure of anti-Gal B cells to α-gal epitopes on syngeneic (i.e., wild type) lymphocytes or endothelial cells.[80-82] These studies further indicate that exposure of anti-Gal B cells to the α-gal epitope on syngeneic cells in the complete absence of T cell help results in the deletion of anti-Gal B cells. In contrast, if such exposure occurs in the presence of T cells that are activated by xenopeptides, the anti-Gal B cells are rescued from deletion and are activated to produce anti-Gal. Future studies based on the observations on tolerance induction by bone marrow cells or by other syngeneic cells expressing the α-gal epitope, may provide for new effective methods for overcoming the anti-Gal/α-gal epitope barrier in case it is impossible to raise homozygous α1,3GT KO pigs.

References

1. Galili U, Rachmilewitz EA, Peleg A et al. A unique natural human IgG antibody with anti-α-galactosyl specificity. J Exp Med 1984; 160:1519-1531.
2. Galili U, Macher BA, Buehler J et al. Human natural anti-α-galactosyl IgG. II. The specific recognition of α(1-3)-linked galactose residues. J Exp Med 1985; 162:573-582.
3. Galili U, Buehler J, Shohet SB et al. The human natural anti-Gal IgG. III. The subtlety of immune tolerance in man as demonstrated by crossreactivity between natural anti-Gal and anti-B antibodies. J Exp Med 1987; 165:693-704.
4. Towbin H, Rosenfelder G, Wieslander J et al. Circulating antibodies to mouse laminin in Chagas disease, American cutaneous leishmaniasis and normal individuals recognize terminal galactosyl (α1-3)-galactose epitopes. J Exp Med 1987; 166:419-432.
5. Thall A, Galili U. Distribution of Galα1-3Galβ1-4GlcNAc residues on secreted mammalian glycoproteins (thyroglobulin, fibrinogen, and immunoglobulin G) as measured by a sensitive solid-phase radioimmunoassay. Biochemistry 1990; 29:3959-3965.
6. Galili U, Matta KL. Inhibition of anti-Gal IgG binding to porcine endothelial cells by synthetic oligosaccharides. Transplantation 1996; 62:256-262.

7. Galili U, Mandrell RE, Hamadeh RM et al. Interaction between human natural anti-α-galactosyl immunoglobulin G and bacteria of the human flora. Infect Immun 1988; 56:1730-1737.
8. Galili U, Anaraki F, Thall A et al. One percent of circulating B lymphocytes are capable of producing the natural anti-Gal antibody. Blood 1993; 82:2485-2493.
9. Wang L, Radic MZ, Galili U. Human anti-Gal heavy chain genes. Preferential use of VH3 and the presence of somatic mutations. J Immunol 1995; 155:1276-1285.
10. Eto T, Ichikawa Y, Nishimura K et al. Chemistry of lipids of the posthemolytic residue or stroma of erythrocytes. XVI. Occurance of ceramide pentasaccharide in the membrane of erythrocytes and reticulocytes in rabbit. J Biochem 1968; 64:205-213.
11. Stellner K, Saito H, and Hakomori S. Determination of aminosugar linkage in glycolipids by methylation. Aminosugar linkage of ceramide pentasaccharides of rabbit erythrocytes and of Forssman antigen. Arch Biochem Biophys 1973; 133:464-472.
12. Egge H, Kordowicz M, Peter-Katalinic J et al. Immunochemistry of I/i-active oligo- and polyglycosylceramides from rabbit erythrocyte membranes. Characterization of linear, di-, and triantennary neolactoglycosphingolipids. J Biol Chem 1985; 260:4927-4935.
13. Honma K, Manabe H, Tomita M et al. Isolation and partial structure characterization of macroglycolipids from rabbit red cell membranes. J Biochem (Tokyo) 1981; 90:1187-1196.
14. Chien JL, Li SC, Li YT. Isolation and characterization of a heptaglycosylceramide from bovine erythrocyte membrane. J Lipid Res 1979; 20:669-673.
15. Uemura K, Yuzawa M, Taketomi T. Characterization of major glycolipids in bovine erythrocyte membrane. J Biochem (Tokyo) 1978; 83:463-471.
16. Watanabe K, Hakomori SI, Childs RA et al. Characterization of a blood group I-active ganglioside. Structural requirements for I and i specificities. J Biol Chem 1979; 254:3221-3228.
17. Wood C, Kabat EA, Murphy LA et al. Immunochemical studies of the combining sites of the two isolectins, A4 and B4, isolated from Bandeiraea simplicifolia. Arch Biochem Biophys 1979; 198:1-11.
18. Eckhardt AE, Goldstein IJ. Isolation and characterization of a family of α-D-galactosyl containing glycopeptides from Ehrlich ascites tumor cells. Biochemistry 1983; 22:5290-5297.
19. Cummings RD, Kornfeld S. The distribution of repeating (Galβ1→4GlcNAcβ1→3) sequences in asparagine-linked oligosaccharides of the mouse lymphoma cell lines BW5147 and PHAR2.1. J Biol Chem 1984; 259:6253-6260.
20. Santer UV, DeSantis R, Hard KJ et al. N-linked oligosaccharide changes with oncogenic transformation require sialylation of multiantennae. Eur J Biochem 1989; 181:249-260.
21. Spiro RG and Bhoyroo VD. Occurance of α-D-galactosyl residues in the thyroglobulin from several species. Localization in the saccharide chains of the complex carbohydrate units. J Biol Chem 1984; 259:9858-5966.
22. Galili U, Clark MR, Shohet SB et al. Evolutionary relationship between the anti-Gal antibody and the Galα1-3Gal epitope in primates. Proc Natl Acad Sci USA 1987; 84:1369-1373.
23. Galili U, Shohet SB, Kobrin E et al. Man, apes, and Old World monkeys differ from other mammals in the expression of α-galactosyl epitopes on nucleated cells. J Biol Chem 1988; 263:17755-17762.
24. Basu M, Basu S. Enzymatic synthesis of blood group B-related pentaglycosyl ceramide by an α-galactosyltransferase from rabbit bone marrow. J Biol Chem 1973; 248:1700-1706.
25. Betteridge A, Watkins WM. Two α-3-D-galactosyltransferases in rabbit stomach mucosa with different acceptor substrate specificities. Eur J Biochem 1983; 132:29-35.
26. Blake DD, Goldstein IJ. An α-D-galactosyltransferase in Ehrlich ascites tumor cells. Biosynthesis and characterization of a trisaccharide (α-D-galacto(1-3)-N-acetyllactosamine). J Biol Chem 1981; 256:5387-5393.
27. Blanken WM, van den Eijnden DH. Biosynthesis of terminal Galα1-3Galß1-4GlcNAc-R oligosaccharide sequence on glycoconjugates. Purification and acceptor specificity of a UDP-Gal: N-acetyllactosamine α1,3galactosyltransferase. J Biol Chem 1985; 260:12927-12934.
28. Thall A, Etienne-Decerf J, Winand R et al. The α galactosyl epitope on mammalian thyroid cells. Acta Endocrin 1991; 124:692-699.
29. Larsen RD, Rajan VP, Ruff M et al. Isolation of a cDNA encoding murine UDP galactose: ßD-galactosyl-1,4-N-acetyl-D-glucosaminide α1,3-galactosyltransferase: expression cloning by gene transfer. Proc Natl Acad Sci USA 1989; 86:8227-8231.
30. Joziasse DH, Shaper JH, Van den Eijnden DH et al. Bovine α1-3galactosyltransferase: Isolation and characterization of a cDNA clone. Identification of homologous sequences in human genomic DNA. J Biol Chem 1989; 264:14290-14297.
31. Larsen RD, Rivera-Marrero CA, Ernst LK et al. Frameshift and nonsense mutations in a human genomic sequence homologous to a murine UDP-Galß-D-Gal(1,4)-D-GlcNAc α(1,3) galactosyltransferase cDNA. J Biol Chem 1990; 265:7055-7061.

32. Galili U, Swanson K. Gene sequences suggest inactivation of α1,3 galactosyltransferase in catarrhines after the divergence of apes from monkeys. Proc Natl Acad Sci USA 1991; 88:7401-7404.
33. Koike C, Fung JJ, Geller DA et al. Molecular basis of evolutionary loss of the α1,3-galactosyltransferase gene in higher primates. J Biol Chem 2002; 277:10114-10120.
34. Galili U, Andrews P. Suppression of α-galactosyl epitopes synthesis and production of the natural anti-Gal antibody: A major evolutionary event in ancestral Old World primates. J Hum Evol 1995; 29:433-442.
35. Repik PM, Strizki JM, Galili U. Differential host-dependent expression of α-galactosyl epitopes on viral glycoproteins: A study of eastern equine encephalitis virus as a model. J Gen Virol 1994; 75:1177-1181.
36. Takeuchi Y, Porter CD, Strahan KM et al. Sensitization of cells and retroviruses to human serum by (α 1-3) galactosyltransferase. Nature 1996; 379:85-88.
37. Rother RP, Galili U. α-Gal epitopes on viral glycoproteins. Subcell Biochem 1999; 32:143-172.
38. Avila JL, Rojas M, Galili U. Immunogenic Gal α1-3Gal carbohydrate epitopes are present on pathogenic American Trypanosoma and Leishmania. J Immunol 1989; 142:2828-2834.
39. Almeida IC, Milani SR, Gorin PA et al. Complement-mediated lysis of Trypanosoma cruzi trypomastigotes by human anti-α-galactosyl antibodies. J Immunol 1991; 146:2394-2400.
40. Krivan HC, Clark GF, Smith DF et al. Cell surface binding site for Clostridium difficile enterotoxin: Evidence for a glycoconjugate containing the sequence Gal α1-3Gal β1-4GlcNAc. Infect Immun 1986; 53:573-581.
41. Galili U. Interaction of the natural anti-Gal antibody with α-galactosyl epitopes: A major obstacle for xenotransplantation in humans. Immunol Today 1993; 14:480-482.
42. Good H, Cooper DKC, Malcolm AJK et al. Identification of carbohydrate structures which bind human anti-porcine Abs: Implications for discordant grafting in man. Transplant Proc 1992; 24:559-562.
43. Sandrin M, Vaughan HA, Dabkowski PL et al. Anti-pig IgM antibodies in human serum react predominantly with Galα1-3Gal epitopes. Proc Natl Acad Sci USA 1993; 90:11391-11395.
44. Collins BH, Cotterell AH, McCurry KR et al. Cardiac xenografts between primate species provide evidence for the importance of the α-galactosyl determinant in hyperacute rejection. J Immunol 1995; 154:5500-5510.
45. Simon PM, Neethling FA, Taniguchi S et al. Intravenous infusion of Galα1-3Gal oligosaccharides in baboon delays hyperacute rejection of porcine heart xenografts. Transplantation 1998; 56:346-353.
46. Xu T, Lorf T, Sablinski T et al. Removal of anti-porcine natural antibodies from human and nonhuman primate plasma in vitro and in vivo by a Galα1-3Galβ1-4Glc-R immunoaffinity column. Transplantation 1998; 65:172-179.
47. Kozlowski T, Ierino F, Lambrigts D et al. Depletion of anti-Galα1-3Gal antibody in baboons by specific α-Gal immunoaffinity columns. Xenotransplantation 1998; 5:122-131.
48. Cozzi E, White DJG. The generation of transgenic pigs as potential organ donors for humans. Nat Med 1995; 1:964-966.
49. Byrne GW, McCurry KR, Martin MJ et al. Transgenic pigs expressing human CD59 and decay accelerating factor produce an intrinsic barrier to complement-mediated damage. Transplantation 1997; 63:149-155.
50. Schaapherder AFM, Daha MR, te Bulte MT et al. Antibody-dependent cell-mediated cytotoxicity against porcine endothelium induced by a majority of human sera. Transplantation 1994; 57:1376-1382.
51. Watier H, Guillaumin J-M, Vallee I et al. Human NK cell-mediated direct and IgG dependent cytotoxicity against xenogeneic porcine endothelial cells. Transpl Immunol 1996; 4:293-299.
52. Satake M, Kawagishi N, Rydberg L et al. Limited specificity of xenoantibodies in diabetic patients transplanted with fetal porcine islet cell clusters. Xenotransplantation 1994; 1:89-96.
53. Galili U, Tibell A, Samuelsson B et al. Increased anti-Gal activity in diabetic patients transplanted with fetal porcine islet cell clusters. Transplantation 1995; 59:1549-1556.
54. Groth CG, Korsgren O, Tibell A et al. Transplantation of fetal porcine pancreas to diabetic patients: Biochemical and histological evidence for graft survival. Lancet 1994; 344:1402-1404.
55. Link Jr CJ, Moorman D, Seregina T et al. A phase I trial of in vivo gene therapy with the herpes simplex thymidine kinase/ganciclovir system for the treatment of refractory or recurrent ovarian cancer. Hum Gene Ther 1996; 7:1161-1179.
56. Galili U, Chen ZC, Tanemura M et al. Understanding the induced antibody response in xenograft recipients. GRAFT 2000; 4:32.
57. Yu PB, Parker W, Everett ML et al. Immunochemical properties of anti-Gal α1-3Gal antibodies after sensitization with xenogeneic tissues. J Clin Immunol 1999; 19:116-126.

58. Galili U, LaTemple DC, Walgenbach AW et al. Porcine and bovine cartilage transplants in cynomolgus monkey. II. Changes in anti-Gal response during chronic rejection. Transplantation 1997; 63:646-651.
59. Stone KR, Ayala G, Goldstein J et al. Porcine cartilage transplants in cynomolgus monkey. III. Transplantation of α-galactosidase treated porcine cartilage. Transplantation 1998; 65:1577-1583.
60. Palmetshofer A, Galili U, Dalmasso AP et al. α-galactosyl epitope-mediated activation of porcine aortic endothelial cells: Type I activation. Transplantation 1998; 65:844-853.
61. Palmetshofer A, Galili U, Dalmasso AP et al. α-galactosyl epitope-mediated activation of porcine aortic endothelial cells: Type II activation. Transplantation 1998; 65:971-978.
62. Tanemura M, Yin D, Chong AS et al. Differential immune responses to α-gal epitopes on xenografts and allografts: Implications for accommodation in xenotransplantation. J Clin Invest 2000; 105:301-310.
63. Speir JA, Abdel-Motal UM, Jondal M et al. Crystal structure of an MHC class I presented glycopeptide that generates carbohydrate-specific CTL. Immunity 1999; 10:51-61.
64. Sandrin M S, Fodor WL, Mouhtouris E et al. Enzymatic remodelling of the carbohydrate surface of a xenogenic cell substantially reduces human antibody binding and complement-mediated cytolysis. Nat Med 1995; 1:1261-1267.
65. Koike C, Katayama A, Kadomatsu K et al. Reduction of α-Gal epitopes in transgenic pig by introduction of human α1-2 fucosyltransferase. Transplant Proc 1997; 29:894.
66. Sharma A, Okabe J, Birch P et al. Reduction in the level of Gal(α1,3)Gal in transgenic mice and pigs by the expression of an α(1,2)fucosyltransferase. Proc Natl Acad Sci USA 1996; 93:7190-7195.
67. Tanemura M, Miyagawa S, Koyota S et al. Reduction of the major swine xenoantigen, the α-galactosyl epitope by transfection of the α2,3-sialyltransferase gene. J Biol Chem 1998; 273:16421-16425.
68. Tanemura M, Miyagawa S, Ihara Y et al. Significant downregulation of the major swine xenoantigen by N-acetylglucosaminyltransferase III gene transfection. Biochem Biophys Res Commun 1997; 235:359-364.
69. Tanemura M, Miyagawa S, Ihara Y et al. Reduction of the major swine xenoantigen Gal α (1,3)Gal by transfection of N-acetylglucosaminyl transferase III (GnT-III) gene. Transplant Proc 1997; 29:891-892.
70. Osman N, McKenzie IF, Ostenried K et al. Combined transgenic expression of α-galactosidase and α1,2-fucosyltransferase leads to optimal reduction in the major xenoepitope Galα (1,3)Gal. Proc Natl Acad Sci USA 1997; 94:14677-14682.
71. Thall AD, Maly P and Lowe JB. Oocyte Galα1-3Gal epitopes implicated in sperm adhesion to the zona pellucida glycoprotein ZP3 are not required for fertilization in the mouse. J Biol Chem 1995; 270:21437-21440.
72. Tearle RG, Tange MJ, Zannettino ZL et al. The α-1,3-galactosyltransferase knockout mouse. Implications for xenotransplantation. Transplantation 1996; 61:13-19.
73. Joziasse DH, Shaper NL, Kim D et al. Murine α 1,3-galactosyltransferase. A single gene locus specifies four isoforms of the enzyme by alternative splicing. J Biol Chem 1992; 267:5534-5541.
74. LaTemple DC, Galili, U. Adult and neonatal anti-Gal response in knockout mice for α-galactosyltransferase. Xenotransplantation 1998; 5:191-196.
75. Lai L, Kolber-Simonds D, Park KW et al. Production of α-1,3-galactosyltransferase knockout pigs by nuclear transfer cloning. Science 2002; 295:1089-1092.
76. Phelps CJ, Koike C, Vaught TD et al. Production of α1,3-galactosyltransferase-deficient pigs. Science 2003; 299:411-414.
77. Yang YG, deGoma E, Ohdan H et al. Tolerization of anti-Galα1-3Gal natural antibody-forming B cells by induction of mixed chimerism. J Exp Med 1998; 187:1335-1342.
78. Ohdan H, Yang YG, Shimizu A et al. Mixed chimerism induced without lethal conditioning prevents T cell- and anti-Galα1,3Gal-mediated graft rejection. Transplantation 2000; 69:910-913.
79. Bracy JL, Sachs DH, Iacomini J. Inhibition of xenoreactive natural antibody production by retroviral gene therapy. Science 1998; 281:1845-1847.
80. Ogawa H, Yin DP, Shen J et al. Tolerance induction to a mammalian blood group-like carbohydrate antigen by syngeneic lymphocytes expressing the antigen. Blood 2003; 101:2318-2320.
81. Mohiuddin M, Ogawa H, Yin DP et al. Tolerance induction to a mammalian blood group like carbohydrate antigen by syngeneic lymphocytes expressing the antigen: II. tolerance induction on memory B cells. Blood 2003; 102:229-236.
82. Ogawa H, Mohiuddin MM, Yin DP et al. Mouse heart grafts expressing an incompatible carbohydrate antigen: II. transition from accommodation to tolerance. Transplantation, in press.

Index

F

G

H

I

L

M

N

P

S

T

X